Flora of Telangana
The 29ᵗʰ State of India

THE AUTHOR

Prof. T.Pullaiah obtained his M.Sc. (1973) and Ph.D. (1976) degrees in Botany from Andhra University. He was a Post Doctoral Fellow at Moscow State University, Russia during 1976-78. He traveled widely in Europe and USA and visited Universities and Botanical Gardens in about 17 countries. Professor Pullaiah joined Sri Krishnadevaraya University as Lecturer in 1979 and became Professor in 1993. He held several positions in the University, which include Dean, Faculty of Life Sciences, Head of the Department of Botany, Chairman, BOS in Botany, Head of the Department of Sericulture, Co-ordinator and Chairman, BOS in Biotechnology, Vice Principal and Principal, S.K.University College. He retired from active service on 31st May 2011. He was selected by UGC as UGC-BSR Faculty Fellow and is working in Sri Krishnadevaraya University. Prof. Pullaiah has published 60 books, 303 research papers and 35 popular articles. He is Principal Investigator of 20 major Research Projects totaling more than a Crore of Rupees funded by DBT, DST, CSIR, UGC, BSI, WWF, GCC etc. Under his guidance 53 students obtained their Ph.D. degrees and 35 students their M.Phil. degrees. He is recipient of Best Teacher Award from Government of Andhra Pradesh, Prof. P. Maheswari Gold Medal and Prof. G.Panigrahi Memorial Lecture of Indian Botanical Society and Prof. Y.D.Tyagi Gold medal of Indian Association for Angiosperm Taxonomy. He was Past President of Indian Association for Angiosperm Taxonomy and presently he is President of Indian Botanical Society. He was Member of Species Survival Commission of International Union for Conservation of Nature and Natural Resources (IUCN).

Flora of Telangana

The 29th State of India

Volume 1

Author

T. Pullaiah

Department of Botany
Sri Krishnadevaraya University
Anantapur – 515 003
Andhra Pradesh
India

2015

Regency Publications

A Division of

Astral International Pvt. Ltd.

New Delhi – 110 002

Flora of Telangana- *The 29ᵗʰ State of India*
(3 Volume Set)

Volume 1: Page 0001-0446
Volume 2: Page 0447-0892
Volume 3: Page 0893-1306

© 2015 AUTHOR

Cataloging in Publication Data--DK
   Courtesy: D.K. Agencies (P) Ltd. <docinfo@dkagencies.com>

Pullaiah, T., author.
Flora of Telangana : the 29th state of India / author, T. Pullaiah.
   3 volumes   cm
   Includes bibliographical references and index.
   ISBN 9789351308676 (Vol.1)
   ISBN 9789351305460 (Set)

   1. Plants--India--Telengana--Classification.   2. Flowers--India--Telengana--Classification.   I. Title.
   DDC 582.13095485   23

Published by   :   **Regency Publications**
                             A Division of
                             Astral International Pvt. Ltd.
                             – ISO 9001:2008 Certified Company –
                             4760-61/23, Ansari Road, Darya Ganj
                             New Delhi-110 002
                             Ph. 011-43549197, 23278134
                             E-mail: info@astralint.com
                             Website: www.astralint.com

Laser Typesetting   :   **Classic Computer Services**, Delhi - 110 035

Printed at   :   **Replika Press Pvt. Ltd.**

PRINTED IN INDIA

Preface

Flora of Former Andhra Pradesh was published in 1997 and Additions in 2008. The book went out of print in 2010. The long awaited decision to divide the state of Andhra Pradesh coincidentally happened when I had started revising the book sitting in Missouri Botanical Garden, U.S.A. It was then the idea of separate Flora for Telangana and Andhra Pradesh began. I continued my work using the facilities of Missouri Botanical Garden Herbarium and Library. I would like to highlight the warmth and affection which I received there and was inspired by their hard work and dedication. I thank authorities of Missouri Botanical Garden, Herbarium and Library for the permission to consult the Library and Herabrium.

Out of ten districts of Telangana we already published Flora of Five districts and I have taken liberally information from these books. The contributions of the scholars in these projects who are presently well placed in various esteemed institutions include Dr. P.V. Prasanna, Prof. B. Ravi Prasada Rao, Dr. C. Prabhakar, Dr. M.Silar Mohammad and Dr. B.V.Raghava Swamy. I thank all these scholars for their contributions which enabled me to write this book.

I thank the Scientists and scholars of Botanical Survey of India, Deccan Circle, Hyderabad, especially Dr. P.Venu, Dr. P.V.Prasanna, Scientists, Dr. L.Rasingam, Dr. Shankara Rao, Alok Chorghe, P.S.Annamma, besides ministerial staff. Many a times I had stimulating discussion with them. I am confident that this centre will grow. Dr. G.V.S.Murthy, Joint Director of Botanical Survey of India, Sourthern Circle, Coimbatore was always encouraging, supporting and extending facilities at Madras Herbarium for which I am grateful to him. Dr. M.V.Ramana, Assistant Professor, Osmania University gave his thesis on Flora of Hyderabad and many a times had animated discussion on the subject. This young boy will contribute in future much to the science of Plant Taxonomy. I also thank Prof. K. Seshagiri Rao, of University of Hyderabad, for giving the details of Flora of their campus. Librarian of Kakatiya University extended Library facilities during my visit and I am grateful to them for the same. Dr.

M.Sanjappa, former Director of Botanical Survey of India was critical about my work and always wanted best from my hands and I am grateful to him for the same. May his tribe grow. Taxonomists at Calicut University helped me in nomenclature and other aspects of different groups of plants, I thank especially Prof. M.Sabu, Prof. Santhosh Nampy and Dr. A.K.Pradeep for their help in various ways.

Authorities of Sri Krishnadevaraya University were graceful enough to provide facilities for carrying out this work. I thank Prof. R.R.Venkata Raju, Prof. C.Sudhakar, Prof. B.Ravi Prasad Rao, Prof. S. Thimma Naik, Heads of Department during different periods for their support and extending the facilities in the Department. Dr. S. Sandhya Rani, the latest entrant in the Botany Department needs special mention who read the manuscript, gave suggestions for its improvement, corrected the manuscript and contributed in many ways to bring out this book. Dr. S,.Karuppusamy Assitant Professor at The Madura College and Dr. K.Sri Rama Murthy, Professor and Head, Department of Biotechnology, Montessori Mahila Kalasala, Vijayawada have also contributed much for this book.

University Grants Commission gave me BSR Faculty Fellowship which enabled me to carry out this work after superannuation and the same is gratefully acknowledged.

I thank Anil Mittal and Dinesh Arora at Astral International, New Delhi for their patience, support and encouragement to bring out his book.

I tried to make this work as perfect as possible. Had I tried for perfection I could never complete this work. I request the readers of this book to point out errors and omissions which can be rectified in future.

T. Pullaiah

Contents

Preface   *v*

Introduction   1

Artificial Key to the Families of Flowering Plants in Telangana   31
B. Ravi Prasad Rao

Systematic Enumeration   49

Gymnosperms   1190

Bibliography   1191

Index   1217

Introduction

In a developing country like India, soil and land resources surveys play an important role in economic development. Ramachandran (1978) says "the importance of survey, exploration and exploitation of the floral wealth of the country needs to be reiterated because of its obvious significance in the country's overall socio-economic and cultural development."

The importance of survey of plant resources had also been emphasized by Jain (1978) who says "After independence our planners realized that in an agricultural country like India where the flora is so varied and rich, a proper consensus of the flora of the country and its evaluation for economic exploitation is very important." Such a precise data for the newly carved out state of Telangana is very important for the economic planning and development of the state.

Telangana state is one of the 29 states of India. It was formerly part of Hyderabad State (Medak and Warangal divisions) which was ruled by the Nizams. When India became independent from the British Empire in 1947, the Nizam of Hyderabad did not want to merge with Indian Union and wanted to remain independent under the special provisions given to princely states. The Government of India annexed Hyderabad State on 17 September 1948 in Operation Polo.

When India became independent, the Telugu speaking population was distributed in about 20 districts, 9 of them in the Nizam's Dominion and 11 in the Presidency of Madras. In 1953, first Andhra State was carved out of the erstwhile Presidency of Madras so as to include predominantly Telugu speaking areas, Kurnool as its capital in response to the desire of the Telugu speaking people. On November 1st, 1956, according to the recommendations of the States reorganisation Commission, Andhra Pradesh was formed, by the addition of nine districts, which were formerly in the Nizam's Dominion. Later on 3 more new districts were carved out making 23 districts in the State as a whole.

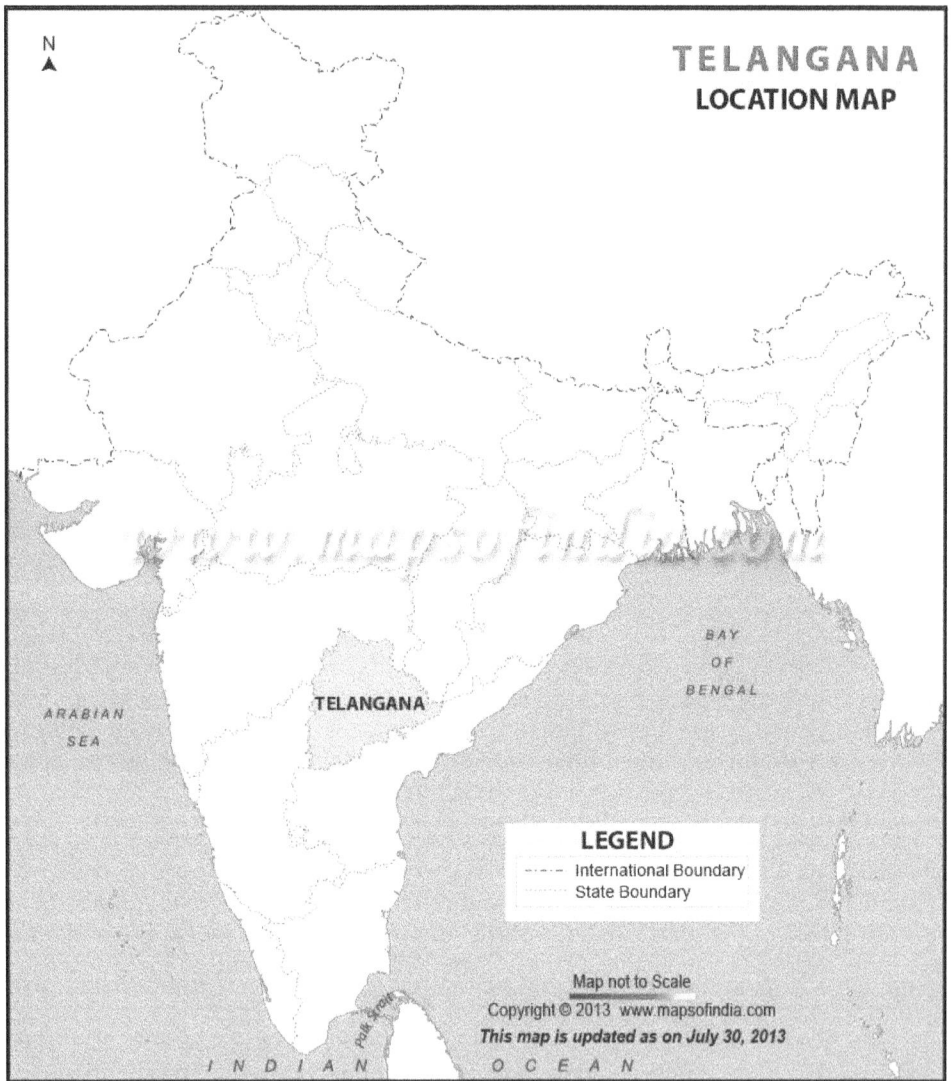

Figure 1: Map of India Showing the Position of Telangana.

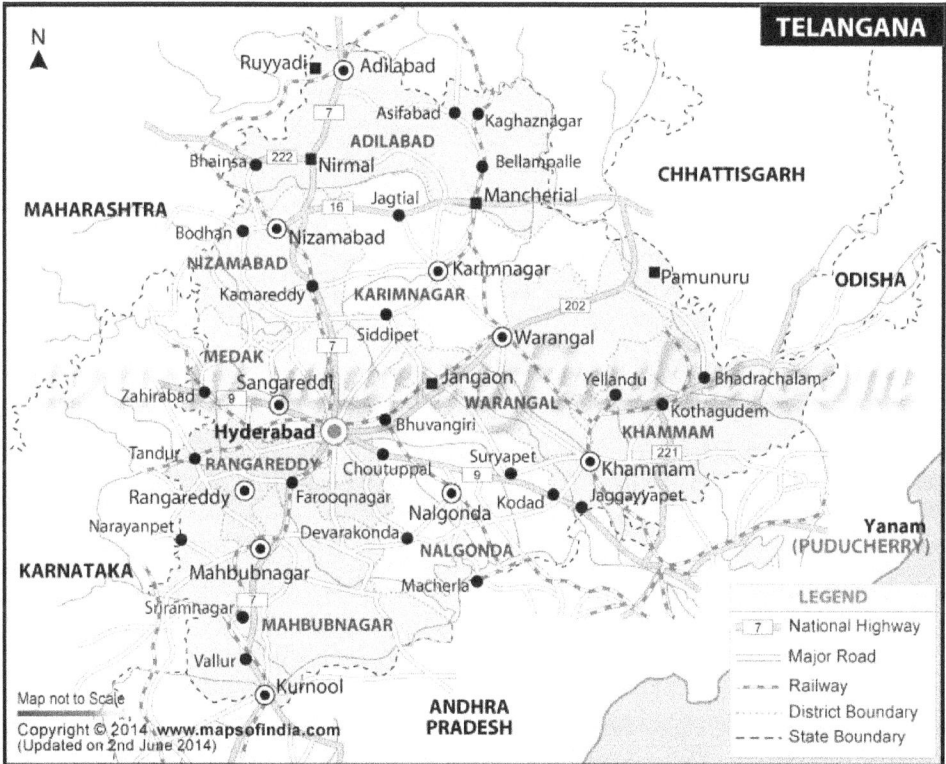

Figure 2: Map of Telangana with District Boundaries.

On 2ⁿᵈ June 2014 Central government has decided to form a separate state of Telangana (the 29ᵗʰ State of India). The city of Hyderabad would serve as the capital of Telangana.

Telangana and the language spoken in that region, "Telugu", is thought to have been derived from *trilinga*, as in *Trilinga Desa*, "the country of the three lingas". According to a Hindu legend, Shiva descended as Linga on three mountains namely, Kaleshwaram, Srisailam and Draksharama, which marked the boundaries of the Trilinga desa This is roughly the region between Krishna and Godavari rivers or modern Telangana region.

The term "Telangana" was designated to distinguish the Telugu region from Marathwada as part of Hyderabad State.

Telangana comprises 10 districts: Adilabad, Nizamabad, Medak, Ranga Reddi, Hyderabad, Karimnagar, Warangal, Khammam, Nalgonda and Mahabubnagar. Bhadrachalam and Nuguru Venkatapuram Taluks of East Godavari district (part of coastal Andhra Pradesh), which are on the other side of the river Godavari were merged into Khammam district on grounds of geographical contiguity and administrative viability. Earlier Ashwaraopeta was also part of West Godavari district and added to Khammam in the year 1959. Similarly, Munangala mandal was added to Nalgonda district from Krishna district in 1959.

Geography

Telangana State is situated in the central stretch of the eastern seaboard of the Indian Peninsula. Telangana state has an area of 114,840 square kilometres (44,300 sq mi). The area is divided into two main regions, the Eastern Ghats and the plains. The surface is dotted with low depression. Eastern Ghats are seen in Mahabubnagar and Khammam districts.

Telangana lies between 15°50′ –19°55′ North latitudes and 77°14′–78°50′ East longitudes. Telangana is bordered by the states of Maharashtra to the north and north-west, Karnataka to the west, Chattisgarh to the north-east and Odisha to the east and Andhra Pradesh to the south.

The state is drained by two major rivers, with about 79 per cent of the Godavari river catchment area and about 69 per cent of the Krishna catchment area, but most of the land is arid.

It is an extensive plateau with an average elevation of about 400 m above sea level. This plateau consists mainly of the ranges of erosional surface: (i) above 600 mt, (ii) from 300 – 450 mt and (iii) fromn 150 – 300 mt. First erosional surface lies around Hyderabad and it extends into north-west and north-east of Telangana. Another small patch lies along the northern bank of the river Krishna, where it is known as Amrabad plateau. Second erosional surface extends over a large area of the plateau in the middle of Telangana covering rounded flat hills and vast rolling plains. Granite is the chief rock in this range. Third erosional surface area is in between the Godavari trough on the east and on the west. It also extends from north to south.

There are extensive coal deposits, which are excavated by the Singarenic Collieries Company, for power generation and industrial purposes. There are limestone deposits in the area, which are utilised by cement factories. Telangana also has deposits of bauxite and mica.

Climate

Telangana is a semi-arid area and has a predominantly hot and dry climate. Summers start in March, and peak in May with average high temperatures in the 42 °C range. The monsoon arrives in June and lasts until September with about 755 mm of precipitation. A dry, mild winter starts in late November and lasts until early February with little humidity and average temperatures in the 22–23 °C range.

The State Telangana has the monsoon type of tropical climate. On the whole State enjoys warm climate. In northern Telangana tropical rainy type of climate prevails. Hot Steppe type of climate is noticed in the southern parts of the State. In Tropical Rainy type, the mean daily temperature is above 20°C with an annual rainfall of 150 to 200 cms, mostly in summer and South-West monsoon. In the Hot Steppe type, the mean daily temperature is 18°C and less. Maximum temperature in the summer season varying between 37°C and 44°C and minimum temperature in the winter season ranging between 14°C and 19°C.

In Telangana temperature increases from West to east during the month of December. The western part of Nizamabad, Medak, Hyderabad and Mahabubnagar districts are of low temperature, the temperature is below 20 °C during December. The central zone of Telangana and western part of Karimnagar, Adilabad and Warangal districts experience temperatures varying between 20-22 °C, the rest of Telangana shows temperatures above 22 °C.

The hottest month is May when the maximum temperature touches 45°C in the northeast and eastern areas of Telangana. In Ramagundam, Bhadrachalam and Khammam the temperature in May reach up to 45 °C very frequently. The central and southern areas of Telangana covering Nalgonda, Nizamabad and eastern areas of Medak, Hyderabad and Mahabubnagar districts along with the western areas of Karimnagar, Warangal and Khammam districts experience temperature between 32°C and 34 °C. The extreme western areas of Telangana covering Medak, Hyderabad, Ranga Reddi and Mahabubngar districts have slightly less severe summers comparatively due to their higher altitudes.

Monsoons

In Telangana south west monsoon is the major source of rainfall giving 80 per cent of annual rainfall. The northeast monsoon is significant only in the southern and southeastern corners of Telangana. The annual rainfall is between 900 to 1500 mm in northern Telangana and 700 to 900 mm in southern Telangana, from the southwest monsoon.

Southern areas of Telangana receives lower rainfall than the western, northern and eastern areas. Northern and north-eastern parts of Adilabad and western areas of Nizamabad receives high rainfall. The lowest rainfall zone of Telangana situated south and sourtheast, *i.e.,* Nagarkurnool and Alampur areas of Mahabubnagar district.

Table 1: District-wise Annual Average Rainfall (in mm)

Sl.No.	District	2004-05	2005-06	2006-07	2007-08	2008-09
1.	Mahaboobnagar	413	973	484	844.9	457.6
2.	Ranga Reddi	598	1136	631	913.8	762.9
3.	Hyderabad	629	1170	744	952.8	972.3
4.	Medak	529	896	735	807.3	708.4
5.	Nizambad	767	1172	961	962.5	840.5
6.	Adilabad	758	1210	1139	909.6	886.7
7.	Karimnagar	557	1024	1072	892.9	784.5
8.	Warangal	695	1191	1020	1107.7	1031.4
9.	Khammam	1058	1540	1080	1271.2	1326.9
10.	Nalgonda	519	1000	547	817.4	686.1

Geology

Telangana contains a wide variety of Geological formations ranging from among the oldest Dharwar schists to the recent alluvium. The Archaen gneissic and granite complex dominates the rock formation in Telangana. Adilabad, Nizamabad, Medak, Warangal and Khammam districts are having different types of rock systems. These are ignaceous rocks, partly metamorphosed. These rock systems contain good quantities of iron and copper ore, asbestos and barites, apart from these lime stone, steatite etc., are also available from rock systems in Telangana.

The Dharwar formations extend in Mahabubnagar district. The Dharwar formations contain important minerals. The Archaean or Peninsular gneisses dominate the rock formations in Telangana. The rock formation consists of Granites, Granodiorides and branded gneisses.

The purenas are found in in Mahabubnagar, Nalgonda, Ranga Reddi and Adilabad districts. These rocks include limestones, sand stones, slates and shales.

The Gondwanas occupied by the Pranahita and Godavary rivers along the eastern margin of Telangana is the main repositary of the Gondwana rock formations. The lower Gondwana contains coal seams among sandstones and shales. The coal seams are being exploited at Kothagudem and Yellandu in Khammam district, Bellampalli in Adilabad district and Godavari Khani in Karimnagar district. The upper Gondwanas consists mostly of Shales and Sandstones extending along Godavari trough in eastern Adilabad and Karimnagar districts.

The Deccan trap formations are found in the western and north-western parts of the State. They are composed of mostly greenish basaltic rocks, with occasional limestone beds interbedded with the basalt.

The laterites occur as caps over Deccan traps in Western Telangana (Vikarabad, Zaheerabad, Narayakhed mandals).

Soils

The State has a wide variety of soils and they form into three broad categories. They are red, black, laterite (Table 2).

Table 2: Soils in Telangana

Sl.No.	Traditional Name	USDA Taxonomy Order	Places of Occurrence in Telangana
1.	Red Soils	Association of	The great part of the Telangana
	Red sandy soils:	ENTISOLS	
	Red earths	INCEPTISOLS	
	Red loamy soils	ALFISOLS	
2.	Black Cotton Soils:		
	Black Soils	VERTISOLS	Adilabad, Warangal and Khammam
	Black Cotton Soils		

Red Soil

They are poor in nitrogen content (0.2 to 0.3 per cent) and plant nutrients but low to medium in available phosphates and medium to high in potash. The moisture holding capacity of the red soils is also poor. Red soils are derived from the weathering of gneisses and granite. The red sandy soils are locally known as "chalka" soils, cover the largest area in Telangana. A large area in Karimnagar district, the entire Warangal district, south-east and eastern parts of Nizamabad district, except the western part of Medak and Ranga Reddi district, the southern part of Mahabubnagar and Nalgonda districts and the eastern part of Khammam, all the remaining districts are extensively covered with this type of soils This soil is mainly supporting the cultivation of jowar, rice, chilly and sugarcane. The red soils around Hyderabad have given rise to an extensive development of grape cultivation.

Black Soils

Black soils are the second important soil group in area extent. Among the black soils the deep and medium black soils, which are also known as black cotton soils, are found in western and north -western portions of the State. The black soils are rich in calcium and potash but poor in nitrogen.

The deep black soils are found along the Pranahita, the Godavary and the Krishna rivers in broad belt, ranging from 10 to 26 kms on either side of these rivers. Deep black soils can be seen in Adilabad district and parts of Mahabubnagar district.

The light black soils are developed from the Deccan trap rocks, found in the north-western part of the State in Adilabad, Ranga Reddi, Medak and Nizamabad districts.

The mixed black and red soils are found in wide patches between Krishna river. In the elevated plateaus, the black soils are thin with light colour and low fertility, whereas in the low lying areas, they are thick with deep colour and high fertility. It is distinctive for its shrinkage and cracking when dried. Mixed red and black soils occur mainly in Adilabad distrct, north of the Godavari river and also found in small patches in Mahabubnagar district.

Laterite Soils

These soils are formed by the composition of gneisses, the colour of these soils vary from deep reddish to brown or black. They are rapidly permeable and well drained. They are generally poor in organic matter and plant nutrients like nitrogen, phosphate and potash. Laterite pockets occur in parts of Zaheerabad and Narayankhed mandals in Medak district, small patches around Vikarabad in Ranga Reddi district and Hyderabad districts.

Rivers

Telangana is popularly and rather appropriately called a "river State", nearly 75 per cent of its territory is covered by the basins of two big rivers, *viz.* Godavary and Krishna and their tributaries *viz.*, Kadam, Pranahita, Peddavagu, Manair, Bhima, Dindi, Kinnerasani, Manjira, Munneru, Moosi, Penganga, and Taliperu. General terrian of the State slope downwards from the west to east.

The Godavary

The Godavary, which is the largest river in South India and the second largest in India, takes its origin in the Western Ghats at Triambak near Nasik in Maharastra State. Its total length is about 1584 kms of which about 720 kms lies within this State. Its total catchment area is 312,812 sq.kms spread in Maharashtra, Madhya Pradesh, Orissa, Telangana and Andhra Pradesh. Entering the State near Basar village of Adilabad district, it flows through Nizamabad, Karimnagar, Warangal and Khammam before entering Andhra Pradesh. The river receives more than 2/3 of its water from its tributaries in this State, *viz.*, the Kadam river, the Pranahita, the Manjira, the Penganga, the Manair, the Wardha, the Wainganga, the Sabari, the Indravathi and the Purana. Nizamsagar and Sriramsagar are the important projects in this river basin.

The Krishna

Krishna, one of the mighty rivers in Peninsular India, and the second longest river in the State, rises about 64 kms from the Arabian sea (17° 58′ E) in the Western Ghats, north of the hill station of Mahabaleswar in Maharashtra state. Its catchment area of 258,938 sq.kms is spread over Maharashtra, Karnataka, Telangana and Andhra Pradesh. The river enters the State near Tangadigi village in Mahabubnagar district. The total length of the river is 1400 kms. It courses through a plateau, 338 kms long, covering Mahabubnagar and Nalgonda districts of which about 193 kms from Siddeswaram to Nandikonda lie in a narrow and continuous gorge and emerges into plains at Pulichintala in Krishna district of Andhra Pradesh. The important tributary of this river is the Moosi.

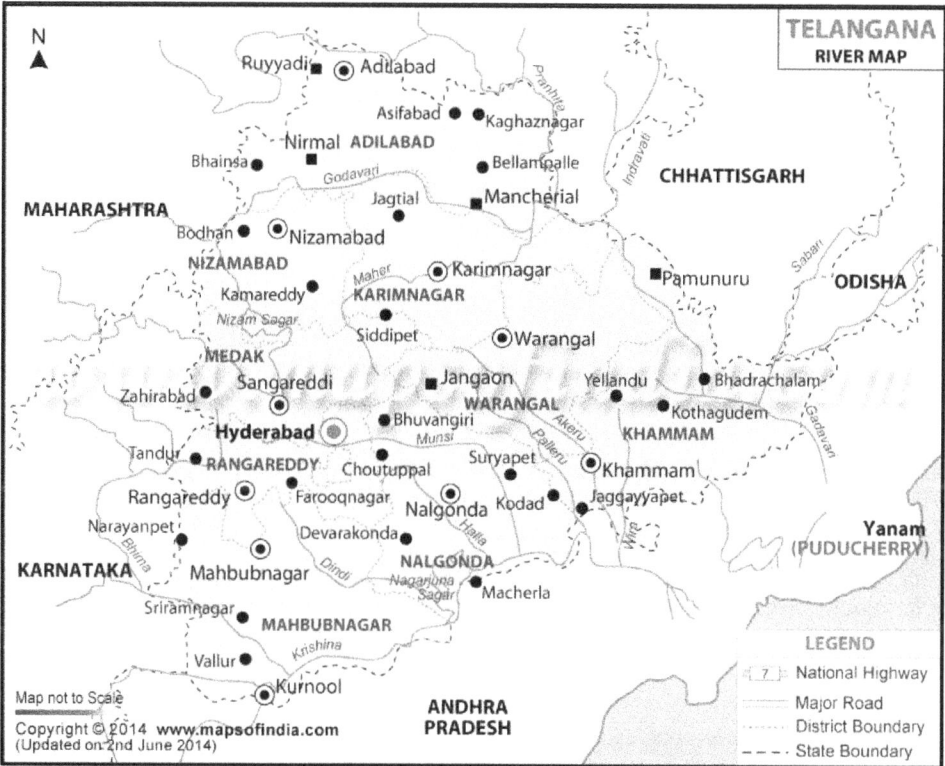

Figure 3: Map of Telangana – Rivers and Drainage.

The Manjira

This is the chief tributary to Godavari. It rises from 823 m height in Pathoda taluk of Bir district in Maharashtra state. It flows through or along the districts Osmanabad and Bidar and enters the Medak district of Telangana in South-East direction. It crosses the district from the south-west and joins the Godavary near Kandakurthy in Bodhan mandal of Nizamabad, after a course of 100 kms. The total catchment area is about 30,821 sq.kms.

Tanks and Lakes

Huge tanks and lakes, most of them fed by river channels, dot the countryside. To cite only a few, the the Pakhal lake, Ramappa lake, Lakkavaram lake all in Warangal district cater to irrigation needs.

Water Falls

The main water falls in Telangana are, Potchera and Kunthala in Adilabad.

Forests and Vegetation

Forests

Humankind depends in a myriad ways on plants and plant products. The quest for new plant products, and new methods of using them to satisfy the emerging need is an ever expanding enterprise. An account of the plants that serve the varied human needs for food, clothing, shelter, medicines and drugs, and of the innumerable ways in which they are put to use is a fascinating story of human ingenuity and innovation. So old saying "Wood is required from cradle to grave" is true in every sense.

The depletion of forests all over the world at an estimated rate of 245,000 sq.kms per year is one of the most alarming aspects of present day biospheral tendencies. Due to such an extensive and unabated destruction of forests, mankind is loosing many valuable plants even before we come to know that they exist. Causes of threats to the flora have generally been grouped in two categories, *viz.* (1) natural and (2) man-made. The natural causes include floods, earthquakes, landslides, natural competition between species, biology of the species such as lack of pollination and natural regeneration, diseases etc. In man-made threats, could be included deliberate destruction of habitat (such as by mining, dam constructions, conversion of forests etc.) excessive grazing, over-exploitation etc.

The State of Telangana receives handsome revenue from its forest wealth. Major forest products of the State include timber, especially teakwood, rose wood, fuel wood, charcoal, bamboos, beedi leaf, tanning barks, myrobalans, park mohwa, lac and gums, soap nuts etc. Bamboo and beedi leaf support paper and beedi industries respectively. These two items together contribute more than 80 per cent of the total revenue.

The forest area of Telangana extends over 67.5 lakh acres forming 23.8 per cent of its geographical area. Most of the forest cover in the State is reserved where it is protected from over exploitation for the purposes of fire wood, grazing, cultivation etc. High percentage of forest cover is found in the northern part of the State comprising the whole of Adilabad and Khammam districts and parts of Karimnagar, Warangal districts. Nearly 51 per cent of area is in Adilabad and Khammam districts only. The Eastern Ghats in the southern parts of State comprising parts of Khammam and Mahaboobnagar districts too have a dense forest cover.

The study of the district-wise forest area shows that Khammam district with 52.2 per cent, Adilabad district with 43.9 per cent, Warangal district 28.8 per cent have higher percentage of forest cover than the state's average forest cover. Districts with low forest cover are Medak (9.8 per cent), Nalgonda (8.4 per cent), Ranga Reddi (7 per cent), and Hyderabad (Nil).

Effect of Biotic Factors on the Vegetation

The vegetation is often considerably modified by the biotic factors operating on it. It took nature ages to produce the great forests but it takes man only moments to decimate them. Biotic interference has had a marked effect on the vegetation of Telangana leading to the extinction of rare and valuable plants. Luxurious flora and fauna is on the verge of disappearance. Even some of the evergreen areas once present some areas of Khammam are being converted into deciduous patches. In these areas drought indicating plants like thorny bushes are appearing indicating the impact of various biotic factors on the once luxurious forests.

The herbivors indiscriminately graze on the seedlings of tree and shrub species and herbaceous vegetation, with the result many herbaceous species, especially rare ones, are seldom seen in all the stages of development. The seedlings of shrubs and tree species often do not attain their full growth.

With the increase in the population, the demand of land for agricultural purposes has been increasing, thus resulting in the encroachment of large forest areas by the people of plains. Wavering policy of the Government regarding the protection of forests on one side and the implementation of plan of supply of raw materials to the industries and fire wood and other necessities has ultimately resulted in great loss of forest cover and the denudation of soil.

Forest fire is an important factor which occurs either accidentally or by negligence of man which causes large scale damage to the forest cover and forest soil. Urban development also enhanced the effect of deforestation.

The indiscriminate cutting of trees for the past several years has thinned down many forests. This has reduced abundancy and also intensity of rains. For example, the complete extermination of teak can be seen in large areas of Nallamalai forest and also in some areas of Adilabad district.

Many of the ongoing projects and the proposed ones will certainly alter the natural habitats of plants.

Due to the construction of many reservoirs and Hydro-electric projects on rivers like Srisailam, Nagarjunasagar, Pochampadu, Manjeera etc. large forest areas are submerged. In addition to this, the settlements around these projects increase the exploitation of forests. For example Srisailam project area which was once surrounded by dense forests not so long ago, now vast stretches of this area is barren.

Forest based industries such as (1) Paper mills, (2) Saw mills (3) Plywood industries, (4) Rayon mills and a few other minor industries which are managed either by government or private sector have over-exploited the forests. Some paper mills have unscrupulously denuded the Bamboo forests. The A.P. Rayons with its chemical effluents has polluted the Eturunagaram affecting its ecology.

The open cast coal mining at Manugur in Khammam district resulted in complete destruction of the open typical dry deciduous forests. Large forest cover is cut down for laying power-lines, new railway tracks, roads etc.

The ecodegradation and wood famine triggered by wanton destruction of forests necessitated the realignment of strategies and policies of forest department. Emphasis has been given to the conservation and consolidation of the existing forests to ensure ecological security.

Biotic factors also operate in a different way, that is to say, by introduction of exotic species. A number of exotic plants have been introduced in this country and in the State of Telangana also. Such plants are naturalised in many parts of the State and they are now the integral members of the indigenous flora. Some such exotic plants are *Acanthospermum hispidum, Ageratum conyzoides, Alternanthera pungens, Argemone mexicana, Casuarina equisetifolia, Chromolaena odorata, Lantana camara, Passiflora foetida, Parkinsonia aculeata, Parthenium hysterophorus* and *Prosopis chilensis.* The last species has become an obnoxious weed by its prolific distribution and acclimatization.

Wild Life Sanctuaries and Parks are dedicated to the preservation of wild life and to represent ecological units in Andhra Pradesh. There are sixteen sanctuaries in the State. Some of them are given in Table 3.

Table 3: Wildlife Sanctuaries and Parks

Sl.No.	Name	Area in sq. km.	Location (District)
1.	Kawal	893	Adilabad
2.	Eturnagaram	803	Warangal
3.	Pakhal	860	Warangal
4.	Pocharam	130	Medak and Nizamabad
5.	Kinnerasani	635.41	Khammam
6.	Papikonda	591	Khammam and (East and West Godavari in A.P.)
7.	Nagarjunasagar-Srisailam Tiger Reserve	3,568	Mahaboobnagar, Nalgonda (and Guntur, Kurnool, and Prakasam of A.P.)
8.	Pranahita	136	Adilabad
9.	Lanjamadugu, Sivaram	38.66	Adilabad, Karimnagar
10.	Manjira (Crocodile)	20	Medak

General Vegetation Types

The types vegetation in any area are determined by climatic, edaphic and biotic factors coupled with altitude. The wide variation in climate and topography of the

State have resulted in various types of forest growth ranging from Moist Deciduous to pure dry and barren lands. The major concentration of forests is found in Eastern Ghats. Champion and Seth (1968) while presenting Revised classification of forest types of India indicated that their differentiation of plant communities for presenting classification of vegetation is of general nature and pointed out that their knowledge of the various floristic components of different types of vegetation covering limited areas of tropical forest is far too incomplete to present the distinct vegetation type based on floristic analysis of limited area. The type of forests met within Telangana, as per the classification of Champion and Seth (1968) are

1. Tropical moist deciduous forests,
2. Southern dry deciduous forests,
3. Northern mixed dry deciduous forests,
4. Dry savannah forests,
5. Tropical dry evergreen scrub.

1. Tropical Moist Deciduous Forests
Southern Tropical Moist Deciduous Riverian Forests
Along the courses of rivers and streams in the plains, where alluvial soil is deposited, there are many plants predominantly exclusive to these areas. This riparian vegetation is maintained by the interaction of constant erosion and redeposition of the soil going on the banks of rivers. It generally forms a very narrow belt along the banks; sometimes it may extend to the higher elevations. The riparian trees may be evergreen or deciduous depending upon the region. This type of forest is present along the banks of river Godavari and other hill streams in a narrow belt. The most common trees in these forests are *Terminalia arjuna, Mitragyna parviflora, Tamarindus indica, Bombax ceiba, Barringtonia acutangula, Butea monosperma, Strychnos nux-vomica, Pongamia pinnata, Syzygium cumini, Oroxylum indicum, Trema orientalis, Memecylon umbellatum* etc.

Homonoia riparia, Tamarix ericoides, Rotula aquatica, Syzygium heyneanum, Combretum albidum, Vitex leucoxylon etc. are some of the representative shrubby elements one will come across in the rocky and sandy 'banks' of Godavari.

Most of the common herbs that occur in this type are *Indigofera linnaei, Pedalium murex, Polygonum* spp. *Typha angustata, Saccharum spontaneum* etc.

2. Dry Deciduous Forests
In this type of forests, the trees begin to shed their leaves by about December and between February and May the forest looks very open and at times eye-soring, but no area is completely leafless at any one time of the year. Flowering and fruiting are generally far advanced before the first flush of new leaves appears with the conventional showers in April-May. These forests are widely spread in almost all the districts of the State, where the soil conditions are poor. The forest composition does not show zonations.

Tectona grandis and *Anogeissus latifolia* are perhaps the commonest trees in these forests. *Boswelia serrata, Cochlospermum religiosum, Diospyros melanoxylon, Gardenia latifolia, Givotia rottleriformis, Gyrocarpus americanus, Lannea coromandelica (=Kavalama urens), Sterculia urens, Strychnos potatorum, Ziziphus xylopyrus, Terminalia* spp., *Chloroxylon swietenia, Pterocarpus marsupium, Albizia odoratissima, Haldina cordifolia, Cassia fistula, Diospyros melanoxylon* etc. are some of the typical trees. The orange blossoms of *Firmiana colorata* are less common. *Balanites aegyptiaca, Gmelina asiatica* and *Naringi crenulata* are armed trees present, at the edges of the forest. In Deccan plateau *Tectona grandis-Terminalia alata* are the dominant species, while in Eastern Ghats of Khammam district shows predominance of *Xylia xylocarpa* with *Terminalia alata.*

Holarrhena antidysenterica, Wrightia tinctoria, Alangium salvifolium, Bauhinia racemosa, Tarenna asiatcia, Flacourtia indica, Helicteres isora, Nyctanthus arbor-tristis, Woodfordia fruticosa, Grewia hirsuta etc. are some of the common shrubs found in this type of forests.

A number of climbers and twiners are found in this type of forests. To mention, some of them are *Aspidopterys cordata, Butea superba, Cansjera rheedii, Celastrus paniculatus, Combretum ovalifolium, Paracalyx scariosus, Dioscorea* spp., *Pueraria tuberosa, Ventilago denticulata, Mucuna* spp. etc.

The ground flora is mostly seasonal. *Achyranthes aspera, Aerva sanguinolenta, Leea edgeworthii, Sida mysorensis, Solanum indicum, Scilla hyncinthina, Habenaria roxburghii* etc. are some of the common examples.

3. Northern Mixed Dry Deciduous Forests

These generally occur at about and above 400 m in shallow soils of well drained hill sides. The canopy is closed though uneven and not dense. Most of the species are deciduous. The under- growth is usually dense since enough light penetrates through the upper canopy. Epiphytes and ferns are very rare. This type of forests are confined to Cherukupalli of Nalgonda district, parts of Medak and Ranga Reddi districts.

Common among the canopy trees are *Albizia amara, A. odoratissima, Anogeissus latifolia, Hardwickia binata, Terminalia chebula, T. tomentosa, T. paniculata, Kavalama urens (=Sterculia urens), Bauhinia racemosa, Butea monosperma, Cassia fistula, Dalbergia* spp., *Phyllanthus emblica* L., *Lannea coromandelica, Mangifera indica, Pterocarpus marsupium* etc.

The middle storey comprises small trees such as *Chloroxylon swietenia, Dalbergia paniculata, Vitex altissima, Dolichandrone atrovirens, Gardenia gummifera, G. latifolia, Strychnos potatorum* etc.

The common shrubs are *Acacia* spp. *Dodonaea viscosa, Ixora arborea, Securinega virosa, Helicteres isora, Tarenna asiatica, Combretum albidum, Hiptage benghalensis, Ventilago madraspatana* etc. The bamboo, *Dendrocalamus strictus* is often found.

The ground layer comprises the species of following genera: *Abutilon, Achyranthus, Aristida, Bulbostylis, Cleome, Cymbopogon, Cyperus, Digitaria, Glinus, Polygala, Tragus* and *Tribulus.* The grass growth is usually heavy. These forests are often subjected to the interference of man as well as fire.

4. Dry Savannah Forests

These forests, formed as a result of intense biotic interference, are scattered in Mahabubnagar and Khammam districts. The stunted trees are *Phyllanthus emblica* L., *Phoenix humilis, Pterocarpus marsupium, Terminalia chebula* are associated with grasses like *Aristida setacea, Bothriochloa pertusa, Brachiaria ramosa, Themeda triandra, Cymbopogon flexuosus, Chrysopogon aciculatus, Panicum* spp., *Setaria* spp., etc.

5. Tropical Scrub Forests

Scrub forests are widely distributed on the arid and semiarid zones of earth, where the rainfall is scanty. The vegetation presents a very open appearance so that the trees and shrubs are widely spaced. The bulk of the vegetation consists of co-dominant, spinous shrubs and trees capable of great drought resistance. In this type there are two categories, *viz.* (a) the permanent vegetation occurring throughout the year and (b) the temporary vegetation consisting of the annuals growing mainly during short rainy season. Corresponding to this, the area represent two distinct seasonal variations. (1) The permanent xerophytic vegetation consists of trees and shrubs, which flower in the summer and winter seasons, when the soil is devoid of the ground cover (2) in the rainy season, the vegetation will be at its best and the soil which is otherwise barren between the trees and shrubs, is covered by a vivid-green carpet of a temporary vegetation. This flowers and fruits in a short time and disappears soon after the surface layer of the soil dries up as winter sets in. These are mainly present in almost all the drier parts of Telangana like the districts of Nalgonda, Ranga Reddi, Medak and peripheries of forests in other districts. The main species are *Acacia chundra, Albizia amara, Balanites aegyptiaca, Anisochilus carnosus, Canthium parviflorum, Erythroxylum monogynum, Flacourtia indica, Premna tomentosa, Ziziphus* sps., *Dodonaea viscosa, Euphorbia antiquorum, Dichrostachys cinerea, Capparis brevispina, Maytenus emarginata, Carissa spinarum, Grewia tenax* etc.

Aquatic Vegetation

The State of Telangana is quite rich in streams, ponds, ditches and rivers, which harbour a large number of hydrophytic plants (including aquatic and marshy-wetland plants). Most of the ditches and temporary ponds are filled up with water during monsoon, in the second half of which a number of plants of the hydrophytic vegetation appear. These hydrophytes can be classified as 1. Floating hydrophytes 2. Submerged hydrophytes 3. Emergent hydrophytes 4. Wetland hydrophytes.

1. Floating Hydrophytes

There are three types of plants in this division basing on the relationship between the plant and the substratum. They are (a) Free floating on the surface of water : In this subtype the plants have no contact with the soil. They float freely on the surface of water and are in contact with air and water. *Eichhornia crassipes, Lemna perpusilla, Pistia stratiotes* and *Spirodela polyrhiza* are common examples for this type. (b) Attached hydrophytes with floating shoots : These plants are attached to the muddy floor by their roots, but their shoots come out and float on the surface of water. The principal examples of this category are *Hygrorhiza aristata, Ipomoea aquatica, Ludwigia adscendens* and *Neptunia oleracea.* (c) Attached hydrophytes with floating leaves : In this category the plants are attached to the sub-stratum and their stems (mostly rhizome) remain under water in contact with soil and water while the leaves float on the surface of the water. *Aponogeton natans, Limnophyton obtusifolium, Monochoria vaginalis, Nelumbo nucifera, Nymphaea pubescens, N. nouchali, N. rubra, Nymphoides cristatum, N. indicum, Ottelia alismoides, Potamogeton nodosus* and *Tenagoncharis latifolia* are common examples.

2. Submerged Hydrophytes

These plants always remain under water surface and can be grouped into two categories *viz.*, suspended submerged hydrophytes and attached submerged hydrophytes. (a) Suspended submerged hydrophytes : These plants remian submerged in water but have no contact with the soil. Their flowers may or may not come above the water level *e.g.*, *Ceratophyllum demersum, Utricularia aurea* and *U. exoleta.* (b) Attached submerged hydrophytes : These plants remain in contact with soil and water. Their vegetative portion remains completely submerged in water, while the flowers may come out of water surface. *Aponogeton crispus, Cryptocoryne retrospiralis, Hydrilla verticillata, Najas graminea, Lagarosiphon alternifolia, Polypleurum stylosum, Potamogeton crispus, P. pectinatus* and *Vallisneria natans* are found in this type.

3. Emergent Hydrophytes

Plants which are attached to soil covered with water but most of their vegetative parts come out of water surface, *e.g.*: *Aeschynomene aspera, A. indica, Ammannia baccifera, Bacopa monnieri, Cyperus distans, C. pangorei, Echinochloa colona, Fimbristylis* spp., *Hygrophila auriculata, Ischaemum rugosum, Limnophila indica, Polygonum barbatum, Phragmites karka* and *Typha angustata.*

4. Wetland Hydrophytes or Marshy Plants

The plants included in this category are rooted to the soil saturated with water, which may also survive in dried conditions too in the later part of their life cycle. A large number of species are found in this habitat. Some typical ones are *Phyla nodiflora, Alternanthera sessilis, Polygonum plebeium, Commelina* spp., *Glinus lotoides, Caesulia axillaris, Eclipta prostrata, Melochia corchorifolia, Sphaeranthus indicus, Ipomoea carnea, Cynodon dactylon, Murdannia nodiflora, Justicia betonica* etc.

Weeds

Centaurium roxburghii, Chenopodium album, C. murale, Cyperus rotundus, Eclipta prostrata, Echinochloa colona, E. crusgalli, Fuirena ciliaris, Ischaemum rugosum, Panicum repens, Paspalum spp., *Paspalidium* spp., and *Plectranthus japonicus* are common weeds of irrigated fields. *Argemone mexicana, Aristolochia bracteolata, Celosia argenta, Chrozophora rottleri, Corchorus trilocularis, Cynodon dactylon, Dactyloctenium aegyptium, Digitaria* spp., *Dinebra retroflexa, Elytrophorus spicatus, Euphorbia leata, E. prostrata, Goniocaulon glabrum, Justicia* spp., *Leucas aspera, Luffa tuberosa, Phyllanthus amarus, Portulaca quadrifida, Sida acuta, Sphaeranthus indicus, Trianthema portulacastrum,* and *Trichurus monsoniae* are common weeds of dry fields. Other common weeds encountered in cultivated fields are *Cardiospermum halicacabum, Catharanthus pusillus, Cleome* spp., *Crotalaria laburnifolia, Malvastrum coromandelianum, Melilotus alba, M. indica, Melochia corchorifolia, Portulaca oleracea, Psoralea corylifolia, Rorippa indica, Vigna* spp., etc.

Parasitic Plants

A few parasitic plants have also been recorded from the State of Telangana. *Dendrophthoe falcata, Scurrula parasitica, Taxillus bracteatus* are seen in dry deciduous forests. *Cassytha filiformis* is a partial stem parastie on a number of species like *Carissa spinarum, Ziziphus mauritiana, Hardwickia binata, Erythroxylum monogynum, Vitex negundo, Azadirachta indica* etc. This particular species is found both in the forest and non-forest areas. *Striga angustifolia, S. asiatica,* and *S. densiflora* are frequent root parasites on grasses including *Sorghum* and Bajra, *Striga gesnerioides* is a root parasite on *Lepidagathis cristata. Orobanche cernua* is a complete root parasite on tobacco and brinjal. *Cuscuta reflexa* is a common stem parasite on a variety of plants.

Economically Important Plants

The information about the vegetation and flora of the State is not complete without mention of various economically important plants. For that a list of the common crop plants and other important plants is appended.

Cereals

Oryza sativa (vari) is the prinicpal cereal. *Zea mays* (mokka jonnalu) is cultivated extensively in the State. *Sorghum vulgare* (jonnalu), *Panicum miliaceum* (varigalu), *Pennisetum americanum* (sajjalu), *Panicum miliare* (samalu) are millets grown in the State.

Pulses

Cicer arietinum (sanagalu), *Cajanus cajan* (kandulu), *Vigna mungo* (minimulu), *Vigna radiata* (pesalu), *Vigna unguiculata* (Alasandulu) are the pulse species common in the State.

Spices

Foeniculum vulgare (somp), *Trigonella foenum-graecum* (menthulu), *Cuminum cyminum* (jilakara), *Tamarindus indicus* (chinta), *Trachyspermum ammi* (vamu), *Murraya koenigi* (karivapak) are the spices and condiments cultivated here.

Vegetables

Brassica oleracea var. *capitata* (cabbage), *Lycopersicon esculentum* (Tomato), *Solanum melongena* (Vankai), *Cucurbita maxima* (Gummadi), *Cucumis sativus* (Dosa), *Abelmoschus esculentus* (Benda), *Luffa cylindrica* (Beera), *Momordica charantia* (Kakara), *Lagenaria vulgaris* (Sorakai), *Raphanus sativus* (Mullangi), *Hibiscus cannabinus* (Gongura), *Cyamopsis tetragonaloba* (Motika or goruchikkudu) are commonly grown in the State.

Fruits

Mangifera indica (Mamidi), *Citrus* sps. (Nimma), *Vitis vinifera* (Draksha), *Carica papaya* (Boppai), *Punica granatum* (Danimma), *Psidium guajava* (Jama), *Annona squamosa* (Seetapalam), *A. reticulata* (Ramapalam), *Ziziphus mauritiana* (Regu), *Aegle marmelos* (Maredu), *Feronia limonia* (Velaga) and *Syzygium cumini* (Neredu) are encountered.

Medicinal Plants

Eucalyptus citriodora, Azadirachta indica (Vepa), *Butea monosperma* (Moduga),

Terminalia chebula (Karaka), *T. bellarica, Pterocarpus marsupium* (Vegisa) are some of the examples for medicinal plants.

Timber Trees

The major revenue for the forest department comes from the timber of some species. They are *Tectona grandis* (Teak), *Bombax malabaricum* (Buruga), *Boswellia serrata* (Anduga), *Erythrina suberosa* (Mula moduga), *Albizzia odoratissima* (Chinduga), *Calophyllum inophyllum* (Ponna), *Chloroxylon swietina* (Billudu), *Dalbergia latifolia* (Jittegi), *Pterocarpus marsupium* (Yegisa), *Acacia nilotica* (Thumma), *Anogeissus latifolia* (Velama or Chirumenu), *Soymida febrifuga* (Somi), *Garuga pinnata* etc.

Oil Yielding Plants

Arachis hypogaea (Verusenaga), *Brassica* sps., *Azadirachta indica* (Vepa), *Sesamum indicum* (Nuvvulu), *Ricinus communis* (Amudalu) etc. are plants from which oil of commerce is obtained.

Minor Forest Products Gum and Resin Yielding Plants

Azadirachta indica, Acacia leucocephala, A. nilotica, Anogeissus latifolia Boswellia serrata, Bombax ceiba, Commiphora caudata, Cochlospermum religiosum, Hardwickia binata, Lannea coromandelica, Pterocarpus marsupium, Kavalama urens (=Sterculia urens) etc.

Dye Yielding Plants

Acacia catechu, A. arabica, A. chundra, Punica granatum, Terminalia chebula, T. bellirica, Aegle marmelos, Butea monosperma, Woodfordia fruticosa etc.

Match Wood Species

Ailanthus excelsa, Bombax ceiba, Garuga pinnata, Kydia calycina, Lannea coromandelica etc.

Avenue Trees

Some species are planted along road sides for shade and such tree species commonly found in the State are *Albizia lebbeck, Azadirachta indica, Bauhinia purpurea, Cassia fistula, C. roxburghii, Senna siamea, Dalbergia sissoo, Delonix regia, D. elata, Erythrina variegata, Gliricidia sepium, Melia azedarach, Millingtonia hortensis, Peltophorum pterocarpum, Polyalthia longifolia, Syzygium cumini, Tamarindus indicus, Albizia saman, Mangifera indica, Ficus benghalensis, Pongamia pinnata* etc.

Past and Present Work

Past Work

Systematic studies of the erstwhile Hyderabad state came from two principal sources, the State Forest Department and the Department of Botany, Osmania University, Hyderabad. Before independence Telangana region was in Hyderabad State. The study of the flora of Hyderabad State dates back to the 19th century when Walker (1849) and Bradley (1849) published their pioneer work, which included agricultural, medicinal and other economically important species of Daulatabad and Warangal districts of the State. Campbell in 1898 included a list of forest plants of Hyderabad State in his "Glimpses of the Nizams Dominions". During the same period Bisco, a Forest officer listed 128 chief timber-yielding and other economically important plants of the state. Patridge (1911) published a book entitled "Forest Flora of Hyderabad State". He described 450 species belonging to 69 families and provided keys to taxa, information on local names and economic importance of plants. The book was later revised by Khan (1953), who added some more information on the vegetation and described 567 wild and cultivated species. Sayeeduddin (1935, 1938, 1941a, b, 1954) published a series of papers on the flora of Hyderabad State and reported a total of 370 species. Suxena (1947) listed 115 grasses from Hyderabad State. Other works on the flora of Telangana region include Santapau (1954), Sebastine and Henry (1966) etc.

Botanical explorations were revived with the reorganisation of Botanical Survey of India in 1955. Collections were made and interesting results were published. G.V. Subba Rao, K. Subramanyam, N.P. Balakrishnan, K. Thothathri, K.M. Sebastine are some of the important contributors from Botanical Survey of India to the flora of Telangana.

Sebastine *et al.* (1960) enumerated 268 species from Medak distrct. In 1966 Sebastine and Henry studied the Flora of Pakhal and surrounding regions of Narasampet taluk in Warangal district and reported 254 species of 198 genera belonging to 70 families. Thothathri (1964) studied Nagarjunakonda and surroundings and noted 251 species. Subba Rao and Kumari (1967) published a small account of 434 species from Kodimial, Manthani and Raikal of Karimnagar district. S.L. Kapoor and L.D. Kapoor (1973) enumerated an additional list of 66 species from Karimnagar district. Seshagiri Rao (2012) reported 734 species of flowering plants in University of Hyderabad campus.

Floristic Wealth of different Districts of Telangana

The importance of district floras has been recognised by funding agencies and the botanists. District Flora projects were started in 1977 by different funding agencis like Botanical Survey of India, University Grants Commission and Council of Scientific and Industrial Research. Several of these projects have been completed. An analysis of district floras is given in Table 4.

The dominant families and their species number in each district have given in Table 5. Fabaceae (Leguminosae) is the dominant family in all the districts with largest number of species. Family Poaceae is coming under second position in most of the districts. Next order families are Euphorbiaceae, Cyperaceae, Acanthaceae, Asteraceae and Rubiaceae.

The district-wise floristic wealth is discussed briefly as follows:

Adilabad District

Adilabad district lies within the tropical deciduous belt which occupies 43.9 per cent of total geographic area of the district. In this district Pullaiah *et al.* (1992) reported 673 species belonging to 422 genera and 118 families. Fabaceae (Leguminosae) is the dominant family, represented by 88 species, followed by Poaceae (85 species),

Cyperaceae (39) and Euphorbiaceae (34). The dominant genera are *Fimbristylis* (12 speices), *Cyperus* and *Eragrostis* (9 species each), *Euphorbia, Ficus* and *Acacia* (8 species each). *Dactyloctenium aristatum* and *Digitaria radicosa* were recorded from this district as new reports to South India, *Argyreia sericea, Asphodelus tenuifolius, Brachiaria milliformis, Curcuma decipiens, Dichanthium filiculme, Kyllingia hyalina, Fimbristylis tetragona* and *Rhynchopora wightiana* are additions to the Flora of Andhra Pradesh. Prabhakar Raju and Venkata Raju (1999) described a new species *Cyathocline manilaliana* from Pochera fields.

Table 4: Analysis of District Flora of Andhra Pradesh

District	Investigator	Species	Genera	Families
Mahaboobnagar	S.R.R. Rao	1174(132)	615	137
Ranga Reddi	T. Pullaiah, M.S. Mohamad	698	414	110
Hyderabad	T. Rajagopal	951	583	124
Medak	T. Pullaiah, C. Prabhakar, B.R.P. Rao	708	414	119
Nizamabad	T. Pullaiah, B.R.P. Rao	700	439	123
Adilabad	T. Pullaiah, P.V. Prasanna and G. Obulesu	673	422	118
Karimnagar	A.H. Naqvi	1055	601	135
Warangal	C.S. Reddy	715	431	111
Khammam	V.S. Raju			
Nalgonda	P.N. Rao, B.V.R. Swamy and T. Pullaiah	506	329	96

Hyderabad District

Rajagopal (1973) enumerated 951 species under 583 genera and 124 families in Hyderabad district. The largest families are Poaceae (108) followed by Fabaceae (including Faboideae, Caesalpinioideae and Mimosoideae 103 species), Asteraceae (38), Cyperaceae (32), Euphorbiaceae (30 species), Malvaceae (28), Verbenaceae (28) Acanthaceae (22), Convolvulaceae (22) and Rubiaceae (20). The largest genera are *Cyperus* (16 species), *Cassia* (15 species), *Ipomoea* (11), *Euphorbia* (10) and *Crotalaria* (9). In addition he investigated the epidermal features in relation to taxonomy.

Ramana (2010) reported 1335 species (including cultivated ornamentals) in Greater Hyderabad, of which 536 species had appeared new and 77 species had disappeared. These 1335 species are spread over 724 genera and 160 families. The dominant families are Fabaceae (Leguminosae – 164 species), Poaceae (118), Arecaceae (78), Asteraceae (65), Cyperaceae (59), Euphorbiaceae (48), Malvaceae (38), Acanthaceae (36), Verbenaceae (28) and Rubiaceae (27). The dominant genera in Flora of Hyderabad are *Cyperus* (19 species), *Fimbristylis* (15), *Crotalaria* (15), *Eragrostis* (13), *Hibiscus* (12), *Ficus* (12), *Senna* (12), *Acacia* (11), *Indigofera* (11) and *Ipomoea* (11 species).

Karimnagar District

Naqvi (2001) worked on Flora of Karimnagar district for his Ph.D. thesis. He reported 1055 species of Angiosperms (including cultivated plants) belonging to 601 genera and 135 families. Fabaceae (Leguminosae) is the dominant family with 161 species followed by Poaceae (100 species), Euphorbiaceae (57), Cyperaceae (45), Asteraceae (44), Acanthaceae (35), Malvaceae (35) and Lamiaceae (23). Dominant genera include *Crotalaria* (20 species), *Euphorbia* (18), *Cyperus* (15), *Cassia* (13), *Indigofera* (12), *Ficus* (11), *Grewia* (9), *Acacia* (8) and *Phyllanthus* (7). Naqvi and Raju (1998) gave additions to the flora of Karimnagar district.

Khammam District

There is no systematic study on flora of Khammam district except for stray collections and reports by V.S. Raju.

Medak District

In Flora of Medak district Pullaiah *et al.* (1998) reported 708 wild naturalized species belonging to 414 genera and 119 families. Family Fabaceae (Leguminosae) is the dominant family which comprised about 104 species followed by Poaceae (83), Cyperaceae (49), Asteraceae (37), Euphorbiaceae (31) and Acanthaceae (22), Genus wise dominance include *Cyperus* and *Eragrostis* (12 species), *Crotalaria* and *Fimbristyllis* (11), *Indigofera* (10), *Cassia* and *Ipomoea* (9), and *Desmodium, Phyllanthus, Euphorbia* and *Schoenoplectus* (7 species). Some rare taxa for Telangana recorded from this district are *Plantago asiatica, Asparagus laevissimus, Elytrophorus spicatusi, Leersia hexandra, Pseudoraphis spinescens, Sehima sulcatum* and *Neanotis montholoni.* Floristic studies in the Narsapur taluk of Medak district by Narasimha Rao (1985) yielded 616 taxa of Angiosperms including cultivated species.

Mahabubnagar District

Ramachandrachary (1980) studied the flora of Achampet taluk in Mahabubnagar district. A floristic study of Mahabubnagar district was undertaken by Raghava Rao during 1983-89. A total of 1042 species occurring in the wild were recorded for the district (Raghava Rao, 1989). Based on his collections one new species has been described *Alysicarpus mahabubnagarensis* (Raghava Rao, 1990). *Habenaria ramayyana,* a new species was described from this district by Ramachandrachary and Wood (1981).

Nalgonda District

Flora of Nalgonda district was carried out by Rao *et al.* (2001). They recorded 506 species under 329 genera and 96 families. Fabaceae (Leguminosae) is the dominant family in this district which comprises 46 species followed by Poaceae (44), Euphorbiacae (34), Cyperaceae (25) and Asteraceae (22). *Cyperus* and *Crotalaria* are dominant genera represented by 10 species each, followed by *Indigofera, Cleome, Corchorus, Acacia, Chamaescye, Phyllanthus* and *Fimbristyllis* (6 species each). *Abrus fruticulosus, Corchorus urticifolius, Fuirena wallichiana, Ludwigia hyssopifolia, Mariscus sumatrensis, M. tenuifolius, Rhynchopora rubra* and *Seseli diffusum* are the new records to the State of Andhra Pradesh from this district. Reddy (2001) described a new species *Hybanthus vatsavayi* from Nalgonda district.

Table 5: Comparative Analysis of Dominant Families in different Districts in Telangana

Name of the District	Name of the Family	Total Number of Species
Adilabad district	1. Fabaceae (Leguminosae)	88
	2. Poaceae	85
	3. Cyperacae	39
	4. Euphorbiaceae	34
	5. Asteracae	31
	6. Rubiaceae	25
	7. Acanthaceae	21
	8. Convolvulaceae	17
	9. Tiliaceae	15
	10. Lamiaceae	13
Hyderabad District	1. Fabaceae	109
	2. Poaceae	34
	3. Euphorbiaceae	34
	4. Asteraceae	32
	5. Acanthaceae	29
	6. Malvaceae	22
	7. Convolvulaceae	22
	8. Lamiaceae	21
	9. Rubiaceae	18
	10. Cyperaceae	18
	11. Verbenaceae	18
	12. Amaranthaceae	16
Karimnagar District	1. Fabaceae (Leguminosae)	161
	2. Poaceae	100
	3. Euphorbiaceae	57
	4. Cyperaceae	45
	5. Asteraceae	44
	6. Acanthaceae	35
	7. Malvaceae	35
	8. Lamiaceae	23
Medak	1. Fabaceae (Leguminosae)	104
	2. Poaceae	83
	3. Cyperaceae	49
	4. Asteraceae	37
	5. Euphorbiaceae	31
	6. Acanthaceae	22

Contd...

Table 5–*Contd...*

Name of the District	Name of the Family	Total Number of Species
Nalgonda	1. Fabaceae (Leguminosae)	46
	2. Poaceae	44
	3. Euphorbiacae	34
	4. Cyperaceae	25
	5. Asteraceae	22
Nizamabad District	1. Fabaceae (Leguminosae)	108
	2. Poaceae	60
	3. Euphorbiaceae	35
	4. Asteraceae	34
	5. Cypeaceae	34
	6. Acanthaceae	24
	7. Rubiaceae	19
	8. Lamiaceae	18
	9. Scrophulariaceae	17
	10. Malvaceae	17
Ranga Reddi District	1. Fabaceae (Leguminosae)	109
	2. Poaceae	70
	3. Cyperaceae	56
	4. Euphorbiaceae	33
	5. Asteraceae	32
	6. Acanthaceae	24
	7. Rubiaceae	19
	8. Convolvulaceae	17
	9. Lamiaceae	17
	10. Asclepiadaceae	14

Nizamabad District

The district is with good irrigation resulting in nearly 40 per cent of the total geographical area under cultivation. The cultivated fields harbour good number of weeds. A total of 708 species wild and naturalized species were reported in the district, belonging to 436 genera and 123 families (Pullaiah and Ravi Prasad Rao, 1995). The family Leguminosae with 108 species is dominant family in the district. An analysis of the dominant genera in the district indicates that as many as 20 genera are represented by 5 or more than 5 species. *Cyperus* is the largest genus with 12 species, followed by *Euphorbia* (11). *Crotalaria* and *Ipomoea* (10 species each). *Chrysanthellum americanum, Rumex dentatus* and *Potamogeton crispus* were recorded in this district as new reports for Andhra Pradesh. Some species recorded as rare include *Malachra capitata, Crotalaria hirta, Indigofera nummularifolia, Acacia polyacantha, Sutera dissecta, Rotala serpyllifolia, Euphorbia laeta, Homonoia retusa* and *Tenagocharis latifolia.*

Ranga Reddi District

Pullaiah and Silar Mohammed (2000) explored Ranga Reddi district and brought out the District Flora. A total of 698 wild and naturalized species belonging to 414 genera and 110 families have been reported in Ranga Reddi district. The family Fabaceae (Leguminosae) with 109 species is dominant followed by Poaceae (70), Cyperaceae (56), Euphorbiaceae (33) and Asteraceae (32). *Cyperus* is the largest genus with 16 species, followed by *Fimbristylis* (14), *Cassia* (10), *Eragrostis, Euphorbia, Indigofera* and *Ipomoea* (9 species each). Some rare taxa found in this district are *Utricularia stellaris, Drosera burmannii, D. indica* and *Arisaema leschenaultia.* From this district some taxa are additions to Flora of Andhra Pradesh such as *Fimbristylis alboviridis, F. dichotoma* subsp. *podocarpa* and *Alysicarpus ovalifolius.*

Warangal District

In 1966, Sebastine and Henry studied the Flora of Pakhal and surrounding regions of Narasampet taluk in Warangal district and reported 254 species of 198 genera belonging to 70 families. Later Reddy (1985) made floristic study of Warangal and also investigated the epidermology of Cyperaceae.

Present Work

The present work is based on the specimens collected by the author and his associates during the last 34 years (1980-2013). The abbreviations of the collectors of Sri Krishnadevaraya University are as follows:

AJR: A.J.Ram

AMR: A. Madhusudana Reddy

ANS: A. Narayana Swami

AU: A. Ugraiah

BR: B. Ravi Prasad Rao

BSS: B. Sadasivaiah

CP: C. Prabhakar

CPR: C. Prabhakar Raju

DAM: D. Ali Moulali

DMR: D. Muralidhara Rao

DV: D. Veeranjaneyulu

EC: E. Chennaiah

GO: G. Obulesu

GS: G. Sivaram

KH: K. Hanumanthappa

KI: K.Indira

KP: K. Prasad

KSM: K. Sri Rama Murthy

MB: M.Bheemalingappa

MCK : M. Chennakesavulu

MHR : M. Hemambara Reddy

MSG : M.S. Gayathri

MSM : M. Silar Mohammed

MV : M.Venkateswarlu

MVS : M.V.Suresh

NV : N. Venkatappa

NVN : N.V. Nagalakshmi

PRSR : P. Rajasekhara Reddy

PSPB : P. Surya Prakash Babu

PVP : P.V.Prasanna

RVR : R.R. Venkata Raju

SKB : S. Khader Basha

SS : S. Sunitha

SSR : S.Sandhya Rani

TDCK : T. Dharma Chandra Kumar

TP : T. Pullaiah

TSS : T. Shali Saheb

VSL : V. Sreenivasulu

The specimens collected by the above botanists are deposited in the Herbarium of department of Botany, Sri Krishnadevaraya University, Anantapur. The dry method of collection has been followed. Author has also consulted various National and Regional herbaria besides herbaria of the universities in Andhra Pradesh. Specimens examined by the authors at various herbaria are given in the enumeration and the herbaria acronyms are given in the brackets. The abbreviations of plant collectors from other herbaria are as follows.

ANH : A.N. Henry

AR : A. Ragan

BVRS : B.V. Raghava Swamy

CAB : C.A. Barber

CSR : C.S. Reddy

DN : D. Narasimhan

GNR : G. Narasimha Rao

GVS : G.V. Subba Rao

JSG : J.S. Gamble

KCJ : K.C. Jacob

KMS : K.M. Sebastine

KNR : K.N. Reddy

KS : K. Subramanyam
KTH : K. Thothathri
LJGV : L.J.G. Van der Maesen
MRS : M.R. Suxena
MVR : M.Venkata Ramana
NPBK : N.P. Balakrishnan
NRR : N.Rama Rao
PVS : P.V.Sreekumar
RC : R. Chandrasekhar
RG : R.Gopalan
RKP : R.K. Premnath
RR : R. Rajan
SLK : S.L. Kapoor
SRR : S.R.Raghava Rao
SRS : S.R. Srinivasan
STR : S.T. Ramachandrachari
TR : T. Rajagopal
TRS : T. Ravi Shankar
VBH : V.B. Hosagondar
VNS : V. Narayana Swami
VSR : Vatsavaya S. Raju

Abbreviations – Districts

ADB : Adilabad
HYD : Hyderabad
KRN : Karimnagar
KMM : Khammam
MBNR : Mahaboobnagar
MDK : Medak
NLG : Nalgonda
NZB : Nizamabad
RR : Ranga Reddi
WGL : Warangal

Other Abbreviations

auct non : auctorum non; not of authors
auct. pl. : auctorum plurimorum; of most authors
BSI : Botanical Survey of India

ca :	*circa*; about
CAL :	Central National Herbarium, Botanical Survey of India, Howrah
cm :	centimeter
ed. :	editor, edition
emend :	*emedavit*; emeded
& *al.* :	and others
etc. :	et cetera; and other things
ex :	Published by
f. :	*filius*; son
Fig. :	Figure
Fl. & Fr.: :	Flowering and fruiting
Ic. :	Icone; illustration
loc. cit. :	*loco citato;* at the place cited
m :	meter
MH :	Madras Herbarium, Botanical Survey of India, Coimbatore
nom. illegit. :	*nomen illegitimum*; illegitimate name
nom. nud. :	*nomen nudum;* name published without description and without reference to previous published description
NU :	Acharya Nagarjuna University Herbarium
p.p. :	*pro parte*; in part
RF :	Reserved Forest
SKU :	Sri Krishnadevaraya University Herbarium
sp :	species
spp :	species (plural)
syn :	Synonym
t. :	tabula; plate
var. :	variety
Vern. :	Vernacular name; local name

Plan of Enumeration

Bentham and Hooker's System (1862-83) of classification has been followed in the enumeration of families with certain exceptions to accommodate recent changes. A dichotomous, indented key is given for the genera and species. The genera under each family and species under each genus are arranged in alphabetical sequence. Nomenclature of each species has been checked according to International code of Nomenclature for Algae, Fungi and Plants-Melbourne Code – 2011 (McNeill., 2012). Conserved genera are indicated by *nom. cons.* after the generic name.

Under each species the original citation is given. The relevant basionyms are given. Synonyms if necessary to connect the Hooker's Flora of British India, Gamble's

Flora of Presidency of Madras, some authenticated references that appeared in the latest floras and monographic works are also given in the citation wherever necessary. The citation is followed by a detailed discription of the species. At the end of description, its distribution and relative abundance in the State of Telangana is given. Vernacular names if any are also given. The flowering and fruiting period either observed by the author or noted from collections seen by us from earlier work is given. The field numbers of the specimens collected by the authors and also the field numbers of earlier collectors from Telangana present in different herbaria are given. The name of districts in which the particular species is available has been given in the brackets, in the abbreviated form. The names of herbaria where the specimens have been housed are given in the brackets, except for the collections that are housed in the Sri Krishnadevaraya University. Ecological notes and notes on nomenclatural changes are given wherever necessary.

Plants Endemic to Peninsular India found in the State of Telangana

1. *Adenostemma lavenia* (L.) Kuntze
2. *Aglaia elaeagnoidea* (A. Juss.) Benth. var. *beddomei* (Gamble) K.K.N.Nair
3. *Andrographis serpyllifoila* (Rottl. ex Vahl) Wight
4. *Argyreia cuneata* (Willd.) Ker.Gawl.
5. *Argyreia kleiniana* (Roem. & Schult.) Raizada
6. *Argyreia pilosa* Arn.
7. *Aspidopterys indica* (Roxb.) Hochr.
8. *Bridelia retusa* (L.) A.Juss
9. *Caralluma adscendens* (Roxb.) R.Br. var. *adscendens*
10. *Caralluma adscendens* (Roxb.) R.Br. var. *attenuata*
11. *Caralluma adscendens* (Roxb.) R.Br. var. *fimbriata*
12. *Ceropegia candelabrum* L. var. *candelabrum*
13. *Ceropegia spiralis* Wight
14. *Crotalaria epunctata* Dalzell
15. *Crotalaria paniculata* Willd. var. *nagarjunakondensis* Thoth.
16. *Crotalaria willdenowiana* DC. subsp. *willdenowiana*
17. *Cyperus clarkei* Cooke
18. *Decalepis hamiltonii* Wight & Arn.
19. *Deccania pubescens* (Roth) Tirveng. var. *candolleana* (Wight & Arn.) Tirveng.
20. *Desmodiastrum racemosum* (Benth.) A. Pramanik & Thoth., var. *racemosum* (Syn.: *Alysicarpus racemosus* Benth.)
21. *Dicliptera cuneata* Nees
22. *Dimeria orissae* Bor

23. *Dolichandrone atrovirens* (Heyne ex Roth) Sprague
24. *Dyschoriste vagans* (Wight) Kuntze
25. *Eriolaena lushingtonii* Dunn
26. *Habenaria roxburhgii* Nicolson
27. *Hemigraphis latebrosa* (Heyne ex Roth) Nees
28. *Indigofera mysorensis* Rottl. ex DC
29. *Lepidagathis mitis* Dalzell
30. *Maerua apetala* (Roth) Jacobs
31. *Mucuna pruriens* (L.) DC. var. *hirsuta* (Wight & Arn.) Wilmot-Dear, (Syn.: *M. hirsuta* Wight & Arn.)
32. *Ochna gamblei* King ex Brandis
33. *Piper hymenophyllum* Miq.
34. *Polyalthia cerasoides* (Roxb.) Bedd.
35. *Polycarpaea aurea* Wight & Arn.
36. *Rostellularia crinita* (Nees) Nees
37. *Senna montana* (Heyne ex Roth) V. Singh
38. *Sesbania procumbens* (Roxb.) Wight & Arn.,
39. *Sophora interrupta* Bedd.
40. *Taxillus heyneanus* (Schult.) Danser
41. *Tephrosia strigosa* (Dalzell) Santapau & Maheshw.
42. *Tribulus subramanyamii* P.Singh
43. *Tricholepis radicans* (Roxb.) DC.
44. *Wendlandia gamblei* Cowan

Floristic Analysis

Total taxa 1945 (including 163 cultivated taxa) were recorded in the Telangana State. These 1945 taxa belong to 1891 species spread over 794 genera and 147 families.

	No. of Families	*No. of Genera*	*No. of Species*
Dicotyledons	118	624	1412
Monocotyledons	29	170	479
Total	147	794	1891

Largest families are Fabaceae (Leguminosae) (273 species; 191+40 +42), Poaceae (208 species), Cyperaceae (126 species), Euphorbiaceae (118), Asteraceae (84), Acanthaceae (60), Rubiaceae (50), Malvaceae (47), Lamiaceae (42), Convolvulaceae

(39), Asclepiadaceae (36) and Scrophulariaceae (29). Orchidaceae, one of the top ten families in Flora of India is represented by only 12 species in the State of Telangana.

Largest genera are *Cyperus* (42 species), *Euphorbia* (29), *Crotalaria* (28), *Fimbristylis* (25 species), *Indigofera* (20), *Ficus* (18), *Ipomoea* (18), *Eragrostis, Acacia* and *Phyllanthus* (17 species each).

Artificial Key to the Families of Flowering Plants in Telangana

B. Ravi Prasad Rao

Note

1. When a character applies to 'up to' two genera/species in a family, these are given in brackets at the end of the concerned lead.
2. The number of times a family keyed out is indicated by roman numerals in bracket after family name.

KEY TO MAIN GROUPS

1. Stem with central pith, surrounded by concentric rings of vascular bundles with cambium; leaves usually net-veined, stipulate or exstipulate flowers 2-polymerous, cotyledons usually 2, lateral; tap root usually persistent (DICOTS):

   2. Perianth in one whorl, apetalous OR perianth in two whorls, monochlamydeous OR perianth absent .... **GROUP 1**

   2. Perianth in 2-many whorls, differentiated into petals and sepals;

      3. Petals free .... **GROUP 2**

      3. Petals connate .... **GROUP 3**

1. Stem without a central pith, vascular bundles scattered: without cambium; leaves usually parallel-veined, exstipulate; flowers usually 3-merous; cotyledons 1, terminal or undifferentiated; mature root system adventitious (MONOCOTS) .... **GROUP 4**

GROUP 1

Dicots with one whorl of perianth, apetalous OR with 2 whorls of perianth, monochlamydeous OR without perianth.

1. Plants parasitic or semi-parasitic:

    2. Root parasites .... **Santalaceae**

    2. Stem parasites:

        3. Branchlets flattened .... **Viscaceae**

        3. Branchlets not flattened:

            4. Branchlets wiry; perianth limb
campanulate .... **Lauraceae (I) (Cassytha)**

            4. Branchlets not wiry; perianth limb tubular .... **Loranthaceae**

1. Plants autotrophic:

    5. Climbers or twiners:

        6. Perianth absent .... **Piperaceae**

        6. Perianth present:

            7. Ovary inferior; stamens adnate to stylar column ...... **Aristolochiaceae**

            7. Ovary superior; stamens free:

                8. Leaves succulent; fruit enclosed in fleshy perianth ... **Basellaceae**

                8. Leaves not succulent; fruit not as above:

                    9. Carpels free; fruits feathery .... **Ranunculaceae**

                    9. Carpels 1, if more, syncarpous;
fruits not feathery .... **Euphorbiaceae (I) (Tragia)**

    5. Other than climbers or twiners:

        10. Aquatic .... **Ceratophyllaceae**

        10. Terrestrial:

            11. Stems with jointed branchlets; leaves reduced .... **Casuarinaceae**

            11. Stems without jointed branchlets; leaves reduced
or well developed:

                12. Inflorescence a syconium .... **Moraceae (I) (Ficus)**

                12. Inflorescence other than a syconium:

                    13. Flowers unisexual:

                        14. Ovary inferior .... **Begoniaceae**

                        14. Ovary superior, if half-inferior,
tepals 6-10, shortly connate:

                            15. Ovary 1-locular with erect ovule .... **Urticaceae**

15. Ovary 1-many locular with pendulous ovules or ovules on axile placenta:

    16. Flowers aggregated in umbels .... **Lauraceae (II) (Litsea)**

    16. Flowers not aggregated into umbels:

        17. Inflorescence a cyathium .... **Euphorbiaceae (II)**

        17. Inflorescence other than a cyathium:

            18. Cells of ovary 2-ovuled, if 1-ovuled, then stamens alternating with sepals:

                19. Unarmed, styles 3 .... **Euphorbiaceae (III)**

                19. Armed, if unarmed then styles 1 .... **Flacourtiaceae (I) (Flacourtia)**

            18. Cells of ovary 1-ovuled and stamens opposite to sepals:

                20. Trees:

                    21. Leaves 2-pinnatifd .... **Proteaceae**

                    21. Leaves not pinnatifid:

                        22. Plants with milky juice .... **Moraceae (II)**

                        22. Plants without milky juice .... **Ulmaceae (I)**

                20. Herbs or undershrubs ... **Cannabaceae**

13. Flowers bisexual:

    23. Leaves opposite or whorled:

        24. Stamens attached to perianth lobes:

            25. Ovary 1-celled .... **Combretaceae (I) (Calycopteris & Terminalia)**

            25. Ovary 4-celled .... **Lythraceae (I) (Ammannia & Nesaea)**

        24. Stemens free from perianth:

            26. Fruit an anthocarp .... **Nyctaginaceae**

            26. Fruit a capsule:

                27. Tepals free .... **Aizoaceae**

                27. Tepals connate .... **Molluginaceae**

    23. Leaves alternate:

        28. Flowers bracteate and 2-bracteolate with scarious perianth .... **Amaranthaceae**

28. Flowers otherwise:

 29. Perianth many-seriate .... **Magnoliaceae**

 29. Perianth 1-3-seriate:

  30. Stipules ochraceous .... **Polygonaceae**

  30. Stipules not ochraceous:

   31. Leaves 2-foliolate; fruit a
   legume .... **Fabaceae (Caesalpinioideae (I) (Hardwickia)**

   31. Leaves other than 2-foliolate; fruit not a legume:

    32. Ovary half-inferior ....
    .... **Combretaceae (II) (Anogeissus, Terminalia** *p.p.)*

    32. Ovary superior:

     33. Stamens on staminal column ....
     .... **Sterculiaceae (I) (Sterculia, Firmiana)**

     33. Stamens not on staminal column:

      34. Stamens numerous .... **Annonaceae (I)**

      34. Stamens 4-10:

       35. Staminodes present:

        36. Stamens 6-8, connate basally
        with staminodes ....
        .... **Flacourtiaceae (II) (Casearia)**

        36. Stamens 4-7, free from
        staminodes .... **Hernandiaceae**

       35. Staminodes absent:

        37. Stems armed ....
        .... **Rhamnaceae (I) (Ziziphus rugosa)**

        37. Stems unarmed:

         38. Leaves even pinnate ....
         .... **Sapindaceae (I) (Schleichera)**

         38. Leaves simple:

          39. Trees .... **Ulmaceae (II)**

          39. Herbs or shrubs:

           40. Ovary 1-celled
           1-ovuled .... **Chenopodiaceae**

           40. Ovary 2-4-celled, ovules
           2 per cell .... **Sapindaceae (II)**
           .... **(Dodonaea)**

GROUP 2

Dicots with perianth in 2-many whorls, differentiated into petals and sepals; petals free

1. Plants insectivorous, leaves with glandular hairs, styles 3-5 .... **Droseraceae**

1. Plants otherwise:

    2. Fruit a legume or lomentum, very rarely indehiscent:

        3. Corolla papilionaceous, descendingly imbricate, the posterior petal outermost .... **Fabaceae (Faboideae)**

        3. Corolla otherwise:

            4. Flowers actinomorphic; perianth lobes valvate in bud .... **Fabaceae (Mimosoideae (I)**

            4. Flowers zygomorphic; perianth lobes imbricate in bud .... **Fabaceae** (**Caesalpinioideae (II)**

    2. Fruit otherwise:

        5. Ovary half-inferior or inferior:

            6. Plants with tendrils .... **Cucurbitaceae (I)**

            6. Plants without tendrils:

                7. Plants aquatic, with dimorphic leaves or peltate leaves

                    8. Leaves pelate; carpels sunk in a fleshy to rub .... **Nymphaeaceae**

                    8. Leaves dimorphic; carpels not sunk in a fleshy to rub .... **Trapaceae**

                7. Plants terrestrial, if aquatic, then leaves not dimorphic or peltate:

                    9. Leaves opposite or whorled:

                        10. Sepals 2 .... **Portulacaceae (I) (Portulaca)**

                        10. Sepals 4-14:

                            11. Anthers dehisce by terminal pores .... **Melastomataceae (I) (Osbeckia)**

                        11. Anthers dehisce longitudinally:

                            12. Small herbs .... **Vahliaceae**

                            12. Large erect or climbing shrubs or trees:

                                13. Fruit 4-5-winged .... **Combretaceae (III)** .... **(Combretum)**

                                13. Fruit not winged:

             14. Stamens 8 .... **Malastomataceae (II)**
                        .... **(Memecylon)**

             14. Stamens numerous .... **Myrtaceae**

  9. Leaves alternate or in basal rosettes:

      15. Ovary semi-inferior .... **Rosaceae**

      15. Ovary inferior:

          16. Ovary 2-celled crowned by disk;
              fruit of 2 mericarps .... **Apiaceae**

          16. Ovary 1-many-celled, not crowned
              by disk; fruit otherwise:

             17. Plants herbaceous .... **Onagraceae**

             17. Plants woody:

                18. Petals crumpled in bud .... **Punicaceae**

                18. Petals not crumpled in bud:

                    19. Stamens 4-10 .... **Alangiaceae**

                    19. Stamens many .... **Lecythidaceae**

  5. Ovary superior:

    20. Stamens more than 15; rarely 11-12, then placentation axile:

      21. Ovary apocarpous:

          22. Plants aquatic with floating leaves .... **Nelumbonaceae**

          22. Plants terrestrial

             23. Flowers 3-merons .... **Annonaceae (II)**

             23. Flowers 5-merons .... **Dillimaceae**

      21. Ovary monocarpous or syncarpous :

          24. Stamens monadelphous with
             monothecous anthers .... **Malvaceae**

          24. Stamens and anthers otherwise:

             25. Leaves opposite:

                26. Fruit a drupe .... **Clusiaceae**

                26. Fruit a capsule .... **Lythraceae (II)**
                    .... **(Lagerstroemia, Woodfordia)**

             25. Leaves alternate:

                27. Flowers unisexual .... **Euphorbiaceae (IV)**

27. Flowers bisexual:

    28. Sepals 3:

        29. Placentation parietal .... **Papaveraceae**

        29. Placentation free-central ....
          .... **Portulacaceae (II) (Talinum)**

    28. Sepals more than 3:

        30. Placentation parietal:

            31. Stamens inserted at the base or
                top of the gynophore .... **Capparaceae (I)**

            31. Stamens not on
                gynophore .... **Bixaceae II (Bixa)**

        30. Placentation other than parietal:

            32. Leaves 3-9-foliolate:

                33. Leaves 3-foliolate,
                    gland-dotted;
                    stamens free .... **Rutaceae (II) (Aegle)**

                33. Leaves 5-9-foliolate, not gland
                    dotted; stamens mono or
                    polyadelphous .... **Bombacaceae**

            32. Leaves simple or palmately lobed :

                34. Anthers poricidal:

                    35. Petals fimbriate .... **Elaeocarpaceae**

                    35. Petals entire:

                        36. Fruit a capsule .... **Bixaceae**
                          .... **(Cochlospermum)**

                        36. Fruit of drupelets .. **Ochnaceae**

                34. Anthers dehisce by longitudinal
                  slits or lateral clefts:

                  37. Stamens monadelphous ....
                    .... **Sterculiaceae (II)**

                  37. Stamens free, if connate at
                    base, then filaments in bundles .....
                    .... **Tiliaceae (I)**

20. Stamens 10 or fewer:

    38. Plants with tendrils:

        39. Flowers with distinct corona between petals
          and stamens .... **Passifloraceae**

39. Flowers otherwise:

 40. Fruit an inflated trigonous capsule .... **Sapindaceae (III) (Cardiospermum)**

 40. Fruit a berry, often watery .... **Vitaceae**

38. Plants without tendrils:

 41. Leaves minute, scale like, imbricating the stem .... **Tamaricaceae**

 41. Leaves well-developed:

  42. Climbers or twiners:

   43. Climbing by means of spiral hooks .... **Linaceae (I) (Hugonia)**

   43. Climbing or twining otherwise:

    44. Fruit enclosed in accrescent calyx .... **Olacaceae (I) (Olax)**

    44. Fruit not enclosed in accrescent calyx:

     45. Calyx segments 6 .... **Menispermaceae**

     45. Calyx segments 4-5:

      46. Petals cucullate, often clawed .... **Rhamnaceae (II)**

      46. Petals otherwise:

       47. Armed:

        48. Leaves opposite .... **Malpighiaceae**

        48. Leaves alternate .... **Rutaceae (III) (Toddalia)**

       47. Unarmed:

        49. Fruit is a drupe .... **Opiliaceae (I) (Opilia)**

        49. Fruit is other than a drupe ...... **Celastraceae (I)**

  42. Non-climbers or twiners:

   50. Trees:

    51. Leaves opposite:

     52. Calyx persistent .... **Celastraceae (II) (Pleurostylia)**

     52. Calyx not persistent .... **Rutaceae (IV)**

    51. Leaves alternate:

     53. Leaves 2-foliolate .... **Balanitaceae**

     53. Leaves 1 or 3-many foliolate:

      54. Leaves simple, 1-foliolate:

       55. Stamens 10, monadelphous .... **Erythroxylaceae**

       55. Stamens otherwise:

        56. Fruit a capsule or indehiscent with 15-20 bony pyrenes:

57. Stamens on a gonophore ....
.... **Euphorbiaceae (V) (Cleistanthus)**

57. Stamens otherwise ....
.... **Celastraceae (Meytenus rufa)**

56. Fruit a drupe:

58. Ovules erect .... **Rhamnaceae (III)**

58. Ovules pendulous:

59. Disc with fimbriate corona ....
.... **Euphorbiaceae (VI) (Bridelia)**

59. Disc without corona ....
.... **Anacardiaceae (I)**

54. Leaves 3-many-foliolate :

60. Leaves 2-3-pinnate:

61. Fruit a berry .... **Leeaceae (I)**

61. Fruit a drupe or capsule:

62. Fruit an elongated trigonus
capsule .... **Moringaceae**

62. Fruit a drupe .... **Meliaceae (I) (Melia)**

60. Leaves 3-foliolate or 1-pinnate:

63. Stamens monadelphous,
staminal tube cupular,
apically lobed .... **Meliaceae (II)**

63. Stamens free:

64. Leaves gland-dotted:

65. Fruit a capsule .... **Flindersiaceae**

65. Fruit other than capsule ....
.... **Rutaceae (V)**

64. Leaves not gland-dotted:

66. Style simple:

67. Bark resinous .... **Burseraceae**

67. Bark not resinous ....
.... **Sapindaceae (III)**

66. Styles 4-5:

68. Leaflets 3-4 pairs ....
.... **Simaroubaceae**

68. Leaflets more than 7 pairs ....
.... **Anacardiaceae (II)**

50. Herbs or shrubs:

  69. Leaves 2-3 pinnate .... **Leeaceae(II)**

  69. Leaves simple or 3-7-foliolate or simple pinnate:

    70. Leaves pinnatisect, stamens 6, diadelphous or tetradynamous:

      71. Stamens tetradynamous .... **Brassicaceae**

      71. Stamens diadelphous .... **Fumariaceae**

    70. Leaves and stamens otherwise:

      72. Leaves simple:

        73. Flowers spurred:

          74. Lower sepal spurred .... **Balsaminaceae**

          74. Sepals over petal spurred; stamens 5 not spurred .... **Violaceae**

        73. Flowers not spurred:

          75. Leaves opposite:

            76. Placentation free central .... **Caryophyllaceae**

            76. Placentation axile:

              77. Stipules paired .... **Elatinaceae**

              77. Stipules absent, if present not paired .... **Lythraceae (III)**

          75. Leaves alternate:

            78. Pedicels adnate to the petiole with 2 apical bracteoles .... **Turneraceae**

            78. Pedicels otherwise:

              79. Stamens forming a column or united into a sheath:

                80. Petals 3 .... **Polygalaceae**

                80. Petals 5 .... **Sterculiaceae (III)**

              79. Stamens free or united at base or on a gynophore:

                81. Stamens connate at base or adnate to petals:

                  82. Fruit a drupe .... **Olacaceae (II) (Ximenia)**

                  82. Fruit a capsule ..... **Linaceae (II)**

81. Stamens free or on a gynophore:

    83. Stamens unilaterally arranged on a gynophores .... **Capparaceae (II) (Cadaba)**

    83. Stamens free:

        84. Armed .... **Celastraceae (IV)** .... **(Maytenus)**

        84. Unarmed .... **Tiliaceae (II)**

72. Leaves compound:

    85. Leaves even-pinnate .... **Zygophyllaceae**

    85. Leaves odd pinnate or 3-7-foliolate:

        86. Herbs:

            87. Placentation axile... .... **Oxalidaceae (I) (Oxalis)**

            87. Placentation parietal .... **Capparaceae III (Cleome)**

    86. Shrubs:

        88. Stamens monadelphous .... **Meliaceae (III) (Cipadessa)**

        88. Stamens free:

            89. Leaves gland-dotted ...... **Rutaceae (VI)**

            89. Leaves not gland-dotted .... **Anacardiaceae (II) (Rhus)**

GROUP 3

(Dicots with 2 whorls of perianth, differentiated into petals and sepals; petals united)

1. Plants parasitic or insectivorous with 2 stamens :

    2. Plant aquatic, marshy, insectivorous .... **Lentibulariaceae**

    2. Plants terrestrial, parasitic:

        3. Stems wiry, twining .... **Cuscutaceae**

        3. Stems otherwise:

            4. Leafless root parasites .... **Orobanchaceae**

            4. Leafy semi parasites .... **Scrophulariaceae (I)**

1. Plants otherwise:

    5. Stems fleshy, columnar, articulated, leaves caducous .... **Cactaceae**

    5. Stems and leaves otherwise:

6. Stamens free from corolla:
  7. Leaves simple:
    8. Flowers polygamous or unisexual:
      9. Fruit a berry with enlarged accrescent calyx .... **Ebenaceae**
      9. Fruit otherwise:
        10. Seeds curved .... **Menispermaeae (III)**
        10. Seeds not curved .... **Euphorbiaceae (VII) (Jatropha & Givotia)**
    8. Flowers bisexual:
      11. Stragglers .... **Opiliaceae (II) (Cansjera)**
      11. Non-stragglers:
        12. Flowers actinomorphic; filaments free .... **Campanulaceae**
        12. Flowers zygomorphic; filaments connate in the upper half .... **Lobeliaceae**
  7. Leaves compound .... **Campanulaceae**
    13. Fruit a legume or lomentum .... **Fabaceae (Mimosoideae (II)**
    13. Fruit other than a legume or lomentum:
      14. Stamens 10 .... **Oxalidaceae (II) (Biophytum)**
      14. Stamens 5 .... **Leeaceae (III)**
6. Stamens epipetalous:
  15. Ovary inferior or half-inferior:
    16. Inflorescence capitulum .... **Asteraceae**
    16. Inflorescence other than capitulum:
      17. Flowers unisexual; plants with tendrils ..... **Cucurbitaceae (II)**
      17. Flowers uni- or bisexual; plants without tendrils .... **Rubiaceae**
  15. Ovary superior:
    18. Corolla actinomorphic:
      19. Plants scapigerous .... **Plantaginaceae**
      19. Plants non-scapigerous:
        20. Leaves alternate:
          21. Plants aquatic, stoloniferous with peltate leaves .... **Menyanthaceae**
          21. Plants otherwise:
            22. Carpels 4, with hypogynous scales .... **Crassulaceae**

22. Carpels otherwise:

    23. Stamens more than the number of corolla lobes .... **Sapotaceae**

    23. Stamens equal or less than the number of corolla lobes:

        24. Ovary 4-locular, if 2, then ovule solitary in each locule, basal or pendulous:

            25. Corolla plicate, fruit a capsule or berry .... **Convolvulaceae**

            25. Corolla not plicate, fruit a pyrene or drupe:

                26. Flowers in paniculate corymbose cymes; shrubs or trees .... **Cordiaceae**

                26. Flowers solitary or in spikes or racemes; subshrubs or herbs .... **Boraginaceae**

        24. Ovary 2 or rarely 5-6 locular; ovules 2 in each locule, axile:

            27. Styles 2-5:

                28. Styles 2 .... **Hydrophyllaceae**

                28. Styles 5 .... **Plumbaginaceae**

            27. Style solitary:

                29. Fruit a berry, if capsule, then spinescent .... **Solanaceae**

                29. Fruit a capsule, not spinescent ... **Scrophulariaceae** .... **(Verbascum) (II)**

20. Leaves opposite (except *Cascabela* & *Plumeria*)

    30. Latex milky; seeds with coma, anthers connate or connivent around the stigma:

        31. Corona present; pollen masses collected in pollinia .... **Asclepiadaceae**

        31. Corona absent; pollen masses not collected in pollinia .... **Apocynaceae**

    30. Latex watery or absent; seeds without coma; anthers free:

32. Stamens 2 .... **Oleaceae**

32. Stamens more than 2:

    33. Herbs:

        34. Ovary 2-4-celled .... **Gentianaceae**

        34. Ovary 1-celled .... **Primulaceae**

    33. Shrubs or trees:

        35. Fruit a berry .... **Loganiaceae**

        35. Fruit drupaceous:

            36. Ovary monolocular ...... **Salvadoraceae**

            36. Ovary 2-4-locular ....
            .... **Verbenaceae I (Tectona)**

18. Corolla zygomorphic:

    37. Fruit opening elastically from the apex of
    2 loculicidal valves .... **Acanthaceae**

    37. Fruit not opening elastically from the apex
    of 2 loculicidal valves:

        38. Seeds winged; leaves pinnately
        compound .... **Bignoniaceae**

        38. Seeds and leaves otherwise:

            39. Leaves alternate .... **Pedaliaceae**

            39. Leaves opposite or whorled:

                40. Fruit a capsule .... **Scrophulariaceae (III)**

                40. Fruit other than a capsule:

                    41. Style terminal; fruit a drupe
                    or of pyrenes .... **Verbenaceae (II)**

                    41. Style gynobasic; fruit of 4 nutlets ... **Lamiaceae**

GROUP 4
(MONOCOTS)

1. Flowers in spikelets with bristly or scaly perianth; ovary 1-celled:

    2. Stem triquetrous, solid, leaves more or less 3-ranked, with
    closed sheaths; flowers in the axils of single bract .... **Cyperaceae**

    2. Stem terete, hollow, leaves 2-ranked with open sheaths;
    flowers in between lemma and paleas .... **Poaceae**

1. Flowers not in spikelets; ovary 1-many-celled:

    3. Plants aquatic, marsh or riparian:

        4. Plant body thalloid, not exceeding 1 cm across .... **Lemnaceae**

4. Plant body otherwise:

   5. Totally submerged aquatics (with submerged flowers):

      5. Flowers bisexual .... **Potamogetonaceae**

      6. Flowers unisexual .... **Najadaceae**

   5. Plants submerged or not; if submerged,
       flowers always at or above the water surface:

      7. Flowers unisexual, in involucrate capitula;
         perianth scarious or membranous .... **Eriocaulaceae**

      7. Flowers uni-or bisexual, not in involucrate capitula:

         8. Trees or shrubs, dioecious; leaves armed .... **Pandanaceae**

         8. Herbs or subshrubs; leaves unarmed:

            9. Perianth absent:

               10. Flowers in 2 cylindrical super-imposed
                   brownish spikes .... **Typhaceae**

               10. Flowers otherwise .... **Araceae (I)**

            9. Perianth present:

               11. Perianth 1-seriate, 2-lobed .... **Aponogetonaceae**

               11. Perianth 2-seriate, 6-lobed; if 1-seriate,
                   perianth 3-lobed:

                   12. Flowers in umbels, involucrate with
                       spathaceous bracts:

                       13. Fruit indehiscent, of 3 or more
                           achenes; stamens 6 .... **Alismataceae**

                       13. Fruit dehiscent, of 6-7 follicles;
                           stamens 8-12 .... **Butomaceae**

                   12. Flowers otherwise:

                       14. Ovary inferior .... **Hydrocharitaceae**

                       14. Ovary superior:

                           15. Ovary 1-celled .... **Xyridaceae**

                           15. Ovary 3-celled:

                               16. Leaves radical, floating
                                 or submerged .... **Pontederiaceae**

                             16. Leaves cauline not floating
                                 or submerged .... **Commelinaceae (I)**

3. Plants terrestrial or epiphytic:

  17. Perianth absent .... **Araceae (II)**

  17. Perianth present:

18. Leaves compound; flowers enclosed in spathe like bracts .... **Arecaceae**

18. Leaves simple; if compound then flowers not enclosed in spathe like bracts:

    19. Flowers prominently zygomorphic:

        20. Fruit fleshy, indehiscent; fertile stamens 5 .... **Musaceae**

        20. Fruit not fleshy, dehiscent; fertile stamens 1 or 2:

            21. Median petal produced into a lip; stamens 2 or 1 united with the stigma bearing column; pollen in pollinia .... **Orchidaceae**

            21. Floral characters not as above:

                22. Leaves distichous or spirally arranged .... **Zingiberaceae**

                22. Leaves alternate .... **Cannaceae**

    19. Flowers actinomorphic:

        23. Climbers or twiners:

            24. Tendrils present; if absent, then leaves scaly:

                25. Petioles with 2 tendrils; leaves reticulately veined .... **Smilacaceae**

                25. Petioles without any tendrils; leaves parallel-veined .... **Liliaceae (I)**

            24. Tendrils absent; leaves well-developed

                26. Flowers unisexual; capsules winged ... **Dioscoreaceae**

                26. Flowers bisexual; capsules not winged .... **Stemonaceae**

        23. Non-climbers or twiners:

            27. Ovary inferior:

                28. Bracts dimorphic; outer ones leafy, inner ones filiform, pendent .... **Taccaceae**

                28. Bracts not as above:

                  29. Rootstock a tunicated bulb .... **Amaryllidaceae**

                  29. Rootstock otherwise:

                      30. Arborescent shrubs or stout herbs, leaves armed .... **Agavaceae**

                      30. Herbs, leaves unarmed .... **Hypoxidaceae**

27. Ovary superior:

    31. Stamens 6, all perfect .... **Liliaceae (II)**

    31. Stamens 3, if 6 then 1-3 reduced to staminodes or inflorescence enclosed in leaf sheaths .... **Commelinaceae (II)**

Systematic Enumeration

RANUNCULACEAE
CLEMATIS L.

Clematis gouriana Roxb. ex DC., Syst. Nat. 1: 138. 1817; Hook. f. & Thomson in Hook.f., Fl. Brit. India 1 : 5. 1872; Gamble, Fl. Madras 1: 3. 1915; Rau in B.D.Sharma & *al.*, Fl. India 1: 64. 1993.

Vern: *Gowri-kuntala.*

Glabrous climbers, up to 8 m long, stems and branches ribbed, ribs 6-12, sparsely hairy. Leaves opposite, 2-3-pinnate, leaflets ovate, ovate- lanceolate, 4.5-8 × 1.5-3 cm, pubescent below, base obtuse or cordate, margin entire, apex acuminate, glabrous or pubescent along nerves, petiole twining, up to 6 cm long, petiolules slender, up to 2.5 cm long. Flowers greenish-white in axillary or terminal, crowded panicles, apetalous; sepals 4, ca 1.1 cm long, oblong; stamens 30-35, filaments linear, flat, glabrous, connectives scarcely produced beyond the anther lobes; carpels 10-15, oblong or linear, hairy, styles up to 2.5 mm long, stigma clavate. Achenes 0.2-0.4 cm long, ovoid or narrowly oblong, flat, with to 5 cm long feathery style.

Occasional in hilly regions in Hyderabad district (Rajagopal, 1973).

Fl. & Fr.: November-March.

DILLENIACEAE
DILLENIA L.

Dillenia pentagyna Roxb.,Pl. Coromandel 1: 21, t. 20. 1795; Hook.f. & Thomson in Hook. f., Fl. Brit. India 1: 33. 1872; Gamble,Fl. Madras 1: 6. 1915; Majumdar in

B.D. Sharma & *al.*, Fl. India 1: 156. 1993.

Vern.: *Chinna kalinga, Parudu.*

Tall deciduous trees, up to 20 m high; trunk straight, cylindrical, crown rounded; bark grey or pale brown with shallow depressions. Leaves at the extremities of the branches, large, oblong-obovate, elliptic to oblanceolate, 15-70 × 7-25 cm, (the leaves of younger trees larger than those of older, often attaining a length of 120 cm or more), smooth and shining when old and downy when young, apex obtuse, margin crenate, base cuneate, lateral nerves 30-50 pairs, often forking towards the margins; petioles 2-6 cm long, glabrous above, densely strigose below. Flowers yellow, 3 cm across, fragrant, in fascicles on naked branches; sepals 5, ovate, acute or obtuse at apex, ciliate on margins, glabrous on both sides, ca 7 mm long; petals 5, obovate, 15-20 mm long, bright yellow, rounded at apex; stamens in 2 series, outer series with 60-90 stamens, erect, each 2.5-4 mm long, inner series up to 10 stamens, each 6-9 mm long with reflexed apex in bud; carpels 5 (-6), arranged on a narrow conical receptacle, oblong-lanceolate, glabrous, each with 5-20 ovules. Fruit subglobose, pendulous, fleshy, yellow, orange or red, 1.5 cm across1(-2)-seeded; seeds ovoid, 3 x 2 mm, glabrous, black, exarillate.

Frequent in Tadwai and Mallur in Warangal district in deciduous forests. .

Fl. & Fr.: February-March.

Tadwai (WGL), *CSR* 1025 (KU).

MAGNOLIACEAE

MAGNOLIA L.

Magnolia champaca (L) Baill. ex Pierre in Fl. Forest. Cochinch. t. 3. 1880. *Michelia champaca* L., Sp. Pl. 536. 1753; Hook.f. & Thomson, Fl. Ind. 79. 1855; Hook.f., Fl. Brit. India 1 : 42. 1872; Gamble, Fl. Madras 1: 9. 1915; D.C.S. Raju in B.D.Sharma & *al.*, Fl. India 1: 175. 1993; Pullaiah & Chennaiah, Fl. Andhra Pradesh 1: 66. 1997.

Vern.: *Sampenga, Champakamu.*

Densely foliaceous, medium sized, evergreen trees, up to 20 m high, bark grey, thick, smooth, branches ascending, spreading, form a close crown, young branches appressed pubescent. Leaves ovate-lanceolate, thinly coriaceous, glabrescent, 10-30 × 5-10 cm, base subactue or cuneate, margin entire, apex long-acuminate, minutely puberulous beneath, secondary nerves 10-18 pairs, petiole 1-8 cm long. Flowers solitary, axillary, pale yellow becoming orange, fragrant, 5 cm across; bracts spathaceous, broadly ovate, silky outside, caducous; perianth parts 12-15, oblanceolate, oblong-lanceolate, fleshy; stamens numerous, 6-8 mm long, connective appendages to 1 mm long; carpels numerous, ovoid-oblong, stigmatic crest recurved. Fruit an aggregate of follicles, laxly arranged, fruitlets warty, 2 cm across, dehiscing by 2-valves, seeds 5, subglobose, scarlet red, arillate.

Cultivated especially in temples for the sake of its flowers (Ramana, 2010).

Fl. & Fr.: March-August.

ANNONACEAE

1. Stragglers or lianas .... **Artabotrys**

1. Trees or shrubs :

    2. Inner petals scaly or 0; ripe carpels connate into a large,
        fleshy syncarpium .... **Annona**

    2. Inner petals prominent, not scaly; ripe carpels free or distinct
        and from stalked monocarps:

        3. Connectives not produced above the anther lobes;
            monocarps few, tomentose or pubescent .... **Miliusa**

        3. Connectives produced above the anther lobes;
            monocarps numerous, glabrous:

            4. Seeds 1-5; leaf base not oblique .... **Polyalthia**

            4. Seeds 6-12 in 2 rows; leaf base oblique .... **Cananga**

ANNONA L.

1. Petals 6, the inner 3 conspicuous .... **A. muricata**

1. Petals 3, inner rudimentary or absent:

    2. Leaves elliptic, velvety outside; fruits smooth .... **A. cherimola**

    2. Leaves oblong or lanceolate, almost glabrous; fruits not smooth:

        3. Leaves obtuse, glaucous beneath; flowers greenish;
            fruit green, ovoid, with projecting, ovoid areoles .... **A. squamosa**

        3. Leaves acuminate, green beneath; flowers pale green;
            fruit orange, subglobose, with flat, 5-cornered areoles .... **A. reticulata**

Annona cherimola Mill., Gard. Dict. ed. 8. 5. 1768; Debika Mitra in B.D.Sharma & *al.*,
Fl. India 1: 205. 1993.

Small trees, young branches fulvous tomentose. Leaves elliptic, velvety tomentose
beneath. Flowers solitary, sometimes 2-3 together, extra-axillary or leaf-opposed;
sepals 3, valvate; petals 3+3, inner petals minute and scaly. Fruit a fleshy syncarp of
carpels and receptacle, ovoid; seeds numerous.

Cultivated for its edible fruit.

Fl. & Fr.: March – August.

Hyderabad, *PVP* & *MVR* 2289 (BSID).

Annona muricata L., Sp. Pl. 536. 1753; Debika Mitra in B.D.Sharma & *al.*, Fl. India 1:
206. 1993.

Vern.: *Lakshmana phalamu*; English: Prickly custard apple.

Evergreen small tree, 3-7 m high, branchlets pubescent when young and glabrous
when old. Leaves obovate-oblong, 11-14 × 4-5 cm, glabrous above, sparsely pubescent

beneath, base acute, margin entire, apex shortly acuminate, petiole 4-5 mm. Flowers yellowish, solitary, axillary, greenish yellow; sepals 3, valvate, triangular, ca 4 mm long, puberulous outside, glabrous inside, persistent; petals 6 (3+3), broadly ovate, subcordate at base, short acuminate at apex, sparsely puberulous out side, glabrous inside; stamens numerous, linear, 4-5 mm long, filaments broad at base, connective capitate at apex; carpels many, oblong, linear, 4 mm long. Fruit 6-8 cm long, ovoid, carpels dull green, tuberculate, covered with long, curved, fleshy spinules; seeds many, reddish-browh.

Planted in gardesns for edible fruits and evergreen foliage.

Fl. & Fr.: April – October.

Annona reticulata L., Sp. Pl. 537. 1753; Hook. f. & Thomson in Hook.f., Fl. Brit. India 1: 78. 1872; Gamble, Fl. Madras 1: 20. 1915; Debika Mitra in B.D.Sharma & *al.*, Fl. India 1: 207. 1993.

Vern. : *Rama phalamu*

Small, evergreen trees, up to 7 m high; branchlets grey-sericeous when young, later glabrous. Leaves oblong-lanceolate, 10-18 × 2-5 cm, base cuneate or rounded, margin entire, apex acuminate, nerves 15-18 paris, petiole to 1.8 cm. Flowers 2-4 on lateral, extra axillary cymes, greenish; pedicel up to 2 cm long; sepals 3, broadly ovate, pubescent outside, glabrous inside; petals apparently 3, linear oblong, but 3 minute ones represent the inner whorl, puberulous on both sides; stamens numerous, 1-1.3 mm long, connective ovoid at apex; carpels many, ovoid to linear, pubescent. Fruit subglobose or ovoid, 10-16 cm in diam., nearly smooth, with impressed lines, orange when ripe; seeds numerous, blackish, smooth, arillate.

Commonly planted in gardens and homesteads for its edible fruits.

Fl. & Fr.: April-June.

Tekrial (NZB), *TP* & *BR* 6420.

Annona squamosa L., Sp. Pl. 537. 1753; Hook.f. & Thomson in Hook.f., Fl. Brit. India 1: 78. 1872; Gamble, Fl. Madras 1: 20. 1915; Debika Mitra in B.D.Sharma & *al.*, Fl. India 1: 207. 1993.

Vern.: *Seethaphalamu*. Eng.: Custard apple.

Shrubs or small trees, up to 8 m high, bark grey, thin. Leaves elliptic or oblong-lanceolate, 4-8 × 2-3 cm, thin-coriaceous, glabrescent, glaucous beneath, lateral nerves 5-10 pairs, pellucid dotted with a peculiar smell, base obtuse, margin entire, apex obtuse, petiole to 1.5 cm. Flowers greenish, solitary, on leaf-opposed peduncles or 2-4 in a cluster; sepals broadly ovate-triangled, 2-3 x 0.7 mm, glabrous or pubescent; petals in two whorls, outer linear obong, 15-30 x 3-7 mm, pubescent, green with purple base, inner petals reduced into scales or absent; carpels ovoid, free or subconnate, up to 1 mm long, pubescent, ovules 1, style oblong, stigma entire. Fruit a fleshy syncarp, cordate or ovoid, 5-10 cm diam., pulp sweet; seeds many, 14 x 8 mm, angular, oblong, black, shining, arillate.

Occasional in forests and wastelands in all districts and planted in gardens for its fruits.

Fl. & Fr.: April-November.

Saikara (ADB), *GO* & *DAM* 4987; Laxmapur (NZB), *TP* & *BR* 6269; Ramayampet (MDK), *BR* & *CP* 11510; Amaragiri (MBNR), *BR* & *TSS* 29066; Mohammadabad (RR), *MSM* & *KH* 10206; Mannanur (MBNR), *TDCK* 16811; Kodimial (KRN), *GVS* 20034 (MH); Yenkur (KMM), *RR* 107928 (BSID).

ARTABOTRYS R. Br.

Artabotrys hexapetalus (L.f.) Bhandari, Baileya 12: 149. 1964; Debika Mitra in B.D.Sharma & *al.*, Fl. India 1: 251. 1993. *Annona hexapetala* L.f., Suppl. Pl. 270. 1781. *Artabotrys odoratissimus* (Roxb.) R. Br., Bot. Reg. 5: 423. 1820; Hook.f. & Thomson in Hook.f., Fl. Brit. India 1: 54. 1872; Gamble, Fl. Madras 1: 14. 1915. *Uvaria odoratissima* Roxb., Fl. Ind. 2: 666. 1832.

Vern.: *Tiga sampangi, Manoranjitham.*

Straggling shrubs, young branches puberulous. Leaves oblong-lanceolate, 8-14 × 2.5-4 cm, glabrous, base cuneate, margin entire or undulate, apex acuminate, lateral nerves 6-18 pairs, petiole ca. 5 mm long. Flowers solitary or paired on leaf-opposed or terminal peduncle which is curved and hooked, dark green when unripe and yellow when mature, very fragrant; sepals 3, ovate, connate at base, acute and reflexed at tip, puberulous; petals 6 (3+3), sometimes up to 9, ovate-lanceolate, saccate or concave below, appressed villous; stamens many, anthers beaked concealing anther lobes; carpels few, sickle-shaped, pubescent, stigma blunt. Fruit an aggregate of orange-colored drupelets; seeds 1-2, oblong, deeply grooved on one side, brown.

Cultivated for its strongly fragrant flowres.

Fl. & Fr.: Round the year.

CANANGA (DC.) Hook.f. & Thomson, *nom. cons.*

Cananga odorata (Lam.) Hook.f. & Thomson, Fl. Ind. 130. 1855 & in Hook.f., Fl. Brit. India 1: 56. 1872; Debika Mitra in B.D.Sharma & *al.*, Fl. India 1: 254. 1993. *Uvaria odorata* Lam., Encycl. 1: 595. 1789.

Evergreen tree, 10-25 m high, trunk straight, branched at the top, young branches minutely pubescent. Leaves elliptic to oblong, 10-17 × 4-7 cm, glabrous, base obtuse or rounded, oblique, margin entire, apex narrowly acuminate, secondary nerves 10 pairs, petiole to 1 cm long. Flowers greenish-yellow, very fragrant, few to numerous in shortly pedunculate, pendulous racemes; sepals 3, ovate or triangular, shortly connate at base, acute, reflexed at apex, 8 × 5 mm, pubescent, valvate in bud; petals 6 (3+3), linear lanceolate, grey pubescent; stamens numerous, 3 mm long, closely arranged, connective conic, acuminate at apex; carpels many, linear-oblong, 2-3 mm long, glabrous except at base, ovules many, styles short, slender. Monocarps 10-12, black, oblong-ovoid or globose, pulpy, 1-2.5 cm long; seeds 6-12 in 2 rows, transversely flat, pale brown with pitted surface.

Grown as an ornamental (Rajagopal, 1973; Prasanna & *al.,* 2012).

Fl.: July; Fr.: October – November.

MILIUSA Lesch. ex DC.

1. Sepals 5-10 mm long, sepals and outer petals linear, inner petals saccate at the base; fruit brown when ripe. .... **M. tomentosa**
1. Sepals 2-4 mm long, sepals and outer petals ovate, inner petals flat at base; fruit purple when ripe. .... **M. velutina**

Miliusa tomentosa (Roxb.) J.Sinclair, Gard. Bull. Singapore 14: 378. 1955; Debika Mitra in B.D.Sharma & *al.,* Fl. India 1: 221. 1993. *Uvaria tomentosa* Roxb., Pl. Coromandel 1: t. 35. 1795. *Saccopetalum tomentosum* (Roxb.) Hook. f. & Thomson in Hook.f., Fl. Brit. India 1: 88. 1872; Gamble, Fl. Madras 1: 22. 1915.

Vern.: *Barre duduga, Pedda chilka duduga, Budda duduga.*

Large deciduous trees, up to 20 m high, bark blackish-brown, deeply longitudinally fissured. Leaves elliptic or oblong-ovate, 6-15 × 4-8 cm, thick coriaceous, tomentose above when young, glabrous with age except midrib, tomentose beneath, base rounded, margin entire, apex acute, petiole 2-5 mm long. Flowers terminal or axillary, solitary or in leaf-opposed or terminal cymes, greenish, 1.5 cm across; sepals 3, linear-lanceolate, 0.5-1 cm long, pubescent; petals 6 (3+3), outer petals sepaloid, linear-lanceolate, inner ones oblong-ovate; stamens many, ovate, 1 mm long; carpels many, tomentose, ovules many, stigma capitate. Fruit a ring of subglobose, stalked carpels, 0.8-1 cm across, tomentose, brown; seeds 4-5, greysih brown.

Occasional in dry deciduous forests in most districts.

Fl. & Fr.: February-June.

Inchapalli RF (ADB), *GO* & *DAM* 5154; Manchippa (NZB), *BR* & *GO* 9214; Rudraram RF (RR), *MSM* & *TP* 12707; Mohammadabad RF (RR), *MSM* & *KH* 10213; Borapur (MBNR), *SKB* & *BSS* 32390; Eturunagaram (WGL), *SSR* 18552.

Miliusa velutina Hook. f. & Thomson, Fl. Ind. 151. 1855 & in Hook.f., Fl. Brit.India 1 : 87. 1872; Gamble, Fl. Madras 1: 21. 1915; Debika Mitra in B.D.Sharma & *al.,* Fl. India 1: 222. 1993. *Uvaria velutina* Dunal, Monogr. Annon. 91. 1817.

Large deciduous trees, up to 12 m high, bark greyish, fairly smooth, often fluted, young branches densely yellowish tomentose. Leaves coriaceous, ovate or oblong, 5-15 × 3-7 cm, tomentose on both surfaces, base rounded or slightly cordate, margin entire, apex acute or obtuse, secondary nerves 9-11 pairs, petiole to 5 mm long. Flowers yellow in leaf-opposed cymes, sepals 3, ovate, 2-4 mm long, densely pubescent; petals 6 (3+3), outer petals sepaloid, inner ones broadly ovate, tomentose inside, glabrous outside; stamens many, oblong, 2 mm long; carpels many, oblong, densely pubescent, 2-ovuled, style short, slightly bent, stigma capitate. Fruitlets subglobose with woody stalks, purple when ripe, tomentose; seeds 1-2.

Rare, confined to Mahadevpur forests in Karimnagar district (Naqvi, 2001).

Fl. & Fr.: February-July.

POLYALTHIA Blume

1. Flowers in fascicles or umbels; crown conical; leaves glossy, undulate-margined. .... **P. longifolia**

1. Flowers solitary, occasionally many but not closely fascicled; crown other than conical; leaves not glossy, entire-margined:

    2. Leaves glabrous or nerves puberulous beneath, obtuse at apex; petals unequal in 2 series; stamens 1 mm long .... **P. suberosa**

    2. Leaves pubescent beneath, acute at apex; petals equal in 2 series; stamens 1-1.5 mm long .... **P. cerasoides**

Polyalthia cerasoides (Roxb.) Benth. & Hook.f. ex Bedd., Fl. Sylv. t. 1. 1869; Hook.f. & Thomson in Hook.f., Fl. Brit. India 1: 63. 1872; Gamble, Fl. Madras 1: 17. 1915; Debika Mitra in B.D.Sharma & *al.*, Fl. India 1: 270. 1993. *Uvaria cerasoides* Roxb., Pl. Coromandel. t. 33. 1795.

Vern.: *Gutti.*

Small semi-evergreen trees, up to 13 m high with light grey bark, branchlets densely tomentose. Leaves chartaceous, oblong or elliptic-lanceolate, 4-18 × 2.5-4 cm, glabrous above, densely pubescent below, base obtuse, margin entire, apex acute, petiole to 4 mm long. Flowers greenish, axillary or extra-axillary, solitary or in clusters; bracts 1-2, basal or submedian, leafy, oblong-ovate; sepals 3, free, ovate-lanceolate, 3 × 2 mm, densely pubescent; petals 6 (3+3), subequal, outer petals ovate-oblong, 7 x 5 mm, inflexed, puberulous outside, inner one slightly longer; stamens numerous, cuneate, 1-1.5 mm long, anthers extrorse, connective flat at apex; carpels many, oblong-linear, covered with stiff hairs, ovules 2, stigma clavate, slightly curved. Monocarps 30-40 in a cluster, globose, reddish brown when ripe, 1-seeded; seed globose, brown.

Common in forests of most districts.

Fl. & Fr.: January-October.

Pakhal (WGL), *SSR* & *KSM* 18550; Eturnagaram (WGL), *SSR* 18525; Pakhal (WGL), *SSR* & *KSM* 18550; 18550; Mallelatheertham (MBNR), *TDCK* 15390; Near Mannanur (MBNR), *VBH* 86606 (BSID).

Polyalthia longifolia (Sonner) Thwaites, Enum. Pl. Zeyl. 398. 1864; Hook.f. & Thomson in Hook.f., Fl. Brit. India 1: 62. 1872; Gamble, Fl. Madras 1: 16. 1915; Debika Mitra in B.D.Sharma & *al.*, Fl. India 1: 274. 1993. *Uvaria longifolia* Sonner, Voy. Aux. Indes 2: 233. Pl. 131. 1782.

Vern. : *Ashoka.*

Evergreen trees, 15-20 m high, branchlets sparsely puberulous when young. Leaves narrowly lanceolate, 15 × 20 cm, shining, glabrous, base cuneate, undulate, tip acuminate. Flowers yellowish-green in axillary or terminal fascicles or umbels; sepals 3, connate at base, broadly ovate-triangular, 1-2 × 1-1.5 mm, tomentose; petals 6 (3+3), subequal, linear, outer petals 6-7 × 2 mm, inner ones 10-15 × 2.5 mm, stamens numerous, 1 mm long; carpels many, linear, puberulous, ovule 1, stigma sessile.

Fruits in a cluster of one-seeded berries; seed almost ovoid, smooth or slightly longitudinally grooved, pale brown, shiny.

Common, planted in gardens, along avenues and parks.

Fl.: March- April, Fr.: May-September.

Polyalthia longifolia (Sonner) Thwaites var. **pendula** with straight stem and pendulous branches and **Polyalthia longifolia** (Sonner) Thwaites var. **angustifolia** with spreading, erect branches and linear leaves are planted in gardens and avenue.

Polyalthia suberosa (Roxb.) Thwaites, Enum. Pl. Zeyl. 398. 1864; Hook.f. & Thomson in Hook.f., Fl. Brit. India 1: 65. 1872; Gamble, Fl. Madras 1: 16. 1915; Debika Mitra in B.D.Sharma & *al.*, Fl. India 1: 278. 1993. *Uvaria suberosa* Roxb., Pl. Coromandel t. 34. 1795.

Vern.: *Chilkadudu.*

Small evergreen trees, up to 5 m high, bark corky, rough, reddish inside; young branches rufous tomentose. Leaves oblong, 3-6 × 1.5-2.5 cm, base slightly oblique, margin entire, apex rounded or obtuse, petiole up to 2 cm. Flowers solitary, rarely paired, extra axillary, pale green or yellow; bracts 2, one basal and the other at middle of pedicel; sepals 3, spreading, ovate, pubescent outside, glabrous inside; petals 6 (3+3), outer petals ovate to oblong-lanceolate, inner ones slightly longer; stamens numerous; carpels many, pubescent, stigma triangular, flat. Ripe carpels many, purple, subglobose; seeds 1-2, globose, smooth.

Occasional in dry forests of Mahabubnagar district.

Fl.: March-July; Fr.: August-December.

MENISPERMACEAE

1. Cotyledons foliaceous; endosperm ruminate:
    2. Bark corky, flaking off with age; leaves broadly ovate-cordate, basal nerves 5-7, palmate .... **Tinospora**
    2. Bark not corky; leaves ovate, basal nerves 3-5, subpalmate .... **Tiliacora**
1. Cotyledons not foliaceous; endosperm non ruminate:
    3. Leaves ovate, ovate-oblong, sagittate; style simple; fruit black purple when ripe .... **Cocculus**
    3. Leaves reniform, orbicular or cordate, peltate; style 2-8-parted; fruit scarlet when ripe .... **Cissampelos**

CISSAMPELOS L.

Cissampelos pareira L., Sp. Pl. 1031. 1753. var. **hirsuta** (Buch.-Ham. ex DC.) Forman, Kew Bull. 22. 356. 1968; Gangopadhyay in B.D.Sharma & *al.*, Fl. India 1: 317. 1993. *C. hirsuta* Buch.-Ham. ex DC., Syst. 1: 535. 1817. *C. pareira* L., Sp. Pl. 1031. 1753 *p.p.*, quoad; Hook.f., Fl. Brit. India 1: 103. 1872; Gamble, Fl. Madras 1: 30. 1915.

Vern. : *Adavi banka teega.*

Slender climbers with hairy branches. Leaves reniform, orbicular or cordate, peltate, 4-5.5 × 4-7 cm, glabrescent above, pubescent below, 5-7-nerved, base cordate, apex retuse or mucronate. Flowers greenish-yellow, minute. Male flowers: in axillary pendulous, to 6 cm long, subcorymbose cymes; bract obovate to orbicular, bracteoles linear, sepals 4, free, pilose above, petals 4, puberulous inside, connate into a 4-lobed cup, stamens 4 in a peltate synandrium, anthers dehiscing transversely; Female flowers: in pendulous racemes in axils of conspicuous bracts; bracts ovate-orbicular, foliaceous, persistent, sepal 1, obovate, petal 1, obovate, carpel 1, stigma 3-fid. Drupes ovoid, scarlet, glabrescent or hairy.

Very common in most districts in forests.

Fl. & Fr.: April-October.

Narsapur RF (MDK), *TP* & *CP* 14021, 14038; Anantagiri RF (RR), *MSM* & *TP* 11060; Ibrahimpet RF (NZB), *TP* & *BR* 6116; Mudheli RF (NZB), *BR* & *CPR* 9265; Gangapur (MDK), *RG* 106743 (BSID).

COCCULUS DC. *nom. cons.*

Cocculus hirsutus (L.) Diels in Engl., Pflanzenr. 46: 236. 1910; Gamble, Fl. Madras 1: 29. 1915; Gangopadhyay in B.D.Sharma & *al.*, Fl. India 1: 318. 1993. *Menispermum hirsutum* L., Sp. Pl. 341. 1753. *Cocculus villosus* (Lam.) DC., Syst. 1: 525. 1817; Hook.f. & Thomson in Hook. f., Fl. Brit. India 1: 101. 1872. *Menispermum villosum* Lam., Encycl. 4: 97. 1797.

Vern. : *Dusaraputiga.*

Climbing or straggling, villous, evergreen shrubs, up to 6 m long. Young stem, leaves and inflorescence clothed with greyish hairs. Leaves variable; ovate, ovate-oblong or sagittate, 3-4 × 2-2.5 cm, pubescent above, velvety tomentose below, base cordate, tip retuse. Flowers greenish-yellow, males in slender panicles, females in fascicles. Male flowers: sepals 3 + 3, free, petals 6, free, oblong-ovate; stamens 6, free. Female flowers: sepals and petals as in male; staminodes ca 0.5 cm long; carpels 3, styles terete. Drupelets globose, black-purple.

Common in most districts straggling over bushes and small trees.

Fl. & Fr.: November-March.

Sattenapally (ADB), *TP* & *PV* 4101; Satnella river (ADB), *GO* & *PVP* 4229; Kammarpally (NZB), *TP* & *BR* 6250; Maddigunta (NZB), *BR* & *SPB* 9556; Mohammadabad RF (RR), *MSM* & *KH* 10241; Kanmankalva RF (RR), *MSM* & *KH* 10306; Mannanur forest (MBNR), *TDCK* 13643; Nagarjuna Sagar (NLG), *KMS* 19165 (MH); Venkatapuram (MDK), *TP* & *CP* 14002; Pakhal (WGL), *KMS* 11626 (MH); Parnasala (KMM), *RCS* 99048 (BSID); Narsapur RF (MDK), *RG* 116450 (BSID).

TILIACORA Colebr.

Tiliacora acuminata (Lam.) Miers, Ann. Mag. Nat. Hist. Ser. 27: 39. 1851; Gamble, Fl. Madras 1: 28. 1915; Pramanik in B.D.Sharma & *al.*, Fl. India 1: 343. 1993. *Menispermum acuminatum* Lam., Encycl. 64:101. 1797. *Tiliacora racemosa* Colebr.,

Trans. Linn. Soc. London 12: 67. 1821; Hook.f. & Thomson in Hook.f., Fl. Brit. India 1: 99. 1872.

Vern.: *Kappa teega, Tivva mushidi, Verri chitramulamu*.

Large evergreen climbing shrubs, with milky latex, 5-8 m long, branches striate. Leaves ovate, 10-12 × 6.5-8.5 cm, coriaceous, pinnately nerved, glabrous, base truncate to subcordate, margin entire, apex acuminate; petiole 1.5-5 cm long. Flowers greenish-yellow, small in axillary panicles, unisexual; Male flowers : sepals 3 + 3; petals 6, obovate, concave; stamens 6, free. Female flowers: sepals and petals as in male; carpels 6. Drupes oblong, red.

Common climbing shrub in Khammam, Mahabubnagar and Nalgonda districts on bushes and trees especially on *Hardwickia binata* Roxb., *Dalbergia paniculata* Roxb. and *Anogeissus latifolia* (Roxb. ex DC.) Wall. ex Guill.

Fl. & Fr.: July-December.

Billakal (MBNR), *TDCK* 15395; Kambalpally (NLG), *BVRS* 314 (NU); Ramanagutta (KMM), *RR* 10795 (BSID); Madisisigutta (KMM), *RR* 106058 (BSID).

TINOSPORA Miers

Tinospora cordifolia (Willd.) Miers ex Hook. f. & Thomson, Fl. Ind. 184. 1855 & in Hook.f., in Fl. Brit. India 1: 97. 1872; Gamble, Fl. Madras 1: 26. 1915; Pramanik in B.D.Sharma & *al.*, Fl. India 1: 347. 1993. *Menispermum cordifolium* Willd., Sp. Pl. 4: 826. 1806.

Vern.: *Tippa teega, Tellatippatige, Jivantika, Koratippatige.*

Deciduous woody climbers, up to 10 m long, branchlets glabrous, sometimes with aerial roots, bark corky, flaking off with age. Leaves broadly ovate-cordate, 4-8 × 4-6.5 cm, chartaceous, 5-7-nerved, with glandular papillose patches on the lower surface in basal nerve-axils, base sinuate, margin entire, apex acuminate, petiole pulvinate, 2-7 cm long. Flowers in pseudoracemes, axillary or on leafless branchlets, usually solitary; Male flowers: yellow, outer 3 sepals ovate, inner 3 elliptic, concave; petals 6, rhombic-unguiculate, stamens 6, Female flowers: solitary; petals spathulate; staminodes 6; carpels 3, ellipsoid, stigma capitate. Drupe globose, red, sessile, ca 6 × 5 mm, ventrally flat, dorsally convex, stylar-scar subterminal, endocarp rugose or tubercled; seed grooved ventrally, curved round a 2-lobed intrusion of the endocarp.

Common in forests of all the districts.

Fl. & Fr.: Through the year.

Wankidi (ADB), *GO* 4475; Laxmapur (NZB), *TP & BR* 6294; Ghanapur (NZB), *BR & KH* 7109; Pathur thanda (MDK), *TP & CP* 12103; Medak (MDK), *BR & CP* 11484; Naskal (RR), *MSM & PRSR* 15122; Rechapalli (KRN), *GVS* 25689 (NH); Tirmalgiri (NLG), *BVRS* 464 (NU); Mamillagudem (KMM), *RR* 105912 (BSID); Mannanur (MBNR), *SRS* 109735 (BSID).

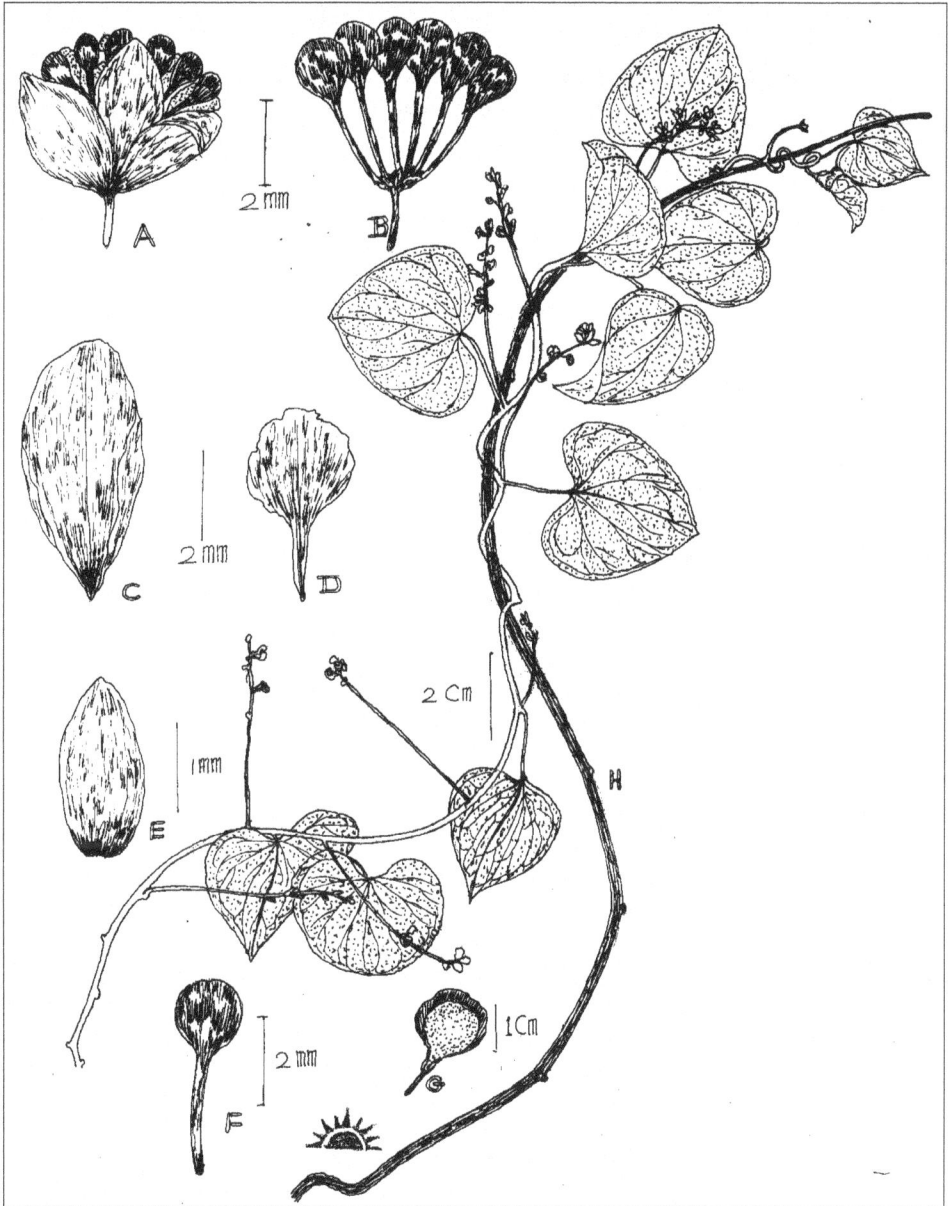

Figure 4: Tinospora cordifolia (Willd.) Miers ex Hook. f. & Thomson
A. Flower, B. Androecium, C. Sepal, D. Petal, E. Bract, F. Stamens, G. Fruit, H. Twig.

NYMPHAEACEAE

NYMPHAEA L. *nom. cons.*

1. Mature leaves entire-undulate, glabrous on both surfaces;
   stamens invariably yellow; anther connectives distinct and caudate;
   sepals persistent in fruit; seeds vertically fine-lined .... **N. nouchali**

1. Mature leaves sinuate-dentate, pubescent beneath; stamens yellow
   or red; anthers without connectives; sepals not persistent in fruit;
   seeds vertically thick-lined .... **N. pubescens**

Nymphaea nouchali Burm.f., Fl. Ind.120.1768; R.L.Mitra in B.D.Sharma & *al.*, Fl.
   India 1: 430. 1993. *N. stellata* Willd., Sp. Pl. 2: 1153. 1799; Hook.f. & Thomson in
   Hook.f., Fl. Brit. India 1: 114. 1872; Gamble, Fl. Madras 1: 33. 1915.

   Vern. : *Kaluva puvvu.*

   Aquatic rhizomatous, perennial herbs. Leaves long petioled, peltate, orbicular,
   sagittate when young, green above, pale purple below, base cordate, margin entire -
   undulate. Flowers solitary, blue, white, rose or purple, 4-17 cm across; sepals 4, free,
   triangular-ovate or lanceolate-oblong, green, often with dark purplish streaks outside;
   petals 8-15, outer ones lanceolate to oblong-lanceolate, mauve coloured; stamens 25-
   40, yellow, anthers brown with creamy appendages; ovary 8-16-locular, urceolate,
   stigmatic appendages bright orange. Berry green, spongy, 1.5-3.5 cm, globose or ovoid,
   seeds numerous.

   Common in ditches, ponds and tanks in all districts.

   Fl. & Fr.: Throughout the year.

   Swarna project (ADB), *MHR* 14567; Lakarapur pond (KMM), *MHR* & *MCK* 16690;
   Malkapur – Nizamabad Road (Mosra RF) (NZB), *TP* & *BR* 6103; Nasarullabad (NZB),
   TP & *BR* 6180; Shivampet (MDK), *TP* & *CP* 14011; Kanmankalva RF (RR), *MSM*
   10469; Dosalgunda lake (MDK), *KMS* 6793 (MH); Narsapur (MDK), *KMS* 6593 (CAL);
   Bourapur (MBNR), *TDCK* 16835; Azampur (NLG), *BVRS* 488 (NU).

Nymphaea pubescens Willd., Sp. Pl. 2: 1154. 1799; Gamble Fl. Madras 1: 34. 1915;
   R.L. Mitra in B.D.Sharma & *al.*, Fl. India 1: 431. 1993. *N. lotus* auct non. L. 1753;
   Hook.f. & Thomson in Hook.f., Fl. Brit. India 1: 114. 1872. *p.p. N. lotus* var. *pubescens*
   (Willd.) Hook.f. & Thomson in Hook.f., Fl. Brit. India 1: 114. 1872.

   Vern.: *Kaluva.*

   Large perennial, rhizomatous herbs. Leaves peltate, orbicular, 6-45 × 5-41 cm,
   sagittate when young, glabrous and often punctulate on both surfaces, green and
   sometimes also blotched purple above, reddish purple beneath, leaf blade sharply
   dentate at margin. Flowers bluish purple, blue or pale bluish-white, solitary, 8-15 cm
   across, arising from the rhizome, fragrant; sepals 4, free, triangular-ovate; petals 8-15,
   outer ones lanceolate to oblong-lanceolate, obtuse or acute at apex, white or rose or
   pink; stamens 25-40, yellow with blue appendages, ovary 8-16-locular, stigmatic
   appendages incurved, orange. Berry 1.5-4 cm across; seeds ellipsoid, reticulate,

longitudinally ribbed, conspicuously ciliate along ribs when immature, becoming glabrate with growth of aril.

Common in Adilabad, Nizamabad, Warangal and Medak districts in ditches and tanks.

Fl. & Fr.: January- November.

Jannaram (ADB), *GO* & *DAM* 5412; Mustapur (NZB), *BR* & *GO* 9072; Rusthempet (MDK), *RG* 96298 (BSID); Medak (MDK), *RG* 1041107 (BSID).

NELUMBONACEAE

NELUMBO Adans.

Nelumbo nucifera Gaertn., Fruct. Sem. Pl. 1: 73. t. 19. f. 2. 1788; Mitra in B.D.Sharma & *al.*, Fl. India1: 441. 1993. *Nymphaea nelumbo* L., Sp. Pl. 51. 1753. *Nelumbium speciosum* Willd., Sp. Pl. 2: 1258. 1799; Hook.f. & Thomson in Hook.f., Fl. Brit. India 1: 116. 1872; Gamble, Fl. Madras 1: 34. 1915.

Vern. : *Thamara puvvu, Kamalamu.*

Large perennial aquatic herbs with milky juice with stout creeping underground rhizomes. Leaves peltate, rotund, 30 cm across, flat or some what hollowed, radiately nerved, waxy, glabrous and glaucous on both surfaces, entire, petioles up to 2 m long. Flowers 8-25 cm across, rose-red or white, fragrant; peduncles up to 2 m long, sepals ovate or elliptic, concave, green (in white flowers) or pinkish green (in rose-pink flowers); petals ca 20 (single form) or ca 110 (double form), each elliptic gradually becoming obovate or spathulate; stamens up to 225, outer most ones in double form staminodal, connective appendages recurved, white or yellow; receptacles 2-4 cm across, spongy, yellow during anthesis turning green and finally becoming dark brown and 5-10 cm across in fruit; carpels 8-10 × 2-3 mm, cylindric, stigmas protruding from receptacle; seeds ovoid-oblong,black.

Common in tanks and ponds of most districts.

Fl. & Fr.: Throughout the year.

Banjapally (NZB), *BR* & *CPR* 9249; Gunavaram (KMM), *RCS* 98942 (BSID).

PAPAVERACEAE

ARGEMONE L.

1. Flower buds subglobose, body length ± equal to breadth; petals mostly bright yellow .... **A. mexicana**
1. Flower buds oblong, body length 1.5-2 times breadth; petals white to pale lemon yellow .... **A. ochroleuca**

Argemone mexicana L., Sp. Pl. 508. 1753; Hook.f. & Thomson in Hook.f., Fl. Brit. India 1: 117. 1872; Gamble, Fl. Madras 1: 35. 1915; Debnath & Nayar in B.D.Sharma & *al.*, Fl. India 2: 2. 1993.

Vern.: *Yerrikusuma.*

Annual prickly herbs, 30-125 cm high, stem greyish-white with yellow sap. Leaves both radical and cauline, variegated, white spiny on margins and veins, sessile, amplexicaul, oblong-ovate, 4-7 × 3-5.5 cm, glaucous, sinuate-pinnatifid. Flowers bright yellow, solitary, terminal; sepals 3, caducous, elliptic, 8-15 × 5-10 mm, petals 4-6, imbricate, obovate, cuneiform, crumpled in aestivation; stamens many, 8-10 mm long, filaments yellow; ovary ovoid, style very short, obsolete, stigma sessile, radiating, 3-7-lobed. Capsule ellipsoid or oblong, 4-6-ribbed, prickly or not, deshicing along sutures by 3-6 short valves in upper part; seeds deeply reticulate, blackish-brown.

A common weed in open wastelands and cultivated fields and road sides. Native of Central America (Ridley, 1930; Baker and Brink, 1963) introduced into India during 17th century.

Fl. & Fr.: Through the year.

Sone (ADB), *TP* & *PVP* 4063; Mingaram (NZB), *TP* & *BR* 6316; Mombojipally (MDK), *BR* & *CP* 11449; Mohammadabad RF (RR), *MSM* & *KH* 10204; Srinivasapur (NLG), *BVRS* 406; Narsapur (MDK), *KMS* 7975 (CAL); Kodimial (KRN), *GVS* 20106 (MH); Koida (KMM), *PVS* 84068 (BSID); Koilkonda road (MBNR), *SRS* 104517 (BSID); Mannanur (MBNR), *SRS* 109123 (BSID).

Argemone ochroleuca Sweet, Brit. Fl. Gard. 3: t. 242. 1829; Nair, Rec. Bot. Surv. India 21(1): 7-8. 1978; C.S.Reddy & C.Pattanaik, Zoos Print Jour. 22(12): 2949. 2007.

Herbs, up to 60 cm high; young stems whitish green with reddish tinge. Leaves oblanceolate, semi-amplexicaul, sessile at base, sinuate to pinnatifid, basal leaves deeply lobed, lobes oblong, glaucous. Flowers whitish or pale lemon yellow, 2.5-3.5 cm across, sessile, sepals 3, 8-12 × 5-7 mm, petals 28-30 × 16 – 19 mm, stamens many, 8-10 mm long, filaments pale yellow, anthers dark yellow, stigmas 5-lobed, deeply dissected, dark red. Capsules lanceolate-ovoid, 1-4 cm long excluding style, 4-17 mm thick, with 12-26 erecto-patent spines per valve, the larger spine up to 10 mm long; seeds 1.5-2 mm in diam., finely reticulate, black.

A weed found along railway tracks and in waste lands in Hyderabad (C.S.Reddy and Pattanaik, 2007). It is associated with *Argemone mexicana* and *Alternanthera tenella*. Native of Mexico.

Fl. & Fr.: February – June.

FUMARIACEAE
FUMARIA L.

Fumaria indica (Hausskn.) Pugsley, J. Linn. Soc. Bot. 44: 313. 1919; Ellis & Balakrishnan in B.D.Sharma & *al.*, Fl. India 2: 84. 1993. *F. vaillantii* Lois. var. *indica* Hausskn., Flora 56: 443. 1873. *F. parviflora* Lam. subsp. *vaillanti* sensu. Hook. f. & Thomson in Hook. f., Fl. Brit. India 1: 128. 1872. *F. vaillantii* Lois., Desv. Journ. Bot. 2: 358. 1809.

Vern. : *Chatarashi.*

Annual branched sarmentose herbs, 15-60 cm long, branchlets grooved, puberulous. Leaves 2-3-pinnatisect, 5-7 cm, segments 5-15 mm, membranous, base obtuse, margin entire, apex acute, mucronulate. Flowers small, in terminal or leaf-opposed racemes, white or pink with purple tips; sepals 2, lanceolate, minute, caducous; petals 4, crested at back with ascending spur, upper petal emarginated. Fruit indehiscent, globose, 1-seeded nutlet.

A common weed in cultivated lands (Sayeeduddin, 1938).

Fl. & Fr.: October- January.

BRASSICACEAE (*nom. alt.* CRUCIFERAE)

1.   Fruit as long as broad :

   2.   Fruit globular, indehiscent .... **Coronopus**

   2.   Fruit flat, dehiscent :

      3.   Fruit 2-seeded .... **Lepidium**

      3.   Fruit 10-many-seeded .... **Schouwia**

1.   Fruit 2-3 times as long as broad :

   4.   Plants aquatic, rooting at nodes .... **Nasturtium**

   4.   Plants terrestrial, not rooting at nodes:

      5.   Pods dehiscing and bearing seeds throughout their
      whole length; sepals not pouched:

         6.   Leaves lyrate to pinnatifid; stamens 6 .... **Rorippa**

         6.   Leaves odd-pinnate; stamens 4 .... **Cardamine**

      5.   Pods with seedless indehiscent beak; sepals pouched at base:

         7.   Siliqua deeply segmented, breaking into 1-seeded segments;
         tap root tuberous .... **Raphanus**

         7.   Siliqua neither segmented nor breaking into one-seeded
         segments; tap root not tuberous:

            8.   Beak of pod cylindric or conical; seeds 1-seriate,
            flowers yellow .... **Brassica**

            8.   Beak flattened; seeds 2-seriate, flowers lilac or
            yellow with lilac veins .... **Eruca**

BRASSICA L.

1.   Leaves thick, glaucous blue or blue green; flowers mostly
light yellow or white .... **B. oleracea**

1.   Leaves thin, dark or light green; flowers bright yellow:

   2.   Siliqua 0.8-3 cm long, more or less appressed to the rachis;
   seeds usually 5-12 .... **B. nigra**

   2.   Siliqua 3-5 cm long, widely spreading or divaricate at
   maturity; seeds usually 9-45 .... **B. juncea**

Brassica juncea (L.) Czern. & Coss in Czern., Consp. Fl. Chark. 8. 5. 1859; Hook.f. & T.Anderson in Hook.f., Fl. Brit. India 1 : 157. 1872; Gamble, Fl. Madras 1 : 38. 1915; Hajra & *al.*, in B.D.Sharma & *al.*, Fl. India 2: 134. 1993. *Sinapsis juncea* L., Sp. Pl. 668. 1753.

Vern. : *Avalu.*

Annual erect herbs, up to 1 m high. Leaves lyrate to pinnatifid, with 2 or 3 pairs of lobes, lateral lobes broad, terminal ones obovate, irregularly dentate or lacerate, upper leaves often simple, oblanceolate or obovate, glabrous, attenuate or decurrent at base, margin irregularly dentate, apex obtuse or subacute. Flowers yellow in terminal racemes; sepals 4, ca 5 mm long, spreading; petals 4, obovate, clawed, 8 mm long, bright yellow; stamens 6; 5 mm long, anthers oblong. Siliqua subterete, 3-5 cm x 2.5 mm, glabrous, torulose, beak often seedless, septum white, membranous, seeds in one row, ovoid, finely alveolate.

Cultivated for seeds, also in the fallow fields or as an escape.

Fl. & Fr. : September-March.

Kadthala (ADB), *GO* & *DAM* 4996; Kammarpalli (NZB), *TP* & *BR* 6241; Lingampalle (RR), *MSM* 19220; Medak (MDK), *BR* & *CP* 11457; Bourapur forest (MBNR), *TDCK* 13647; Narsapur (MDK), *KMS* 7975 (CAL).

Brassica nigra (L.) Koch. in Roehling's Deutschland Fl. ed. 3, 4 : 713. 1833; Hook.f. Fl. Brit. India 1 : 156.1872 Hajra & *al.*, in B.D.Sharma & *al.*, Fl. India 2: 136. 1993.

Vern.: *Avalu, Nalla avalu.* English: Black mustard.

Much-branched, erect, annual herb, 0.3-1.5 m high. Leaves variable, radical ones pinnatifid or lobed, 10-12 x 2-3.5 cm, terminal lobe very large; apical leaves ovate, entire or pinnatifid, all petioled, glabrous or slightly hairy. Flowers in terminal, usually much branched racemes; sepals 5-6 mm long; petals bright yellow, 8-10 mm long. Siliqua subtetragonous, 0.8-3 cm long, on short pedicels which are close to the rachis; seeds ovoid, dark brown or black.

Cultivated and escaped weed of wet lands.

Fl. & Fr.: September- February.

Lingampalli (RR), *MSM* 19220.

Brassica oleracea L., Sp. Pl. 667. 1753.

Key to Varieties
1. Low and stout stock, bearing a dense terminal head .... **B. oleracea** var. **capitata**
1. Low with stout, short stock, bearing a dense,
   terminal teratological head .... **B. oleracea** var. **botrytis**

Brassica oleracea L. var. **botrytis** L., Sp. Pl. 667. 1753; Bailey, Man. Cult. Pl. 435.1949.

English: Cauliflower.

It is largely cultivated occasionally.

Brassica oleracea L. var. **capitata** L., Sp. Pl. 667. 1753; Bailey, Man. Cult. Pl. 435.1949.

English: Cabbagae.

It is largely cultivated occasionally.

CARDAMINE L.

Cardamine trichocarpa Hochst. ex A. Rich., Tent. Fl. Abyss.1:18.1847; Gamble, Fl. Madras 1:38. 1915; Hajra & *al.*, in B.D.Sharma & *al.*, Fl. India 2: 117. 1993. *C. subumbellata* Hook. ex Hook.f. & T. Anderson in Hook.f., Fl. Brit. Ind. 1:138. 1872.

Slender annual herbs, to 20 cm high. Younger shoots pubescent. Leafy shoots many arising from base. Leaves odd-pinnate, lobes 3 pairs, opposite or spiral, end ones larger, ovate or elliptic-ovate, chartaceous, base obtuse or cuneate, margin sinuate, apex subacute, 1-2 × 1-1.2 cm. Flowers in terminal corymb, few-flowered, white, 2 mm across, petals 0; stamens 4. Siliqua linear, 2 cm, erect, compressed, pubescent, midrib obscure, seeds 9-12, suborbicular.

Winter season weed of margins of tanks and gardens, rare (C.S. Reddy & *al.,* 2000).

Fl. & Fr.: November -January.

Hyderabad-Hussain sagar tank, *CSR* 1562 (KU).

Note: It is an invasive alien species, native to Tropical America (C.S.Reddy & *al.,* 2008).

CORONOPUS Zinn. *nom. cons.*

Coronopus didymus (L.) Smith, Fl. Britan 2: 691. 1800; Bhaumik in B.D.Sharma & *al.*, Fl. India 2: 192. 1993. *Lepidium didymum* L., Syst. Nat. 2: 433. 1759. *Senebiera didyma* (L.) Pers. Syn. 2: 185. 1806; Gamble, Fl. Madras 1: 39. 1915.

Annual, small, pubescent ascending or procumbent herbs, 10-40 cm high. Radical leaves pinnate, 7-10 × 1.5-2 cm, pinnae incised; cauline leaves much smaller, pinnatifid. Flowers greenish- white in 30-60-flowered racemes; sepals 4, ovate-rounded; petals 4, linear, 0.5 mm long, greenish white; stamens 4-6, fertile stamens 2. Silicula emarginate, reticulate; seeds reniform, striate, light brown.

Rare weed in Warangal and Hyderabad districts (Rajagopal, 1973).

Fl. & Fr.: Almost throughout the year.

ERUCA Adans.

Eruca sativa Mill., Gard. Dict. ed. 8. n. 1. 1768; Hook.f. & T. Anderson in Hook.f. Fl. Brit. India 1: 158. 1872; Gamble, Fl. Madras 1 : 30. 1915; Hajra & *al.* in B.D.Sharma & *al.*, Fl. India 2: 143. 1993. *Brassica eruca* L., Sp. Pl. 667. 1753.

Annual erect herbs, up to 25 cm high. Basal leaves lyrately pinnatifid, 7-15 × 3-5 cm, petiolate, upper ones lyrate-pinnatifid with entire to subulate-dentate segments, sessile or subsessile. Flowers in terminal racemes; sepals 4, oblong; petals 4, obovate-cuneate, lilac or yellow with lilac veins; stamens 1-1.5 cm long. Pod turgid, glabrous, beaked; seeds subglobose, 2-seriate, smooth, brownish-black.

A rare weed in gardens in Telangana.

Fl. & Fr.: October-January.

LEPIDIUM L.

Lepidium sativum L., Sp. Pl. 644. 1753; Hook. f. & T. Anderson in Hook.f., Fl. Brit. India 1: 159. 1872; Gamble, Fl. Madras 1: 39. 1915; Bhaumik in B.D.Sharma & *al.*, Fl. India 2: 206. 1993.

Annual, erect, glabrous herbs, up to 75 cm high. Radical leaves long petioled, usually pinnatifid, elliptic-ovate or oblong-linear, 5-7 × 1-2 cm, cauline leaves sessile and usually entire. Flowers blue or white in terminal spikes; sepals 4, 1.5 mm long, sparsely pubescent out side; petals 4, spathulate, somewhat clawed, 2.5 mm long, distinctly nerved. Silicula flat, dehiscent, valves smooth, falling away; seeds ovoid, reddish brown.

Rare as an escape near human habitations in Adilabad district.

Fl. & Fr.: July-December.

Kadthala (ADB), *GO* & *DAM* 4995.

NASTURTIUM R. Br. *nom. cons.*

Nasturtium officinale R. Br. in Ait., Hort. Kew. ed. 2: 4: 110. 1812; Hook.f. & T. Anderson in Hook. f., Fl. Brit. India 1: 133. 1872; Gamble, Fl. Madras 1: 37. 1915; Hajra & Chowdhery in B.D.Sharma & *al.*, Fl. India 2: 125. 1993. *Rorippa nasturtium-aquaticum* (L.) Hayek. in Sched. Fl. Stir. Ex 5. 22. 1905. *Sisymbrium nasturtium-aquaticum* L., Sp. Pl. 657. 1753.

Delicate, glabrous, aquatic herbs, stem creeping and floating, rooting at nodes. Leaves pinnate, the upper with 3-7 pinnules and a terminal one, the lower cut into 3 repand segments. Flowers white in short racemes, 4-10-flowered; sepals oblong, white; petals obovate, narrowly clawed at base, 5 x 2 mm, white; stamens 3 mm long; Pod stalked, spreading or bent upwards; seeds 2-seriate, ovoid-rounded, red.

Rare weed in fields. Probably an introduced plant.

Fl. & Fr.: September-December.

RAPHANUS L.

Raphanus sativus L., Sp. Pl. 669. 1753; Gamble, Fl. Madras 1: 39. 1915; Hajra & *al.*, in B.D.Sharma & *al.*, Fl. India 2: 145. 1993.

Vern.: *Mullangi.* English: Radish.

Herbs, to 1 m high; tap root tuberous, fusiform. Lower leaves lyrate-pinnatisect; upper ones lobed, terminal lobes larger, margin dentate to crenate; uppermost ones simple with entire margin. Flowers in terminal racemes; sepals 4, obovate, caducous; petals 4, pinkish white with darker veins; stamens 6, tetradynamous. Siliqua cylindrical, beaked, deeply segmented, breaking into 1-seeded segments; seeds subglobose.

Cultivated in many parts as a subsidiary crop.

Fl. & Fr.: November – February.

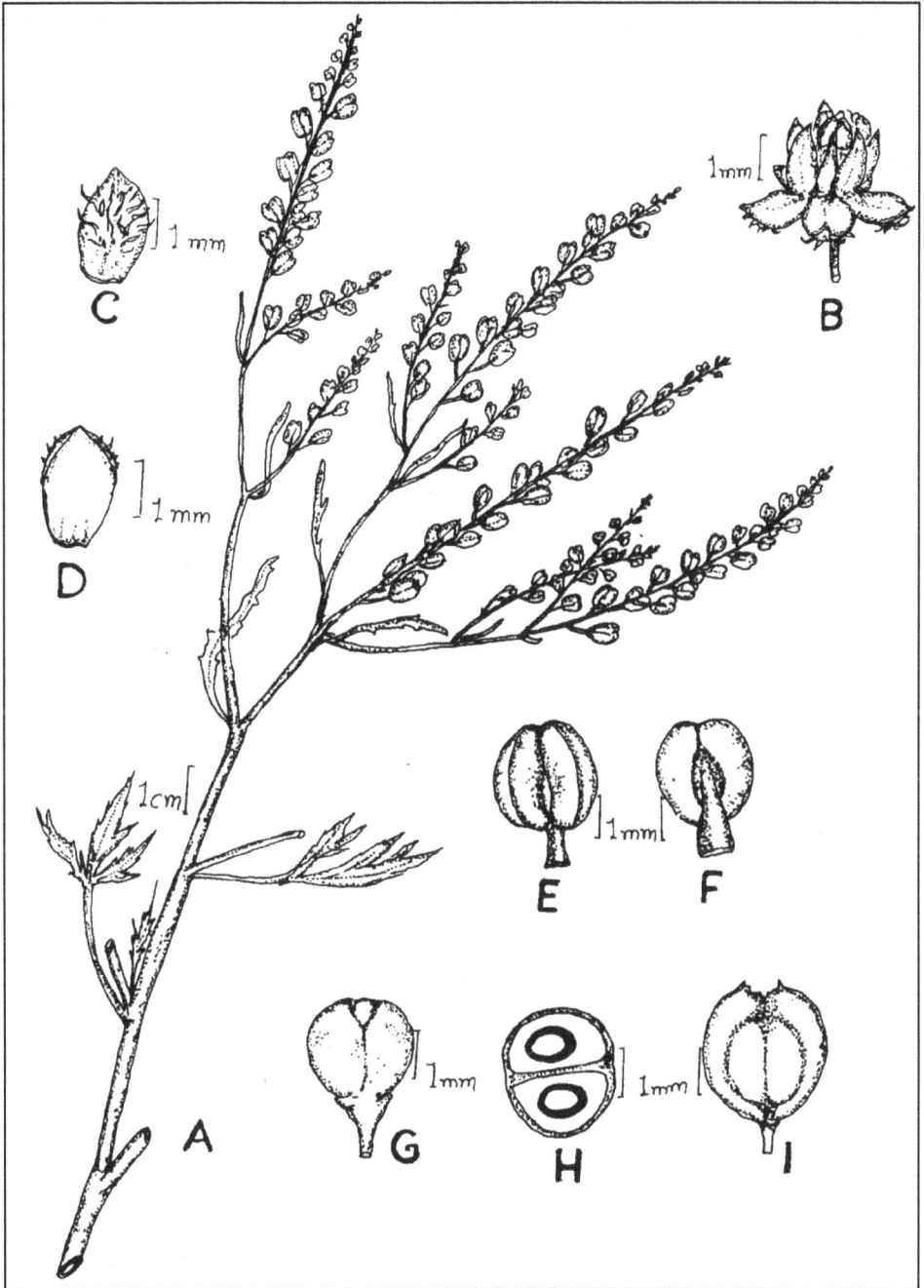

Figure 5: Lepidium sativum L.

A. Twig, B. Flower-entire, C & D. Sepals, E. & F. Stamens, G. Pistil, H. Ovary c.s, I. Fruit.

RORIPPA Scopoli

1. Petals usually absent .... **R. dubia**

1. Petals always present :

    2. Upper and middle leaves obovate, coarsely blunty toothed .... **R. indica**

    2. Leaves all bipinnatifid, teeth sharp .... **R. madagascariensis**

Rorippa dubia (Pers.) Hara, J. Jap. Bot. 30: 196. 1955. *Sisymbrium dubium* Pers., Syn. Pl. 2: 199. 1806. var. *apetala* (DC.) Hochr., Candollea 2: 370. 1925. *Nasturtium indicum* DC. var. *apetalum* DC., Prodr. 1: 139. 1824.

Annual erect, slender, glabrous herbs, up to 20 cm high. Basal leaves many, crowded, long-petioled, simple to lyrate, with 1-2 basal segments on either side, upper leaves subsessile, obovate oblanceolate, entire, irregularly serrate-dentate. Flowers greenish yellow in terminal racemes; sepals 4, erect, green, oblong-ovate, 2-3 mm long; petals often absent; stamens 2 mm long. Siliqua erect, thin, cylindrical, 1.6-3.5 cm long, seeds 2-seriate, smooth, yellowish brown, more than 20 in each row.

Occasional along river banks in Karimnagar district.

Fl. & Fr.: September – January.

Mahadevpur (KRN), *CSR* & *VSR,* 1568 (MH & KU).

Rorippa indica (L.) Hiern., Cat. Afr. Pl. Welw. 1: 26. add & Corr. 1896 & 2: 481, errata 1899; Hajra & Chowdhery in B.D.Sharma & *al.*, Fl. India 2: 129. 1993. *Sisymbrium indicum* L., Mant. 1: 93. 1763. *Nasturtium indicum* (L.) DC., Prodr. 1: 139. 1824; Hook.f. & T.Anderson in Hook.f., Fl. Brit. India 1: 134. 1872; Gamble, Fl. Madras 1: 37. 1915.

Trailing or suberect herbs, up to 35 cm high, branching from the base. Leaves pinnatipartite, 4-8 cm long, membranous, terminal lobe largest, gradually becoming smaller towards base, pinnae ovate-lanceolate, acute, coarsely, bluntly toothed. Flowers yellow in 6-10 cm long, many-flowered, terminal racemes; sepals 4, 2-3 mm long, petals 4, 3-4 mm long, stamens ca. 2 mm long, stigma subsessile. Pods 1 cm long, straight, linear, shortly beaked; seeds numerous, reddish-brown, smooth, glabrous, spherical or nearly so.

Common in moist locations in forests and rice fields in Adilabad and Nizamabad districts.

Fl. & Fr.: February-June.

Utnoor (ADB), *GO* & *PVP* 4605; Bhasar (ADB), *GO* & *PVP* 4648; Potchera (ADB), *GO* & *DAM* 5081; Kandakurthi (NZB), *BR* 7253; Rudroor farm (NZB), *BR* & *KH* 9640.

Rorippa madagascariensis (DC.) Hara, J. Jap. Bot. 30: 198. 1955; Hajra & Chowdhery in B.D.Shharma & *al.*, Fl. India 2: 129. 1993. *Nasturtium madagascariense* DC., Syst. 2: 192. 1821; Gamble, Fl. Madras 1: 37. 1915.

Herbs, up to 35 cm high. Leaves sessile, obovate, oblong, deeply bipinnatifid, teeth sharp, base auriculate-amplexicaul, margins toothed, sparsely hairy, sessile.

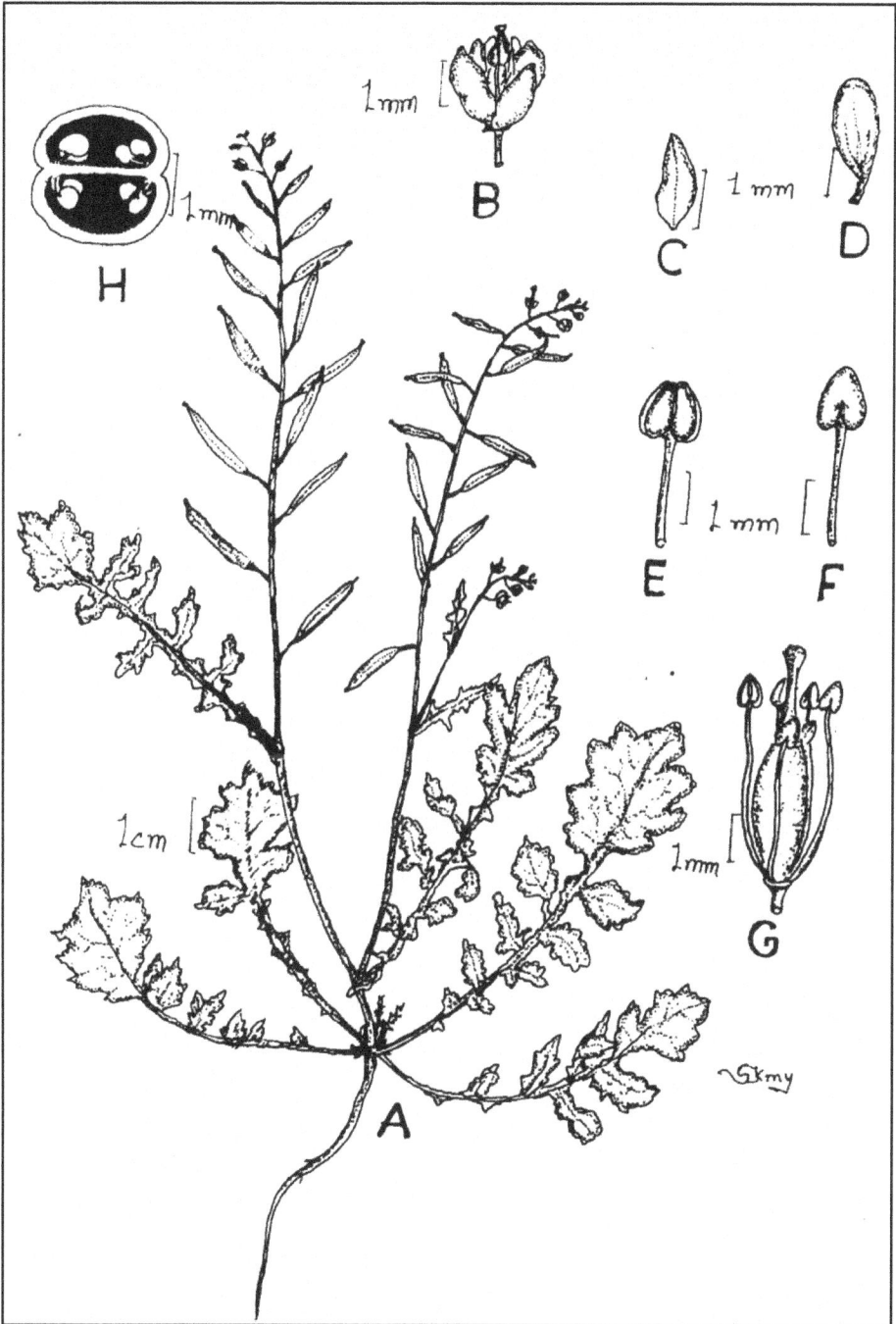

Figure 6: Rorippa indica (L.) Hiern.

A. Habit, B. Flower-entire, C. Sepal, D. Petal, E & F. Stamens, G. Stamens & Pistil, H. Ovary c.s.

Flowers yellow in racemes; sepals 4, oblong, petals 4, narrowly oblong, stamens 6, filaments 1.6 mm long, anthers 0.5 mm long; ovary ellipsoid, style stout, stigma capitate. Pod cylindric, glabrous, upcurved; seeds 2-seriate, lenticular.

Hills of Karimnagar and Medak districts.

Fl. & Fr.: November-April.

Rechapalli (KRN), *GVS* 22266 (MH); Narsapur Ooracheru (MDK), *KMS* 7988 (MH).

SCHOUWIA DC. *nom. cons.*

Schouwia purpurea (Forssk.) Schweinf., Bull. Herb. Boiss. 4. App. 2: 183. n. 486. 1896; Hajra & Chowdhery in B.D.Sharma & *al.*, Fl. India 2: 147. 1993. *Subulaia purpurea* Forssk., Fl. Aegypt-Arab. 117. 1775. *Thlaspi arabicum* Vahl, Symb. Bot. 2: 76. 1791 (excl. syn. L.). *Schouwia arabica* (Vahl) DC., Syst. Nat. 2: 644. 1821 &. Prodr. 1: 224. 1824; W.J. Hook. Icon. Pl. 3. t. 223. 1840 (excl. Pl. Seib.); Rao, Bull. Bot. Surv. India. 5: 265. 1963.

Annual, glabrous herbs, up to 75 cm high; stem erect, dichotomously branched from the base. Leaves 1.5-9 × 1-4 cm, simple, alternate, sessile, basal leaves obovate or oblanceolate, tapering towards base, acute at apex, crenately lobed or wavy-dentate to almost entire, upper leaves oblong-ovate or obovate, deeply cordate-auricled at base. Flowers purple, in terminal and leaf-opposed racemes, elongating up to 30 cm long in fruits; sepals 4, 4-5 mm long; petals 4, unguiculate, purple. Siliqua 1.5-2 cm in diameter, sub-ellipic to ovate- orbicular, laterally compressed with a cordate base, reticulate and broadly winged, replum linear, membranous, style persistent, pyramidal, 5-7 mm long, seeds many, 2-seriate, globose, reddish- brown, mucilaginous.

In Mahabubnagar district (Raghava Rao, 1989)

Fl. & Fr.: November-January.

CAPPARACEAE

1. Herbs .... **Cleome**
1. Shrubs or trees:
    2. Calyx lobes fused below .... **Maerua**
    2. Calyx lobes divided to base :
        3. Stamens up to 6, inserted on the middle of the gynophore .... **Cadaba**
        3. Stamens more than 10, inserted on the base of the gynophore :
            4. Leaves simple; sepals unequal .... **Capparis**
            4. Leaves compound (digitately 3-foliolate); sepals equal .... **Crataeva**

CADABA Forssk.

Cadaba fruticosa (L.) Druce, Bot. Exch. Club. Soc. Brit. Isles 3: 415. 1914; Sundara Raghavan in B.D.Sharma & *al.*, Fl. India 2: 250. 1993. *Cleome fruiticosa* L., Sp. Pl. 671. 1753. *Cadaba farinosa* Forssk., Fl. Aegypt.-Arab. 68. 1775. *C. indica* Lam.,

Encycl. 1: 554. 1785; Hook.f. & Thomson in Fl. Brit. India 1: 172. 1872; Gamble, Fl. Madras 1: 43. 1915.

Vern.: *Adamorinika, Vutharasi chettu, Chikondi.*

An unarmed straggling much branched shrubs, up to 3 m high. Leaves simple, elliptic-oblong, 2-3.5 × 1-1.5 cm, base cuneate, margin entire, apex obtuse-apiculate; petioles 2-8 mm long. Flowers greenish-white, in few-flowered, terminal corymbs, bracts subulate; sepals 4, greenish, outer sepals ovate or obovate, boat-shaped, inner ones elliptic-lanceolate or oblong-ovate; petals 4, white, spathulate, claw very narrow, as long as the limb; gynandrophore elongated; stamens 4-6, spreading; ovary 3-5 mm long, stigma sessile. Berry cylindrical, torulose; seed striate, clothed by a red aril.

Common in forests, waste lands and hedges in most districts.

Fl. & Fr.: April-November.

Anand Nagar (NZB), *BR* & *KH* 9652; Rayalatepenta (MBNR), *TDCK* 16808; Elagundlagunta (NLG), *BVRS* 131 (NU); Mannanur (MBNR), *SRS* 109023 (BSID).

CAPPARIS L.

1. Mature branches leaf-less .... **C. decidua**
1. Mature branches leafy:
    2. Berry up to 8 mm across .... **C. sepiaria**
    2. Berry more than 1 cm across:
        3. Flowers axillary, solitary .... **C. divaricata**
        3. Flowers other than axillary in clusters or terminal corymbs :
            4. Flowers in supra-axillary rows .... **C. zeylanica**
            4. Flowers in terminal racemes or umbels .... **C. grandis**

Capparis decidua (Forssk.) Edgew., J. Linn. Soc. Bot. 6: 184. 1862; Sundara Raghavan in B.D.Sharma & *al.*, Fl. India 2: 265. 1993. *Sodada decidua* Forssk., Fl. Aegypt.-Arab. 81. 1775. *Capparis aphylla* Roth, Nov. Pl. Sp. 238. 1821; Hook.f. & Thomson in Fl. Brit. India 1: 174. 1872; Gamble, Fl. Madras 1: 45. 1915.

Shrubs or small trees, up to 7 m high, branchlets glabrous, striate, spines straight, 3-6 mm long, rarely absent, cataphylls confined to base. Leaves only on young branchlets, linear-oblong, 6-20 × 1-3 cm. Flowers in axillary and terminal short racemes; sepals 4, unequal, tomentose, adaxial sepal ovate-oblong, saccate, hooded, others petaloid, elliptic to linear-lanceolate; petals 4, almost equaling sepals, puberulous, anterior pair triangular-ovate, included within hooded sepal, posterior pair slightly smaller, oblong-lanceolate; stamens 10-16; gynophores 1.2-1.6 cm long, slender; ovary globose. Berry globose, beaked, smooth, red when ripe, 1.5 cm across; seeds 1-4, reniform.

Occasional in scrub jungles of Telangana (Khan, 1953; Ramana, 2010).

Fl. & Fr.: February-August.

Capparis divaricata Lam., Encycl. 1: 606. 1785; Hook.f. & Thomson in Hook. f., Fl.
   Brit. India 1: 174. 1872; Sundara Raghavan in B.D.Sharma & *al.*, Fl. India 2: 266.
   1993. *C. stylosa* DC., Prodr. 1: 246. 1824; Gamble, Fl. Madras 1: 45. 1915.

Vern. : *Remidi, Badreni.*

Armed shrubs or small trees, up to 5 m high, branches glabrous or shoots hoary;
stipular thorns variable, divaricate, straight or curved upwards or downwards;
cataphylls absent. Leaves thick, coriaceous, variable, dimorphic, either linear to linear-
oblong or elliptic to oblong, 4-6 × 1-2.5 cm, base cuneate, margin entire, apex mucronate;
petioles 2-4 mm long. Flowers yellow, axillary, solitary; sepals 4, elliptic-orbicular,
thick, pubescent on both sides, outer pair boat-shaped, inner pair petaloid, equaling
petals; petals 4, greenish-yellow, creamy or white, linear, oblong, obovate-spathulate
or strap-shaped; stamens 45-65, filaments yellowish, purplish red at base, longer
than petals; gynophore 2- 2.5 cm long, ovary ovoid, beaked. Berry globose, 5 × 2 cm,
ribbed apiculate, reddish; seeds 6-8, embedded in white or creamy pulp.

Occasional in Karimnagar (Naqvi, 2001), Khammam, Warangal, Hyderabad
(Ramana, 2010) and Mahabubnagar (Raghava Rao, 1989) districts in scrub or
deciduous forests, on black cotton or laterite soil

Fl.: March – April; Fr.: June- September.

On the way to Gandikudiai (KMM), *RR* 113841 (BSID).

Capparis grandis L.f., Suppl. Pl. 263. 1781; Hook.f. & Thomson in Hook. f., Fl. Brit.
   India 1: 176. 1872; Gamble, Fl. Madras 1: 46. 1915; Sundara Raghavan in
   B.D.Sharma & *al.*, Fl. India 2: 274. 1993.

Vern. : *Guli, Ragota, Nalluppi.*

Erect shrubs or crooked trees, up to 10 m high, bark yellowish to dark brownish,
irregularly cracked with short straight spines; branchlets velvety-tomentose, stipular
thorns straight or rarely recurved, 5-6 mm long in older branches, but often lacking in
flowering or tender shoots. Leaves obovate, 3-7 × 2-5 cm, base cuneate or acute,
margin entire, apex acute. Flowers white in terminal, 25-30-flowered corymbs or
subumbels; sepals 4, outer pair boat-shaped, orbicular, inner pair linear-oblong or
obovate, sparsely puberulous; petals 4, oblong to narrowly obovate; stamens 35-50;
gynophores 1.3-2.8 cm long, hairy towards base, incrassate in fruit; ovary ovoid.
Berry globose, 2.5 cm across, rusty brown, seeds 2, embedded in pink pulp.

Occasional in forests in most districts.

Fl.: Throughout the year, with peak period in February – July and November –
December; Fr.: July- January.

Dharur RF (RR), *MSM* & *TP* 11095; Gowraram RF (NZB), *TP* & *BR* 6379;
Pocharam RF (MDK), *BR* & *CP* 11604; Mannanur (MBNR), *TDCK* 16806.

Capparis sepiaria L., Syst. Nat. (ed. 10) 2: 1071. 1759; Hook.f. & Thomson in Hook.f.,
   Fl. Brit. India 1: 177. 1872; Sundara Raghavan in B.D.Sharma & *al.*, Fl. India 2:
   289. 1993.

Vern.: *Nalluppi.*

Scandent prickly shrubs, rarely trees, 2-6 m high, stems zig-zag, terete, branchlets densely fulvous or grey-pubescent. Leaves oblong, lanceolate or elliptic, rarely obovate, 1.5-3 × 0.8-1.5 cm, thin, coriaceous, base cuneate, margin entire, apex notched. Flowers white in terminal corymbose umbels; sepals 4, subequal, ovate or suborbicular, inner pair smaller than outer, margin ciliate; petals 4, obovate or oblong-spathulate, puberulous at base; stamens 25-40, filaments 10-14 cm long, exerted, anthers brown; gynophores 5-6 mm long, slender, ovary ovoid. Berry globose, ca. 0.8 cm across, blue when ripe; seeds 2, embedded in a sticky pulp.

Common in dry forests and hedges and scrub jungles of most districts.

Fl.: March – July; Fr.: October - November.

Kagajnagar road (ADB), *GO* 4367; Sirpur RF (ADB), *GO* & *PVP* 4910; Mohammadabad RF (RR), *MSM* 10481; Ranjhole (MDK), *CP* 11599; Nagarjunakonda (NLG), *BVRS* 138 (NU); Amrabad (MBNR), *BR* & *SKB* 32546 (BSID).

Capparis zeylanica L., Sp. Pl. ed. 2: 270. 1762; Gamble, Fl. Madras 1: 46. 1915; Sundara Raghavan in B.D.Sharma & *al.*, Fl. India 2: 250. 1993. *C. horrida* L. f., Suppl. Pl. 264. 1781; Hook.f. & Thomson in Hook.f., Fl. Brit. India 1 : 178. 1872.

Vern. : *Adavinimma.*

Large, climbing shrubs, up to 8 m long, with hooked spines, stems woody, rough, young parts green, rusty tomentose, with pungent smell. Leaves ovate or elliptic, 3.5-6.5 × 2.5-4 cm, rusty-tomentose when young, glabrous at maturity, base cuneate, margin entire, apex mucronate. Flowers yellowish-white or white in supra-axillary, solitary, 2-3 pedunculate; sepals 4, subequal, outer pair elliptic or orbicular, inner pair elliptic-oblong, slightly smaller than outer; petals 4, white with a reddish spot within, fading to purple, tomentose inside; stamens 30-50, filaments creamy white, turning to pink or purple before dusk; gynophores glabrous, except at base, ovary ellipsoid or ovoid. Berries globose, scarlet red; seeds many, embedded in scarlet pulp.

Common in forests and hedges of all districts.

Fl.: February – April; Fr.: August - October.

Kunthala (ADB), *GO* & *DAM* 5121; Mingaram RF (NZB), *TP* & *BR* 6301; Mamidipally RF (NZB), *BR* & *KH* 9018; Mombojipally (MDK), *BR* & *CP* 11453; Mohammadabad RF (RR), *MSM* & *KH* 10215; Nagarjunasagar (NLG), *BVRS* 169 (NU); Gandikudisai (KMM), *RR* 113832 (BSID); Way to Nawabpet (MBNR), *SRS* 10906(BSID).

Note: The nomenclatural problem in this species has been dealt with by Jacobs (in Blumea 12: 505-508. 1965). Unaware that Linnaeus had already described the species based on Herman's collection from Sri Lanka, the younger Linnaeus described *C. horrida* based on Koenig's collection, also from Sri Lanka. Wight & Arnott (Prodr. Fl. Pen. Ind. Orient. 25. 1834, excl. synonym *C. pyrifolia* Lam.) had correctly interpreted both the species but failed to recognize that both are related to one and the same species.

The description of *C. zeylanica* Hook.f. (Fl. Brit. India 1: 174. 1872) pertains to *C. brevispina* DC., an error repeated by many other Indian botanists, except by Dunn who could distinguish *C. zeylanica* from *C. brevispina*.

CLEOME L.

1. Leaves simple:
   2. Stamens 6 .... **C. monophylla**
   2. Stamens 8-16 .... **C. simplicifolia**
1. Leaves usually compound:
   3. Androgynophore present .... **C. gynandra**
   3. Androgynophore absent:
      4. Plants viscous with stalked glands .... **C. viscosa**
      4. Plants mostly glabrous, rarely with scale like hairs :
         5. Stamens 6:
            6. Flowers pinkish, in terminal racemes .... **C. rutidosperma**
            6. Flowers yellow, axillary, solitary .... **C. aspera**
         5. Stamens more than 10:
            7. Siliqua cylindric, glandular-hairy; stamens 35-60 .... **C. chelidonii**
            7. Siliqua flat, glabrous; stamens less than 30 .... **C. felina**

Cleome aspera Köen. ex DC., Prodr. 1: 241. 1824; Hook.f. & Thomson in Hook. f., Fl. Brit. India 1: 169. 1872; Gamble, Fl. Madras 1: 41. 1915; Sundara Raghavan in B.D.Sharma & *al.*, Fl. India 2: 303. 1993.

Annual, much branched, prostrate or ascending, minutely prickly herb. Leaves apically simple, basally 3-foliolate, leaflets obovate or oblanceolate, 1-2 × 0.3-0.5 cm, middle leaflets bigger than the lateral leaflets, base cuneate, entire, tip obtuse, surface hairy; petioles 2-10 mm long, almost absent in uppermost leaves; leaflets subsessile. Flowers yellow, axillary, solitary, sepals 4, linear-lanceolate, obtuse, glandular pubescent; petals 4, elliptic or oblanceolate, rounded at apex; stamens 6, anthers linear; gynophores ca 1 mm long, elongating to 3 mm in fruits, ovary linear. Capsule slender, torulose, beaked; seeds 12-20, suborbicular, yellowish to dark brown.

Occasional in wastelands and forests in most districts.

Fl. & Fr.: July-January.

Jaipur RF (ADB), *GO & PVP* 4802; Sirnapally RF (NZB), *BR & PSPB* 9584; Pochammaralu (MDK), *CP* 11759; Cherukupally (NLG), *BVRS* 350; Nagarjunakonda (NLG), *BVRS* 85 (NU); Vijayapuri (NLG), *KMS* 9829 (CAL); Aklaspur (KRN), *GVS* 20240.

Cleome chelidonii L. f., Suppl. Pl. 300. 1781; Hook. f. & Thomson in Hook. f., Fl. Brit. India 1: 170. 1872; Gamble, Fl. Madras 1: 41. 1915; Sundara Raghavan in B.D.Sharma & *al.*, Fl. India 2: 306. 1993. *Polanisia chelidonii* (L. f.) DC., Prodr. 1: 242. 1824.

Annual strigose suffrutescent herbs, up to 1.2 m high. Leaves 3-9-foliolate, leaflets 5-9 on lower leaves, 3 or rarely one in upper leaves; basal leaflets obovate, base cuneate, margin entire, apex obtuse to rounded, middle one 2-3 × 0.5-1.5 cm, laterals 0.8-1 × 0.3-0.6 cm, appressed glandular. Flowers pink in leafy racemes; sepals 4, narrowly imbricate, elliptic to obovate, sparsely pubescent outside; petals 4, elliptic, ovate to obovate, narrow at base, rounded at apex, glabrous; stamens 35-60, filaments 8-12 mm long, swollen at tips, greenish white or purple tinged, anther 1 mm long, white; ovary sessile, linear. Siliqua cylindric, glandular hairy; seeds 40-50, orbicular on dorsal side, reddish brown, drying black.

Occasional in sandy soils of most districts.

Fl. & Fr.: October- May.

Sone (ADB), *GO* & *BR* 9424; Adluryellareddy (NZB), *TP* & *BR* 6418; Gannaram (NZB), *TP* & *BR* 6405; Kodimial (KRN), *GVS* 20097 (MH); Pakhal (WGL), *KMS* 13106 (MH); Rayagir (NLG), *BVRS* 12 (NU).

Cleome chelidonii L.f. var. **pallai** C.S. Reddy & V.S. Raju, J. Econ. Taxon. Bot. 25: 217-218.

This variety differs from typical variety in the following characters.

Characters	var. *chelidonii*	var. *pallai*
Habit	Erect herbs to 0.5m high	Erect herbs to 1.5 m high
Petiole	Basal leaves 5-13 cm; upper leaves 2-4 mm	Basal leaves 4-8 cm, upper leaves 1.5-3.5 cm
Basal leaflets	Obovate, 3-5 × 1.2-2 cm; margin entire	Lanceolate, 4-7 × 0.3-0.7 cm, margin crenate to crenate-serrate
Upper leaflets	Oblanceolate to lanceolate; margin entire or crenate	Linear; margin crenate to serrate
Internode	Internodes to 5 cm	Internodes to 12 cm
Pedicel	Pedicel 7 cm in fruit	Pedicel 4 cm in fruit
Sepal	Obovate or ovate, 3-4 mm	Elliptic, 1.5- 2 mm
Petal	Obovate or elliptic–ovate, 1.2-1.4 cm	Ovate, 1.4-1.7 cm
Filaments	1.2-1.4 cm	1.5-1.8 cm
Capsule	beak 2-4 mm	beak 5-10 mm
Seeds	Spheroidal to ovoidal, compressed, 1.5-2.5 mm across; cleft fairly open, deep, about 0.7-1.0mm. Testa with blunt tubercles of various heights; waxy coating absent	Ovoidal 1.3-1.8 mm across, cleft narrow, deep about 0.5-0.7 mm. Testa with scattered uneven scales; appendages with waxy coating

Fl. & Fr.: August - December.

Pakhal RF (WGL), *KMS* 13106 (MH, Holotype); Pakhal RF (WGL), *CSR & VSR* 876 (KU, Paratype).

Cleome felina L. f., Suppl. Pl. 300. 1781; Hook. f. & Thomson in Hook. f., Fl. Brit. India 1: 170. 1872; Gamble, Fl. Madras 1: 41. 1915; Sundara Raghavan in B.D.Sharma & *al.*, Fl. India 2: 308. 1993.

Annual prostrate-decumbent herbs, up to 50 cm high, stems and leaves entirely covered with stiff appressed scale like haris. Leaves apically simple or 1-foliolate, basally 3-foliolate, leaflets obovate, 0.7-1.5 × 0.5-1 cm, glandular, base cuneate, margin entire, apex obtuse. Flowers pink, axillary, solitary; sepals 4, linear, lanceolate, oblong; petals 4, elliptic-obovate, spathulate, pinkish; stamens less than 30, anthers yellow; ovary 5-6 mm long. Capsule linear, glabrous, compressed, acute at both ends, striate; seeds many, reniform, yellowish brown.

Common in forests and non-forest areas during rainy season in most districts.

Fl. & Fr.: July-January.

Mancherial (ADB), *GO* & *PVP* 4936; Tiryani RF (ADB), *PVP* 9803; Nagarjunakonda (NLG), *BVRS* 105 (NU); Chandoor (NLG), *BVRS* 357 (NU); Vempalli (KRN), *GVS* 20179 (NH); Molachintapalli (MBNR), *TSS* & *BR* 29827.

Cleome gynandra L., Sp. Pl. 6710 1753 var. **gynandra**; Sundara Raghavan in B.D.Sharma & *al.*, Fl. India 2: 309. 1993. *Gynandropsis pentaphylla* (L.) DC., Prodr. 1: 238. 1824; Hook.f. & Thomson in Hook.f., Fl. Brit. India 1: 171.1872.

Vern.: *Vaminta.*

Annual, erect, branched, hispid herbs, up to 80 cm high, strongly foetid. Leaves long-petioled, digitately 3-5-foliolate, leaflets sessile, obovate, 2-5.5 × 1.5-3 cm, hispid, base cuneate, margin entire or crenate, apex acute. Flowers white, cream or pale pink, in corymbose racemes; sepals 4, ovate, obovate, elliptic or oblanceolate, caducous; petals 4, obovate to oblanceolate, subequal; stamens 6, purple, filaments subequal, anthers linear, 2 mm long; gynophores 0-2 cm long, elongating to 3.5 cm in fruits, ovary sessile among stamens or on up to 2 cm long gynophores. Capsule cylindrical, 3-12 cm long, 4-8 mm thick, sticky glandular-pubescent, stalk 2-4 cm long; seeds 15-40, reniform, compressed, black.

Common weed in waste places, cultivated fields and forests in all districts.

Fl. & Fr.: May-February.

Sattenapalli (ADB), *TP* & *PVP* 4154; Tekrial (NZB), *TP* & *BR* 6422; Narsingi (MDK), *BR* & *CP* 11532; Molachintapally (MBNR), *TDCK* 15306; Sarlapally (MBNR), *GS* 18211; Kodimial (KRN), *GVS* 20050 (MH); Tirmalagiri (NLG), *BVRS* 465 (NU); Kinnerasani Wild Life sanctuary (KMM), *RR* 112697 (BSID).

Cleome monophylla L., Sp. Pl. 672. 1753; Hook. f. & Thomson in Hook. f., Fl. Brit. India 1: 168. 1872; Gamble, Fl. Madras 1: 41. 1915; Sundara Raghavan in B.D.Sharma & *al.*, Fl. India 2: 312. 1993.

Erect, annual, sparsely branched hispid herbs, up to 60 cm high. Leaves simple, linear, linear-lanceolate, oblong or lanceolate, 2-4 × 0.4-0.6 cm, sparsely hairy, base truncate, margin entire, apex acute. Flowers white or pink with blue tinge in terminal

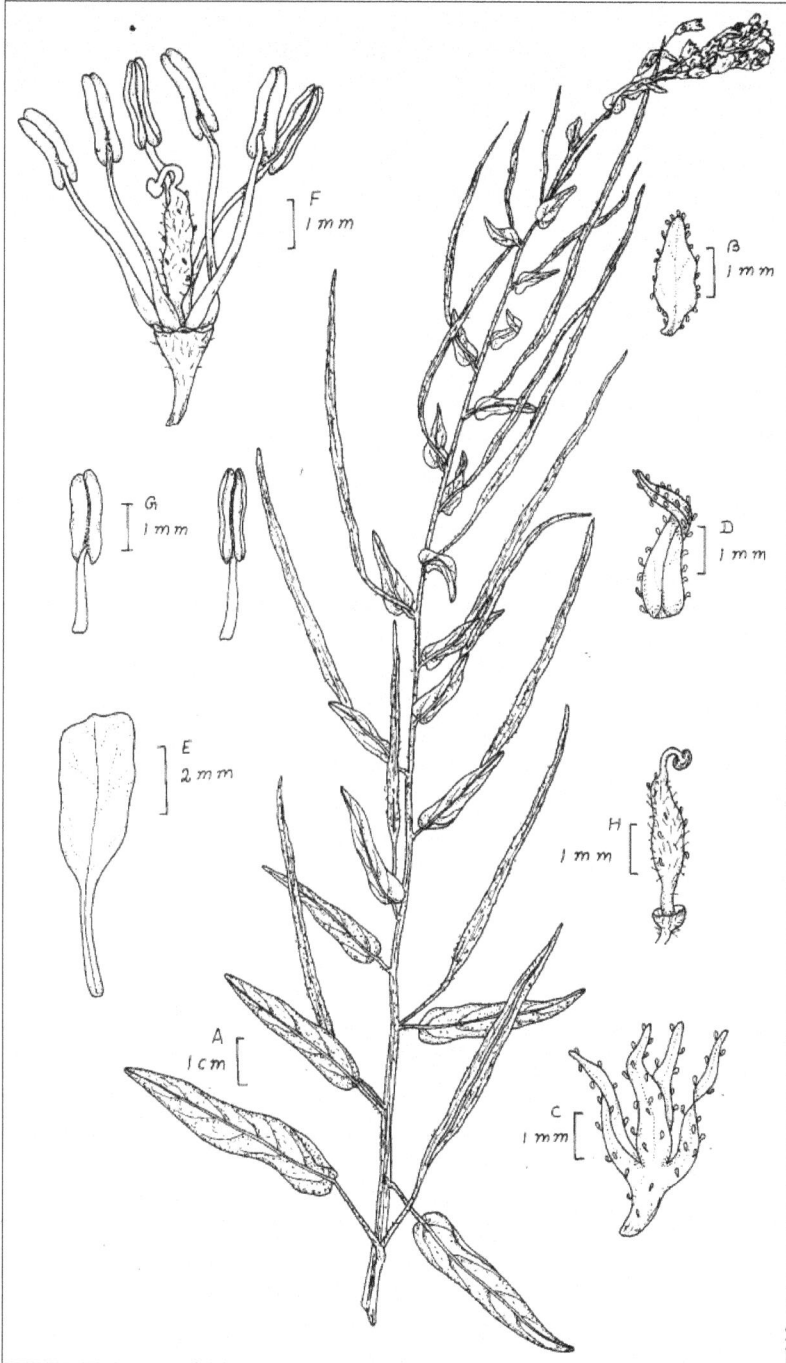

Figure 7: Cleome monophylla L.

A. Habit. B. Bract, C. Calyx, D. Sepal, E. Petal, F. Stamens with pistil, G. Anthers, H. Pistil.

racemes; sepals 4, linear or lanceolate; petals 4, oblong, obovate or spathulate, tapering and clawed at base, rounded at apex; stamens 6, filaments unequal, 5-8 mm long, anthers 2 mm long; ovary linear, sessile or on a short gynophore. Capsule slightly curved, striate, beaked; seeds numerous, rugose, dark brown.

A weed of fields and waste places in all districts.

Fl. & Fr.: July-December.

Laxmapur (NZB), *TP* & *BR* 6286; Marrikunta (MDK), *CP* 11386; Mombojipally (MDK), *BR* & *CP* 11657; Dharmaram (RR), *MSM* & *KH* 10285; Venkatreddipalli (RR), *MSM* & *KH* 10307; Kodimial (KRN), *GVS* 25677, 20067 (MH); Durachipalli (NLG), *BVRS* 25 (NU).

Cleome rutidosperma DC., Prodr. 1: 211. 1824; C.S.Reddy & *al.*, J. Econ. Taxon. Bot. 24: 291. 2000.

Annual erect herbs, to 30 cm high, stems ribbed, weak and pilose. Leaves 3-foliolate, floral leaves subsessile, leaflets rhomboid – elliptic or obovate, 1-5 × 0.2-1 cm, glabrous, base attenuate or acuminate, margin obscurely crenulate-serrulate, apex acute or acuminate. Flowers in terminal racemes, few - flowered, pinkish; sepals 4, linear-lanceolate; petals 4, oblanceolate to elliptic, clawed at base, apiculate at apex, showy pink, bluish violet, rarely white with pink streaks; stamens 6, filaments 6-9 mm long, anthers linear, recurved after anthesis; gynophores 1.5-2 mm long elongating to 8 mm in fruits, ovary linear, stigma sessile, capitate. Capsule linear – cylindric, compressed, alternate at both ends, ribbed, seeds many, suborbicular to reniform with prominent concentric and transverse ridges and open cleft, black, elaiosome massive.

Weed of damp, shady localities in Ranga Reddi and Khammam districts, rare (C.S. Reddy & *al.*, 2000).

Fl. & Fr.: June – December.

Achutapuram (KMM), *CSR* & *KNR* 1469 (KU).

Cleome simplicifolia Hook. f. & Thomson in Hook. f., Fl. Brit. India 1: 169. 1872; Sundara Raghavan in B.D.Sharma & *al.*, Fl. India 2: 314. 1993.

Annual scabrid, erect herbs, up to 60 cm high, branches many from the base, stem furrowed. Leaves simple, obovate, oblong or elliptic-oblong, scabrid, base truncate, margin entire, apex acute, densely strigose with scaly hairs; petiole up to 2 cm long but almost absent upwards. Flowers solitary in the axils of leafy bracts on elongate racemes; sepals 4, triangular or ovate; petals 4, oblong, obovate, oblanceolate or spathulate, pink, mauve, lilac or purplish violet; stamens 8-16; ovary cylindric. Capsule 4-6 cm long, beaked, glabrous, torulose; seeds many, pale-brown.

Rare weed of cultivated lands, waste places and also in forest areas in Karimnagar district.

Fl. & Fr.: July-December.

Bhupatipur (KRN), *GVS* 20173 (MH).

Note: The resemblance of *C. simplicifolia* to *C. monophylla* in having simple leaves and pink flowers, is often misleading to plant collector. It is better to count the number of stamens which are 10-12 in the former and 6 (7) in the later. Further more elaisome is absent in *C. monophylla*.

Cleome viscosa L., Sp. Pl. 672. 1753; Hook.f. & Thomson in Hook. f., Fl. Brit. India 1: 170. 1872; Gamble, Fl. Madras 1: 41. 1915; var. **viscosa**; Sundara Raghavan in B.D.Sharma & *al.*, Fl. India 2: 317. 1993.

Vern. : *Kukkavaminta.*

Annual erect, branched, viscid pubescent herbs, up to 1 m high. Leaves 3-5-foliolate, leaflets obovate or elliptic, 1.5-3.2 × 0.7-1.5 cm, hairy, base cuneate, margin ciliate, apex acute. Flowers yellow in axillary or terminal racemes; sepals 4, oblong, acute at apex, glabrous inside, glandular hairy outside; petals 4, subequal, obovate or oblanceolate to oblong-spathulate, orange yellow or yellow or creamy white, distinctly nerved; stamens 12-18, often intermingled with staminodes, filaments 4-8 mm long, broad at tips, anthers linear, 2.8 mm long, bluish; ovary sessile, oblong-cylindric or linear-oblong. Capsule terete, sticky- pubescent, short beaked, up to 6 cm long; seeds numerous, dark- brown, sub-globose.

A common weed, in fields and waste places in all districts. Occasional in open forests.

Fl. & Fr.: June-December.

Bersaipet (ADB), *GO* & *PVP* 4616; Dharmapuram (RR), *MSM* & *KH* 10285; Mudheli RF – Road side (NZB), *BR* & *KH* 7135; Sivanoor (MDK), *BR* & *CP* 11546; Narsapur (MDK), *KMS* 6109 (CAL); Pakhal (WGL), *KMS* 13103 (MH); Tirmalagiri (NLG), *BVRS* 473 (NU); Rathamhutta hills (KMM), *RCS* 104235 (BSID).

Key to the Varieties
1. Stamens 36-40; styles 5-8 mm long in flower, elongating to 20 mm long in fruits .... var. **nagarjunakondensis**
1. Stamens 12-18; styles 1-3.5 mm long in flower, 4-5 mm long in fruits .... var. **viscosa**

Cleome viscosa L. var. **nagarjunakondensis** Sundara Raghavan in Bull. Bot. Surv. India 28: 187-191. t. 1-8. 1986 (1988) & in B.D.Sharma & *al.*, Flora of India 2: 318.1993.

Annual herbs, up to 2 m high, woody and sparsely branched from base. Leaves 3-5 (7)-foliolate, 0.6-3 × 0.2 – 1.5 cm. Flowers cream-coloured, 2-2.5 cm across, stamens 36-40, intermixed with staminodes. Capsules linear to oblong, 6-8.5 cm long.

Very rare, occurs in dry deciduous forests of Nagarjunakonda.

Fl. & Fr.: July – September.

CRATEVA L.

1. Mostly leafless when flowering, flowers many (up to 100) in corymbs; rachis 2-3 cm long, fruit orange yellow, turning to reddish at maturity .... **C. adansonii**

1. Mostly leaf – bearing when flowering, flowers few (12 – 30) in corymbs, rachis 10 cm long, fruit yellowish – grey to brown on maturity .... **C. magna**

Crateva adansonii DC., Prodr. 1:243. 1824 subsp. **odora** (Buch.-Ham.) Jacobs, Blumea 12: 198. 1964; Sundara Raghavan in B.D.Sharma & *al.*, Fl. India 2: 322. 1993. *C. odora* Buch.-Ham., Trans. Linn. Soc. London 15: 188. 1827. *C. religiosa* var. *roxburghiana* (R.Br.) Hook.f. & Thomson in Hook. f., Fl. Brit. India 1: 172. 1872. *C. religiosa* sensu Dunn in Gamble, Fl. Madras 1: 47. 1915 *p.p.* non Forst.f. 1782.

Vern.: *Mugalina, Muvva.*

Small, deciduous trees, 3-10 m high, bark greyish with yellow specks; branchlets glabrescent. Leaves 3-foliolate, petiole 7-10 cm, glandular at tip; leaflets subcoriaceous, ovate to elliptic – lanceolate, central leaflets up to 12 × 7 cm with abruptly acuminate at apex; lateral leaflets 6-8 × 3.5-5 cm with acute tip, base oblique and tapered, glabrous. Flowers greenish – white, fragrant, 2.5 – 3.5 cm across in terminal corymbs; sepals petaloid, elliptic, acuminate; petals initially greenish white turning yellowish, finally fading to pink, clawed; stamens 15-26, filaments 4.5 cm long, pinkish; gynophores pinkish, ovary ovoid or ellipsoid. Berries orange – yellow turning to reddish at maturity, globose, 2.5 – 4 × 2 cm, seeds brown, broadly elliptic, 5-6 × 2 cm, smooth, embedded in yellow pulp.

Occasional, in deciduous forests of Warangal, Khammam and Hyderabad districts (Rajagopal, 1973; Ramana, 2010).

Fl.: February – April; Fr.: May – November.

Khammam (KMM), *PVS* 84049 (MH, BSID).

Note: *C. religiosa* sensu Dunn (in Gamble's Fl. Madras 1:47.1915) partly reflects subsp. *odora* and partly *C. magna*, both common in Peninsular india (Sundara Raghavan *l.c.*).

Crateva magna (Lour.) DC., Prodr. 1: 243. 1824; Jacobs in Steenis, Fl. Males. (Ser. 1) 7: 822. 1976 (in addendum) & Blumea 12: 206. 1964; Sundara Raghavan in B.D.Sharma & *al.*, Fl. India 2: 324. 1993. *Capparis magna* Lour., Fl. Cochinch. 1: 331. 1790. *Crateva nurvala* Buch.-Ham., Trans. Linn. Soc. London 15: 121. 1827. "*naravala*". *C. religiosa* Forst. f. var. *nurvala.* (Buch. Ham.) Hook. f. & Thomson in Hook. f., Fl. Brit. India 1: 172. 1872. *C. religiosa* sensu Dunn in Gamble, Fl. Madras 1: 47. 1915. *p.p.* non. Forst. f. 1786.

Vern.: *Uskia-man, Voolemara, Vulimiri-chettu.*

Moderate sized deciduous trees, up to 6 m high, bark grey, fairly smooth with horizontal wrinkles. Leaves digitately 3-foliolate, leaflets ovate or ovate-lanceolate, 5-8.5 × 2-5 cm, glabrous above, pale beneath, base attenuate or oblique, margin entire, apex acuminate; petioles 4-12 cm long with glands at apex. Flowers yellow, in many-flowered, lax, terminal corymbs, tetramerous; sepals 4, oblong to ovate-oblong; petals 4, obovate, obtuse, clawed; stamens more than 24, filaments lilac or purple; gynophores 4-6 cm long, ovary oblong-ellipsoid. Berry oblong-ellipsoid or ovoid, ca 5 cm across; seeds many, reniform, embedded in creamy plp.

Occasional in forest outskirts and villages in Nizamabad district.

Fl.: January – April; Fr.: April-July.

Manchippa (NZB), *BR* & *KH* 9024.

MAERUA Forssk.

1. Petals 0, berry ovoid; leaves 3-5-foliolate; small tree .... **M. apetala**
1. Petals 4; berry moniliform; leaves simple; stragglers .... **M. oblongifolia**

Maerua apetala (Roth) Jacobs, Blumea 12: 207. 1964; Sundara Raghavan in B.D.Sharma & *al.*, Fl. India 2: 330. 1993. *Capparis apetala* Roth, Nov. Pl. Sp. 238. 1821. *Niebuhria linearis* DC., Prodr. 1: 244. 1824; Hook.f. & Thomson in Hook. f., Fl. Brit. India 1: 171. *N. apetala* Dunn in Gamble, Fl. Madras 1: 41. 1915.

Vern.: *Pilli adugu, Lukki-chettu, Nemali-adugu.*

Small, straggling, unarmed, glabrous trees, up to 4 m high. Leaves present on young shoots only, older branches leafless, leaves 3-5-foliolate, leaflets ovate, linear or linear-oblong, 3.5-6 × 1.5-2.5 cm, base cuneate, margin entire, apex acute, spinous pointed; petioles 4-12 cm long, petiolules 6-8 mm long. Flowers creamish in terminal lax, corymbs; calyx-tube greenish, lobes ovate; petals absent; stamens ca 30, filaments pinkish, shorter than gynophores; gynophores 2-2.8 cm long, elongating up to 4 cm while fruiting; ovary ovoid. Berry ovoid, oblong; seed solitary, embedded in scarlet pulp, white.

Occasional in forests in Nalgonda and Mahabubnagar districts.

Fl. & Fr.: August-December.

Farahabad (MBNR), *BR, VSR* & *KP* 39269; Appapur (MBNR), *SKB* & *BSS* 30592 (BSID); Nagarjunakonda (NLG), *BVRS* 136, 137 (NU).

Maerua oblongifolia (Forssk.) A. Rich in Guill. & Perr. Fl. Seneg. Tent. 1: 32. t. 6. 1847; Sundara Raghavan in B.D.Sharma & *al.*, Fl. India 2: 250. 1993. *Capparis oblongifolia* Forssk., Fl. Aegypt.-Arab. 99. 1775. *Maerua arenaria* (DC.) Hook. f. & Thomson in Hook. f., Fl. Brit. Inida 1: 171. 1872; Gamble, Fl. Madras 1: 42. 1915. *Niebuhria arenaria* DC., Prodr. 1: 244. 1824.

Unarmed stragglers, up to 4 m long, branches glabrous. Leaves simple, ovate, elliptic-oblong or lanceolate, 3- 4.5 × 2-3 cm, scabrous, margin entire, base and apex obtuse; petioles 6-9 mm long. Flowers greenish-yellow in corymbs, rarely flowers solitary, axillary, mildly fragrant; bracts small, ovate; sepals petaloid, united near base or up to one-third from base, calyx-tube 3-8 mm long, lined by a tubular truncate disc, lobes elliptic-oblong; petals on cup-shaped disc, ovate-lanceolate to obovate; stamens 20-26, filaments greenish or white, brownish or purple on drying; gynophores

1.5-2.5 cm long, ovary cylindrical, stigma sessile. Berry fleshy, elongate, twisted and deeply constricted between the seeds; seeds globose, minutely echinate-tuberculate.

Occasional in forests in Karimnagar, Mahabubnagar, Warangal, Hyderabad and Nalgonda districts.

Fl.: January – March, August – October; Fr.: February-June, September-December.

Nimmagudem (KRN), *SLK* 70912 (MH); Jangaon (NLG), *BVRS* 388 (NU).

VIOLACEAE

HYBANTHUS Jacq. *nom. cons.*

1. Annual herb; stem unbranched; flowers bright orange .... **H. stellarioides**
1. Perennial herb; stem branched; flowers pink:
    2. Flowers conspicuous; fruit stalked; seeds many .... **H. enneaspermus**
    2. Flowers inconspicuous; fruit without stalk; seeds 5-8 .... **H. vatsavayii**

Hybanthus enneaspermus (L.) F.V. Muell., Fragm. Phyt. Austr. 10: 81. 1872; Banerjee & Pramanik in B.D.Sharma & *al.*, Fl. India 2: 343. 1993. *Viola enneasperma* L., Sp. Pl. 937. 1753. *Ionidium suffruticosum* (L.) Ging. in DC., Prodr. 1: 311. 1824; Hook.f. & Thomson in Hook. f., Fl. Brit. India 1 : 185. 1872; Gamble, Fl. Madras 1: 49. 1915. *Viola suffruticosa* L., Sp. Pl. 937. 1753.

Vern.: *Ratnapurusha.*

Small perennial much branched, erect, woody herbs, up to 25 cm high. Leaves simple, linearly lanceolate, 2-3.5 × 0.3-0.5 cm, base attenuate, margin serrate, apex acute. Flowers pink, axillary, solitary; sepals 5, subequal, lanceolate; petals 5, unequal, obovate, upper ones oblong, laterals falcate, lower one larger, orbicular, clawed, saccate at base; stamens 5, anterior stamen with a small recurved fleshy appendage; ovary globose, glabrous, style thickened towards tip. Capsule sub-globose, yellowish; seeds many, ovoid, striated.

Common weed of waste lands, road sides, fields and open forest lands in almost all districts.

Fl. & Fr.: July-January.

Bellampalli RF (ADB), *PVP* 9465; Nasarullabad RF (NZB), *TP* & *BR* 6159; Arsapally (NZB), *BR* & *KH* 6491; Pillutla (MDK), *BR* & *CP* 11264; Thornal (MDK), *CP* 11399; Pakhal RF (WGL), *KMS* 6692 (MH); Rangammagudem RF (RR), *MSM* & *KH* 10318; Kusumasamudram RF (RR), *MSM* & *KH* 10345; Lingampalli (RR), *MSM* & *PRSR* 12763; Kannaigudem (KMM), *AJR* & *PAN Reddy* 20468; Kodimial (KRN), *GVS* 20182 (MH); Azampur (NLG), *BVRS* 485 (NU).

Hybanthus stellarioides (Domin) P.I. Forst., Muelleria 8(1): 18. 1993. *H. enneaspermus* var. *stellarioides* Domin, Bibl. Bot. 89(4): 983. 1928. *H. enneaspermus* subsp. *stellarioides* (Domin) E.M. Benn., Nuytsia 1(3): 229. 1972; George, Fl. Australia 8: 103. 1982.

Annual herb, 15 – 30 cm high, stem erect, grooved, unbranched (rarely branched), hairy; hairs antrorse-divaricate. Leaves simple, alternate, clustered at apex, linear or linear to lanceolate, 1.5 x 0.1-0.4 cm, attenuate at base, entire with occasional marginal tooth and ciliate at margins, acute at apex, sparsely hairy; stipules linear, 1 – 2 mm long, densely hairy; hairs subsessile, gland-tipped. Flowers solitary, axillary; peduncle filiform, with scattered to sparse indumentum, ca 7 mm long; pedicels ca 2 mm long; a distinct joint exists between peduncle and pedicel; bracts triangular, ca 1 mm long, ciliate at margins; sepals 5, ovate-lanceolate, subequal, 2.5 – 4 mm long, acuminate, bent backwards at apex, keeled, hairy; petals 5, unequal; upper oblong, 3 – 4 mm long, pale yellow; lateral 2 falcate, 2 – 3 mm long, pale yellow; lower enlarged into a spathulate limb with a claw; limb broader than long, ca 6-10 mm, acute at middle, bright orange; claw ca 4 mm long; stamens 5, ca 3 mm long; filaments free; anterior 2 filaments with hairy appendages; anthers connate, 2 of them villous, others glabrous; connective extended vividly; pistil ca 4 mm long; ovary ovoid, glabrous; ovules 6 – 9; style suberect; stigma enlarged, flat. Capsules 3-angled, 4 – 6 mm long, 3-valved, with remnant petals; seeds 6 – 9, ellipsoid, ca 2 mm long, longitudinally ribbed, glabrous, pale yellow.

Rocky terrains on hill slopes along rock margins and in rocky crevices in Hyderabad (Ramana & *al.,* 2011).

Fl. & Fr. July – August.

Dammaiguda hills, Hydrabad, *M. Venkat Ramana* & *P. V. Prasanna* 1961 (BSID).

Hybanthus vatsavayii C.S. Reddy, J. Econ. Tax. Bot. 25: 219. 2001.

Erect herb, branched from base, up to 20 cm high, stem pale yellow puberulous. Leaves linear to linear – lanceolate, 2-4 × 0.1-0.5 cm, base alternate, margin entire to distantly serrate, apex acute-apiculate. Flowers axillary, solitary, inconspicuous. Capsule 3-lobed, puberulous, seeds 5-8 per capsule, ovoid-ellipsoid

Rare in crevices of rocks in Khammam, Warangal and Nalgonda districts.

Fl. & Fr.: August – December.

Yadagirigutta (NLG), *CSR* 1268 (MH) Holotype; Shobhanadripuram (NLG), *CSR*1674 (KU) Paratype; Pakhal RF (WGL), *CSR* 1302; Vazheedu (KMM), *VSR* 3001 (KU).

BIXACEAE

1. Leaves entire; petals pink; disc annular; ovary and capsules covered densely by thick flexible bristles; ripe seeds with thick red testa .... **Bixa**

1. Leaves palmatipartite; petals yellow; disc absent; ovary and capsules not covered by such bristles; ripe seeds wrapped in long white wool .... **Cochlospermum**

BIXA L.

Bixa orellana L., Sp. Pl. 512. 1753; Hook.f. & Thomson in Hook.f., Fl. Brit. India 1 : 190. 1872; Gamble, Fl. Madras 1: 51. 1915; Balakrishnan in B.D.Sharma *& al.*, Fl. India 2: 381. 1993.

Vern. : *Japhra.*

Small evergreen trees, 2-9 m high, bark greyish with yellow specks; young branches densely dark scaly. Leaves alternate, ovate, up to 14 × 9 cm, 5-nerved, base cordate, margin entire, apex acuminate; petiole 4-10 cm long. Flowers pink in terminal panicles, or corymb, 8-50-flowered, fragrant; sepals concave, broadly ovate to suborbicular, purple; petals 5-7, unequal, obovate, light red, pink to white, rounded at apex; filaments slender, yellow at base, red at apex; ovary densely covered with thick red bristles, style 12-15 mm long, swollen above, red. Capsule 2-valved, globose or broadly ovoid, loculicidal, with dense long stiff but soft and flexible bristles; seeds covered by red powder.

Planted in gardens, house yards in Warangal, Hyderabad and Karimnagar districts (Rajagopal, 1973; Naqvi, 2001; Prasanna & *al.*, 2012).

Fl. & Fr.: August.-December.

COCHLOSPERMUM Kunth *nom. cons.*

Cochlospermum religiosum (L.) Alston in Trimen, Handb. Fl. Ceylon 6 (Suppl.). 14. 1931; Balakrishnan in B.D.Sharma & *al.*, Fl. India 2: 383. 1993. *Bombax religiosum* L., Sp. Pl. 512. 1753. *Cochlospermum gossypium* DC., Prodr. 1: 527. 1824; Hook.f. & Thomson in Hook. f., Fl. Brit. India 1: 190. 1872; Gamble, Fl. Madras 1: 50. 1915.

Vern. : *Kondagogu.*

Deciduous trees, up to 15 m high with yellow or red juice. Leaves orbicular in outline, 3-5-lobed, 5-15 × 7-20 cm, glabrous above, densely brownish- tomentose below, 5-7-nerved at base, lobes ovate-lanceolate, base cordate, margin entire, apex acute to acuminate. Flowers large, 8 cm across, golden-yellow with silky sepals in terminal panicles; bracts triangular, caducous; sepals 5, deltoid-ovate, silky, partly purplish; petals 5, obovate, yellow with distinct odour, thick at base; stamens many, slightly S-curved, unequal, anthers linear, falcate, orange; ovary globose, glabrous, style long, to 2 cm, glabrous. Capsule brown, pear- shaped, striate, leathery, 5-10 cm long, 2.5 – 8 cm broad; seeds numerous, reniform to cochleate, densely woolly.

Common in dry forests, especially on stony hills, in all districts.

Fl.: January – March; Fr.: March-August. It flowers after leaf shedding.

Laxmapur RF (NZB), *TP* & *BR* 6256; Narsapur RF (MDK), *TP* & *CP* 14049; Pillutla RF (MDK), *RG* 106792 (BSID); Farahabad (MBNR), *TDCK* 13666; Pakhal RF (WGL), *ANH* 15948 (MH).

FLACOURTIACEAE
(including SAMYDACEAE)

1. Flowers bisexual .... **Casearia**
1. Flowers unisexual .... **Flacourtia**

CASEARIA Jacq.

1. Leaves tomentose beneath; pedicel 4-5 mm long .... **C. tomentosa**
1. Leaves hairy or glabrous beneath; pedicel 1-2 mm long .... **C. graveolens**

Casearia graveolens Dalzell, Hooker's J. Bot. Kew. Gard. Misc. 4: 107. 1852; C.B. Clarke in Hook.f., Fl. Brit. India 2: 593. 1879; Gamble, Fl. Madras 1: 520. 1919; R.L.Mitra in B.D.Sharma & *al.*, Fl. India 2: 394. 1993.

Small deciduous trees conspicous in forest by its rather large leaves turning red in winter before falling, up to 10 m high. Leaves broadly elliptic, 4.5- 22 × 2.5-11 cm, base rounded, margin shallowly crenate, obtuse or shortly acute at apex, stiff-coriaceous when old, glabrous, lateral nerves 7-9 pairs, raised and prominent beneath, petiole to 6-19 mm long. Flowers pale green in leafless axillary clusters, greenish; calyx 5-lobed, ovate-oblong, gland-dotted, pubescent outside at base; petals absent; stamens 8, alternating with linear-oblong villous, disc scales, staminodes oblong; ovary ovoid, glabrous, styles as long as stamens with discoid stigma. Capsules orange-yellow, ellipsoid or subglobose, 3- valved; seeds 10-12, ovoid, compressed with scarlet aril.

Common in deciduous forests of Mahabubnagar district.

Fl.: March – April; Fr.: April-August.

Farahabad (MBNR), *SKB* & *BSS* 34436; Mannanur RF (MBNR), *MVS* & *MB* 38001.

Casearia tomentosa Roxb., Fl. Ind. 2: 421. 1832; C.B.Clarke in Hook. f., Fl. Brit. India 2 : 593. 1879; Gamble, Fl. Madras 1: 521. 1919; R.L.Mitra in B.D.Sharma & *al.*, Fl. India 2: 394. 1993. *C. elliptica* Willd., Sp. Pl. 2: 628. 1799.

Vern. : *Chilaka duddi.*

Bushy shrubs or small trees, up to 7 m high, young parts tomentose. Leaves ovate-lanceolate 7-14 × 2-5 cm, rounded and asymmetrical at base, margin serrulate or entire, apex acute, obtuse or acuminate; tomentose beneath; lateral nerves 8-10 pairs, distinct beneath; petioles 3-8 mm long. Flowers to 7 mm across, in dense, axillary, somewhat capitate clusters; calyx 5-lobed, to 4 mm long, broadly elliptic, pubescent, persistent; petals absent; stamens 8, filament stout, to 2 mm long, pubescent; ovary ovoid, style short, stigma discoid. Capsule succulent, 3-valved, globose, 6-ribbed, greenish yellow; seeds numerous, yellowish brown with scarlet aril.

Occasional in forests of Adilabad, Nizamabad, Warangal and Karimnagar districts.

Fl.: February-April; Fr.: April – August.

Bheemaram (ADB), *GO* & *DAM* 5171, 5173; Gowraram RF (NZB), *TP* & *BR* 6383; Sirnapally RF (NZB), *BR* & *GO* 9087, 9088, 9089; Gowraram (KMM), *PVS* 76950 (BSID); Way to Khammam (KMM), *PVS* 84055 (BSID); Pillutla (MDK), *RG* 106795 (BSID).

FLACOURTIA L'Herit.

Flacourtia indica (Burm. f.) Merr., Interpr. Rumph. Herb. Amb. 377. 1917; R.L.Mitra in B.D.Sharma & *al.*, Fl. India 2: 402. 1993. *Gmelina indica* Burm. f., Fl. Ind. 132. t. 39. f. 5. 1768. *Flacourtia ramontchi* L'Herit., Strip. Nov. 3: 59. tt. 30 & 30 B. 1786; Hook.f. & Thomson in Hook. f., Fl. Brit. India 1: 194. 1872; Gamble, Fl. Madras 1: 54. 1915. *F. sepiaria* Roxb., Pl. Coromandel t. 68. 1795.

Vern. : *Kanregu.*

Shrubs or small trees, up to 4 m high, juvenile shoots and branches with simple thorns; branchlets glabrous or pubescent. Leaves usually clustered at apex, basaly opposite or alternate, ovate or obovate, 1.5 - 3.5 × 1-1.5 cm, thin, coriaceous, glabrous, base attenuate, margin entire, apex acute to obtuse, retuse, petioles very short, up to 5 mm. Flowers small, white, in short racemes; sepals 4-5, slightly connate at base, ovate, obtuse, to 1.5 mm long; Male flowers: stamens numerous, filaments 2-2.5 mm long, hairy at base, anthers versatile, disc lobed; Female flowers: disc entire, ovary globular, subglandular, style short, connate, stigma faintly 2-lobed. Drupe globose, dark purple or red when ripe, up to 4 mm across; seeds 3-5, obovoid, trigonous, pale yellow to brown.

Occasional in scrub forests in most districts.

Fl.: December – March; Fr.: ripens from May – August.

Way to Kunthala (ADB), *GO* & *DAM* 5122; Bheemaram (ADB), *GO* & *DAM* 5176; Gandivet RF (NZB), *BR* 9548; Gangapur RF (MDK), *BR* & *CP* 11433; Pocharam RF (MDK), *BR* & *CP* 11614; Pakhal (WGL), *ANH* 15940 (MH); Mallelatheertham (MBNR), *SRS* 108986 (BSID).

Note: Sleumer (in Steenis, Fl. Males. 1(5): 76. 1954) in his revision of *Flacourtia* united *F. ramontchi* L' Herit. and *F. indica* (Burm. f.) Merr. because of the intergrading characters displayed in herbarium materials. This has been followed by Saldanha and Nicolson (Fl. Hassan Dist. 163. 1976) and Saldanha (Fl. Karnataka, 269. 1984). Matthew (Fl. Tam. Carnatic 3: 60. 1983) however has preferred to keep them distinct, based on strong field characters. The white-flowered shrubs of open, dry forests with flowering thorns and fruits were placed under *F. sepiaria* and the trees with non-flowering thorns growing in wet forests under *F. ramontchi.* Here Sleumer's view is being followed.

POLYGALACEAE

1. Flowers minute, in long terminal spikes; stamens 4;
    sepals nearly equal .... **Solomonia**

1. Flowers fairly large, stalked, stamens 8; 2 sepals larger
    (wing sepals) .... **Polygala**

POLYGALA L.

1. Wing sepals symmetric, petaloid; lateral petals obliquely quadrangular or oblong .... **P. persicariifolia**

1. Wing sepals asymmetric, herbaceous; lateral petals oblong, obovate, ovate or suborbicular:

  2. Capsules orbicular or suborbicular :

    3. Flowers 2.5-3 mm long; staminal sheath bearing 2 free filaments at the centre and 2 bundles of 3 connate filaments and subsessile anthers on the sides; style broadened like a knob with a tooth at apex .... **P. telephoides**

    3. Flowers more than 4 mm long; staminal sheath bearing 8 free filaments; style horse-shoe-shaped at apex :

      4. Wing sepals glabrous excepting the margins; capsules 5.5 mm broad; caruncle with 2 oblong, membranous appendages covering 1/2 the length of seed and a short tooth .... **P. chinensis**

      4. Wing sepals sparsely ciliate without and along the margins; capsule 3.5 mm broad; caruncle with 3 obtusely toothed appendages .... **P. bolbothrix**

  2. Capsules oblong, obliquely oblong, elliptic-oblong, ovate-oblong or rhomboid;

    5. Wing sepals elliptic-oblong or elliptic obovate, obtuse to subobtuse at apex; style widened at apex like a knob .... **P. erioptera**

    5. Wing sepals obliquely ovate or ovate-elliptic, acute or acuminate at apex; style horse-shoe shaped or hook-like with wings at apex :

      6. Staminal sheath split half way into 3 bundles bearing 3 sessile anthers in each side-bundle and 2 anthers on short free filaments in the mid-bundle; seeds oblong ...... **P. wightiana**

      6. Staminal sheath bearing 8 free filaments; seeds elliptic-obovate or oblong-obovate :

        7. Keel crested with shortly or deeply forked filiform appendages; caruncle with 2 oblong membranous appendages and pointed tooth .... **P. rosmarinifolia**

        7. Keel crested with shortly forked or notched blunt appendages; caruncle 3-toothed :

          8. Racemes up to 2 cm long; wing sepals sparsely pubescent without, style-apex hooked; capsule symmetric, ovate- oblong .... **P. arvensis**

          8. Racemes 4-10 cm long; wing sepals glabrous, sometimes sparsely puberulous only along the margins; style apex horse- shoe-shaped; capsules asymmetric, rhomboid .... **P. elongata**

Polygala arvensis Willd., Sp. Pl. 3(2): 876. 1802; Chandrabose & Nair, Proc. Indian
    Acad. Sci. (Plant Sci.) 90: 123. 1981; Banerjee in B.D.Sharma & *al.*, Fl. India 2: 460.
    1993. *P. chinensis* auct. non. L. 1753; Bennett in Hook.f., Fl. Brit. India 1: 204. 1872
    p.p.; Gamble, Fl. Madras 1: 58. 1915.

Annual herbs, erect, prostrate, procumbent or ascending, 5-30 cm long. Leaves
0.5-5 × 0.2-1.5 cm, orbicular, obovate, elliptic-oblong, linear- oblong, linear or
lanceolate, base narrowed, apex emarginate. Flowers yellow, solitary or in lateral
cymes; sepals 5, persistent, outer sepals broadly ovate to ovate-lanceolate, wing sepals
obliquely ovate, falcate; petals 3, yellow, mid lobe keeled, auricled at base, with a crest
of shortly multifid or notched, lateral lobes of petals suborbicular or broadly ovate;
staminal sheath free, free portion of filaments up to 2 mm long, united at same level;
ovary obovoid, style curved, flat, dilated and hooked at apex forming a hood with
lateral broad capitate stigma at its concavity. Capsules ovate- oblong, narrowly
winged along margins; seeds black, pilose.

Common weed in waste lands and cultivated fields in most districts.

Fl. & Fr.: August-January.

Rebbana (ADB), *GO* & *PVP* 4839; Bhimnagar RF (NZB), *TP* & *BR* 6218;
Kamareddy (NZB), *BR* & *KH* 7150; Narsapur RF (MDK), *TP* & *CP* 14023;
Kusumasamudram RF (RR), *MSM* & *NV* 10371; Gadirayal RF (RR), *MSM* 10520;
Nagarjunakonda (NLG), *KTH* 9772 (CAL), *BVRS* 119 (NU).

Polygala bolbothrix Dunn in Gamble, Fl. Madras 1: 58. 1915; Chandrabose & Nair,
    Proc. Indian Acad. Sci. (Plant. Sci.) 90: 120. 1981; var. **bolbothrix**; Banerjee in
    B.D.Sharma & *al.*, Fl. India 2: 462. 1993. *P. ramaswamiana* Mukherjee, J. Bombay
    Nat. Hist. Soc. 53: 55. 1955.

Diffusely branched herbs, 25-50 cm high, stem brown, bulbous based hairy.
Leaves 0.3-4 × 0.3-1.2 cm oblong-obovate, base narrowed or cuneate, margin entire
and revolute, apex rounded or emarginate and mucronate, sparsely bulbous hairy
especially on nerves. Flowers pale pink, axillary in 5-9 cm long racemes; bracts 3,
opposite, deltoid, lanceolate; sepals 5, outer sepals ovate-lanceolate, wing sepals
obliquely elliptic-ovate; petals 3-lobed, pink or yellow, lobed up to base, middle lobe
auricled, keeled, with a crest of filiform forked appendages, lateral lobes oblong,
suborbicular, emarginated, clawed and ciliate-margined at base; staminal sheath
free, free portion of filaments ca 1.7 mm long, connate at slightly different levels;
ovary orbicular, slightly oblique, styles curved, widened and horse-shoe shaped at
apex, with lateral hooded stigma. Capsule 4 x 3 mm, obliquely oblong, notched at
apex, bulbous ciliate on margins; seeds 3 mm long, black, obovate, woolly pubescent,
caruncle cup like, three-toothed, membranous, appendaged.

Rare among grasses in Mahabubnagar district.

Fl. & Fr.: Almost throughout the year.

Polygala chinensis L., Sp. Pl. 704 1753; Chandrabose & Nair, Proc. Indian Acad. Sci.
    (Plant Sci.) 90: 118. 1981; Banerjee in B.D.Sharma & *al.*, Fl. India 2: 464. 1993. *P.*
    glomerata Lour., Fl. Cochinch 426. 1790; Bennett in Hook.f., Fl. Brit. India 1: 206.
    1872.

Perennial erect herbs, up to 50 cm high. Leaves 1-5 × 0.2-1 cm, elliptic, elliptic-oblanceolate or obovate, glabrous or sparsely pubescent, base cuneate or obtuse, margin entire, apex acute or obtuse, apiculate, sometimes emarginate. Flowers pale blue, few, in short axillary racemes; sepals 5, outer sepals 1.5-3 × 0.8-1 mm, unequal, ovate to ovate- lanceolate, acute, long-mucronate to aristate, ciliate along the margins, wing sepals 6 × 3 mm, obliquely ovate, acute or acuminate, 5-7-nerved; petals 3-lobed, lateral petals almost as long the keel, obovate; keel crested with filiform appendages; staminal sheath free or adnate to base of petal, 2.5-3mm long, hairy, free portion of filaments arising at different levels on sheath; ovary orbicular, style slightly bent horse-shoe shaped at apex, stigma bilobed. Capsules suborbicular, notched at apex; seeds obovoid or ellipsoid, black, white silky hairy, caruncle hood-shaped with 2 narrowly oblong membranous appendages and a short tooth.

Common in grasslands in all districts.

Fl. & Fr.: April-December.

Aklaspur (KRN), *GVS* 22500 (MH); Amrabad (MBNR), *BR* & *BSS* 32576; Madikonakatta- Narsapur (MDK), *KMS* 6739 (MH); Janakalancha RF (WGL), *RKP* 108201 (BSID).

Polygala elongata Klein ex Willd., Sp. Pl. 3. 879. 1802; Bennett in Hook.f., Fl. Brit. India 1: 203. 1872; Gamble, Fl. Madras 1: 58. 1915; Chandrabose & Nair, Proc. Indian Acad. Sci. (Plant Sci.) 90: 125. 1981; Banerjee in B.D.Sharma & *al.*, Fl. India 2: 466. 1993.

Annual erect herbs, up to 60 cm high. Leaves 1-7 × 0.1-1 cm, oblong, linear to lanceolate, glabrescent, base cuneate, margin entire, apex obtuse; petiole up to 1.5 mm long. Flowers yellow in terminal up to 10 cm long racemes; bracts ca 1 mm long, persistent; sepals 5, outer sepals ovate, elliptic, wing sepals obliquely falcate, ovate or elliptic; petals 5-6 mm long, 3-lobed, midlobe keeled with a crest of branched and shortly forked and notched appendages, lateral lobes obovate, suborbicular, clawed at base; staminal sheath free, up to 3 mm long, free portion of filaments up to 2 mm long, joined at different levels; styles curved, hooked at apex, with stigma inside. Capsule rhomboid; seeds black, ciliate.

Common in waste lands and in open forests in all districts.

Fl. & Fr.: August-January.

Neelwai RF (ADB), *TP* & *GO* 5428; Mudheli RF (NZB), *BR* & *KH* 7120; Arsapally (NZB), *BR* & *KH* 6484; Velutla RF (NZB), *BR* 9516; Venkatapuram (MDK), *TP* & *CP* 14009; Narsapur (MDK), *TP* & *CP* 14022; Pillutla RF (MDK), *BR* & *CP* 11262; Nagulabanda (MDK), *CP* 11329; Kusumasamudram RF (RR), *MSM* & *NV* 10370; Kodimial (KRN), *GVS* 20119 (MH); Pakhal (WGL), *KMS* 13164 (MH); Mannanur (MBNR), *SRS* 109742 (BSID).

Polygala erioptera DC., Prodr. 1: 326. 1824; Bennett in Hook.f., Fl. Brit. India 1: 203. 1872; Gamble, Fl. Madras 1: 59. 1915; Chandrabose & Nair, Proc. Indian Acad. Sci. (Plant Sci.) 90: 121. 1981; Banerjee in B.D.Sharma & *al.*, Fl. India 2: 467. 1993.

Annual erect or diffuse herbs, up to 60 cm long. Leaves 1-3 × 0.3-0.6 cm, linear, linear-oblong or obovate, glabrous to densely pubescent, narrow towards base, subacute, obtuse or emarginate at apex. Flowers pink or purple, solitary or many in axillary or extra axillary racemes, up to 3.5 cm long; outer sepals unequal, lanceolate to elliptic ovate; wing sepals elliptic-obovate or elliptic-oblong, 3-7-nerved; lateral petals shorter than keel, ovate; keel cresed with filiform appendages; staminal sheath free or slightly adnate to middle lobe of petal, free portion of filaments ca 1.5 mm long; ovary ovoid, style curved, broadened at apex like knob. Capsule oblong, obliquely emarginate; seeds black, oblong, pilose, caruncle galeate with 3 short appendages.

Common in scrubs and wastelands in most districts.

Fl. & Fr.: Through the year.

Bellampalli (ADB), *PVP*9460; Bijjur (ADB), *GO* & *PVP*4854; Kamareddy (NZB), *BR* & *KH* 7144; Nagarjunakonda (NLG), *BVRS* 120; Kodimial (KRN), *GVS* 20087 (MH); Aklaspur (KRN), *GVS*25651 (MH).

Polygala persicariifolia DC., Prodr. 1: 326. 1824; Bennett in Hook.f., Fl. Brit. India 1: 202. 1872; *"persicariaefolia"*; Gamble, Fl. Madras 1: 58. 1915; Chandrabose & Nair, Proc. Indian Acad. Sci (Plant Sci.) 90: 115. 1981; Banerjee in B.D.Sharma & *al.*, Fl. India 2: 480. 1993.

Annual erect herbs, up to 65 cm high. Leaves alternate, 1.5-6.5 × 0.2-1.2 cm, linear-lanceolate, base subacute, margin entire, apex acute. Flowers purple in terminal or extra-axillary racemes, up to 14 cm long; bracts 3, lanceolate, persistent; sepals 5, persistent, outer sepals elliptic-oblong, lanceolate, wing sepals rounded, broadly obliquely obovate to suborbicular, petaloid, reddish; petals pink, purple or yellowish with purple streaks, mid lobe keeled with a crest of 2-3 bundles of linear membranous forked appendages, lateral lobes obliquely quadrangular or oblong, smaller; staminal sheath adherent to petals, free portion of filaments ca 1.2 mm long; ovary obovoid, style curved, widened towards middle, broadened at apex, stigma oblique, lateral. Capsules elliptic-oblong, winged and cilate along the margins; seeds black-brown hirsute.

Occasional in forests of Hyderabad district (Ramana, 2010).

Fl. & Fr.: June- March.

Polygala rosmarinifolia Wight & Arn., Prodr. Fl. Ind. Orient. 1: 37. 1834; Bennett in Hook.f., Fl. Brit. India 1: 204. 1872; Gamble, Fl. Madras 1: 58. 1915; Chandrabose & Nair, Proc. Indian Acad. Sci. (Plant Sci.) 90: 122. 1981; Banerjee in B.D.Sharma & *al.*, Fl. India 2: 480. 1993.

Annual herbs, erect, ascending or prostrate, branches up to 45 cm long. Leaves 0.5-5.5 × 0.1-0.6 cm, linear-oblong, elliptic-oblong, glabrous or sparsely pubescent, base narrowed, margin entire, apex acute or obtuse; petioles up to 0.5 mm long. Flowers yellow in lateral or extra- axillary racemes, up to 3 cm long, sometimes solitary; sepals 5, outer sepals with pale yellow or chocolate-coloured tips, ovate to obovate or lanceolate, wing sepals obliquely oblong or elliptic, persistent; petals yellow with red markings, mid lobe up to 3 mm long, keeled with a crest of forked filiform appendages on back, lateral lobes suborbicular, spathulate, prominently nerved, wrinkled; staminal sheath up to 1.5 mm long, lateral pairs of filaments subconnate, terminal filaments free or up to 0.6 mm long; ovary obliquely obcordate, style curved, broadened towards the horse-shoe shaped apex, stigma lateral, capitate. Capsule rhomboid or obliquely oblong, ciliate along the margins.

Occasional in open grasslands in the hills in Warangal, Karimnagar and Mahabubnagar districts.

Fl. & Fr.: August-January.

Warangal, *P.S. Rao s.n.* (CAL).

Polygala telephioides Willd., Sp. Pl. 3(2). 876. 1802; Bennett in Hook.f., Fl. Brit. India 1: 205. 1872; Gamble, Fl. Madras 1: 59. 1915; Chandrabose & Nair, Proc. Indian Acad. Sci. (Plant Sci.) 90: 17. 1981; Banerjee in B.D.Sharma & *al.*, Fl. India 2: 485. 1993. *P. brachystachya* DC., Prodr. 1: 326. 1824 (non Blume 1825).

Annual herbs, prostrate, ascending or erect; branches 3-18 cm long. Leaves 0.2-1.7 × 0.1-0.7 cm, elliptic-obovate, elliptic- oblong or oblanceolate, glabrous, base narrowed, margin entire, apex obtuse or sub-acute; petioles up to 1 mm long. Flowers violet or blue, crowded in lateral or leaf-opposed racemes up to 1.1 cm long; sepals 5, outer sepals linear-lanceolate to ovate-lanceolate, concave, wing sepals obliquely elliptic or ovate-lanceolate; petals blue, mid lobe keeled, crested with stalked, capitate, finger-like appendages on back near apex, lateral lobes obovate; staminal sheath adherent to base of petals, split for two-third the length, filaments 8, arising at different levels from sheath, with 2 free filaments in the middle and 2 bundles of 3 connate filaments on either side with sessile anthers; ovary orbicular, flat, style bent, broadened at middle and apex, stigma hooded. Capsule suborbicular; seeds black, ciliate.

Occasional in open forests of Mahabubnagar district (Raghava Rao, 1989a).

Fl. & Fr.: August-March.

Polygala wightiana Wall. ex Wight & Arn., Prodr. Fl. Ind. Orient. 38. 1834; Gamble, Fl. Madras 1: 59. 1915; Chandrabose & Nair, Proc. Indian Acad. Sci. (Plant Sci.) 90: 121. 1981; Banerjee in B.D.Sharma & *al.*, Fl. India 2: 488. 1993.

Annual erect or ascending herbs, up to 40 cm high. Leaves 0.3-2.2 × 0.1-0.3 cm, linear-oblong or narrowly oblanceolate, glabrous, base narrowed, margin entire, apex obtuse or subacute; subsessile or shortly petioled. Flowers yellow in extra-axillary or lateral racemes; sepals 5, persistent, outer sepals unequal, lanceolate to ovate-lanceolate, wing sepals obliquely ovate, falcate, petaloid; petals greenish, tinged with pink, middle lobe keeled, auricled at base; staminal sheath free, filaments connate into 3 bundles, the lateral ones bearing 3 sessile anthers each and the middle one bearing 2 anthers with filaments united for half way; ovary ovoid, style slender, curved, hooked at apex with membranous wings, stigma dilated. Capsules oblong, overtopped by wing sepals; seeds black, ciliate.

Occasional weed in wastelands and cultivated fields in Mahaboobanagar district.

Fl. & Fr.: August-January.

SALOMONIA Lour. *nom. cons.*

Salomonia ciliata (L.) DC., Prodr. 1: 334. 1824; Raju, Indian J. Bot. 9: 75. 1986; Banerjee in B.D.Sharma & *al.*, Fl. India 2: 490. 1993. *Polygala ciliata* L., Sp. Pl. 705. 1753. *Salomonia oblongifolia* DC., Prodr. 1: 334. 1824; Benth., Fl. Austr. 1: 1328. 1862; Bennett in Hook.f., Fl. Brit. India 1: 207. 1872; Gamble, Fl. Madras 1: 56. 1915.

A small slender, simple or much branched herb with angular stems, 6-26 cm high. Leaves sessile, elliptic or ovate lanceolate, 4-14 × 2-8 mm, cuneate at base, margins entire, occasionally with a few hairs, branches and leaves glabrous. Flowers minute, pink in long thin terminal spikes; sepals 5, unequal, linear-lanceolate, persistent, 2 inner somewhat larger, acute at apex, ciliate margins; petals 3, pink or white and purplish tipped, median petal keeled, lateral ones lanceolate, adnate to keel petal; stamens 4, monadelphous, anthers 2-loculed, undifferentiated; ovary compressed, sessile, obcordate or orbicular, style curved, stigma capitate. Capsule sessile with setose-dentate margins; seeds rounded, gelatinously strophiolate, 0.5 mm across, black.

Rare in Khammam district.

Fl. & Fr.: November-February.

Vazheedu (KMM), *VSR* 684 (MH).

CARYOPHYLLACEAE

1.   Stipules scarious :
    2.   Stamens 5; sepals other than keeled, white to rose,
       membranous throughout .... **Polycarpaea**
    2.   Stamens 2 or 3; sepals keeled, green, membranous along
       margins .... **Polycarpon**
1.   Stipules absent:
    3.   Bracteoles (epicalyx) absent .... **Vaccaria**
    3.   Bracteoles (epicalyx) present .... **Dianthus**

DIANTHUS L.

Dianthus chinensis L. Sp. Pl. 411. 1753; Hook.f. Fl. Brit. India 1 : 213. 1874.

Erect, glabrous, annual herbs, 20-45 cm high; stems corymbosely branched in upper part. Leaves narrowly lanceolate, entire, with smooth margins, sessile. Flowers solitary or 2-3 together; bracts shorter or longer than calyx, 1.5-2 cm long, inner ones form a broad base, narrowed towards apex, or acuminate; calyx 1.6-2.4 cm long; lobes erect, ciliate, sharply acute; petals 2-3.5 cm long; limb coarsely dentate, pink, red or white. Capsule oblong, apically 4-valved, 2.25 cm long; seeds smooth, many, dark brown.

Cultivated as a bedding garden ornamental for showy flowers (Ramana, 2010).

Fl. & Fr.: Februar- May.

POLYCARPAEA Lam. *nom. cons.*

1. Flowers in terminal and axillary paniculate cymes; leaves more than 8 per node, flat, to 2.5 x 0.3 cm, stipules much fimbriate, internodes with scattered hairs .... **P. corymbosa**

1. Flowers in terminal compound cymes; leaves less than 6 per node, with inrolled margins, stipules hardly fimbriate, internodes densely white tomentose .... **P. aurea**

Polycarpaea aurea Wight & Arn., Ann. Nat. Hist. Ser. 1. 3: 91. 1893; Gamble, Fl. Madras 1: 65. 1915. *P. corymbosa* Lam. var. *aurea* (Wight & Arn.) Wight, Ill. 2: 44. t. 110. 1850; Edgew. & Hook.f. in Hook.f., Fl. Brit. India 1 : 245. 1874; Majumdar in B.D.Sharma & *al.*, Fl. India 2: 551. 1993.

Perennial, erect under shrubs, up to 25 cm high, internodes densely white tomentose. Leaves opposite, linear, 1-1.5 × 0.2 cm, leaves with inrolled margins, few leaves in axils or O, minutely hairy, entire, tip acute. Stipules hardly fimbriate. Flowers pinkish-violet in terminal compound cymes, bracts scarious; sepals 5, free, ovate-lanceolate, entire, acute, 2 x 0.8 mm, exceeding petals and capsules, scarious, reddish brown; petals 5, free, oblong-obovate, obtuse, partly enclosing capsule, shining, brown; stamens 5, forming a cup at base with petals and encircling ovary. Capsules oblong or ovate-elliptic, 3-valved, tip faintly incurved when young, recurved after dehiscence; seeds 3-5, compressed, reniform, brown with radiating lines at groove.

Occasional in rocky places in Karimnagar, Warangal, Mahabubnagar and Medak districts.

Fl. & Fr.: August-February.

Polycarpaea corymbosa (L.) Lam., Tabl. Encycl. 2: 129. 1797; Edgew. & Hook.f. in Hook.f., Fl. Brit. India 1: 245. 1874; Gamble, Fl. Madras 1: 65. 1915; var. **corymbosa**; Majumdar in B.D. Sharma & *al.*, Fl. India 2: 549. 1993. *Achyranthes corymbosa* L., Sp. Pl. 205. 1753.

Annual, erect, much branched herbs, up to 30 cm high, branchlets densely villous or glabrescent. Leaves decussate or in false whorls, linear to subulate, 1-2.5 × 0.1-0.3 cm, minutely hairy, base cuneate, entire, apex acute, stipules much fimbriate. Flowers light pink, at length white, in terminal and axillary paniculate cymes; sepals 5, lanceolate; acuminate, 1.5-5 mm long, scarious at margin, glabrous; petals 5, ovate or linear-oblong, subentire, obtuse, 0.5-1.2 mm long, dull yellow, persistent; stamens 5, filaments 0.5-1 mm long, anthers oblong, 0.3 mm long, acuminate, 1.5-5 mm long, scarious at margin, glabrous; petals 5, ovate or linear-oblong, subentire, obtuse, 0.5-1.2 mm long, dull yellow, persistent; stamens 5, filaments 0.5-1 mm long, anthers oblong, 0.3 mm long. Capsules oblong, 3-valved, brown; seeds numerous, compressed, reniform, yellowish-brown.

Common in fields, waste places and forests, in most districts.

Fl. & Fr.: August-December.

Gadirayal (RR), *MSM* 10521; Nagaram (NZB), *BR* 9706; Narsapur RF (MDK), *BR* & *CP* 11228; Pochammaralu (MDK), *BR* & *CP* 11476; Kusumasamudram (RR), *MSM* & *NV* 10344; Bingidoddi (MBNR), *BR* & *MV* 35856; Vijayapuri (NLG), *KMS* 9831 (CAL); Aklaspur (KRN), *GVS* 25664 (MH); Thaskapally (MBNR), *SRS* 109079 (BSID); Koriguttalu (KMM), *RR* 106042 (BSID).

Key to Varieties

1. Stipules fimbriate; sepals more than 2.5 mm long; capsules ellipsoid .... var. **corymbosa**

1. Stipules lanceolate, hardly fimbriate; sepals 2.3-2.6 mm long; capsules ellipsoid .... var. **yadgiriense**

Polycarpaea corymbosa (L.) Lam. var. **yadagirense** C.S.Reddy & *al.*, J.Econ. Taxon. Bot. 32(3): 519. 2008.

Rare in Nalgonda district (C.S.Reddy & *al.*, 2008).

POLYCARPON L.

Polycarpon prostratum (Forssk.) Asch. & Schweinf, Oesterr. Bot. Z. 39: 128. 1889. Majumdar in B.D.Sharma & *al.*, Fl. India 2: 553. 1993. *Alsine prostrata* Forssk., Fl. Aegypt.-Arab. 207. 1775. *Polycarpon loeflingiae* (Wall. ex Wight & Arn.) Benth. ex Edgew. & Hook. f. in Hook.f., Fl. Brit. India 1: 245. 1874; Gamble, Fl. Madras 1: 64. 1915.

Annual prostrate herbs, 15-25 cm long. Leaves decussate, subsessile, spathulate or oblanceolate, 2.5 × 0.5 mm, chartaceous, base narrowed, apex acute or obtuse; stipules scarious. Flowers in terminal dichasial paniculate cymes; sepals 5, linear oblong or ovate, subequal, obtusely keeled, broadly scarious-margined; petals 5, linear-lanceolate, rarely absent; stamens 2 or 3; filaments 0.5 mm long, anthers ovoid. Capsule ovoid globose, 3-valved; seeds numerous, subcylindric, 0.3-0.6 mm long, minutely tubercled, pale brown.

Occasional by the river side in Hyderabad and Mahabubnagar districts.

Fl. & Fr.: August-January.

Gandipet lake (HYD), *TR* 642 (HH).

VACCARIA Medik.

Vaccaria pyramidata Medic., Phil. Bot. 1: 96. 1789; Majumdar in B.D.Sharma & *al.*, Fl. India 2: 593. 1993. *Saponaria vaccaria* L., Sp. Pl. 407. 1753; Edgew. & Hook. f. in Hook.f., Fl. Brit. India 1 : 217. 1874; Gamble, Fl. Madras 1: 61. 1915.

Annual erect glabrous herbs, up to 60 cm high. Leaves sessile, 5- 8.5 × 0.6-1 cm, elliptic, lanceolate, base cordate, margin entire, apex acute, lowest leaves somewhat petiolate, the rest sessile. Flowers rose, in dichotomous cymes which are combined into a terminal corymbose panicle; calyx tube inflated, with 5 sharp angles or wings, teeth 5, triangular; petals rose-coloured, limb cuneate or rounded or somewhat emarginated; coronal scales absent. Capsules ovoid- globose, included within the calyx; seeds some what angled, brown-black.

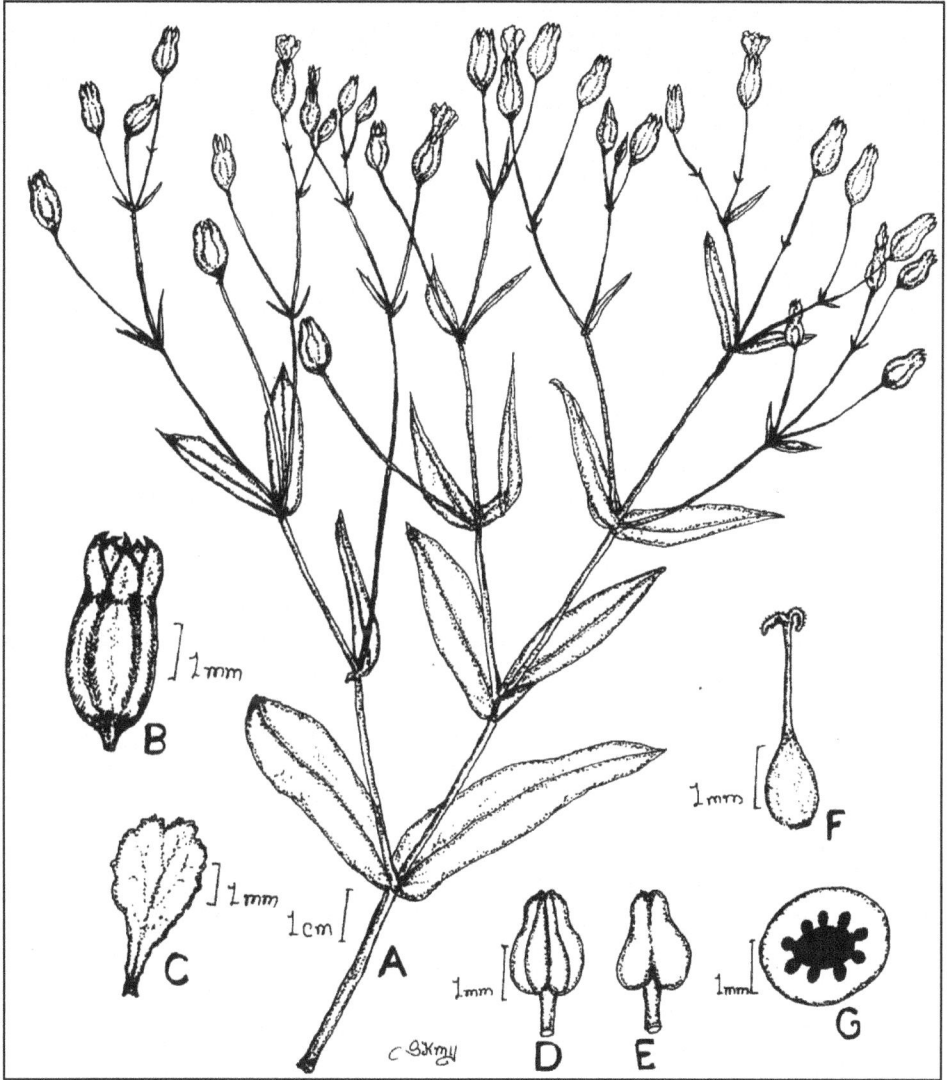

Figure 8: Vaccaria pyramidata Medic.

A. Twig, B. Calyx, C. Petal, D & E. Stamens, F. Pistil, G. Ovary c.s.

Occasional weed of cultivated fields in Adilabad, Nizamabad, Warangal and Mahabubnagar districts.

Fl. & Fr.: July-February.

Kadthala (ADB), *GO* & *DAM* 4992; Adlur yellareddy (NZB), *TP* & *BR* 6419.

PORTULACACEAE

1. Ovary superior; capsule valvular; leaves more than 2 cm broad .... **Talinum**
1. Ovary inferior; capsule circumscissile; leaves less than 2 cm broad .... **Portulaca**

PORTULACA L.

1. Roots tuberous .... **P. tuberosa**
1. Roots not tuberous:
    2. Flowers more than 2.5 cm across .... **P. grandiflora**
    2. Flowers less than 1 cm across:
        3. Nodes without scales or hairs; sepals green, fleshy, carinate, persistent .... **P. oleracea**
        3. Nodes with scales or hairs; sepals membranous, not carinate, marcescent :
            4. Nodes surrounded by lanceolate scales .... **P. wightiana**
            4. Nodes surrounded by few or abundant hairs:
                5. Herbs rooting at nodes, leaves opposite, flat .... **P. quadrifida**
                5. Erect herbs, leaves alternate, spiral, subterete .... **P. pilosa**

Portulaca grandiflora Hook. in Bot. Mag. n.s. 3: t. 2885. 1829; Vasudeva Rao in B.D.Sharma & Sanjappa, Fl. India 3: 3. 1993.

Diffuse, decumbent herbs, up to 30 cm high. Leaves alternate or subopposite, 12-25 × 1-4 mm, linear-subulate, often curved, terete, obtuse or acute at both ends, with ca 5 mm long, axillary hairs. Flowers 2-8 in capituli, flowering successively, subtended by 5-8 involucral leaves, deltoid bracteoles and ca 10 mm long hairs; sepals 2, ovate; petals 5 (many in cultivated forms), pink, red, orange or yellow, obovate; stamens numerous; styles 7-13 mm long with 5-10 arms. Capsules globose; seeds many.

Cultivated as an ornamental and occasionally occurs as an escape.

Fl. & Fr.: Through out the year.

Portulaca oleracea L., Sp. Pl. 445. 1753. var. **oleracea**; Dyer in Hook.f., Fl. Brit. India 1: 246. 1874; Gamble, Fl. Madras 1: 66. 1915; Vasudeva Rao in B.D.Sharma & Sanjappa, Fl. India 3: 4. 1993.

Vern. : *Payalaku.*

Annual prostrate succulent herbs, stem glabrous, tinged with pink. Leaves fleshy, alternate or subopposite or whorled, sub-sessile, obovate or oblong 1-3 × 0.5-1.5 cm, base attenuate, entire, apex obtuse. Flowers yellow in terminal clusters; surrounded by a cluster of crowded leaves; sepals 2, carinate, connate at base, lobes oblong-ovate, keeled or slightly winged on back, 2-4 mm long; petals 4 or 5, yellow, connate at base, broadly obovate-oblong, rarely emarginate at apex; stamens 7-10, filaments to 4 mm long; ovary ovoid, styles up to 5 mm long. Capsules 0.6-0.8 cm long, ovoid, enveloped by marcescent corolla, dehiscing transversely in middle; seeds many, reniform, 0.5-1mm across, granular, dull black.

Common weed in fields and wastelands of most districts.

Fl. & Fr.: Through the year.

Neredugunda (ADB), *GO* & *DAM* 5109; Laxmapur (NZB), *TP* & *BR* 6288; Navipet (NZB), *BR* & *PSPB* 9593; Central University campus (RR), *MSM* 19201; Mombojipally (MDK), *BR* & *CP* 11666, 11666; Ramancheru (MDK), *RG* 96285 (BSID); Medak (MDK), *RG* 104164 (BSID).

Key to Varieties
1. Leaves linear, subterete; seeds 55-75 per capsule .... var. **linearifolia**

1. Leaves spathulate, obovate, dorsi-ventral; seeds 20-25 per capsule .... var. **oleracea**

Portulaca oleracea L. var. **linearifolia** Sivarajan & Manilal, New Botanist 4: 30. 1977; Vasudeva Rao in B.D.Sharma & Sanjappa, Fl. India 3: 4. 1993.

Annual prostrate glabrous herbs. Leaves linear, 1-2 × 0.2-0.5 cm, fleshy, alternate, entire, tip obtuse. Flowers yellow. Capsules oblong; seeds numerous, black, minute, granulate.

Common weed of wastelands, fields and moist localities of most districts.

Fl. & Fr.: Round the year.

Vempalli (ADB), *GO* & *PVP* 4743; Ramayampet (MDK), *BR* & *CP* 11512.

Portulaca pilosa L., Sp. Pl. 445. 1753; Geesink, Blumea 17: 294. 1969; Vasudeva Rao in B.D.Sharma & Sanjappa, Flora India 3: 6. 1993.

Perennial, fleshy, radially spreading herbs. Leaves spiral, crowded at apices of branches, linear, 1-2 × 0.1-0.2 cm, sessile, succulent, base cuneate, margin entire, apex subacute, axillary hairs sparse. Flowers solitary or clustered, pink or yellow, sepals 2, ovate, ecarinate or sometimes inconspicuously hooded at apex, acute, 2-5 x 1-3.5 mm, glabrous; petals 4-6, shortly connate at base, obovate, obtuse, 5-8 x 2-10 mm; stamens 9-14, filaments 1.5 mm long, connate or adnate to petals at base, anthers globose to elliptic; ovary ovoid; styles 2-8 mm long, 3-7-armed. Capsules ovoid, setose appendages equalling about 1/2 of the length of the capsule, dull green, dehiscing horizontally in middle; seeds very small, many, reniform, dull black, with concentric rows of minute tubercles.

Occasional in Hyderabad and Nalgonda districts in sandy soil.

Fl. & Fr.: April-November.

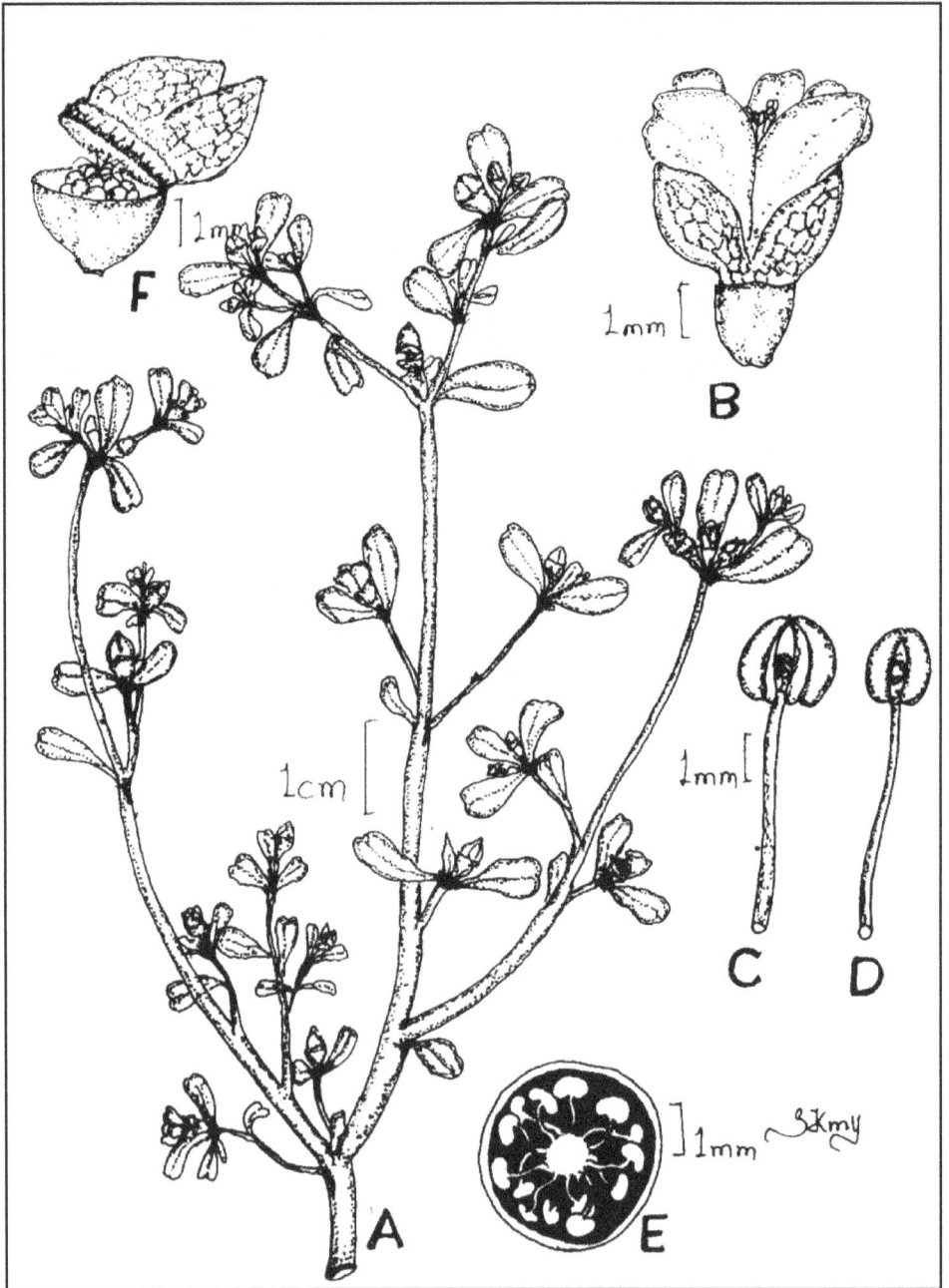

Figure 9: Portulaca oleracea L. var. **oleracea**
A. Twig, B. Flower-entire, C & D. Stamens, E. Ovary c.s., F. Capsule

Nagarjunasagar (NLG), *BVRS* 58 (NU); Bhongir fort (NLG), *BVRS* 18 (NU); Uppal (HYD), *TR* 1007 (HH).

Portulaca quadrifida L., Syst. Nat. ed. 12. 328. 1766 & Mant. Pl. 1: 73. 1767; Dyer in Hook.f. Fl. Brit. India 1: 247. 1874; Gamble, Fl. Madras 1: 66. 1915; Vasudeva Rao in B.D.Sharma & Sanjappa, Fl. India 3: 6. 1993.

Vern. *Sannapayalaku.*

Small, diffuse, annual herbs, branches up to 10 cm long, creeping, rooting at nodes, nodes covered with white silky hairs. Leaves fleshy, sessile, ovate-lanceolate or oblong, 0.3-0.6 × 0.6-0.3 cm, dull brownish-green, sometimes reddish green, base cuneate, margin entire, apex acute. Flowers yellow, solitary, terminal on an infundibular receptacle, subtended by four leaves and encircled by hairs; sepals 2, connate at base, obtuse, 2-4 x 1.5-3 mm, hyaline; petals 4, free, elliptic-oblong to ovate, 5 x 3 mm, yellow; stamens 8 or 12, filaments hairy at base; ovary conic-ovoid; styles cylindrical with 4 arms. Capsules ovoid, shining, straw yellow, dehiscing horizontally; seed numerous, reniform, verrucose with dark-brown or blackish-brown, minutely tubercled.

Common weed of cultivated fields and open forests in most districts.

Fl. & Fr.: Through the year.

Bijjur (ADB), GO & PVP 4974; Neredugunda (ADB), *GO* & *DAM* 5110; Kammarpalli (NZB), *TP* & *BR* 6240; Tirmalagiri (NLG), *BVRS* 470 (NU).

Portulaca tuberosa Roxb., Fl. Ind. 2: 464. 1832; Vasudeva Rao in B.D.Sharma & Sanjappa, Fl. India 3: 7. 1993. *P. suffruticosa* Wall. ex Wight & Arn., Prodr. 356. 1834; Dyer in Hook.f., Fl. Brit. India 1: 246. 1834.

Perennial, erect or decumbent herb, up to 30 cm high, with thick tuberous root. Leaves spiral, oblong, 4-28 × 0.5-5 mm, apex obtuse or rounded, axillary hairs 1-18 mm long. Flowers solitary or 2-4 in capitula, subtended by an involucre of 3-8 cauline leaves, sepals 2, 2-6 × 1-4 mm, petals 4-6, bright yellow, 2.5-12 × 1.8-11 mm, obovate, spreading, stamens 10-25, styles 3-5-armed. Capsule ovoid-globose, 2-3 mm in diam.

On rocky laterite soils in Medak, Karimnagar, Khammam and Ranga Reddi districts.

Narsapur tank (MDK), *TP* & *CP* 12010; Kusumasamudram (RR), *MSM* & *NV* 10376; Ramanagutta (KMM), *RR* 1017981 (BSID).

Portulaca wightiana Wall. ex Wight & Arn., Prodr. Fl. Ind. Orient. 356. 1834; Dyer in Hook.f., Fl. Brit. India 1: 247. 1874; Gamble, Fl. Madras 1: 66. 1915; Vasudeva Rao in B.D.Sharma & Sanjappa, Fl. India 3: 7. 1993.

Small annual herbs, ca 10 cm high, branchlets with closely jointed nodes, nodal appendages scarious, persistent; sparingly leafy above. Leaves alternate, ovate or lanceloate, 4-6 × 1-2.5 mm, membranous, base cuneate, margin entire, apex acute. Flowers creamish, terminal, solitary, surrounded by a few leaves and tufted appendages, sepals 2; petals 4, elliptic; stamens 10; styles ending in 3-6 arms. Capsule globose, straw yellow; seeds numerous.

Occasional in Medak, Warangal, Hyderabad and Mahabubnagar districts.

Fl. & Fr.: September-December.

Osmania University campus (HYD), *TR* 605 (HH); Medak (MDK), *RG* 104152 (BSID).

TALINUM Adans. *nom. cons.*

1. Leaf apex apiculate, obtuse with mucro .... **T. portulacifolium**

1. Leaf apex emarginate or mucronate .... **T. triangulare**

Note: According to Hutchinson and Dalziel (Fl. W. Tropical Africa I, 1: 136. 1954) these two species are very similar and almost indistinguishable morphologically and the only character that distinguishes them is the leaf apex. Nyanayo and Olowokudeju (Willdenowia 15: 455-463.1986) have found differences on the basis of palynology, seed morphology and leaf anatomy.

Talinum portulacifolium (Forssk.) Asch. & Schweinf., Bull. Herb. Boissier 4. App. 2: 172. 1896; Vasudeva Rao in B.D.Sharma & Sanjappa, Fl. India 3: 9. 1993. *Orygia portulacifolia* Forssk., Fl. Aegypt.-Arab. 103. 1775. *Talinum cuneifolim* (Vahl) Willd., Sp. Pl. 2: 864. 1799; Dyer in Hook.f., Fl. Brit. India 1: 247. 1874; Gamble, Fl. Madras 1: 66. 1915. *Portulaca cuneifolia* Vahl, Symb. Bot. 1: 33. 1790.

Annual erect robust herbs, up to 75 cm high, branchlets glabrous. Leaves subsessile, obovate, 6-8 × 2-3 cm, fleshy, glabrous, base attenuate, margin entire, apex obtuse with mucronate tip. Flowers pink in terminal lax panicles; sepals 2, ovate-lanceolate; petals 5, pink, purple or white, obovate to ovate-rotund; stamens many, basally connate; ovary 1-loculed, styles 3-armed. Capsule globose, 0.7 cm across; seeds black, radiate-striate.

Occasional as an escape in Hyderabad (Ramana, 2010), Karimnagar (Naqvi, 2001) and Mahabubnagar districts.

Fl. & Fr.: August-November.

Talinum triangulare (Jacq.) Willd., Sp. Pl. 2: 862. 1799; Vasudeva Rao in B.D.Sharma & Sanjappa, Fl. India 3: 10. 1993. *Portulaca triangularis* Jacq., Enum. Pl. Carib. 22: 1760.

A glabrous, succulent, erect herb, up to 1 m high; tuber clavate and woody, branches transluscent, brown. Leaves elliptic or obovate, 3-6 cm long, fleshy, secondary nerves obscure, margins revolute when young, apex emarginate or mucronate. Flowers pink in corymbiferous cymes or racemes; bracts and bracteoles subulate; sepals 2, obovate, emarginated; petals 5, obovate; stamens 20-40, style 1.5-2.5 mm long, 2-3-armed. Capsules ellipsoid, 2-3-valved, pinkish; seeds many, finely ribbed, with radially elongated cells on testa.

A weed of gardens; elsewhere an escape (C.S. Reddy and Raju, 2002).

Fl. & Fr.: Throughout the year.

A pantropical weed; Native of tropical America.

TAMARICACEAE

TAMARIX L.

1. Stamens 10 .... **T. ericoides**

1. Stamens 5 :

   2. Leaves not auricled .... **T. indica**

   2. Leaves auricled .... **T. dioica**

Tamarix dioica Roxb., Fl. Ind. 2: 101. 1820; Dyer in Hook.f. Fl. Brit. India 1: 249. 1874; Gamble Fl. Madras 1: 67. 1915; Shetty & Pandey in B.D.Sharma & Sanjappa, Fl. India 3: 24. 1993.

A gregarious shrub or small tree, branches with drooping extremities. Leaves sheathing, glabrous, green, keeled, apex acuminate. Flowers pink in narrow terminal spikes, spikes panicled, Male flowers: sepals 5, trullate-ovate to broadly ovate or suborbicular, petals 5, obovate or oblong-obovate; stamens 5, anthers apiculate; ovary abortive or absent. Female flowers: appear as bisexual but stamens functionally sterile; sepals as in male; petals 5, obtriangular-obovate to elliptic-obovate; staminodes 5, ovary narrowed upwards. Capsules with persistent sepals and staminodes and subpersistent petals; seeds ca 0.5 mm long, coma 2.75-3 mm long.

Sandy-river-beds in Hyderabad district (Ramana, 2010).

Fl.: April – December; Fr.: July – January.

Tamarix ericoides Rottl. in Ges., Naturf. Freunde Berlin Neue Schriften 4: 214. 1803; Dyer in Hook.f., Fl. Brit. India 1: 249. 1874; Gamble, Fl. Madras 1: 69. 1915; Shetty & Pandey in B.D.Sharma & Sanjappa, Fl. India 3: 25. 1993.

Bushy shrubs, up to 2.5 m high, bark dark brown or blackish, rough. Leaves minute, ovate-lanceolate, scaly, sheathing at base. Flowers pale to bright rosy, shortly pedicellate, in 8-30 cm long, terminal branched spikes; sepals 5, ovate-elliptic with membranous margins; petals 5, elliptic-obovate, denticulate, twice as long as sepals, persistent; stamens 10, alternately short, glands of the disc separating the filaments. Capsules 0.8-1.2 cm long, conical, pale-brown, or straw-coloured, glabrous; seeds oblong, pilose, hairs pale white.

Beds of rivers in Nizamabad, Medak, Karimnagar, Warangal and Mahabubnagar districts.

Fl. & Fr.: August-March

Gowraram (NZB), *TP* & *BR* 6395; Salura (NZB), *BR* 7273; Pocharam tank (NZB), *BR* 9503; Pochammaralu (MDK), *BR* & *CP* 11496; Bhupatipur (KRN), *GVS* 22297 (MH); Eturunagaram (WGL), *RKP* 110841 (BSID).

Tamarix indica Willd., Ges. Naturf. Freunde Berlin Neue Schriften 4: 214. 1803. Shetty & Pandey in B.D.Sharma & Sanjappa, Fl. India 3: 25. 1993. *T. gallica* L. var. *indica* (Willd.) Ehrenb., Linnaea 2: 276. 1827; Dyer in Hook. f., Fl. Brit. India 1 : 248. 1874. *T. indica* Koen. ex Roxb. Fl. Ind. 2: 100. 1832, *nom. illegit. T. gallica* auct.

non L.; Dyer in Hook.f., Fl. Brit. India 1: 248. 1874; Gamble, Fl. Madras 1 : 67. 1915. *Tamarix troupii* Hole, Indian For. 14: 248. 1919; Pullaiah & Chennaiah in Fl. Andhra Pradesh 1: 109. 1997.

Vern. *Palivi.*

Shrubs or small trees, 1.5 – 3 m high, bark greenish-brown, rough, faintly striate. Leaves scale like, broadly triangular to ovate-lanceolate, or triangular-ovate, 0.5-3 mm long, abruptly acute to acuminate at apex. Flowers minutely pedicellate, pale to bright rose-coloured, in terminal, paniculate spikes; bracts ovate-lanceolate, trullate-lanceolate or trullate-ovate, 1.5-2.5 mm long, deflexed; sepals 5, almost free, equal, ovate; petals 5, obovate, oblong-elliptic, caducous, stamens exserted, disc 5-lobed. Capsules 3-valved, 3-4 × 0.6-1.5 mm, seeds ellipsoidal, ca 0.6 mm long, coma ca 2.5 mm long.

In river beds in Adilabad district.

Fl. & Fr.: August-March

Penganga river bed (ADB), *GO* & *PVP* 4276.

ELATINACEAE

1. Plants erect, decumbent or ascending; sepals free .... **Bergia**
1. Plants prostrate; sepals connate at base, - embryo curved .... **Elatine**

BERGIA L.

1. Plants glandular pubescent; petals reddish pink;
    stamens 5 .... **B. ammannioides**
1. Plants glabrous or nearly so; petals greenish-white or
    transparent; stamens 10 .... **B. capensis**

Bergia ammannioides Roxb. ex Roth, Nov. Pl. Sp. 219. 1821; Dyer in Hook.f., Fl. Brit. India 1: 251. 1874; Gamble, Fl. Madras 1: 69. 1915; Bhattacharya in B.D.Sharma & Sanjappa, Fl.India 3: 33. 1993.

Annual, erect, much branched herbs, up to 35 cm high, branches wiry, glandular-tomentose. Leaves sessile or subsessile, oblanceolate or elliptic-oblong, 1-1.5 × 0.3-0.4 cm, glabrous or tomentose, base attenuate, margin glandular-ciliate or serrulate, apex acute. Flowers subsessile, red, in axillary fascicles; sepals 3, linear-lanceolate to ovate, keeled, glandular pubescent, often reddish pink; petals 3, reddish pink, ovate, elliptic or oblong, membranous; stamens usually 5, rarely more, 1mm long; ovary subglobose, styles 5, recurved. Capsules globose, up to 2 mm long, glabrous, with 5 longitudinal furrows, dehiscing into 5 valves at maturity; seeds many, minute, oblong, slightly curved, dark-brown, reticulate.

Common in marshy places in most districts.

Fl. & Fr.: November-April.

Kadam Dam bank (ADB), *TP* & *PVP* 4140; Mudheli RF (NZB), *TP* & *BR* 6372; Janakampally (NZB), *TP* & *BR* 6320; Sarvapur (NZB), *TP* & *BR* 6372; Mamidipally RF

(NZB), *BR* & *PSPB* 9606; Gangapur (MDK), *TP* & *CP* 12097; Medak fields (MDK), *BR* & *CP* 11424; Kaleswaram (KRN), *SLK* 70892 (MH); Kambalpally (NLG), *BVRS* 231 (NU); Laxmipuram (WGL), *PVS* 76947 (BSID); Pakhal (WGL), *RKP* 108291 (BSID).

Bergia capensis L., Mant. Pl. 2: 241. 1717; Gamble, Fl. Madras 1: 69. 1915; Bhattacharya in B.D. Sharma & Sanjappa, Fl. India 3: 34. 1993. *B. verticillata* Willd., Sp. Pl. 2 : 770, 1799, *nom. illeg.* Dyer in Hook.f. Fl. Brit. India 1: 252. 1874.

Succulent, glabrous, marshy, perennial herbs, erect or ascending, rooting at nodes. Leaves oblanceolate or elliptic-lanceolate, 2-4 × 0.8-1.5 cm, glabrous, base attenuate, somewhat decurrent, margin serrulate, apex obtuse-subacute; petioles stout, 1-5 mm long; stipules 2-3 mm long, ovate-triangular. Flowers red, many in dense axillary fascicles; sepals 3, light green with red tips, ovate lanceolate, entire at margins, acuminate at apex; petals 3, greenish-white or transparent, oblong or subspathulate, suberect or spreading, obtuse at apex; stamens 10, 1.5 mm long; ovary ellipsoid or globose, stigmas 5-notched. Capsule furrowed, 2 mm, globose, glabrous, with 5 longitudinal furrows, break into 5 valves; seeds numerous, oblong-obovoid, brownish, strongly reticulate.

Occasional in rice fields and on the margins of tanks in Khammam, Hyderabad, Mahabubnagar and Karimnagar districts.

Fl. & Fr.: November-January.

Nirmal (ADB), *MHR* 14420; Sagar left canal (KMM), *MHR* & *MCK* 17007; Nirmal-Adilabad Road (ADB), *MHR* 14449, 14420; Sardhana (MDK), *CP* 11776; Rechapalli (KRN), *GVS* 22276 (MH).

ELATINE L.

1. Flowers subsessile; stamens exceeding the sepals .... **E. triandra**
1. Flowers shorter than their pedicels; stamens shorter than
   the sepals .... **E. ambigua**

Elatine ambigua Wight, Bot. Misc. 2. Suppl. 103. t. 5. 1831; Dyer in Hook.f., Fl. Brit. India 1: 251; Gamble, Fl. Madras 1 : 68. 1915; Bhattacharya in B.D.Sharma & Sanjappa, Fl. India 3: 39. 1993.

Prostrate herbs, radially branched, soft, glabrous, stems terete, rooting at nodes. Leaves oblong-lanceolate, 2-5 × 0.5-2 mm, elliptic to oblong-lanceolate, narrowed into a flattened short petiole at base, margin entire, apex obtuse; stipules ca 1 mm long, ovate-triangular. Flowers axillary, solitary; sepals 3, oblong; petals 3, pinkish-white or white, ovate-oblong; stamens 3, alternipetalous; ovary globose, stigmas 3, sessile. Capsule membranous, globose; seeds numerous, oblong, light brown.

Growing in patches on mud in Warangal district (C.S. Reddy, 2001).

Fl.: February.

Elatine triandra Schkuhr, Bot. Handb. ed. 2, 1: 345. t. 109 b. f. 2. 1787. 1791; Bhattacharya in B.D.Sharma & Sanjappa, Fl. India 3: 42. 1993. *Peplis americana* Pursh, Fl. Am. Sept. 1: 238. 1814. *Elatine americana* (Pursh) Arn., Edinburgh J. Nat. Geogr. Sci. 1: 431. 1830; Dyer in Hook.f., Fl. Brit. India 1 : 250. 1874; Gamble, Fl. Madras 1: 68. 1915; Ramayya & Rajagopal, Bull. Bot. Surv. India 13: 330. 1974.

Prostrate herbs, stems 3-12 mm long, creeping, rooting at nodes. Leaves opposite decussate, oblong-lanceolate or ovate, 4-8 × 1-2 mm, glabrous, base attenuate, margin distantly serrulate, apex obtuse-subacute. Flowers axillary, solitary, subsessile; sepals 3, triangular translucent; petals 3, whitish-pink, broadly ovate, 1-1.4 mm long; stamens 3, alternipetalous; ovary globose, stigmas punctiform. Capsules subglobose, opening into 3 valves at apex; seeds numerous, minute, slightly recurved, reticulate, light brown.

Growing in patches on muds in Warangal and Nizamabad districts.

Fl. & Fr.: April-May.

Mustapur tank (NZB), *BR* 9073.

CLUSIACEAE (*nom. alter* GUTTIFERAE)

CALOPHYLLUM L.

Calophyllum inophyllum L., Sp. Pl. 513. 1753; T. Anderson in Hook.f., Fl. Brit. India 1: 273. 1874; Gamble, Fl. Madras 1: 76. 1915; Singh in B.D.Sharma & Sanjappa, Fl. India 3: 92. 1993.

Vern. : *Punna.*

A moderate-sized tree, up to 15 m high, leaf scars persistent, bark brown to pale grey (blackish) and fawn, smooth, often mottled with wide boat-shaped fissures; exudates milky or yellow, clear, very sticky. Leaves opposite, decussate, oblong-obovate, variable in size, 8-14 × 4-9 cm, base acute, margin entire, apex rounded; petioles 1-1.5 cm long, sometimes longer. Flowers polygamous, white, fragrant, numerous in axillary racemes; bracts 3-4 mm long, ovate; sepals 4, fragrant, reflexed, 2 outer ones ovate to suborbicular, concave, inner petaloid, subelliptic; petals usually 4, reflexed, obovate to elliptic or oblong, spreading, concave; stamens numerous, connate up to 2 mm into 4-6 bundles; anthers elliptic or oblong, 0.4-1 mm long; ovary depressed globose, stipitate, styles much longer than ovary, twisted, stigmas peltate. Drupe globose, blunt or short acuminate, smooth; seed solitary, ellipsoid or ovoid.

Occasional in forests of Hyderabad (Ramana, 2010), Mahabubnagar and Ranga Reddi districts.

Fl. & Fr.: November-April.

MALVACEAE

1. Styles twice as many as carpels:

   2. Flowers in condensed racemes; bracts foliaceous .... **Malachra**

   2. Flowers solitary; bracts not foliaceous:

    3. Petals auricled .... **Malvaviscus**

    3. Petals not auricled, fruit neither berry like nor fleshy:

        4. Fruit glochidiate; leaves glandular on midrib beneath .... **Urena**

        4. Fruit muricate or with 1-3 awns at apex; leaves
            without glands .... **Pavonia**

1. Styles as many as carpels:

    5. Fruit a capsule:

        6. Epicalyx segments winged in fruit – Trees .... **Kydi**a

        6. Epicalyx segments not winged in fruit:

            7. Style unbranched; stigma ribbed or lobed:

                8. Epicalyx lobes 3-8, linear to lanceolate,
                    mostly caducous .... **Thespesia**

                8. Epicalyx lobes 3, cordate, persistent .... **Gossypium**

            7. Style distinctly 5-branched; stigma capitate or globose:

                9. Calyx irregularly 2-3-lobed, spathaceous,
                    deciduous .... **Abelmoschus**

                9. Calyx regularly lobed, not spathaceous, persistent:

                    10. Capsule winged .... **Fioria**

                    10. Capsule not winged .... **Hibiscus**

    5. Fruit a schizocarp, breaking into mericarps at maturity:

        11. Stigmas decurrent on adaxial side of style .... **Alcea**

        11. Stigmas apical or nearly so, capitate, discoid or obliquely truncate:

            12. Seeds 2 or more in each mericarp:

                13. Stem erect, stout; mericarps wall thin and papery,
                    rounded at apex.. **Herissantia**

                13. Stem ascending, weak, mericarps wall thick,
                    stout, pointed, mucronate or aristate at apex .... **Abutilon**

            12. Seeds 1 in each mericarp:

                14. Epicalyx present .... **Malvastrum**

                14. Epicalyx absent .... **Sida**

ABELMOSCHUS Medik.

1. Capsules 10-12 cm long .... **A. esculentus**

1. Capsules less than 10 cm long:

    2. Epicalyx lobe ovate, enclosing calyx:

        3. Epicalyx segments free, shorter than capsule .... **A. manihot**

3. Epicalyx segments connate at base, as long as capsule or longer .... **A. angulosus**

2. Epicalyx lobes linear to linear-lanceolate, subtending calyx :

4. Epicalyx lobes 5-6, deciduous before the opening of the flower; leaf lobes obovate .... **A. ficulneus**

4. Epicalyx lobes 7-16, long persistent :

5. Epicalyx lobes 7-10,< 2 cm long, much shorter than capsule .... **A. moschatus**

5. Epicalyx lobes 10-16, linear-filiform, 2.5-5 cm long, usually longer than the capsule .... **A. crinitus**

Abelmoschus angulosus Wall. ex Wight & Arn., Prodr. 53. 1834; Paul in B.D.Sharma & Sanjappa, Fl. India 3: 301. 1993. *Hibiscus angulosus* (Wall. ex Wight & Arn.) Steudel, Nomen. Bot. ed. 2, 1: 758. 1840; Masters in Hook.f., Fl. Brit. India 1: 341. 1874, incl. vars.

Undershrubs, up to 2 m high; young branches covered with simple and stellate hairs. Leaves 3-15 cm across, elliptic to orbicular, cordate and 5-7-nerved at base, palmatilobed to parted, segments 3-7, triangular ovate to lanceolate, acute, crenate to serrate, adpressed, stiff hairy on both surfaces, ultimately glabrescent; petioles longer than lamina; stipules 10-15 × 3-5 mm. Flowers solitary, pedicels 2-6 cm long, accrescent, up to 10 cm; epicalyx 4-5-parted, segments 2-3 × 1-2 cm; calyx ca 3 cm long, hairy; corolla yellow with deep purple centre, rarely white, sometimes ultimately deep pink, petals ca 8 x 6 cm, obovate; staminal column antheriferous throughout. Capsules 3-4.5 × 1.5-2 cm, ovoid to oblong, densely hispid; seeds 3-4 mm long, minutely hairy, blackish.

Rare occurs in deciduous forests in Warangal district (C.S.Reddy & *al.,* 2008).

Fl. & Fr.: November- March.

Tadwai to Eturunagaram (WGL), *CSR* & *VSR* 2643 (KU).

Abelmoschus crinitus Wall., Pl. Asiat. Rar. 1: 39. t. 44. 1830; Borss., Blumea 14: 103. 1966; Paul in B.D.Sharma & Sanjappa, Fl. India 3: 302. 1993. *Hibiscus cancellatus* Roxb., Fl. Ind. 201. 1832; non L. f. 1782; Masters in Hook.f., Fl. Brit. India. 1: 342. 1874.

Annual erect, very bristly herbs, up to 50 cm high with fusiform roots; branches, petioles and pedicels hirsute by shiny simple and stellate hairs, ultimately glabrescent. Leaves cordate, 10-15 × 7-12 cm, margin crenate-toothed, sometimes lobed, upper sagittate; petioles 0.5-24 cm long; stipules linear to filiform. Flowers yellow with purple centre in terminal racemes, rarely solitary; epicalyx segments linear, ciliate; calyx spathaceous, 2-5 cm long, densely puberulous - tomentose; corolla yellow with purple centre, ca 6 cm across, petals broadly obovoid, 4-9 x 2-4 cm, glabrous; staminal column ca 2 cm long, glabrous, antheriferous throughout. Capsule ovoid, furrowed, short acuminate, hirsute; seeds numerous, reniform, ca 4 mm across with concentric rings of hairs, rarely glabrous.

Recorded by Sayeduddin (1940) from Sirnapalli RF.

Fl. & Fr.: July-September.

Abelmoschus esculentus (L.) Moench., Meth. 617. 1794. *Hibiscus esculentus* L., Sp. Pl. 696. 1753; Masters in Hook.f., Fl. Brit. India. 1: 343. 1874; Gamble, Fl. Madras 1 : 99. 1915.

Vern.: *Benda.* English: Lady's finger.

Stout, glabrous, annual herb, 1-1.5 m high. Leaves 3-9 lobed. Flowers axillary, solitary or in racemes; epicalyx segments 5, free; calyx spathaceous, lobes 5, falling with corolls; petals 5, yellow with reddish centre, staminal-tube shorter than corolla. Capsule 5-angled, beaked, hairy, 10-12 cm long; seeds pale brown, globose.

Extensively cultivated as a vegetable for its unripe fruits.

Fl. & Fr.: ThroUGOut the year.

Abelmoschus ficulneus (L.) Wight & Arn., Prodr. 53. 1834; Borss., Blumea 14: 106. 1966; Paul in B.D.Sharma & Sanjappa, Fl. India 3: 304. 1993. *Hibscus ficulneus* L., Sp. Pl. 695. 1753; Masters in Hook.f., Fl. Brit. India. 1: 340. 1874; Gamble, Fl. Madras 1 : 97. 1915.

Annual erect herbs to undershrubs, 0.5-2 m high, branchlets stellate or simply hairy, often minutely prickly. Leaves cordate-rounded when young, adult hispid, orbicular, palmately 3-5-lobed; lobes obovate, 4-7 × 2-3 cm; petioles 1.5-20 cm long. Flowers creamish or white, solitary or in terminal racemes; epicalyx segments 5-6, linear to lanceolate, caducous; calyx in bud lageniform with 3 mm long linear lobes; corolla white becoming pink with deep purple centre, petals obovate. Capsules oblong-ovoid, 5-angled, shortly beaked, tomentose; seeds globose, sulcate, slightly pilose, blackish.

Occasional in wastelands and cultivated fields in most districts.

Fl. & Fr.: September-February.

Sone (ADB), *TP* & *PVP* 4014; Wankidi (ADB), *GO* 4469; Sirnapally RF (NZB), *BR* & *SPB* 9569; Pathur (MDK), *TP* & *CP* 12126; Pargi (RR), *MSM* & *PRSR* 15104; Pakhal (WGL), *KMS* 11578 (MH); Rechapalli (KRN), *GVS* 22246.

Note: This species can be easily recognized by its flask-shaped calyx in buds and blunt, hairy capsules. The leaves resemble those of *Ficus cairica* L., hence the specific epithet.

Abelmoschus manihot (L.) Medik., Malv. 46. 1787, ampl. Hochr. Candollea 2: 87. 1924; Paul in B.D.Sharma & Sanjappa, Fl. India 3: 304. 1993. subsp. *manihot Hibiscus manihot* L., Sp. Pl. 696. 1753; Masters in Hook.f., Fl. Brit. India. 1 : 341. 1874; Gamble, Fl. Madras 1 : 97. 1915.

Perennial, simple or sparingly branched, strigose and weakly prickly undershrubs, up to 3 m high. Leaves extremely variable in shape and size, broadly ovate, 4-15 × 6-20 cm, cordate at base, shallowly to deeply 5-7-lobed, lobes triangular, ovate-lanceloate or linear-spathulate, margin dentate to serrate, glabrous or stellate-pubescent, petioles 2.5-23 cm long, stipules linear to filiform or lanceolate. Flowers yellow with a deep purple or chocolate base within, in terminal racemes; basal flowers axillary, solitary; epicalyx lobes 6-8, free, ovate or oblong, tomentose with stellate haris; calyx spathaceous, 2.5 x 1.5 cm, velutinous out side, sericeous inside; corolla white or yellow with a purple centre, ca 5 cm across, petals obovate to orbicular, glabrous; ovary ovoid, styles yellow or white, stigma hairy. Capsule 3-6 cm long, 4-5-valved, ovoid-oblong, with 5 prominent costae, stellate-tomentose; seeds globose or reniform, stellate hairy in concentric rings, brown or black.

Key to the Subspecies
1. Stems without prickly hairs .... subsp. **manihot**
1. Stems more of less densely covered with prickly hairs .... subsp. **tetraphyllus**

Abelmoschus manihot (L.) Medik., subsp. **tetraphyllus** (Roxb. ex Hornem) Borss. Blumea 14: 97. 1966; Paul in B.D.Sharma & Sanjappa, Fl. India 3: 306. 1993. Stems prickly hairy.

Key to the Varieties
1. Epicalyx segments hispid along margins with stiff hairs;
   seeds almost globose .... var. **megaspermus**
1. Epicalyx segments densely clothed with soft hairs;
   seeds reniform with broad sinus .... var. **pungens**

Abelmoschus manihot (L.) Medik., subsp. **tetraphyllus** *Hibiscus tetraphyllus* sensu Masters in Hook. f., Fl. Brit. India 1 : 341 1874 *p.p.,* non Roxb. ex Hornem. 1815. var. **megaspermus** Hemadri, Bull. Bot. Surv. India 11: 338. 1972.

Rare, occurs along streams of Tadvai RF and Bhupalpalli RF of Warangal district.

Fl. & Fr.: October- November

Regonda, *CSR* 1107 (KU).

Abelmoschus manihot (L.) Medik., subsp. **tetraphyllus** var. **pungens** (Roxb.) Hochr., Candollea 2: 87. 1924. *Hibiscus pungens* Roxb., Fl. Ind. 3: 213. 1832; Masters in Hook.f. Fl. Brit. India 1: 341. 1874.

Occasional, occurs along forest paths.

Fl. & Fr.: October- December

Bayyakkapet (WGL), *CSR* 1456 (KU).

Abelmoschus moschatus Medik., Malv. 46. 1787; Borss., Blumea 14 : 90. 1966; subsp. *moschatus* Paul in B.D.Sharma & Sanjappa, Fl. India 3: 308. 1993. *Hibiscus abelmoschus* L., Sp. Pl. 696. 1753; Masters in Hook.f., Fl. Brit. India. 1 : 342. 1784; Gamble, Fl. Madras 1: 97. 1915.

Suffrutescent herbs, up to 1.5 m high, stems woody, densley retrorsely hirsute, tap root tuber-like. Leaves extremely variable in shape and size, 6-10 × 8-18 cm, palmately 3-5-lobed, lobes oblong-ovate, cordate at base, margin crenate-serrate, densely velvety pubescent to sparingly hairy; petioles 2-20 cm long; stipules linear-filiform. Flowers yellow with maroon centre, axillary, solitary and in terminal racemes; epicalyx lobes 7-10, free, subulate, 10 -13 × 1.2-2 mm, persistent in fruits, hispid; calyx 1.5-3 mm long, splitting on one side, stellate tomentose out side, sericeous inside; corolla yellow with dark purple centre, ca 10 cm across, petals obovate, rounded at apex, fleshy and ciliate at base; staminal column ca 1.5 cm long, glabrous, antheriferous throughout; ovary ovoid, 6-7 mm long, densely sericeous, hairy at apex, stylar branches 5, glabrous, stigmas hairy, purple. Capsules 8 x 4 cm, ovoid, shortly acuminate, faintly angular, hispidly hairy; seeds many, reniform, 3-4 mm, striate, glabrous, often musk-scented.

Rare in Hyderabad, Nallamalais in Mahabubnagar and Khammam districts.

Fl. & Fr.: August-December.

Dhaniyayigudem (KMM), *RR* 108095 (BSID).

Key to Subspecies

1. Stem retrorsely bristly; epicalyx segments appressed
   in fruit .... subsp. **moschatus**

1. Stem with soft spreading hairs; epicalyx segments
   spreading in fruit .... subsp. **tuberosus**

Abelmoschus moschatus Medik. subsp. **tuberosus** (Span.) Borss., Blumea 14: 93. 1966. *Hibiscus longifolius* Willd.var. *tuberosus* Span., Linnaea 15: 170. 1814. *Abelmoschus rugosus* Wall. ex Wight & Arn., Prodr. Fl. Ind. Orient. 53. 1834. *Hibiscus rugosus* (Wight & Arn.) Mast. in Hook. f., Fl. Brit. India 1: 342. 1875.

Herbs with fusiform root stock, stem with spreading hairs, bristly at apex. Leaves orbicular, 3-7-lobed. Flower buds up to 2 cm long; epicalyx segments linear-filiform, 1.5 cm long, prominently setaceous; petals yellow turning pink, with purple centre. Seeds tomentose.

Rare, occurs along streams in dry deciduous forests.

Fl. & Fr.: September-November.

Tadvai (WGL), *CSR* 1433 (KU).

ABUTILON Mill.

1. Mericarps rounded at apex .... **A. pannosum**

1. Mericarps acute at apex :

   2. Stem patently long hairy; corolla with purple centre;
      mericarps 20-25 .... **A. hirtum**

   2. Stem appressed densely hairy; corolla without purple centre;
      mericarps to 18 .... **A. indicum**

Abutilon hirtum (Lam.) Sweet, Hort. Brit. ed. 1. 53. 1826; Gamble, Fl. Madras 1: 91. 1915; Borss., Blumea 14: 168. 1966; var. **hirtum**; Paul in B.D.Sharma & Sanjappa, Fl. India 3: 264. 1993. *Sida hirta* Lam., Encycl. 1: 70 1783. *Abutilon graveolens* (Roxb. ex Hornem.) Wight & Arn. var. *hirtum* (Lam.) Masters in Hook. f., Fl. Brit. India 1: 327. 1874.

Annual, erect, hirsute undershrubs, up to 1.25 m high, branches viscid, stems, petioles and pedicels covered with long patent, simple, minute stellate and glandular hairs. Leaves cordate, 6-8 × 5-6.5 cm, 7-9-nerved, woolly below, tomentose above, base cordate, margin dentate-crenate, apex acute- acuminate; petioles 3-20 cm long; stipules linear to lanceolate. Flowers yellowish-cream, axillary, solitary; calyx campanulate, divided to the middle, lobes ovate or deltoid, densely stellate-pubescent mixed with simple and glandular hairs; corolla orange-yellow with a purple centre, petals longer than calyx lobes, obcordate; staminal column 5-7 mm, yellow or dark purple. Schizocarp globose, densely stellate, mericarps 20-25; seeds reniform, punctuate with minute stellate hairs.

Common weed of wastelands and forests in almost all districts.

Fl. & Fr.: October – April.

Kunthala Waterfalls (ADB), *GO* & *DAM* 5129; Narsapur (MDK), *BR* & *CP* 14099; Chandoor (NZB), *BR* & *GO* 9212; Narasampet (WGL), *KMS* 11593 (MH); Dhaniyayigudem (KMM), *RR* 108502 (BSID); Mannanur (MBNR), *SRS* 109199 (BSID).

Abutilon indicum (L.) Sweet, Hort. Brit. ed. 1.54. 1826; Masters in Hook.f., Fl. Brit. India 1 : 326. 1875; Gamble, Fl. Madras 1 : 91. 1915; Borss., Blumea 14 : 170. 1966.; Paul in B.D.Sharma & Sanjappa, Fl. India 3: 266. 1993.

Annual or perennial herbs or undershrubs, up to 3 m high, stems, petioles and pedicels densely or sparsely velutinous with minute stellate and simple hairs. Leaves simple, ovate to orbicular, base cordate, margin minutely crenate to irregularly dentate, apex obtusely acuminate, velutinous on both surfaces by minute stellate hairs, also with scattered simple hairs; petioles 2-18 cm long. Flowers yellow, axillary, solitary, calyx much shorter than petals, spreading in fruit, campanulate, accrescent; corolla yellow to orange, petals broadly obovate, 1 cm across, rounded or emarginated at apex; staminal column 5-7 cm long, conic below, tubular above, stellate hairy at base, glabrous above. Schizocarp depressed globose, indented at apex, mericarps flat, reniform, dorsally tomentose by stellate hairs; seeds 2 or more in each mericarp, reniform, covered with stellate hairs, brownish black.

Key to Subspecies

1. Calyx shorter than schizocarp, spreading at maturity;
    mericarps shortly acuminate at apex .... subsp. **indicum**

1. Calyx about as long as the schizocarp, appressed, mericarps
    long acute at apex, obtuse or rounded :

    2. Apex of the leaves obtuse to acute or shortly acuminate,
        coarsely stellate hairy; mericarps long acute at apex .... subsp. **guineensis**

2. Apex of the leaves long acuminate, velutinous with short stellate hairs; mericarps usually rounded or obtuse, rarely long acute at apex .... subsp. **albescens**

Abutilon indicum (L.) Sweet, subsp. **albescens** (Miq.) Borss. var. **australiense** Hochr., Ann. Cons. Jard. Bot. Geneve 6: 20. 1902; Borss., Blumea 14: 175. t. 19(a.b.). 1916; Chandrabose, Bull. Bot. Surv. India 12: 276. 1970; Paul in B.D.Sharma & Sanjappa, Fl. India 3: 266. 1993.

Occasional in Guntur district.

Fl. & Fr. : November-January

Nagarjunasagar, Krishna river bank (NLG), *KMS* 9803 (MH).

Abutilon indicum (L.) Sweet, subsp. **guineensis** (Schumach.) Borss., Blumea 14 : 175. 1966; Paul in B.D.Sharma & Sanjappa, Fl. India 3: 267. 1993. *Sida guineensis* Schumach. in Kongel. Danske Vidensk. Selsk. Ska. 4 : 81. 1829. *Abutilon asiaticum* (L.) Sweet, Hort. Brit. ed. 1 : 53. 1826; Masters in Hook.f. Fl. Brit. India. 1 : 326. 1874; Gamble, Fl. Madras 1 : 91. 1915. *Sida asiatica* L., Cent. Pl. 2 : 26. 1756.

Vern. : *Thuthurubenda, Duvvenakaya.*

Abutilon indicum (L.) Sweet, subsp. *guineense* is allied to subsp. *indicum* but differs in having stem with dense woolly hairs, leaves more or less scabrid on the upper surface, large corolla, mericarps long acute with woolly stellate hairs on the dorsal surface.

Occasional in dry areas.

Fl. & Fr.: October-November

Yellutla (ATP), *BR* & *ANS* 35515; Narasampet (WGL), *KMS* 11595 (MH); Parnapalli (HYD), *KMS* 5989 (MH).

Abutilon indicum (L.) Sweet, subsp. **indicum**; Borss., Blumea 14: 170. 1966; Paul in B.D.Sharma & Sanjappa, Fl. India 3: 267. 1993. *Sida indica* L., Cent. Pl. 2: 26. 1756. *Sida populifolia* Lam., Encycl. 1: 7. 1783. *Abutilon populifoloium* (Lam.) Sweet, Hort. Brit. ed. 1. 53. 1826. *A. indicum* var. *populifolium* (Lam.) Wight & Arn. ex Masters in Hook. f., Fl., Brit. India 1: 326. 1874.

Vern. : *Thuthuru benda, Erribenda.*

Common along road sides, in open places and wastelands.

Fl. & Fr.: March-September

Indervelli (ADB), *GO* & *PVP* 4529; Mosra (NZB), *TP* & *BR* 6087; Mombojipally (MDK), *BR* & *CP* 11635; ICRISAT (MDK), *LJGV* 2687 (CAL); Central University campus (RR), *MSM* 19204; Krishna river (NLG), *KMS* 9803 (MH); Way to Wyra (KMM), *RR* 108057 (BSID); Pasra (WGL), *RKP* 108126 (BSID); Amangal to Shadnagar (MBNR), *SRS* 104570 (BSID).

Abutilon pannosum (Forst. f.) Schlecht., Bot. Zeitung (Berlin) 9: 828. 1851; Paul in B.D.Sharma & Sanjappa, Fl. India 3: 268. 1993. *Sida pannosa* Forst. f., Comm. Phys. Soc. Reg. Goett. 1787: 62. 1789. *Abutilon glaucum* (Cav.) Sweet, Hort. Brit. ed. 1. 54. 1826; Gamble, Fl. Madras 1: 91. 1915. *Sida glauca* Cav. Icon. 1: 8. t. 11. 1791. *Abutilon muticum* (Del. ex DC.) Sweet, Hort. Brit. ed. 2. 65. 1830; Masters in Hook. f., Fl. Brit. India 1: 327. 1874. *Sida mutica* Del. ex DC., Prodr. 1 : 470. 1824.

Undershrubs or shrubs, up to 2 m high, stem slender, tomentose by stellate and some simple hairs. Leaves orbicular to round cordate, 2-8 × 2-7 cm, apex acute, margin irregularly toothed, 5-9-nerved at the base, both the surfaces densely stellate hairy; petioles 1.5-6 cm long. Flowers yellow with deep brown centre, solitary, axillary, sometimes in racemes by replacing the upper leaves; calyx campanulate, divided to the middle, lobes deltoid, densely stellate-hairy outside and with simple hairs inside; corolla yellow with deep brown centre, petals obliquely triangular, glabrous; staminal column 5-8 mm long, not exceeding petals, antheriferous throughout. Schizocarp 0.5-1.0 × 1-2 cm, subglobose, depressed at the top, downy, mericarps 20-25, each 6-10 × 5 mm, reniform with awn, compressed on two sides, hairy on the dorsal margin; seeds 2-3 per mericarp, 1-5 mm across, minutely hairy or glabrous, brownish.

Rare in Hyderabad, Karimnagar, Mahabubnagar and Warangal districts.

Fl. & Fr.: October-March.

Eturunagaram (WGL), *RKP* 110824 (BSID).

ALCEA L.

Alcea rosea L., Sp. Pl. 687. 1753. *Althaea rosea* (L.) Cav., Diss. 2: 91. 1786; Masters in Hook.f., Fl. Brit. India. 1 : 338. 1874.

Under- shrub, 1.5-2 m high; stem stellate-tomentose. Leaves suborbicular, 6-15 x 5-16 cm. 5-7-lobed or angled, cordate, rounded or subacute, crenate-serrate, stellate pubescent on both the surfaces; petioles to 20 cm long; stipules ovate, caducous. Flowers solitary or 2-3 in fascicles aggregated into terminal, spiciform inflorescence; bracteoles 6-7, broadly triangular; calyx lobes ovate-acute, 1.5 cm long; corolla single or sometimes double; petals obovate, 4-6 cm long. Schizocarp depressed, globose, enclosed within the accrescent calyx, mericarps 20-40, horseshoe-shaped, indehiscent; seeds black, sparsely stellate hairy.

Cultivated in gardens as an ornamental.

Fl. & Fr.: Throughout the year.

FIORIA Mattei

Fioria vitifolia (L.) Mattei, Bot. Ort. Bot. Palermo n.s.2: 71. 1916; Paul in B.D.Sharma & Sanjappa, Fl. India 3: 310. 1993. *Hibiscus vitifolius* L., Sp. Pl. 696. 1753; Masters in Hook.f., Fl. Brit. India. 1 : 338. 1874; Gamble, Fl. Madras 1: 98. 1915.

Tel.: *Karu-patti*

Herbs or subshrubs, about 2 m high, branchlets tomentose with simple and stellate hairs. Lower leaves with 3-5, broadly angular lobes, upper leaves angled or ovate-cordate, 4-8 × 3-7.5 cm, base cordate, margin serrate, apex acute, chartaceous,

7-8-nerved at base, minutely stellate tomentose; petioles 2-13 cm long; stipules 2-5 mm long, linear. Flowers yellow with maroon centre, axillary, solitary or 3-5 in a cluster at the ends of branches; epicalyx deeply parted into 7-8 segments; calyx campanulate, 5-lobed, lobes ovate-deltoid; corolla yellow with dark purple centre, petals obovate; staminal column glabrous, antheriferous throughout; ovary ovoid. Capsule about 1.5 cm long, shorter than calyx, prominently winged, depressed at top; seeds about 15, faintly muricate, brownish black.

Common in most districts of the state, both in plains and in forests.

Fl. & Fr.: May-January.

Vempalli RF (ADB), *GO* 4435; Kammarapally (NZB), *TP* & *BR* 6248; Sirnapally RF (NZB), *BR* & *GO* 9091; Kodimial (KRN), *GVS* 25674 (MH); Aklaspur (KRN), *GVS* 25663 (MH); Bhupatipur (KRN), *GVS* 22286 (MH); Pakhal (WGL), *KMS* 11577 (MH), *KMS* 13173 (BSID; Bottichalami (NLG), *KMS* 9863 (MH); Borapuram (MBNR), *BR, SKB* 32355;

Note: This species was treated previously by most of the workers *viz.*, Masters 1874, Borssum Waalkes 1966, under the genus *Hibiscus* L. Mattei (1917) treated this species under the genus *Fioria* based mainly on its conspicuous scarious strongly veined wings on the fruit, a unique character absent in the genus *Hibiscus* L. But Hochreutiner (in Candollea 2 : 83-84. 1924) and Brenan and Excell (in Bot. Soc. Broteriana 32 : 69-74. 1958) do not accept this segregation because in this species the density of the indumentum and leaf lobation are variable even in the same plant. The density of the indumentum depends on the habitat. In open places the hairs are usually denser than the shade. However here we followed Mattei and treated it under genus *Fioria.*

GOSSYPIUM L.

1. Epicalyx lobes entire to serrate:

    2. Epicalyx lobes connate to 1 cm or more at base, entire or 3- or 4-toothed towards apex, closely embracing flowers .... **G. arboreum**

    2. Epicalyx lobes connate at very base only, 7-9-toothed or lobed at apex, not embracing flower .... **G. herbaceum**

1. Epicalyx lobes laciniate:

    3. Stipules auricled at base, capsules elongate .... **G. barbadense**

    3. Stipules not auricled, capsule globos*e* .... **G. hirsutum**

Gossypium arboreum L. (Sp. Pl. 693. 1753) emend. Hutch. & *al.,* Evol. Gossypium 32. t. 4. 1947; Borss., Blumea 14 : 121. 1966. *G. arboretum* sensu Hook.f. Fl. Brit. India 1 : 347. 1874.

Shrub, 1.5-2 m high, sometimes arborescent; stem densely covered with minute stellate hairs. Leaves ovate to orbicular in outline, 5-7-lobed to parted, 5-12 x 9-12 cm, shallowly cordate at base, margins entire, petioles 1.5-10 cm long; stipules linear-lanceolate, usually falcate, 0.4-1.2 cm long. Flowers axillary, solitary; pedicels 0.5-2 cm long; epicalyx segments foliaceous, broadly ovate, 2.5 cm long; calyx cupular, 5 mm long, truncate; corolla pale yellow; petals obovate, 3-4 cm long; staminal column 1.5-2 cm long; filaments 1.5-2 mm long. Capsule ovoid-subglobose, 3.5 cm long with an apical beak; seeds ovoid-globular, 5-6 mm in diam., hairs white.

Grown in gardens and houseyards as an ornmental.

Fl. & Fr .: September-December.

Koriguttalu (KMM), *RR* 106029 (BSID); Rollapadu (KMM), *RR* 108046 (BSID).

Gossypium barbadense L., Sp. Pl. 693. 1753; Gamble, Fl. Madras 1 : 102. 1915; Hutch. & al., Evol. *Gossypium* 48. 1947; Borss., Blumea 14 : 127. 1966.

Shrub or undershrub, 3-4 m high; stem sparsely stellate-pubescent to glabrate. Leaves orbicular to ovate, deeply 3-7-lobed, 8-20 cm long, cordate at base, margins entire; petiole 10-20 cm long, gland dotted; stipules 1-5 cm long, falcate or lanceolate, slightly auricled at base, caducous. Flowers yellow, axillary or in sympodial inflorescence; pedicels 1.5 cm long; bracteoles 3, foliaceous; calyx cupular, truncate at apex; corolla funnel-form; petals obovate with rounded apex; staminal column 2-3 mm long. Capsules narrowly ovoid to elongated, prominently pitted; seeds many per locule, subglobose, hairs usually white.

Grown in gardens and houseyards.

Fl. & Fr .: September-December.

Central University campus (RR), *MSM* 15167; Karkalpadu RF (MBNR), *SRS* 104522 (BSID).

Gossypium herbaceum L., Sp. Pl. 693. 1753; Gamble, Fl. Madras 1: 102. 1915.

Large shrubs, branches and leaves densely hairy. Leaves shallowly lobed, lobes 3-5. Flowers yellow with purple base. Capsules ovoid with prominent shoulders, acute.

Cultivated throughout the State.

Fl. & Fr.: September – March.

Appapur (MBNR), *TDCK* 16845.

Gossypium hirsutum L., Sp. Pl. 975. 1753; Borss., Blumea 14: 123. 1966. *G. herbaceum* L. var. *hirsutum* (L.) Mast. in Hook.f. Fl. Brit. India 1: 347. 1874.

Vern. : *Pathi*; English: Cotton.

Shrub or undershrub, 1-3 m high. Leaves orbicular in outline, usually wider than long, mostly 3-lobed, lower ones sometimes 5-lobed, upper ones occasionally ovate and entire, cordate at base, 5-7-nerved, 3-15 cm in diam.; petiole 3-8 cm long; stipules ovate to lanceolate, often falcate. Flowers on axillary, sympodial shoots; pedicel 1-2.5 cm, with 3 nectaries at apex; epicalyx segments ovate to triangular, deeply cordate and auricled at base, 2-4 x 1.5-3 cm; calyx campanulate, 6-7 mm long; segments 5; corolla pale yellow to white; petals obovate, 4-5.5 mm long; staminal column 1-2 cm long; Capsule ovoid, rostrate, 2-5 x 1-1.5 cm long, coarsely pitted.

Grown through out the state for cotton.

Fl. & Fr .: September–February.

HERISSANTIA Medik.

Herissantia crispa (L.) Medik., Phil. Bot. 1: 90. 1789; Brizicky, J. Arn. Arb. 49: 279. 1968; Paul in B.D.Sharma & Sanjappa, Fl. India 3: 276. 1993. *Sida crispa* L., Sp. Pl. 685. 1753. *Abutilon crispum* (L.) Medikus, Malv. 29. 1787; Masters in Hook.f., Fl. Brit. India. 1 : 327. 1874; Gamble, Fl. Madras 1 : 91. 1915.

Vern. : *Nelabenda.*

Annual erect hairy herbs, up to 75 cm high, stems, petioles and pedicels densely minute, stellate and simple hairy, rarely tomentose. Leaves cordate, 4-7 × 3-6.5 cm, chartaceous, 7-nerved, velutinous, base cordate, margin crenate, apex acuminate; stipules filiform, petioles 0.5-6.5 cm long. Flowers yellow, axillary, solitary; calyx campanulate, 5-lobed, lobes ovate to linear-triangular or lanceolate, densely stellate-hairy; corolla as long as or slightly shorter than calyx, white or pale yellow, petals broadly obovate, glabrous except for ciliate base; staminal column 2-2.5 mm long, glabrous. Schizocarp globose, wrinkled, mericarps 10-15; seeds reniform, covered with curved, appressed, simple hairs, brown.

Occasional in wastelands and forests in most districts.

Fl. & Fr.: July – February.

Mannanur (MBNR), *BSS* & *KP* 30662, 30663; Kambalpally (NLG), *BVRS* 215 (NU).

Note: This genus is allied to the *Abutilon* but can be distinguished by its inflated carpels with rounded apex, thin chartaceous pericarp and vine like weak stems.

HIBISCUS L. *nom. cons.*

1. Epicalyx very short, often caducous .... **H. lobatus**
1. Epicalyx present :
    2. Leaves simple, not lobed, not partite:
        3. Flowers 1.5 cm across, staminal column shorter than petals, wild .... **H. micranthus**
        3. Flowers 5-12 cm across, staminal column longer than petals; cultivated:
            4. Petals entire, spreading .... **H. rosa-sinensis**
            4. Petals laciniate, deeply incised, drooping .... **H. schizopetalus**
    2. Leaves partite or lobed, at least upper leaves:
        5. Epicalyx segments basally connate .... **H. panduraeformis**
        5. Epicalyx segments free :
            6. Epicalyx segments forked at apex; stems, petioles and pedicels armed with prickles:

7. Leaves unlobed to shallowly 3-5-lobed; stipules
    ovate-lanceolate; pedicel very distinct, 1.5-7 cm
    long, rambling undershrubs .... **H. aculeatus**

7. Leaves deeply palmately 3-7-lobed; stipules linear;
    pedicel 2-4 mm long, erect undershrubs .... **H. radiatus**

6. Epicalyx segments simple, not forked; stems, petioles
    and pedicels unarmed:

8. Seeds cottony .... **H. hirtus**

8. Seeds glabrous or hairy, not cottony:

9. Midrib eglandular .... **H. trionum**

9. Midrib with a gland beneath near the lobes:

10. Epicalyx segments spreading or reflexed ... **H. cannabinus**

10. Epicalyx segments closely appressed
    to calyx .... **H. sabdariffa**

Hibiscus aculeatus Roxb., Fl. Ind. 3: 206. 1832; Paul in B.D. Sharma & Sanjappa, Fl. India 3: 323. 1993. *H. furcatus* Roxb. ex DC., Prodr. 1: 449. 1824, *non* Willd. 1809; Masters in Hook.f., Fl. Brit. India. 1: 335. 1874; Gamble, Fl. Madras 1: 97. 1915.

Rambling undershrubs, up to 1 m high, branchlets softly prickly and hairy. Leaves 3-8 cm long, almost as much as broad, unlobed or shallowly 3-5-lobed, base cordate, margin serrate-dentate, apex acute. Flowers bright yellow, inside base pink, axillary and solitary; epicalyx forked at apex with one ovate leafy lobe, another oval-shaped lobe projecting upwards; calyx deeply 5-parted, lobes broadly lanceolate, persistent; corolla yellow with purple centre. Capsule ca 1.5 × 1 cm, globose, not exceeding calyx, silky hairy; seeds 3-gonous, brown, with scattered minute stellate hairs.

Rare in open forests, near villages in Karimnagar district (Naqvi, 2001).

Fl. & Fr.: July-October

Hibiscus cannabinus L., Syst. Nat. ed. 10. 1149. 1759; Masters in Hook.f., Fl. Brit. India. 1: 339. 1874; Gamble, Fl. Madras 1: 71. 1915; Paul in B.D.Sharma & Sanjappa, Fl. India 3: 324. 1993.

Vern.: *Gogu.*

Annual, erect herbs, 0.5-1.5 m high. Leaves deeply 5-lobed, lobes 2-10 × 0.4-1.2 cm, elliptic-lanceolate, base obtuse, margin distantly serrate, apex tapering, midnerve glandular below. Flowers 4-6 cm across, pale yellow with a dark purple throat, axillary, solitary. Capsule 1.5-2 cm across, globose, apiculate, bristly.

Cultivated for its leaf and fibre, occasionally as an escape.

Fl. & Fr.: Through out the year

Kadthala (ADB), *GO & DAM* 5018; Appapur (MBNR), *TDCK* 16846; Rangapur (WGL), *PVS* 76981 (BSID).

Hibiscus hirtus L., Sp. Pl. 694. 1753; Masters in Hook.f., Fl. Brit. India. 1 : 335. 1874; Gamble, Fl. Madras 1 : 98. 1915; Paul in B.D.Sharma & Sanjappa, Fl. India 3: 329. 1993.

Hairy shrubs, woody at base, up to 1 m high. Leaves ovate or ovate-lanceolate, 3-lobed, ca 12 × 6 cm, central lobes oblong- lanceolate, much bigger than short lateral lobes, base rounded or more or less cuneate, margin crenate-serrate or irregularly dentate, apex acute to acuminate, stellately hairy, with a large gland on the under side of the midrib; petioles 0.5-1.5 cm; stipules linear, ciliate. Flowers white, rarely rose-coloured, axillary, solitary and in terminal racemes, pedicels longer than petiole, jointed at or below middle; epicalyx segments 6-9, free, lanceolate to linear; calyx narrowly campanulate, 5-fid or parted, divided nearly to the base, lobes linear-lanceolate, persistent; corolla pink or white, rotate, petals obovate; staminal column shorter or as long as petals, pink, antheriferous throughout. Capsule globose, minutely hairy, included in calyx; seeds reniform, cottony.

Very rare in Hyderabad (Ramana, 2010) and Karimnagar (Naqvi, 2001) districts.

Fl. & Fr.: October-January

Hibiscus lobatus (Murr.) Kuntze, Revis. Gen. Pl. 3 : 19. 1898; Paul in B.D.Sharma & Sanjappa, Fl. India 3: 336. 1993. *Solandra lobata* Murr. in Commentat. Soc. Regiae Sci. Gott. 6 : 20. t. 1. 1785. *Hibiscus solandra* L' Herit., Strip. Nov. 1 : 103. t. 549. 1789; Masters in Hook.f. Fl. Brit. India. 1 : 336. 1874; Gamble, Fl. Madras 1 : 98. 1915.

Vern.: *Atakanara.*

Annual, erect, pubescent herbs, up to 1 m high. Leaves vary in shape, basal leaves orbicular-ovate, upper ones 3-lobed or lanceloate, 3-5 × 0.6-1.5 cm; simple and stellate hairy above, woolly below, 5-nerved, base truncate-cordate, margin crenate-serrate, apex acute-acuminate; petioles 2-9.5 cm long; stipules linear to filiform. Flowers white with pale yellow tinge, axillary, solitary; epicalyx segments 6-8, caducous before anthesis; calyx campanulate to rotate, 5-fid to 5-parted, lobes deltoid to lanceolate; corolla white or yellow, 1-2 cm across, petals obovate, glabrous or nearly so; staminal column 6-12 mm long, antheriferous throughout, ovary ovoid, ca 3 mm across, sparsely glandular, styles glabrous, stigma capitate, pale pink. Capsules ovoid, shortly beaked, hispid, wrinkled; seeds many, reniform or ovoid, glabrous, black.

Occasional in the forests in all districts.

Fl. & Fr. : July-November

Sirpur (ADB), *GO* & *PVP* 4930; Manchippa RF (NZB), *BR* & *PSPB* 9622; Pathur (MDK), *TP* & *CP* 12108; Narsapur RF (MDK), *BR* & *CP* 11227; Medak (MDK), *CP* 11731; Pakhal (WGL), *KMS* 13140 (CAL); Bhupatipur (KRN), *GVS* 25707 (MH); Mamillagudem (KMM), *RR* 105902 (BSID); way to Pasra (WGL), *RKP* 108225 (BSID).

Hibiscus micranthus L.f., Suppl. Pl. 308. 1782; Masters in Hook.f., Fl. Brit. India. 1 : 335. 1874; Gamble, Fl. Madras 1: 97. 1915; Paul in B.D.Sharma & Sanjappa, Fl. India 3: 334. 1993. var. **micranthus.** *H. ovalifolius* (Forssk.) Vahl, Symb. Bot. 1: 50.1970. *Urena ovabifolia* Forssk.Fl. Aegypt, Arab. 124.1775.

Vern.: *Nityamalli.*

Small, erect undershrub, 0.5-1.5 m high, with slender terete branches and scabrid with scattered stellate bristles. Leaves ovate or oblong, 2.5-4 × 1.5 – 3 cm, margin serrate, apex acute or obtuse, without glands, more or less scabrid with stiff stellate hairs, petiole 1.5-2.5 cm, stipules 0.3-1.5 cm, filiform, hairy. Flowers white or pink, axillary, solitary, pedicels up to 3.5 cm long, jointed above or below the middle, scabrid with stellate hairs, epicalyx segments 6-8, filiform, hairy, shorter than calyx, calyx divided up to the middle, lobes ca 5 mm long, stellate hairy outside and on apical portion inside, corolla 0.5-1.3 cm across, purplish-white or pink, petals ca 0.3-1.2 cm, oblong, often reflexed, stellate-pubescent outside, glabrous inside. Capsules globose, dehisces into 5 valves, smooth outside, hairy inside; seeds reniform, black, hirsute with long white silky hairs.

Common in almost all districts, in waste places and in forests.

Fl. & Fr.: May-December

Nirmal (ADB), *GO* & *BR* 9405; Kamareddy (NZB), *BR* & *KH* 9057; Domadugu (MDK), *BR* & *CP* 14125; Mohammadabad RF (RR), *MSM* & *KH* 10221; Kodimial (KRN), *GVS* 20095 (MH); Nagarjunakonda (NLG), *KTH* 9622 (CAL); Ramanagutta (KMM), *RR* 107978 (BSID); Zulurpeda (KMM), *RR* 113821 (BSID).

Hibiscus panduraeformis Burm. f., Fl. Ind. 151. t. 47. 6. 2. 1768; Masters in Hook.f., Fl. Brit. India. 1: 335. 1874; Gamble, Fl. Madras 1: 98. 1915; Paul in B.D.Sharma & Sanjappa, Fl. India 3: 339. 1993.

Large erect herbs, up to 2 m high, tomentose, often with bristles and viscid hairs. Leaves cordate-ovate, obscurely lobed or angular, 2-8 × 2-5 cm, 7-nerved at base, margin crenate-serrate, subentire, apex acute, glabrescent above, white or grey-tomentose beneath; petioles 1-15 cm long; stipules 2-3-parted with filiform segments, deciduous. Flowers yellow with a deep-purple base within, axillary, solitary or in terminal racemes; epicalyx segments 6-10, shorter than calyx, segments shortly connate at base, spathulate, persistent; calyx campanulate, 5-lobed, lobes free up to the middle, ovate, obtuse, 3-nerved; corolla yellow with dark purple centre, petals obovate; staminal column dark purple, antheriferous throughout; ovary ovoid, style arms purple, stigma capitate. Capsule ovoid, ca 1.5 × 1 cm, included in calyx, densely hairy; seeds reniform, numerous, densely silky-brown hairy (villous), brown.

Occasional in Medak, Nizamabad, Karimnagar and Warangal districts, in deciduous and scrub forests.

Fl. & Fr.: October – January.

Jalalpur RF outskirts (NZB), *BR* & *KH* 9667; Patancheru (MDK), *BR* & *CP* 11310; ICRISAT (MDK), *LJGV* 2589 (CAL); Rechapalli (KRN), *GVS* 22284 (MH); Eturunagaram (WGL), *RKP* 110818 (BSID).

Hibiscus radiatus Cav., Diss. 3: 150. t. 54.f.2. 1787; Paul in B.D.Sharma & Sanjappa, Fl. India 3: 327. 1993.

Undershrubs, up to 1.5 m high, stems with some bulbous based retrorse prickles and long simple hairs. Leaves 2-12 × 1.5-12 cm, lower leaves orbicular, deeply palmately 3-7-lobed, lobes variable in shape, acute to acuminate at apex, coarsely or sharply serrate, 3-5-nerved at base, often tinged red; petioles 2-15 cm long; stipules linear to lanceolate. Flowers axillary, solitary, showy; epicalyx segments 8-10, 15-18 × 1.5 – 2 mm, linear, acute, forked towards apex; calyx ca 2 cm long, 1.5 cm across, lobes ovate to deltoid; corolla yellow or purple with dark purple centre; staminal column 1.5-2.2 cm long; ovary globose, stigmas capitellate. Capsules 2-2.5 × 1.5 cm, ovoid, with a short beak, densely hairy with long simple bristles, dehiscing into 5 loculicidal valves; seeds 4 mm across, scabrous, brown.

Rare, weed of wastelands and edges of forests in Warangal and Khammam districts.

Fl. & Fr.: October-February

Hanmakonda (WGL), *CSR* 1667 (KU); Sukkumamidi (KMM), *CSR* & *KRN* 2128 (FRLHT).

Hibiscus rosa-sinensis L., Sp. Pl. 694. 1753; Masters in Hook. f., Fl. Brit. India. 1: 344. 1874; Gamble, Fl. Madras 1: 70. 1915; Paul in B.D.Sharma & Sanjappa, Fl. India 3: 391. 1993.

Vern.: *Mandara.*

Evergreen shrub, 2-4 m high; stem sparsely puberulent to glabrate. Leaves ovate, 12-15 x 7-10 cm. truncate or subcordate at base, acuminate at apex, serrate to dentate, entire towards base, glabrous or minutely stellate-hairy on nerves beneath, petioles 2.5-5 cm long; stipules lanceolate-subulate, 3-4 mm long. Flowers solitary, axillary; pedicels to 10 cm long, articulated below the middle; epicalyx-segments 5-8, 5-15 mm long, lanceolate, connate at base; calyx tubular, 1-3 cm long, divided to middle; corolla single or double in a profusion of color forms; petals obovate or oblong, 7-12 cm long; staminal column antheriferous in upper half; staminal filaments 5-10 mm long (or petaloid in double flowered forms). Fruits not seen.

Cultivated as ornamental plant in gardens and private house compounds.

Fl.: Through out the year.

Mannanur (MBNR), *TDCK* 13626.

Hibiscus sabdariffa L., Sp. Pl. 695; Gamble, Fl. Madras 1: 99. 1915.

Vern.: *Erra gogu.*

Annual, erect, glabrous shrubs, 1-1.5 m high; stem red or purplish. Leaves shallowly 3-lobed, rarely 5-lobed, lower leaves often entire or obscurely lobed, lobes lanceolate, glandular on midrib below, margin serrate; stipules linear, acute, 1 cm long. Flowers axillary, solitary; epicalyx segments 10, adnate to the base of calyx, accrescent; calyx fleshy, enlarging in fruits; corolla purple with dark centre, lobes 5. Capsules ovoid, pubescent, beaked; seeds black, stellate hairy.

Largely cultivated in the state.

Fl. & Fr.: October – January.

Penugunda, Narsapur (MDK), *RG* 96284 (BSID); way to Wyra (KMM), *RR* 108069 (BSID).

Hibiscus schizopetalus (Dyer) Hook.f., Bot. Mag. t. 6524. 1880; Bailey, Man. Cult. Pl. 665. 1949. *Hibiscus rosa-sinensis* L. var. *schizopetalus* Dyer, Gard. Chron. n.5.538.1879.

Subshrub, 2-4 m high; stem glabrous; branches spreading or usually drooping. Leaves ovate-lanceolate, 2-7 x 1-5 cm, truncate at base, acute at apex, margins coarsely serrate; petiole short, 0.5-2 cm long; stipules 2-3 cm long, subulate. Flowers axillary, solitary, pendulous; pedicel 8-15 cm long, articulate nearly in the middle; epicalyx segments 5-8, 1-2 mm long; calyx tubular, 1-1.5 cm long, irregularly 2-5-lobed; corolla pinkish, with pink or red streaks or red with white or pink streaks; petals 4-6 cm long, laciniate, recurved; staminal column 8-10 cm long. Fruit setting not seen.

Occasional in gardens.

Fl. & Fr.: Through out the year.

Hibiscus trionum L., Sp. Pl. 697. 1753; Masters in Hook. f., Fl. Brit. India. 1: 334. 1874; Gamble, Fl. Madras 1: 98. 1915; Paul in B.D.Sharma & Sanjappa, Fl. India 3: 341. 1993.

Erect, stellately hairy herbs, up to 75 cm high. Leaves ovate, ca 5 × 4 cm, deeply palmately 3-lobed, often unlobed or shallowly lobed, middle lobe longer, oblong, pinnatisect, punctate, sparsely simple and stellate-hairy on both surfaces, densely so on lower surface, upper surface rarely glabrous; petioles 2-5 cm long; stipules subulate. Flowers solitary, axillary, covered with bulbous-based hairs except for corolla; epicalyx segments 8-12, linear; ciliate at margins, spreading upward in fruit surrounding capsule; calyx campanulate, inflated covering capsule, lobes broadly ovate, membranous; corolla yellowish pink with dark purple centre; staminal column antheriferous towards tip; ovary ovate, styles 5. Capsule oblong, hairy, included in inflated calyx; seeds grey, rugose, minutely stellately hairy.

Rare, in Hyderabad district.

Fl. & Fr.: July-January.

KYDIA Roxb.

Kydia calycina Roxb., Pl. Coromandel 3: 11, t. 215. 1811; Masters in Hook. f., Fl. Brit. India. 1: 348. 1874; Gamble, Fl. Madras 1 : 93. 1915; Paul in B.D.Sharma & Sanjappa Fl. India 3: 344. 1993.

Deciduous trees, up to 10 m high, branchlets stellate-woolly. Leaves broad-ovoid, to nearly orbicular, 3-5-angular or lobed, 5- 8 × 4-8 cm, 7-nerved, stellate hairy above, grey tomentose beneath, base cordate-truncate, margin obscurely crenate-dentate, apex round-obtuse; petioles 2-7 cm long; stipules subulate. Flowers polygamous, axillary or in terminal panicles; epicalyx segments 4-6, connate at base, oblong-spathulate or obovate; calyx cup-shaped, connate at base, lobes ovate; corolla white or pink, ca 1.7 cm across, petals longer than calyx but shorter than epicalyx, obcordate; staminal column ca 3 mm long, pistillode absent in male flower; ovary ovoid. Capsules globose, hard, depressed; seeds reniform, brown, minutely hairy, concentrically ribbed.

Occasional in Mahabubnagar district, in forests (Raghava Rao, 1989a).

Fl. & Fr.: September-April

MALACHRA L.

Malachra capitata (L.) L., Syst. Nat. ed. 12. 2. 458. 1767; Masters in Hook.f., Fl. Brit. India 1 : 329. 1874; Gamble, Fl. Madras 1 : 102. 1915; Paul in B.D.Sharma & Sanjappa, Fl. India 3: 367. 1993. *Sida capitata* L., Sp. Pl. 685. 1753.

Herbs or undershrubs, up to 1.8 m high, stems, petioles and floral axes minutely stellate-hairy. Leaves orbicular or slightly ovate, palmately 3- or 5-lobed or angled, 3-14 × 5-16.5 cm, 7-nerved at base, base cordate, margin entire or slightly lobed, apex obtuse, sparsely hairy; petioles 2-8 cm long; stipules 1-2 cm long, filiform, hispid. Flowers yellow and white, axillary or terminal, 2-7 in a cluster or in dense capitate heads; bracts 3-4 per head, broadly ovate to orbicular; calyx cupular, lobes oblong to deltoid, accrescent; corolla bright yellow, petals obovate, densely stellate-hairy outside, glabrous inside; staminal column ca 1 mm long, antheriferous throughout. Schizocarp subglobose, wrinkled, not exceeding calyx, mericarps indehiscent, 3-gonous, prominently nerved, glabrous; seeds reniform, brown, smooth, glabrous.

Common weed of marshy places, especially near sewage drains in Nizamabad, Karimnagar and Warangal districts. Introduced in the Old World, native of tropical America.

Fl. & Fr.: April – December.

Rudroor (NZB), *BR* & *KH* 9635.

MALVASTRUM A. Gray *nom. cons.*

Malvastrum coromandelianum (L.) Garcke, Bonplandia 5: 297. 1857; Gamble, Fl. Madras 1: 54. 1915; Borss., Blumea 14 : 152. 1966; Paul in B.D.Sharma & Sanjappa, Fl. India 3: 277. 1993. *Malva coromandeliana* L., Sp. Pl. 687. 1753. *Malvastrum tricuspidatum* R. Br. in Ait., Hort. Kew. ed. 2.4. 210. 1812.

Small, erect, branched shrubs, up to 1 m high; stems, branches, petioles and pedicels covered with stiff appressed, 4-armed, stellate hairs, two arms of which pointing upwards and two downwards. Leaves ovate, 2.5 - 5.5 × 1-2.5 cm, 5-nerved at base, scattered hairy, base obtuse-truncate, margin crenate-serrate, apex acute, petioles 0.5-4 cm long. Flowers yellow, axillary, solitary or in terminal clusters; calyx campanulate, lobes deltoid to ovate, 4-armed, stellate hairy outside; corolla yellow, petals obliquely obovate; staminal column ca 2.5 mm long, conical. Schizocarp globular, mericarps reniform, awns with paired apical hooks, mericarp 1-seeded; seeds reniform, compressed, glabrous, brownish-black.

Common weed of wastelands, road sides, and open forests.

Fl. & Fr.: Throughout the year, fruits persistent.

Kamareddy (NZB), *BR* & *KH* 7146; Narsapur (MDK), *BR* & *CP* 11236; Central University campus (RR), *TP* & *MSM* 19248; Moosi river bank (HYD), *KSM* 5966 (MH); ICRISAT (MDK), *LJGV* 3192 (CAL), Hyderabad city (HYD), *KMS* 5966 (CAL).

Note: This species is often confused with the *Sida* species especially *S. acuta*, but can be easily distinguished from the latter by its epicalyx and strigose indumentum on stems, petioles and pedicels.

MALVISCUS Fabricius

Malviscus arboreus Dill. ex Cav. var. **penduliflorus** (DC.) Schery, Ann. Missouri Bot.
Gard. 29: 223. 1942; Borss., Blumea 14: 133. 1966. *M. penduliflorus* Mociro & Sessi
ex DC., Prodr. 1: 445. 1824.

Shrubs, to 2 m high. Leaves ovate, acuminate, 4-7 × 2-5 cm, entire, glabrous.
Flowers pendulous; calyx to 1 cm, lobes 5, acute; petals 5, bright orange red, never
spreading; staminal tube equal to petals; carpels 10. Fruits not seen

Cultivated in gardens.

Fl.: Throughout the year.

PAVONIA Cav. *nom. cons.*

1. Leaves shallowly lobed; carpels rounded at the back, pubescent .... **P. odorata**
1. Leaves deep sinuately lobed; carpels 3-angled, keeled at the
    edges, glabrous .... **P. zeylanica**

Pavonia odorata Willd., Sp. Pl. 3: 387. 1800; Masters in Hook.f., Fl. Brit. India 1: 331.
1874; Gamble, Fl. Madras 1: 93. 1915; Paul in B.D.Sharma & Sanjappa, Fl. India
3: 373. 1993.

Viscidly pubescent, perennial herbs, up to 1 m high; stems, petioles and pedicels
covered with simple glandular hairs. Leaves ovate-suborbicular, shallowly 3-lobed,
1.5-3 × 1-2 cm, basally 5- nerved, stellately hairy, base cordate, margin dentate or
entire, apex acute; petioles 1-8 cm long; stipules linear, deciduous. Flowers pink to
white, solitary or paired at the ends of the branches; pedicels 1.5-4 cm, jointed above
the middle; epicalyx segments 10-12, free, linear, persistent; calyx 5-lobed, lobes
connate at base, ovate-lanceolate; corolla 1-3 cm across, pink, petals obovate, 1-2 cm
long, glabrous; staminal column shorter than petals, glabrous; ovary globose, stigmas
capitate. Schizocarp globose, pubescent, mericaps 5, not winged; seeds reniform,
brown, minutely papillose.

Common weed of waste lands and occasional in forests of most districts.

Fl. & Fr.: Throughout the year.

Ankusapuram (ADB), *GO* 4341; Vempalli RF (ADB), *GO* 4797; Aklaspur (KRN),
GVS 25622 (MH); Pakhal (WGL), *KMS* 11606 (MH); Way to Pegarikutta (MDK), *KMS*
6614 (MH); Nayanigutta (RR), *MSM* & *NV* 10327; Nagarjunasagar (NLG), *BVRS* 68
(NU); Kambalpally (NLG), *BVRS* 240 (NU); Vatavarlapally (MBNR), *VBH* 84963 (BSID);
Lawal RF (WGL), *RKP* 108162 (BSID); Kosiguttalu (KMM), *RR* 106027 (BSID).

Pavonia procumbens (Wall. ex Wight & Arn.) Walp., Rep. Bot. Syst. 1: 301. 1842, non
Casaretto 1842; Gamble, Fl. Madras 1: 93. 1915; Paul in B.D.Sharma & Sanjappa,
Fl. India 3: 374. 1993. *Lebretonia procumbens* Wall. ex Wight & Arn., Prodr. 47.
1834.

Procumbent or spreading undershrubs or herbs, up to 60 cm high, with
indumentum of stellate-pubescent hairs. Leaves ovate or orbicular, angular or 3-

lobed, 2-4 × 2-5 cm, 7-nerved at base, stellate-tomentose, base cordate, margin crenate-dentate, apex acuminate; petioles 0.5 -9 cm; stipules 3-5 mm long, linear to filiform. Flowers axillary, solitary, yellow; pedicels 1-6 cm long, jointed above the middle; epicalyx rotate, segments 5, shortly connate at base; calyx campanulate, connate up to the middle, lobes obovate; corolla yellow, 1.5-2 cm across petals obovate; staminal column antheriferous towards apex. Fruit schizocarp, mericarps 5, muricate, minutely stellate-hairy, margins prickled; seeds angular, glabrous, brownish black.

Rare along hedges in Warangal district (C.S.Reddy, 2001).

Fl. & Fr.: July - January.

Pavonia zeylanica (L.) Cav. Diss. 3: 134. 1787; Masters in Hook.f., Fl. Brit. India. 1 : 330. 1874; Gamble, Fl. Madras 1: 93. 1915; Paul in B.D.Sharma & Sanjappa, Fl. India 3: 377. 1993. *Hibiscus zeylanicus* L., Sp. Pl. 697. 1753.

Erect, perennial, branched viscid herbs, up to 2 m high; stems, petioles and pedicels pubescent with stellate and some long, patent, simple hairs. Leaves 3-lobed, broadly ovate or nearly orbicular, 1-2.5 × 1-2 cm, sparsely hirsute, 3-5-nerved at base, base cordate, margin crenate- dentate, apex acuminate; petioles 0.5-8 cm long; stipules ca 2 mm long, linear, deciduous. Flowers pinkish-white, axillary, solitary or in terminal clusters; epicalyx segments 10, linear, persistent; calyx campanulate, 5-lobed, lobes free to the middle, lanceolate or deltoid; corolla rotate; staminal column antheriferous throughout; ovary more or less ovoid, style branches 10, stigma capitate. Schizocarp globular, pubescent; mericarps 5, oblong, slightly winged at angles, awnless, glabrous, dehisces from top; seeds reniform, brown, almost glabrous, distinctly ribbed on dorsal side.

Common weed in waste lands, cultivated fields and rare in forests.

Fl. & Fr.: May-January.

Sirpur (ADB), *GO* 4918; Neela (NZB), *BR* & *GO* 9223; Burugupalli (MDK), *BR* & *CP* 11617; Central University campus (RR), *MSM* & *TP* 15140; Hyderabad city (HYD), *LJGV* 3198 (CAL); Mannanur – Rangapur (MBNR), *AJR* 20487; Nagarjunasagar (NLG), *KMS* 9770 (MH); Kodimial (KRN), *GVS* 20079 (MH); Aklaspur (KRN), *GVS* 22512 (MH); Jakaram (WGL), *RKP* 105305 (BSID); Kinnerasani wild life sanctuary (KMM), *RR* 112688 (BSID).

SIDA L.

1. Leaves cordate at base:
   2. Plants velvety-hairy; carpels 10, awns of carpels longer than calyx segments, retrorsely hairy .... **S. cordifolia**
   2. Plants not velvety-hairy; carpels 5, carpellary awns shorter than calyx segments, not retrorsely scabrid :
      3. Prostrate or ascending herbs; pedicels longer than petiole .... **S. cordata**
      3. Erect rigid herbs; pedicels not exceeding petiole .... **S. mysorensis**

1. Leaves cuneate to truncate at base, not cordate:

    4. Leaves ovate-oblong or ovate, carpellary awns minute or absent:

        5. Mericarps 5 .... **S. alba**

        5. Mericarps 6-12 .... **S. ovata**

    4. Leaves variable; carpellary awns distinct :

        6. Stipular pair dissimilar .... **S. acuta**

        6. Stipular pair similar :

            7. Stem with spiny projections at the base of the petioles; cocci 5 .... **S. spinosa**

            7. Stem not spiny at the base of the petioles; cocci more than 5

                8. Leaf apex retuse .... **S. alnifolia**

                8. Leaf apex obtuse or truncate .... **S. rhombifolia**

Sida acuta Burm. f., Fl. Ind. 147. 1768. emend K. Schum., in Fl. Bros. 12(3): 326. 1891; Gamble, Fl. Madras 1: 90. 1915; Paul in B.D.Sharma & Sanjappa, Fl. India 3: 281. 1993. *S. carpinifolia* sensu Masters in Hook. f., Fl. Brit. India 1 : 323. 1874, non. L. f. 1781.

Annual, erect branched undershrubs, up to 70 cm high, branchlets minutely stellate pubescent with minute stellate hairs mixed with some simple hairs, ultimately glabrescent. Leaves ovate-lanceolate or elliptic, 2-10 × 1-3 cm, basally 3-5-nerved, glabrescent, base obtuse, margin serrate, apex acuminate; petioles 2-6 mm long, pubescent with minute stellate hairs; each pair of stipules different with one lanceolate to linear, 3-6-nerved, another linear to filiform, 1-4-nerved. Flowers creamish or yellow, axillary, solitary or 2-8 in a cluster; calyx campanulate, slightly accrescent, 5-fid, lobes triangular, stellate and simple hairy outside, glabrous inside; corolla light yellow, petals as long as or slightly exceeding calyx, lobes obliquely obovate; staminal column ca 4 mm long. Fruit schizocarp, mericarps usually 6; seeds ovoid, 3-gonous, dark brown.

Very common in waste places and open forests through out the state.

Fl. & Fr. August-January

Sattaenapalli (ADB), *TP* & *PVP* 4160; Mosra (NZB), *TP* & *BR* 6064; Gandivet RF (NZB), *BR* & *CPR* 7195; Narsapur (MDK), *BR* & *CP* 11229; Mohammadabad (RR), *MSM* 10436; Along Moosi river bank (HYD), *KMS* 5934 (MH); Kamapur (WGL), *KMS* 11565 (MH); Pegarikutta-Narsapur (MDK), *KMS* 6651 (MH); Mahadevpur (KRN), *SLK* 70829 (NBG); Azampur (NLG), *BVRS* 483 (NU); Appapur (MBNR), *SKB* & *BSS* 34458; Hanmakonda (WGL), *RKP* 105363 (BSID).

Note: This species can be easily recognised by having stipules of each pair different, one linear to lanceolate, 3-6-nerved, the other linear to filiform and leaves glabrescent.

Sida alba L., Sp. Pl. ed. 2. 966. 1763. *S. alnifolia* L. var. *obovata* Hu, Fl. China. Fam 153. 22, t. 16.f.5. 1955, non *S. rhombifolia* L. var. *obovata* Wall. ex Masters in Hook.f., Fl. Brit. India 1: 324. 1874.

Undershrubs or herbs, up to 1 m high, stellate-pubescent. Leaves 1-2.5 × 0.5-2 cm, elliptic-obovate, base cuneate or obtuse, margin crenate-dentate or serrate, apex acute, glabrescent above, minutely stellate-pubescent beneath; petioles 0.5-1 cm long, stipules ca 5 mm long. Flowers axillary, solitary or paired, pedicel 3-6 mm long; calyx 3-5 mm across, campanulate, free above the middle, lobes 2-4 × 1-2 mm, deltoid, acute; corolla yellow, slightly exceeding the calyx; staminall column ca 3 mm long, basal portion conical, tubular part short. Schizocarps 4-5 mm across, depressed globose; mericarps 5, ca 2 × 15 mm, stellate-pubescent, with 2 convergent apical awns, ca 0.8 mm long, hairy, mericarps dehiscing at base; seeds ca 1.5 mm long, glabrous, browinish black.

In University of Hyderabad campus (Seshagiri Rao, 2012).

Fl. & Fr.: September- December.

Sida alnifolia L., Sp. Pl. 2: 684. 1753; Sivarajan & Pradeep, Sida Contr. Bot. 16: 69. 1994. *Sida rhombifolia* L. subsp. *retusa* (L.) Borss., Blumea 14: 198. 1966. *S. retusa* L., Sp. Pl. ed. 2: 961. 1763. *S. rhombifolia* var. *retusa* (L.) Masters in Hook.f., Fl. Brit. India 1: 324. 1874.

Erect or diffuse herbs or undershrubs, up to 0.5 m high with erect-patent branches. Younger parts stellate pubescent. Leaves obovate to orbicular, 0.5-6 × 4 cm, upper leaves rarely lanceolate, base acute to acuminate, apex mostly retuse, rounded or truncate. Pedicels jointed below middle. Flowers solitary, axillary or in clusters; calyx ca 6 mm across, orange-yellow; corolla yellow, 2-2.5 cm across, obliquely obovate, hairy at base; staminal column 3 mm long, glabrous or sparsely hairy, antheriferous above; ovary globose-depressed, ca 1.5 mm long, glabrous, styles 10. Mericarps 7-10, included in the calyx, mucronate at apex; seeds glabrous, black.

Occasional, occurs in open forests and wastelands.

Fl. & Fr.: September- February

Ibrahimpet (NZB), *BR* & *KH* 9047; Lingampet (NZB), *BR* & *CPR* 9240; Anantasagar RF (RR), *MSM* 10512; Narsapur – Hyderabad Road (MDK), *BR* & *CP* 11212; Lingareddypalli (MDK), *CP* 11713.

Note: This taxon has been treated as a subspecies or variety of *S. rhombifolia* by most of the contemporary taxonomists. After careful reexamination of all relevant materials Sivarajan and Pradeep (*loc.cit.*) have found that they are quite distinct and have reinstated this as a distinct species.

Sida cordata (Burm. f.) Borss., Blumea 14: 182. 1966; Paul in B.D. Sharma & Sanjappa, Fl. India 3: 283. 1993. *Melochia cordata* Burm. f., Fl. Ind. 143. 1768. *Sida veronicifolia* Lam., Encycl. 1: 5. 1783; Gamble, Fl. Madras 1 : 89. 1915. *S. humillis* Cav., Diss. 5. 277. 1788; Masters in Hook.f., Fl. Brit. India. 1: 322. 1874. *S. beddomei* Jacob, J. Bombay Nat. Hist. Soc. 47: 50. 1947; Pullaiah & Chennaiah, Fl. Andhra Pradesh 1: 131. 1997.

Vern.: *Gayapuaku.*

Prostrate or ascending herbs, 60 cm - 1 m high, stems, petioles and pedicels pubescent with scattered, long, patent simple and minute stellate hairs. Leaves cordate, 4-7 × 2-5 cm, hairy 5-7-nerved at base, stellate tomentose, base cordate, margin crenate-serrate, apex acuminate; petioles 1.5-30 mm long; stipules 1-3 mm long, linear, filiform. Flowers yellow, axillary, solitary or twin or sometimes in lax racemes; calyx ca 3 mm across, campanulate, 5-fid, lobes connate up to just above the middle, deltoid to triangular; corolla yellow to light yellow, petals obovate; staminal column ca 3 mm long, basal portion conical, tubular part short; ovary globose, ca 2 mm long, glabrous, styles 5, stigmas 5, pink. Schizocarp globose, enclosed in persistent calyx, brownish black; mericarps 5, sparsely hairy at apex, short beak; seeds ovoid, 3-gonous.

A weed of waste places in all districts.

Fl. & Fr.: September-January

Ankusapuram (ADB), *GO* 4344; Nasrullabad RF (NZB), *TP* & *BR* 6161; Gandhari RF (NZB), *BR* & *CPR* 7244; Rangammagudem RF (RR), *MSM* 10476; Akkannapet (MDK), *BR* & *CP* 11561; Narsapur (MDK), *KMS* 6691 (CAL); Appapur (MBNR), *TDCK* 16844; Tirmalgiri (NLG), *BVRS* 469 (NU); Thumukunta (MBNR), *DV* & *MV* 40728; Pakhal (WGL), *RKP* 108260 (BSID); Khammam hills (KMM), *RR* 113815 (BSID).

Note: This species shows variation in indumentum and inflorescence. It has the general appearance of *S. mysorensis* Wight & Arn. but differs from the later in having semiprostrate habit and absence of gland hairs.

Sida cordifolia L., Sp. Pl. 684. 1753; Masters in Hook.f., Fl. Brit. India. 1: 324. 1874; Gamble, Fl. Madras 1: 89. 1915; Borss., Blumea 14: 199. 1966; Paul in B.D. Sharma & Sanjappa, Fl. India 3: 285. 1993.

Vern.: *Chirubenda.*

Much branched undershrubs, up to 60 cm high, branchlets, petioles and pedicels densely stellate-tomentose. Leaves cordate or ovate, 2-5 × 2-4 cm, tomentose, base 5-7-nerved, base subcordate or rounded, margin serrate-crenate, apex acute-obtuse; petioles 4-5 mm long; stipules 3-10 mm long, filiform. Flowers pale yellow, axillary, solitary or 5-7 in a cluster particularly towards apices of branchlets; pedicels 2-10 mm long, jointed towards apex; calyx campanulate, lobes triangular; corolla yellow or whitish yellow, petals obliquely obovate; staminal column ca 2.5 mm long; ovary conical, stellate hairy, stigma capitate. Fruit schizocarp, mericarps 10; seeds ovoid or 3-gonous, brown or black.

A weed of wastelands, road sides in all districts.

Fl. & Fr.: August-November

Bersaipet (ADB), *GO* & *PVP* 4592; Velutla RF (NZB), *BR* 9523; Narsapur town (MDK), *BR* & *CP* 11238; Kalmankalva RF (RR), *MSM* 10459; Mohammadabad RF (RR), *MSM* & *KH* 10557; Central University campus (RR), *MSM* 15165; Appapur (MBNR), *TDCK* 16844; Nagarjunakonda (NLG), *BVRS* 79 (NU); Jakaram (WGL), *RKP* 105314 (BSID) Lankapally hills (KMM), *RR* 106006 (BSID).

Sida mysorensis Wight & Arn., Prodr. 59. 1834; Masters in Hook.f., Fl. Brit. India 1 : 322. 1874; Borss., Blumea 14 : 180. 1966; Paul in B.D. Sharma & Sanjappa, Fl. India 3: 286. 1993. *S. glutinosa* Roxb., Fl. Ind. 3 : 172. 1832. non Cav. 1785; Gamble, Fl. Madras 1 : 89. 1915.

Erect, rigid, glutinous-hairy herbs, up to 1 m high; stems, petioles and pedicels densely pubescent with minute stellate hairs, mixed with gland-tipped hairs and some long patent simple hairs. Lower leaves ovate, upper ones lanceolate, 3-4.5 × 2-3 cm, basally 5-7-nerved, stellate-tomentose, base subcordate-obtuse, margin crenate or serrate, apex acuminate; petioles 1-7 cm long; stipules 3-6 mm long, filiform. Flowers yellow, solitary or often in 5-7-flowered lax racemes; calyx widely campanulate, 5-fid, lobes connate up to the middle, deltoid; corolla yellow, slightly exceeding calyx, petals obtriangular; staminal column ca 4 mm long, tubular above, glabrous, antheriferous above; ovary subglobose, ca 1 mm long, brownish black, styles ca 4 mm long, connate to middle. Schizocarp smooth, not exceeding calyx, mericarps 5, 3-gonous; seeds 3-gonous, brown-black.

Common weed of waste places and forests in Hyderabad, Karimnagar and Mahabubnagar districts.

Fl. & Fr. : September-April.

Thaskapalli (MBNR), *SRS* 109076 (BSID); on the way to Nilgiri view point (MBNR), *SRS* 109190 (BSID).

Note: Most of the authors like Cooke (1901), Gamble (1915) Haines (1921) etc. considered *Sida mysorensis* Wight & Arn. as synonymous to *Sida glutinosa* Cav. Borssum (1966) placed *S. mysorensis* Wight & Arn. and *S. glutinosa* Cav. in two different sections like *'Nelavaga* and *Sida'* respectively.

Sida ovata Forssk., Fl. Aegypt.-Arab. 124. 1775; Paul in B.D.Sharma & Sanjappa, Fl. India 3: 288. 1993; Thoth., Curr. Sci. 33(19): 593. 1964. *S. grewioides* Guill. & Perr., Fl. Seneg. Tent 1 : 71. 1831; Masters in Hook.f., Fl. Brit. India 1 : 323. 1874.

Erect, stellately hairy under shrubs, up to 1 m high, petioles and pedicels pubescent with stellate hairs. Leaves ovate-oblong or ovate, 1.5-4 × 1.2-2.7 cm, base rounded, margin crenate, apex obtuse, stellately grey tomentose on both surfaces; petioles 0.5-1.5 cm long; stipules up to 1 cm long, linear to lanceolate. Flowers pale-yellow or white, axillary, solitary or clustered at the ends of branches; calyx 5-8 mm across, campanulate, lobes ovate, densely stellate-pubescent outside, glabrous except for stellate and few simple hairs towards apical margins inside; corolla yellow or yellowish white, petals obliquely truncate; staminal column ca 7 mm long, simple and stellate-hairy. Schizocarp subglobose, ripe carpels 6-8, minutely mucronate or without mucro, brown, stellately hairy; seeds reniform, dark-brown, smooth or very minutely pitted.

Very rare in Nagarjunakonda valley in Nalgonda district.

Fl. & Fr.: July – February.

Nagarjunakonda valley (NLG), *KTH* 9724 (CAL).

Note: Thothathri (*loc.cit*) mentioned that the plant is an undershrub, wih thick root-stock and is one of the rare plants collected from the valley. He also mentioned that it is an exotic plant from Arabia and Tropical Africa which has so far been reported from Punjab, Rajasthan and Kutch.

During intensive explorations in the above mentioned locality and its surroundings Rao & *al.* (2001) have not observed even a single plant. As the valley and its surroundings submerged under water after the completion of Nagarjunasagar dam in 1964, the species might have disappeared.

Sida rhombifolia L., Sp. Pl. 684. 1753; emend. Hook. f., Fl. Brit. India 1: 323. 1874; Gamble, Fl. Madras 1: 90. 1915; Borss., Blumea 14 : 193. 1966; Paul in B.D.Sharma & Sanjappa, Fl. India 3: 289. 1993. *S. rhomboidea* Roxb. ex Fleming in Asiat. Res. 6 : 178. 1810; Gamble, Fl. Madras 1 : 90. 1915.

Erect, perennial herbs, stems, petioles and pedicels covered with scattered minute stellate hairs. Leaves ovate-elliptic or rhomboid, 1.5-3 × 0.5-3 cm, basally 3-nerved, stellately hairy above, woolly below, base cuneate, margin basally entire, distally bidentate-biseriate, apex obtuse-truncate. Flowers yellow, axillary, solitary or 5-7 flowers in clusters, pedicel jointed above middle; calyx up to 10 mm; corolla 2-2.5 mm across; staminal column shorter than petals; glabrous; ovary conical, minutely stellate, style 9-12; ovary conic, minutely stellate. Schizocarp rugulose, mericarps 7-10; seed solitary, ovoid, brown-black.

A weed of waste places and occasional in forests in most districts.

Fl. & Fr.: July-December

Lingampet (NZB), *BR* & *CPR* 9240; Mudheli RF (NZB), *BR* & *KH* 9047; Anantasagar RF (RR), *MSM* 10512; Narsapur – Hyderabad road (MDK), *BR* & *CP* 11212; Lingareddypally (MDK), *CP* 11713; Pakhal (WGL), *KMS* 11600 (MH); Mannanur (MBNR), *SRS* 109725 (BSID); Seethaphalmandi (HYD), *A.M.Saibaba* 5 (BSID); Ramavaram (KMM), *RR* 112661 (BSID).

Sida spinosa L., Sp. Pl. 683. 1753; Masters in Hook.f., Fl. Brit. India 1: 323. 1874; Gamble, Fl. Madras 1: 89. 1915; Borss., Blumea 14: 191. 1966; Paul in B.D.Sharma & Sanjappa, Fl. India 3: 292. 1993.

Woody much branched erect herbs-undershrubs, up to 60 cm high, stems and branches rough with spiny tubercles at the nodes below the petioles, stems, petioles and pedicels cinereous with minute stellate hairs. Leaves ovate-elliptic or oblong, 1.5-4 × 1-2.5 cm, basally 3-nerved, minutely stellate-tomentose above, cinereous beneath, base obtuse-rounded, margin serrate, apex acute. Flowers yellow with red tinge, axillary, solitary or 2-5 in clusters; calyx 3-5 mm across, campanulate, lobes free above the middle, triangular; corolla yellow or yellowish-white. Schizocarp wrinkled, exceeding calyx, mericarps 5; seed ovoid, slightly 3- angled.

Occasional weed of waste places and forests in Karimnagar, Warangal, Nalgonda, Mahabubnagar and Hyderabad districts.

Fl. & Fr.: September-March

Rechapalli (KRN), *GVS* 22283 (MH); Srinivasapur (NLG), *BVRS* 404; Pakhal RF (WGL), *KMS* 11600 (MH); way to Erloo (MBNR), *SRS* 104506 (BSID).

Note: Very similar to *Sida alba* L. but differs in having 1-3 spiny structures on the stem at the base of the petiole and mericarps with a pair at divergent apical awns.

THESPESIA Solander ex Correa *nom. cons.*

1. Shrubs, midrib glandular; seeds 8-15 in each locule .... **T. lampas**

1. Trees; midrib eglandular; seeds 2-4 in each locule:

    2. Leaves deeply cordate, green; pedicels 2-5 cm long, jointed near base; capsules indehiscent; seeds covered with simple hairs .... **T. populnea**

    2. Leaves shallowly cordate or subtruncate, copper coloured; pedicels 8-12 cm long, not jointed; outer layer of capsules dehiscent; seeds covered with short clavate or bulbous hairs .... **T. populneoides**

Thespesia lampas (Cav.) Dalzell ex Dalzell & A. Gibson, Bombay Fl. 19. 1861; Masters in Hook.f., Fl. Brit. India 1 : 345. 1874; var. **lampas**; Paul in B.D.Sharma & Sanjappa, Fl. India 3: 350. 1993; Anand Kumar, J. Econ. Taxon. Bot. 7: 665. 1985. *Hibiscus lampas* Cav., Diss. 3: 154. t. 56. f. 2. 1787; Gamble, Fl. Madras 1: 98. 1915.

Vern. : *Kondapatti, Adavi prathi*

Stellately tomentose shrubs, up to 2 m high. Leaves vary in shape, lower ones 3-palmately lobed; upper ones ovate, 5-8 × 3-7 cm, tomentose above, rusty woolly below, base cordate, margin entire, apex acuminate, linear nectar gland at base on midrib beneath, densely stellate-tomentose beneath, sparsely covered with stellate and short simple hairs above; petioles 1-12 cm long; stipules lanceolate to subulate. Flowers axillary, solitary or in racemes; epicalyx segments 5, free, usually subulate, caducous; calyx cupular, with 5 small subulate to deltoid segments; corolla bright yellow with dark purple centre, campanulate, petals obovate; staminal column 1-2 cm long, glabrous; ovary conical, densely hairy. Capsules ovoid, woody, pointed, villous; seeds numerous, ovoid, glabrescent, black.

Occasional in moist deciduous forests in Medak, Warangal and Mahabubnagar districts.

Fl. & Fr.: August-December.

Narsapur RF (MDK), *TP* & *CP* 14046; Appapur (MBNR), *TDCK* 13654; Appapur (MBNR), *TDCK* 13654; north of Pasra (WGL), *RKP* 108195 (BSID).

Thespesia populnea (L.) Sol. ex Correa, Ann. Mus. Natl. Hist. Nat. 9: 290. t. 8. 1807; Masters in Hook.f., Fl. Brit. India 1: 345. 1874. *p.p.;* Gamble, Fl. Madras 1: 101. 1915; Paul in B.D.Sharma & Sanjappa, Fl. India 3: 352. 1993. *Hibiscus populnea* L., Sp. Pl. 694. 1753.

Vern. : *Gangaravi, Munigangaravi, Gangarenu.*

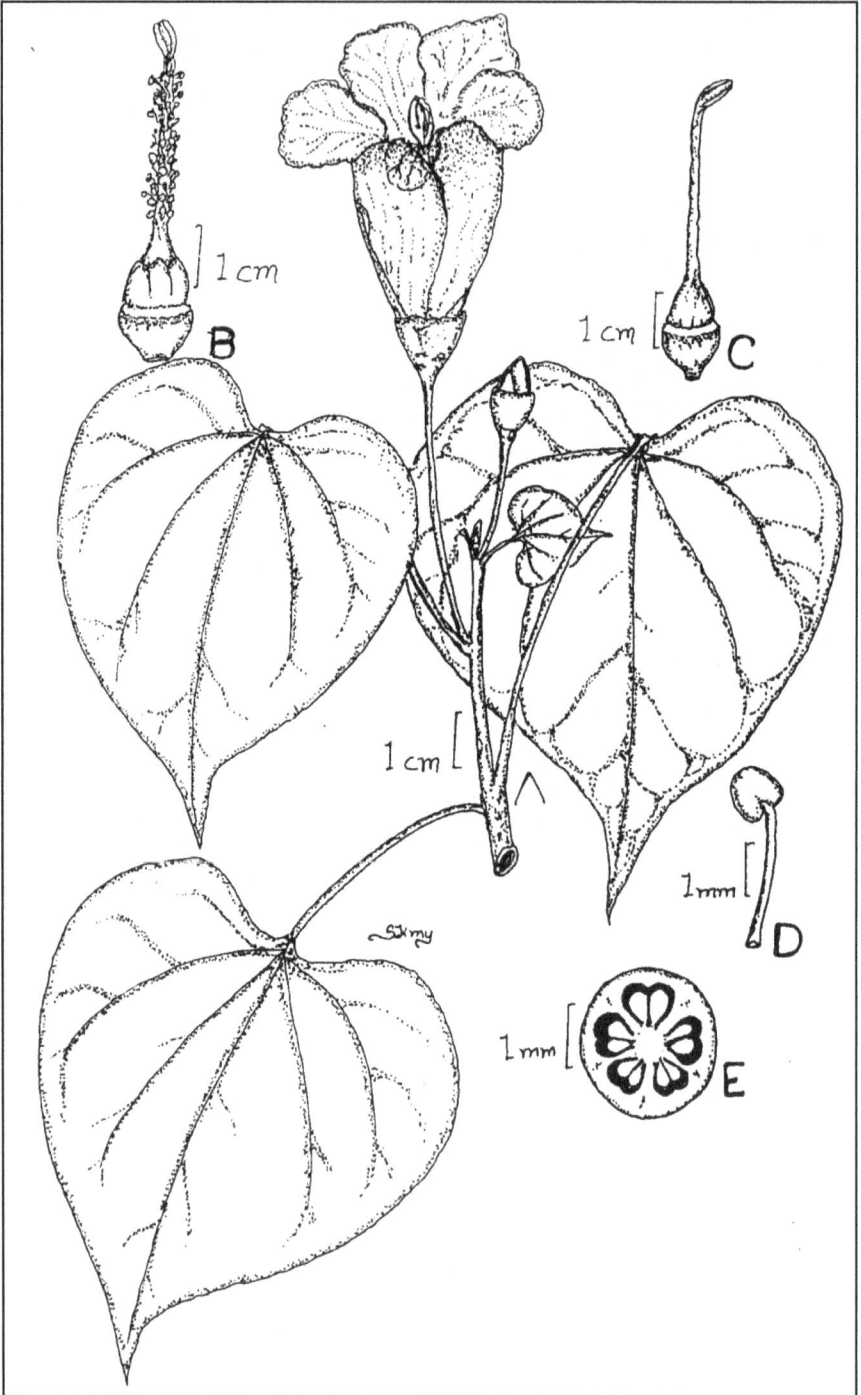

Figure 10: Thespesia populnea (L.) Sol. ex Correa
A. Twig, B. Pistil & Staminal column; C. Pistil, D. Stamen, E. Ovary c.s.

Trees, up to 10 m high; bark greyish black; branchlets clothed with peltate scales. Leaves ovate, not lobed, 6-10 × 4-8 cm, 7-nerved at base, stellate-tomentose beneath, base cordate, entire, apex acuminate; petioles 5-15 cm long; stipules 4-10 mm long, lanceolate to linear, caducous. Flowers solitary, axillary, pedicels 2-5 cm long; epicalyx segments 3, oblong to lanceolate, caducous; calyx cupular, accrescent and flattened in fruit; corolla light yellow with dark purple centre, ultimately reddish, broadly campanulate, petals obliquely obovate; staminal column 1.5-2.5 cm long; ovary globose to ovoid, stigmas connate to a clavate, 5-sulcate body. Capsule indehiscent, apically depressed, mucronate, glabrescent; seeds 2-4 pell cell, ovoid; covered with simple hairs.

Occasional on roadsides and often planted in gardens.

Fl. & Fr.: Through out the year.

Peddagutta (NZB), *BR* & *GO* 9202; Central University campus (RR), *MSM* 19208.

Thespesia populneoides (Roxb.) Kostel., Allg. Med. Pham. Fl. 5 : 1861. 1836; Fosberg & Sachet, Smith. Contrib. Bot 7: 10. 1972; Paul in Sharma & Sanjappa, Fl. India 3 : 353. 1993.

Tree, 10-15 m high; young branches covered with minute brownish peltate scales. Leaves subcordate or cordate with a very shallow broad sinus, 8-15 x 8-12 cm, often truncate, acuminate to subobtuse at apex, margins entire; petiole 6-12 cm long, stipules 10-12 mm long. Flowers drooping, axillary, solitary; pedicels 8-12 cm long, pendant, not jointed; epicalyx segments 3, 1 cm long; calyx with 5 minute teeth or entire, 12-14 mm long; corolla yellow with a dark purple center, turning to deep pink, staminal column 2.5 cm long. Capsule globose, exocarp dehiscing into 4 valves, endocarp hard, fibrous; seeds many, broadly obovoid, angular, with short, rusty, erect bulbous hairs.

Planted in gardens and along waysides as an avenue tree.

Fl. & Fr .: May-January.

URENA L.

Urena lobata L., Sp. Pl. 692. 1753; Masters in Hook. f., Fl. Brit. India 1 : 327. 1874; Gamble, Fl. Madras 1 : 92. 1915 subsp. **lobata**

Annual or perennial, erect, woody undershrubs, up to 1 m high, branchlets, petioles and pedicels stellate- tomentose, intermingled with simple hairs, ultimately glabrescent. Leaves broadly ovate or nearly orbicular, shallowly lobed, glands present on the lower surface, stellate tomentose above, woolly below, base cordate, margin crenate-dentate, apex obtuse-acute; petioles 0.5-12 cm long; stipules 2-4 mm long, linear to lanceolate. Flowers axillary, 2-3 in a cluster or solitary; pedicels 1-5 mm long; epicalyx segments closely enveloping calyx and shortly adnate to it, linear to lanceolate; calyx tubular to somewhat prominent with nectarines at base; corolla pink with a purple centre, 2-3 cm across, petals obovate. Schizocarp globose, mericarps 5, 3-gonous, glochidiate with apical deflexed prongs; seed reniform, angular, glabrescent, brown.

Frequent near villages, often along the agricultural drains in most districts.

Fl. & Fr.: October – February.

Akkannapet to Laxmipuram (MDK), *RG* 106707.

Key to Subspecies

1. Leaves angled or shallowly lobed; epicalyx cupular in fruit,
   appressed to mericarps; segments narrowly triangular .... subsp. **lobata**

1. Leaves palmatilobed or palmatifid; epicalyx spreading or
   reflexed in fruit; segments linear to lanceolate .... subsp. **sinuata**

Urena lobata L. subsp. **sinuata** (L.) Borss., Blumea 14: 142. 1966. var. **glauca** (Blume)
   Borss., Blumea 14: 144. 1966; Paul in B.D.Sharma & Sanjappa, Fl. India 3: 382.
   1993. *Urena sinuata* L., Sp. Pl. 692. 1753; Gamble, Fl. Madras 1: 92. 1915. *U. lappago*
   Smith var. *glauca* Blume, Bijdr. Fl. Ned. Ind. 2: 65. 1825.

Similar to typical variety, but leaves are angular to palmatilobed.

A weed of roadsides, waste places and forests.

Fl. & Fr. : July- December

Jannaram (ADB), *TP* & *PVP* 4179; Bersaipet (ADB), *GO* & *PVP* 4599; Mosra (NZB),
TP & *BR* 6063; Mohammadabad RF (RR), *MSM* & *KH* 10549; Pullaechelma to
Farahabad road (MBNR), *SRS* 109197 (BSID); near Ramappa temple (WGL), *RKP*
105355 (BSID); Pasra (WGL), *RKP* 108146 (BSID); Yenkur (KMM), *RR* 108546 (BSID).

BOMBACACEAE

1. Trunk unarmed; calyx 5-cleft; stamens monadelphous .... **Adansonia**

1. Trunk armed at least when young; calyx campanulateate;
   stamens pentadelphous :

   2. Flowers scarlet red; ultimate filament with a single anther .... **Bombax**

   2. Flowers yellowish-white; filament with 2-3 anthers .... **Ceiba**

ADANSONIA L.

Adansonia digitata L., Sp. Pl. 1150. 1753; Masters in Hook.f., Fl. Brit. India 1 : 348.
   1874; Gamble, Fl. Madras 1 : 102. 1915; Nayar & Biswas in B.D.Sharma & Sanjappa,
   Fl. India 3: 404. 1993.

Tall trees, up to 20 m high, up to 10 m in diam., widely spreading branches, bark
smooth, greyish. Leaves digitate, leaflets 3-9, 2.5-12 × 1.2-5 cm, sessile or subsessile,
obovate-oblong or elliptic-oblong, densely silk- brown-hairy, pubescent beneath,
glabrous above. Flowers white, solitary, axillary, long peduncled, bracteoles 2; calyx
lobes triangular, tomentose outside; petals obovate to flabelliform, adnate to the base
of staminal tube; staminal tube 5-7 cm long, cylindrical to conical, divided into
numerous, slender filaments as along as tube; ovary 5-10-loculed, hirsute, styles 1-
1.5 cm long, stigmas 5-10-lobed, lobes oblong, radiating. Capsule 20-20 cm long,
oblong, woody, indehiscent, with pulp inside, pale brown, densely hairy; seeds
reniform, embedded in fleshy pulp.

Introduced tree, planted near tombs and gardens.

Fl.: April – May; Fr.: September - October.

Hamsapally extension (MDK), *RG* 104078 (BSID).

BOMBAX L. *nom. cons.*

Bombax ceiba L., Sp. Pl. 511. 1753; Nayar & Biswas in B.D.Sharma & Sanjappa, Fl. India 3: 398. 1993. *B. malabaricum* DC., Prodr. 1 : 479. 1824; Masters in Hook.f., Fl. Brit. India 1 : 349. 1879; Gamble, Fl. Madras 1 : 99. 1915. *Salmalia malabarica* (DC.) Schott & Endlicher, Melet. Bot. 35. 1832.

Vern. : *Burugu.*

Deciduous trees, up to 30 m high, trunk straight, older branches horizontal, whorled, prickles conical; bark grey, glabrous, young branches warty, puberulous, glabrescent. Leaves crowded at the ends of the branches, digitately 5-7-foliolate, petioles 12-25 cm long; leaflets elliptic-lanceolate or oblong or ovate-lanceolate, 10-18 × 5-8 cm, terminal one larger, basal one smaller, 10-12-nerved, base cuneate, margin entire, apex acuminate, caudate or acute; petiolules 2-2.5 cm long. Flowers blood red, sessile, crowded at ends of leafless branchlets, solitary, paired or clustered; pedicels 1-2 cm long; calyx campanulate, irregularly 2-5-lobed, lobes coriaceous, glabrous to sparsely puberulous outside, silky inside, falling of with corolla and stamens; petals 5, obovate to elliptic-obovate, rarely oblong, recurved, fleshy, tomentellous outside; stamens 65-80 in 6 bundles in 2 series, the central bundle with 15 stamens of which 5 longer and 10 shorter and 5 bundles in the outer series, each with 10 stamens; staminal tube short; ovary conical, styles ca 6 cm long, stigmas 5-fid, lobes spreading. Capsule oblong, 9-13 x 3-6 cm, loculicidal, 5- valvular; valves silky inside; seeds numerous, pyriform, smooth, dark brown, embedded in creamy white silky fibres.

Occasional in forests of all districts, often cultivated.

Fl.: February - March; Fr.: April – May.

Kadam RF (ADB), *GO* & *DAM* 5193; Gowraram RF (NZB), *TP* & *BR* 6386; Mudheli RF (NZB), *BR* & *KH* 7131; Mohammadabad RF (RR), *MSM* 19228; Narsapur RF (MDK), *BR* & *CP* 14102; Pakhal (WGL), *ANH* 15918; Kadurupalli (KRN), *SLK* 70725 (NBRI); Srinivasapur (NLG), *BVRS* 403 (NU); Saleswaram (MBNR), *BR* & *SKB* 30729 (BSID); way to Gangaram (WGL), *PVS* 84005 (BSID).

CEIBA Mill. emend Gaertn.

Ceiba pentandra (L.) Gaertn., Fruct. Sem. Pl. 2: 244. t. 133. 1791; Nayar & Biswas in B.D.Sharma & Sanjappa, Fl. India 3: 400. 1993. *Bombax pentandrum* L., Sp. Pl. 511. 1753. *Eriodendron pentandrum* (L.) Kurz, J. Asiat. Soc. Bengal 43: 113. 1874; Gamble, Fl. Madras 1: 100. 1915.

Vern. : *Tella buruga.*

Tall trees, up to 20 m high, trunk straight, tapering, prickly when young, up to 30 m high, branches horizontal, verticillate. Leaves digitate, 6-9- foliolate, leaflets usually oblanceolate or elliptic or oblong, 3.5-10.5 × 1.5-2.5 cm, 10-15-nerved, base obtuse-cuneate, margin entire, apex acuminate; petioles 14-20 cm long. Flowers clustered at the ends of branchlets before the leaves or in the axils of leaves, white or yellowish; pedicels 2-3 cm long, stout, glabrous; calyx campanulate, irregularly 5-lobed, glabrous outside, silky inside; petals usually 5, rarely 6, connate at base, oblong to oblanceolate, imbricate; staminal tube divided into 5 phalanges, each dividing again into usually 3 filiform branches bearing 2-3 infractose, 1-locular, twisted anthers; sometimes one petaloid stamen present; ovary globular or ovoid, styles 1 cm long, dilated above the staminal column; stigmas capitate or obscurely 5-lobed. Capsule ellipsoidal, indehiscent or tardily dehiscing into 5 valves; seeds many, subglobose.

Occasional in the forests, often cultivated for its silk cotton.

Fl.: December – February; Fr.: February-May.

Pedduti (MBNR), *TDCK* 16821; Tirmalagiri village (NLG), *BVRS* 463 (NU).

STERCULIACEAE

1. Herbs or subshrubs :
    2. Stamens 15- 20, in clusters of 3 each; flowers red; cultivated .... **Pentapetes**
    2. Stamens 5; flowers other than the red; wild :
        3. Petals appendaged, purple; capsule spinescent .... **Byttneria**
        3. Petals not appendaged, yellow or pink; capsule smooth :
            4. Calyx shallowly or obscurely lobed; flowers pink .... **Melochia**
            4. Calyx deeply lobed; flowers yellow :
                5. Petals clawed; ovary1-celled, bracteoles deciduous ...... **Waltheria**
                5. Petals not clawed; ovary 5-celled; bracteoles persistent .... **Melhania**
1. Shrubs or trees :
    6. Petals absent; flowers polygamous :
        7.   Flowers covered with orange-red tomentum; ovules 2 per cell; follicles thin-walled .... **Firmiana**
        7. Flowers not covered with orange-red tomentum, ovules numerous per cell; follicles woody:
            8. Follicles glabrous, or if villous hairs not irritant .... **Sterculia**
            8. Follicles densely hairy, hairs irritant .... **Kavalama**
    6. Petals present, flowers bisexual :
        9. Follicular fruit spirally twisted; flowers crimson; anthers divergent and superposed .... **Helicteres**
        9. Capsular fruit entire; flowers other than crimson; anthers parallel and collateral:
            10. Flowers yellow :

11. Leaves orbicular-ovate; petals not appendaged;
    stamens numerous; capsules pyriform, not
    tubercled; seeds winged .... **Eriolaena**

11. Leaves oblong-lanceolate; petals appendaged;
    stamens 15; capsule subglobose, tubercled;
    seeds not winged .... **Guazuma**

10. Flowers not yellow:

12. Flowers white .... **Pterospermum**

12. Flowers rosy .... **Kleinhovia**

BYTTNERIA Loefl. *nom. cons.*

Byttneria herbacea Roxb., Pl. Coromandel 1: t. 29. 1795; Masters in Hook.f., Fl. Brit.
India 1 : 376. 1874; Gamble Fl. Madras 1 : 112. 1915; Malick in B.D.Sharma &
Sanjappa, Fl. India 3: 412. 1993.

Prostrate to ascending herbs, up to 70 cm long with perennial root stock;
branchlets minutely stellate-pubescent. Leaves ovate or lanceolate, 1-4 × 1-3 cm,
chartaceous, basally 3-nerved, with a linear gland at base on lower side of midrib,
puberulous above, glabrous below, base cordate or rounded, margin dentate, apex
acuminate, petiole 1 cm long; stipules 2 mm long, subulate. Flowers small, in axillary
and/or terminal umbellate cymes; pedicel ca 4 mm long, subulate; bracts 2-6; sepals
connate at base, narrowly ovate-lanceolate; petals maroon coloured, claw slender,
limb subulate with hood and 2-fid appendages which enter inside the staminal cup
covering the stamens; stamens in inner series of the staminal cup, staminodes ovate;
ovary sparsely hairy, styles 1 mm long, stigmas lobed. Capsule globose, covered with
small subulate prickles with five 1-seeded lobes; seeds ovoid, angular, 5 mm.

Common in Adilabad, Khammam, Mahabubnagar and Warangal districts as
forest undergrowth and shady places.

Fl. & Fr.: Through the year.

Jaipur RF (ADB), *GO* & *PVP* 4957; Westernside of Pakhal lake (WGL), *KMS*
13132 (MH); Mannanur (MBNR), *VBH* 86602 (BSID); Poochavaram RF (KMM), *RCS*
102481 (BSID); way to Sidharam (KMM), *PVS* 84074 (BSID); Pasra (WGL), *RKP* 108227
(BSID).

ERIOLAENA DC.

1. Involucral bracts multisect or laciniate:

2. Flowers 2-3, in racemes .... **E. lushingtonii**

2. Flowers several in cymes .... **E. hookeriana**

1. Involucral bracts entire or nearly so .... **E. quinquelocularis**

Eriolaena hookeriana Wight & Arn., Prodr. Fl. Ind. Orient. 70. 1834; Masters in Hook.f.,
Fl. Brit. India 1 : 370. 1874; Gamble, Fl. Madras 1 : 110. 1915; var. **hookeriana;**
Malick in B.D.Sharma & Sanjappa, Fl. India 3: 415. 1993.

Vern. : *Nar-botku.*

Shrubs or trees, up to 10 m high, bark smooth, greyish, peeling in flakes; herbaceous portions stellate- tomentose. Leaves broadly ovate or cordate, 8-14 × 7-14 cm, basally 7-nerved, thinly stellate hairy or glabrescent above, rusty-tomentose beneath; base cordate, margin irregularly dentate, apex acuminate; petioles 2-11 cm long, rather stout. Flowers yellow, axillary in many- flowered cymes, peduncles up to 15 cm long, stout; involucral bracts shorter than calyx, segments linear; pedicels up to 4 cm long; sepals linear-lanceolate; petals obovate, broad towards the apex, claws densely pubescent within; staminal column 1.5-2 cm long, antheriferous throughout its length; ovary ca 5 mm long, stigmas 8-10-lobed. Capsule 2.5-4 cm long, ovoid, subacute at apex, 10-valved, valves woody, thinly to densely stellate hairy; seed wings papery, brown.

Rare in forests of Adilabad, Warangal, Mahabubnagar and Karimnagar districts.

Fl. & Fr. : March-September

Neelwai RF (ADB), *TP* & *GO* 5423; Mannanur (MBNR), *BR, MSR* & *SKB* 30717; Aklaspur (KRN), *GVS* 25602.

Eriolaena lushingtonii Dunn, Bull. Misc. Inform. Kew 1915: 88. 1915; Gamble, Fl. Madras 1: 110. 1915; Malick in B.D.Sharma & Sanjappa, Fl. India 3: 417. 1993.

Vern.: *Mallabatuku.*

Trees, up to 5 m high, young branches stellate- pubescent. Leaves orbicular, 4-8 cm in diam., cordate at base, margin irregularly dentate, apex acute, regularly shortly dentate, stellately hairy above, white tomentose beneath; petioles half to one-third of the length of blade. Flowers yellow, in axillary, 2-3-flowered racemes, about 8 cm long, flowering buds oblong, constricted at middle; involucral bracts 4-6 mm long, multisect, caducous; calyx 5-partite, 2 cm long, pubescent within, tomentose without; petals 5, obovate, clawed at base, tomentose; stamens 10-12, staminal column with series of fertile stamens, staminodes absent; ovary sessile, ovoid, styles simple, pubescent, stigmas minutely 5-lobed. Capsule about 4 cm long, ovoid, woody, pubescent, loculicidally dehiscent; seeds many, winged on one side top.

Open slopes of moist deciduous forests in Nallamalais.

Fl. & Fr. : June-August.

Mannanur (MBNR), *BS, MSR, SKB* & *BSS* 30660, *VBH* 86610 (BSID); Saleswaram (MBNR), *BR* & *SKB* 30722; Mallelatheertham (MBNR), *MSR, SKB* & *BSS* 30708.

Eriolaena quinquelocularis (Wight & Arn.) Wight, Icon. Pl. Orient. t. 882. 1847; Masters in Hook.f., Fl. Brit. India 1: 371. 1874. *Microchlaena quinquelocularis* Wight & Arn., Prodr. Fl. Ind. Orient. 71. 1834; Malick in B.D.Sharma & Sanjappa, Fl. India 3: 418. 1993.

Small trees, up to 10 m high. Leaves 7-20 × 5-15 cm, orbicular, cordate at base, margin coarsely crenate-serrate, acute to subacuminate at apex, dotted with small tuft of stellate hairs above, softly tomentose beneath, nerves 7 at base, raised beneath;

petioles up to 9 cm long; stipules caducous. Flowers white, 2-3 in axillary peduncled cymes; peduncles often longer than leaves, usually at the ends of the branches; pedicels 2-3 cm long, quadrangular, jointed above the middle; involucral bracts short, entire or rarely lobed, caducous; sepals 15-20 × 3-4 mm, linear-lanceolate, acute, pubescent on both surfaces, glandular at base inside; petals 16-20 × 2-8 mm, obovate-oblong, styles longer than staminal column, stigmas 5-lobed, lobes revolute. Capsules 5-10 × 1-1.5 cm, ovate-lanceolate, beaked, 5-loculed, 5-10-valved, usually silky villous; seeds numerous, winged papery.

Occasional in deciduous forests.

Fl.: February- August; Fr.: April-December

Mahadevpur RF (KRN), *A.H.Naqvi* 696 (KU); Aklaspur (KRN), *GVS* 20226; Medaram RF (WGL), *VSR* 4728 (KU).

FIRMIANA Marsili

Firmiana colorata (Roxb.) R. Br. in Benn. & R.Br., Pl. Jav. Rar. 235. 1844; Gamble, Fl. Madras 1: 107. 1915; Kosterm., Reinwardtia 5. 386. 1961; Malick in B.D.Sharma & Sanjappa, Fl. India 3: 420. 1993. *Sterculia colorata* Roxb., Pl. Coromandel t. 25. 1795; Masters in Hook.f., Fl. Brit. India 1: 359. 1874.

Vern. : *Karaka, Maraka, Karu boppaja.*

Trees, up to 10 m high, trunk erect, buttressed, bark smooth, grey or greyish green, branchlets stellate-tomentose to velvety. Leaves crowded at the ends of branchlets, alternate, palmately 3-lobed, 6-15 × 7-18 cm, basally 7-nerved, woolly below, base cordate, margin entire and faintly serrulate, apex obtuse or acute to acuminate; petioles 10-30 cm long, pulvinate. Flowers orange red, in terminal panicles, densely covered with minute stellate hairs and scales; calyx funnel-shaped, slightly curved, inflated towards apex, teeth acute, with a ring of glistening bundles of long strigose hairs above base inside; staminal column with 10-30 sessile yellow anthers, androgynophore exerted; ovaries 5, flask-shaped. Follicles 5, stipitate, oblong, thin-walled, ca 4.5 × 1.5 cm, strongly nerved; seeds 2 per follicle, ovoid, compressed, attached to the margins of follicles, wrinkled or smooth, yellow.

Occasional in the forests of Mahabubnagar district.

Fl.: February – April; Fr.: April - June.

Appapur (MBNR), *BR* & *SKB* 30594.

GUAZUMA Mill.

Guazuma ulmifolia Lam., Encycl. 2: 52. 1759; Robyns, Ann. Missouri Bot. Gard. 51: 102. 1964; Malick in B.D.Sharma & Sanjappa, Fl. India 3: 424. 1993. *Theobroma guazuma* L., Sp. Pl. 782. 1753. *Guazuma tomentosa* Kunth, Nov. Gen. Sp. 5: 320. 1823; Masters in Hook.f., Fl. Brit. India 1: 375. 1874; Gamble, Fl. Madras 1: 111. 1915.

Large trees, up to 15 m high, branchlets stellate-tomentose to velvety; bark fissured in older parts. Leaves oblong-lanceolate, 10-16 × 3-6 cm, pubescent beneath, scabrid or glabrescent above, base subcordate, oblique, margin serrulate, apex acute to acuminate; petioles 1-2 cm long, terete or subterete. Flowers numerous, yellow in terminal and/or axillary panicles; sepals spathaceous at first, divided into lobes later, lobes oblong-lanceolate, slightly concave, ultimately reflexed, tomentose outside; petals obovate, cucullate, claw narrow, lamina with 2, ca 4 mm long, strap-shaped forked appendages; staminal cup bearing the stamens and petaloid stminodes, fimbriate; ovary sessile, globose with fine tubercles; styles 5, connate at base. Capsules subglobose, woody, tubercled, indehiscent; seeds numerous, small, globose.

An introduced tree, often found run wild, usually in the vicinity of towns and villages.

Fl.: March – September; Fr.: June – February.

Chandoor (NZB), *BR* & *GO* 9204; Central University campus (RR), *MSM* 19207; Hyderabad, *KMS* 5998 (MH).

HELICTERES L.

Helicteres isora L., Sp. Pl. 963. 1753; Masters in Hook.f., Fl. Brit. India 1: 365. 1874; Gamble, Fl. Madras 1: 107. 1915; Malick in B.D.Sharma & Sanjappa, Fl. India 3: 426. 1993.

Shrubs or small trees, up to 5 m high, branchlets apically stellate- tomentose. Leaves obovate or elliptic-oblong, 6-10 × 4-8 cm, scabrous above, pubescent beneath, base roundish or cordate, margin crenate-serrate, apex acuminate, round or subacute; petioles 1-2.5 cm long; stipules up to 1 cm long, subulate, deciduous. Flowers crimson, axillary, solitary or in clusters; calyx gibbous, laterally compressed, somewhat 2-lipped, densely stellate-hairy; petals crimson, reflexed, 2 lower shorter and broader than the 3 upper ones, claws winged; staminal tube 3-4 cm long, slightly bent on one side at the tip, exserted, stamens 10, surrounding ovary and alternating in pairs with 5 minute scaly staminodes attached to the staminal tube; ovary 5-lobed, styles united, as long as the ovary, deflexed. Follicles 5, beaked, spirally twisted or straight, cylindric, pubescent; seeds many, reddish brown, triangular, minutely tuberculate.

Common in forests of all districts.

Fl.: April-December; Fr.: October – January.

Kadam river bank (ADB), *TP* & *PVP* 4134; Sirpur RF (ADB), *GO* & *PVP* 4902; Mamidipally RF (NZB), *BR* & *KH* 6437; Narsapur RF (MDK), *TP* & *CP* 14034; Central University campus (RR), *MSM* & *TP* 15132, *MSM* 15165; Pakhal RF (WGL), *KMS* 11602 (MH); Rechapalli (KRN), *GVS* 20155; Durachipalli (NLG), *BVRS* 37; Nagarjunasagar (NLG), *BVRS* 132 (NU); Bhadrachalam RF (KMM), *RCS* 102429 (BSID); Umamaheswaram RF (MBNR), *SRS* 109142 (BSID); Pasra (WGL), *RKP* 108104 (BSID).

KAVALAMA Raf.

Kavalama urens (Roxb.) Raf., Sylv. Tellur. 72. 1838. *Sterculia urens* Roxb., Pl. Coromandel t. 24. 1795; Masters in Hook.f., Fl. Brit. India 1 : 355. 1874; Gamble, Fl. Madras 1 : 106. 1915; Malick in B.D.Sharma & Sanjappa, Fl. India 3: 470. 1993.

Vern. : *Tabsu, Konda tamara, Kovila chettu, Thapsi chettu, Yerra poliki.* English: Gum Karaya tree

Soft wooded, deciduous trees, up to 15 m high with white papery outer bark, exfoliating in large thin irregular flakes, inner bark fibrous; trunk erect, branches spreading, marked with large scars, wood reddish brown. Leaves palmately 3-or 5-lobed crowded at the tips of the branchlets, 12-15 × 10-15 cm, glabrous above, velvety below, base cordate, margin entire, apex acuminate; petioles 8-9.5 cm long, terete; stipules caducous. Flowers greenish yellow in terminal panicles; panicles clothed with dense sticky tomentum of glandular stellate hairs; male and bisexual flowers are mixed; Male flowers: stamens 10, Female flowers: ovaries 2 mm in diam. on a ca 3 mm gynandrophore with sterile anthers at base; styles as long as ovary, stigmas radiating. Follicles 5, oblong, radiating, yellow, pubescent, with stinging hairs; seeds 3-6, oblong, black, glossy.

A common species in deciduous forests, on the dry, steep, rocky slopes in Adialabad, Nizamabad, Karimnagar, Warangal, Ranga Reddi and Mahabubnagar districts.

Fl.: December – March; Fr.: April onwards dehiscing by June.

Mahaboobghat (ADB), *GO* & *PVP* 4270; Laxmapur (NZB), *BR* 6257; Dharmapur – Anantasagar (RR), *MSM* & *KH* 11014; Mannanur (MBNR), *TDCK* 15371; Pakhal (WGL), *KMS* 11643 (MH); Bhupatipur (KRN), *GVS* 22472 (MH).

KLEINHOVIA L.

Kleinhovia hospita L., Sp. Pl. 1365. 1763; Masters in Hook.f. Fl. Brit. India 1: 364. 1874; Gamble, Fl. Madras 1: 113. 1915.

Small trees. Leaves broadly ovate or suborbicular, base cordate, apex acuminate, glabrous; petioles 6-9 cm long; stipules linear. Flowers rosy, in large terminal panicles; sepals 5, linear, valvate; petals 5, rosy, linear, shorter than sepals; staminal-tube 5-lobed, each with 3 anthers; ovary 5-celled, one ovule per cell. Capsule membranous, inflated.

Cultivated in parks and gardens.

Fl. Fr.: July – October.

MELHANIA Forssk.

1. Leaves lanceolate or elliptic, 2-5 × 0.5-1.5 cm; bracteoles linear; seeds 2-4 in each cell, seeds ovoid .... **M. incana**
1. Leaves ovate, 12-10 x 6-10 cm; bracteoles ovate, seeds many in each cell; seeds oblong .... **M. hamiltoniana**

Melhania hamiltoniana Wall., Pl. Asiat. Rar. 1: 69. t. 77. 1830; Masters in Hook.f. Fl. Brit. India 1: 373. 1874; Gamble, Fl. Madras 1: 113. 1915; Malick in B.D.Sharma & Sanjappa, Fl. India 3: 439. 1993.

Under shrubs, up to 1 m high, with spreading tomentose branches. Leaves ovate, 12-10 × 8-10 cm, pubescent on both surfaces, white beneath, base cordate, margin serrate, unequally toothed, apex acute-acuminate; petioles 1.5-2.5 cm long; stipules linear, setaceous, tomentose. Flowers orange red, axillary, peduncles 3-flowered, bracteoles recurved at the edges; involucral bracts 3, ovate-lanceolate, recurved along margins; sepals linear-oblong to lanceolate, cuspidate, stellate tomentose outside, glabrous inside; petals yellow, obovate, oblique; staminal tube 2 mm long bearing stamens and staminodes; ovary subglobose, hairy, 5-loculed, styles 1 cm long with 5 stigmatic branches. Capsules ovoid, obscurely truncated at the top, villous, shorter than the calyx, many-seeded; seeds oblong, somewhat 4-sided, tubercled.

Occasional in Hyderabad and Nalgonda districts, on rocky hills.

Fl. & Fr.: August-September

Nagarjunakonda (NLG), *KTH* 9642 (CAL); Nagarjunasagar (NLG), *BVRS* 71 (NU); Kambalpally (NLG), *BVRS* 185 (NU).

Melhania incana Heyne ex Wight & Arn., Prodr. Fl. Ind. Orient. 68. 1834; Masters in Hook.f., Fl. Brit. India 1: 372. 1874; Gamble, Fl. Madras 1: 113. 1915; Malick in B.D. Sharma & Sanjappa, Fl. India 3: 439. 1993.

Erect suffruticose herbs, up to 30 cm high, branchlets terete, young ones stellately tomentose. Leaves lanceolate or elliptic, 2- 5 × 0.5-1.5 cm, glabrescent above, grey pilose below, base subcordate, margin crenate-serrate, apex obtuse; petioles 1-15 cm long, slender, stipules 5-7 mm long, filiform. Flowers small, yellow in axillary cymes; involucral bracts 3, linear-oblong, slightly shorter than sepals, tomentose; sepals lanceolate to ovate-oblong, slightly curved, stellate-tomentose outside, glabrous inside; petals yellow, orbicular to obovate, stamens 2 mm long, staminodes up to 4 mm long, staminal tube 1 mm long; ovary 5-loculed, styles 3 mm long with 5 stigmatic branches. Capsules globose-ovoid, hairy; seeds ovoid, angled, tubercled.

Occasional in wastelands and open forests in Karimnagar, Mahabubnagar and Nalgonda districts.

Fl. & Fr.: August-December

Nagarjunakonda (NLG), *KTH* 9696 (CAL); Nagarjunasagar (NLG), *BVRS* 72; Kambalapllay (NLG), *BVRS* 203 (NU).

MELOCHIA L.

Melochia corchorifolia L., Sp. Pl. 675. 1753; Masters in Fl. Brit. India 1: 374. 1874; Gamble, Fl. Madras 1: 110. 1915; Malick in B.D.Sharma & Sanjappa, Fl. India 3: 441. 1993.

Erect branching herbs or undershrubs, up to 70 cm high, thinly stellate-hairy. Leaves variable in shape, ovate, ovate-lanceolate, oblong-ovate or suborbicular, rarely obscurely 3- lobed, 1-5.5 × 0.5-3 cm, glabrous above, thinly stellate-hairy below; base obtuse or truncate, margin serrate, apex acute; petiole up to 3.5 cm long. Flowers pale to deep pink in terminal peduncled, capitate cymes, surrounded by 4-5 bracteoles; calyx ciliate, hairy outside, lanceolate; petals pale or deep pink, obovate to spathulate; staminal cup somewhat spindle-shaped, ovary 5-loculed, hairy. Capsules sub-globose, hispid; seeds angular, black. A weed of roadsides, waste places and open forests in most districts.

Fl. & Fr.: July – April.

Bersaipet (ADB), *GO* & *PVP* 4626; Mosra forest nursery (NZB), *TP* & *BR* 6075; Narsapur RF (MDK), *TP* & *CP* 11994; Dharmapur (RR), *MSM* 10486; Pakhal (WGL), *KMS* 11615 (CAL); Rechapalli (KRN), *GVS* 25695; Banks of Pakhal lake (WGL), *KMS* 11615; Tirmalgiri (NLG), *BVRS* 474 (NU); Pasra RF (WGL), *RKP* 108200 (BSID); Palempet (WGL), *RKP* 105337 (BSID); Bhadrachalam RF (KMM), *RCS* 99074 (BSID).

PENTAPETES L.

Pentapetes phoenicea L., Sp. Pl. 698. 1753; Masters in Hook.f., Fl. Brit. India 1: 371. 1874; Gamble, Fl. Madras 1 : 113. 1915; Malick in B.D.Sharma & Sanjappa, Fl. India 3: 443. 1993.

Herbs, up to 1.5 m high, branches glabrous or with a few scattered stellate hairs. Leaves simple, hastate, lanceolate or oblong, 9- 12 × 1-1.5 cm, glabrescent, base obtuse-rounded; margin 2-serrate or crenate, apex gradually tapering; petioles 1-3 cm long; stipules linear-subulate, equaling petioles. Flowers red, 1-2, axillary; peduncles jointed near the flower, opening at noon and closing at the following dawn; sepals connate at base, lobes lanceolate; petals obovate; staminal cup bearing 5 groups of 3 stamens alternating with 5 staminodes, staminodes as long as petals, linear-spathulate, glandular on inner surface; ovary 5-locular, hairy. Capsule subglobose, 1-2 cm broad, beaked, hairy, half the length of the persistent calyx; seeds angular, rough, 8-12, 2-serrate in each cell.

Ornamental, also found as an escape in waste places.

Fl. & Fr. : August-December

PTEROSPERMUM Schreb. *nom. cons.*

1. Capsules terete .... **P. suberifolium**
1. Capsules angular:
    2. Leaves 5-13× 3-8 cm; flowers up to 6 cm long; capsules pyriform, horned, glabrous or tomentose; locules 8-10-seeded .... **P. xylocarpum**
    2. Leaves 23-38× 14-30 cm; flowers ca 15 cm long; capsules oblong, tubercled, locules more than 10-seeded .... **P. acerifolium**

Pterospermum acerifolium (L.) Willd., Sp. Pl. 3: 729. 1800; Masters in Hook.f., Fl. Brit. India 1: 368. 1874 *p.p.*; Chandra in B.D.Sharma & Sanjappa, Fl. India 3: 448. 1993. *Pentapetes acerifolia* L., Sp. Pl. 3: 698. 1753.

Vern.: *Kanak-champa.*

Trees, 12-15 m high; bark smooth, greyish; branchlets clothed with rusty stellate or floccose pubescence. Leaves 23-38 × 14-30 cm, broadly ovate to elliptic-oblong, cordate and subpeltate or often peltate at base, apex acute, margin coarsely serrate, sinuately lobed at apex, coriaceous, glabrous above, grey or white tomentose beneath, palmately 7-12-nerved at base; petioles 7-9 (-30) cm long. Flowers ca 15 cm long, axillary, solitary or 2-3-flowered cymes, fragrant, 10-15 cm across; sepals 5, fleshy, 8-11 × 0.6-0.9 cm, linear, connate at base, rusty stellate-tomentose outside, silky inside, caducous; petals 5, white, 7-9.5 cm long, linear-revolute; staminodes 6-8.5 cm long, club-shaped; styles 5-6.5 cm long; stigmas club-shaped. Capsules 10-20 × 3-6 cm, woody, 5-angled, covered with brown tubercles; seeds 1-2 × 1-1.5 cm, obliquely ovoid, many, in 2 rows, winged, wings membranous, 4-7 × 1-1.4 cm.

Often planted in gardens; escape elsewhere.

Fl.: March- November; Fr.: July – December.

Pterospermum suberifolium (L.) Lam., Tabl. Encycl. 3 : 136. t. 176. f. 2. 1823; non Willd 1880; Masters in Hook.f., Fl. Brit. India 1 : 367. 1874; Gamble, Fl. Madras 1 : 110. 1915; Chandra in B.D.Sharma & Sanjappa, Fl. India 3: 453. 1993. *Pentapetes suberifolia* L., Sp. Pl. 678. 1753. *Pterospermum canescens* Roxb., Fl. Ind. 3: 162. 1832; Pullaiah & Chennaiah, Fl. Andhra Pradesh, 1: 142. 1997

Vern.: *Lolagu, Tada.*

Moderate sized pretty trees, up to 10 m high, branchlets stellate-tomentose, with light red wood. Leaves oblong-obovate, 4-12 × 2.5-5 cm, coriaceous, prominently nerved, woolly below, glabrous, dark green above, base oblique, rounded or subcordate, margin coarsely toothed or somewhat lobed towards the apex, apex acuminate, petiole 1 cm long. Flowers white, fragrant, axillary, solitary, pedicels 5-10 mm long, stout, jointed; bracteoles 5 mm long, linear, caducous; sepals 5, linear, revolute, stellate-tomentose outside, silky hairy inside; petals 5, linear, revolute; stamens 15, staminodes ca 1 cm long, filiform; ovary ovoid, silky villous. Capsule 3-4.5 × 2.5 cm, persistent, tapering at both ends, 4-5-valved, valves covered with dense fluffy pubescence; seeds 2-4 in each cell, compressed, oboconical, 4 mm, winged.

Occasional in deciduous forest in Mahabubngar district.

Fl. & Fr. : Through the year.

Farahabad (MBNR), *MVS & DV* 38634, *MVS & MB* 38024.

Pterospermum xylocarpum (Gaertn.) Santapau & Wagh, Bull. Bot. Surv. India 5: 108. 1963; Chandra in B.D.Sharma & Sanjappa, Fl. India 3: 454. 1993. *Velaga xylocarpa* Gaertn., Fruct. 2 : 245. t. 133. 1791. *Pterospermum heyneamum* Wall. ex Wight & Arn., Prodr. Fl. Ind. Orient. 69. 1834; Masters in Hook.f., Fl. Brit. India 1 : 369. 1874; Gamble, Fl. Madras 1 : 106. 1915.

Vern.: *Tada, Loluku, Lodgu, Duddika.*

Trees, up to 15 m high; bark greyish brown, smooth; young branchlets rusty tomentose. Leaves variable in shape and size, oblong-obovate, thick, coriaceous, 5-13 × 3-8 cm, glabrous above, rusty below, base oblique, obtuse or truncate, proximally entire, distally dentate, apex acuminate; petioles 6-10 mm long, stout. Flowers large, pure white, fragrant, in axillary clusters; bracteoles broadly ovate; sepals 5, rusty stellate-tomentose, persistent; petals 5, white, obovate; stamens 15, staminodes 5; ovary densely rusty stellate-hairy. Capsule oblong, acute, obscurely 5-angled, scarcely furrowed, rusty, horned; seeds 8-10 in each cell, orbicular, compressed.

Common in dry deciduous forests of Karimnagar, Mahabubnagar and Nalgonda districts.

Fl.: May – January; Fr.: January-July.

Appapur (MBNR), *TDCK* 13632; Kambalpally (NLG), *BVRS* 309 (NU); Umamaheswaram (MBNR), *SRS* 109144 (BSID).

STERCULIA L.

1. Leaves digitately compound; calyx lobes deeply stellate-hairy; follicles glabrous .... **S. foetida**
1. Leaves palmately lobed; calyx lobes glabrous or glandular hairy; follicles hairy .... **S. villosa**

Sterculia foetida L., Sp. Pl. 1008. 1753; Masters in Hook.f., Fl. Brit. India 1: 354. 1874; Gamble, Fl. Madras 1: 105. 1915; Malick in B.D.Sharma & Sanjappa, Fl. India 3: 459. 1993.

Large deciduous trees, up to 30 m high; bark whitish, flaking off; branches whorled, horizontal. Leaves digitately lobed, petioles 14-24 cm long, leaflets 5 or 9, elliptic, 8-13 × 3-5 cm, glabrous, base cuneate, margin entire, apex acute or caudate, pinnately veined; petiolules up to 1 cm long; stipules ensiform, caducous. Flowers dull orange-coloured, in axillary panicles; calyx campanulate, deeply 5-parted, lobes oblong-lanceolate, spreading, villous within, much longer than the tube. Male flowers: anthers 10-15 on ca 1.2 cm long staminal column; Female flowers: carpels 5, ovaries on 5-7 mm long androgynophore, hairy, surrounded by sterile anthers at base, stigmas as many as carpels, radiating. Follicles 1-5, boat-shaped, glabrous, woody, beaked; seeds 10-15, ovoid-ellipsoid, black, smooth.

Planted in gardens, along roadsides, house yards, office fences as an avenue and ornamental tree.

Fl.: March – April; Fr.: May-August.

Sterculia villosa Roxb. ex DC., Prodr. 1: 483. 1824; Masters in Hook.f., Fl. Brit. India 1: 355. 1874; Gamble, Fl. Madras 1: 106. 1915; Malick in B.D.Sharma & Sanjappa, Fl. India 3: 472. 1993.

Vern.: *Gugal, Gogal, Kovila, Vakkunara, Kummari poliki.*

Tall deciduous trees, up to 20 m high, with white bark, branches few, horizontal, spreading with large scars. Leaves suborbicular, 30-45 cm each way, glabrescent or thinly stellate-pilose above, base cordate, deeply palmately 5-7-lobed, lobes acuminate. Flowers in drooping panicles, sepals pale-yelow and hairy outside, deep red or purple within, lobed below middle; Male flowers: staminal column 2-3 mm long, curved, anthers 10; Female flowers: ovaries 5 on 2-3 mm long gynadrophore, globose with sterile anthers at base, style recurved, stigmas 5-lobed. Follicles bright-reddish-purple, oblong, rusty villous, hairs not irritant, 7 × 2 cm; seeds oblong, black.

Occasional in dry rocky hill slopes of deciduous forests in Mahabubnagar district.

Fl.: January – March; Fr .: May-June.

Borapur to Appapur (MBNR), *BR, SKB* & *BSS* 32521.

WALTHERIA L.

Waltheria indica L., Sp. Pl. 673. 1753; Masters in Hook.f., Fl. Brit. India 1 : 374. 1874; Gamble, Fl. Madras 1 : 106. 1915; Malick in B.D.Sharma & Sanjappa, Fl. India 3: 473. 1993.

Erect, perennial herbs, up to 1.5 m high, branchlets stellate- pubescent. Leaves simple, ovate or elliptic, 5-7.5 × 3-4.5 cm, velvety tomentose on both surfaces, base obtuse-subcordate, margin crenate-serrate or dentate, apex obtuse; petioles 0.6-2.5 cm long; stipules subulate. Flowers yellowish in axillary clusters; involucral bracts narrowly lanceolate, villous; calyx campanulate, lobes 5, as along as or shorter than tube, lanceolate; petals spathulate; staminal cup ca 2 mm long, subconical; ovary unilocular, pilose, styles fimbriate at apex, stigma penicillate. Capsule obconical enclosed in calyx; seed solitary, kidney- shaped, black.

Common in all places throughout the State.

Fl. & Fr.: July-January

Ankusapuram (ADB), *GO* 4310; Manchippa RF (NZB), *TP* & *BR* 6004; Mosra forest nursery (NZB), *TP* & *BR* 6079; Ponnala (MDK), *CP* 11369; Narsingi (MDK), *BR* & *CP* 11539; Kanmankalva RF (RR), *MSM* 10463; Amrabad (MBNR), *BSS* & *SKB* 32553; Nagarjunakonda (NLG), *KTH* 9743 (CAL); Chilakur RF (HYD), *Bir Bahadur s.n.;* Pakhal (WGL), *KMS* 11587 (MH); Rechapalli (KRN), *GVS* 22245 (MH); Pasra (WGL), *RKP* 108109 BSID); Kothagudem – Vijayawada road (KMM), *RR* 112515 (BSID).

TILIACEAE

1. Fruit drupe .... **Grewia**
1. Fruit capsule :
    2. Capsule cocci winged .... **Berrya**
    2. Capsule not winged:
        3. Capsule prickled; leaves without filiform processes at base .... **Triumfetta**
        3. Capsule unarmed; leaves with filiform processes at base .... **Corchorus**

BERRYA Roxb. *nom. cons.*

Berrya cordifolia (Willd.) Burret, Notizbl. Bot. Gart. Berlin-Dahlem 9 : 606. 1926; Daniel & Chandrabose in B.D.Sharma & Sanjappa, Fl. India 3: 478. 1993. *Espera cordifolia* Willd., Ges. Naturf. Freunde Berlin Neue Schriften 3 : 450. 1801. *Berrya ammonilla* Roxb., Pl. Coromandel 3: 60, t. 264. 1819; Masters in Hook.f., Fl. Brit. India 1: 383. 1874; Gamble, Fl. Madras 1 : 122. 1915.

Deciduous trees, up to 15 m high; bark pale, smooth. Leaves alternate, ovate, 8-15 × 4-9 cm, basally 5-nerved, glabrous, base truncate, obtuse or subcordate, margin entire or undulate, apex acuminate, acute or obtuse, 5-ribbed; lateral nerves 5-6 pairs above basal nerves, tertiary nerves parallel; petiole to 9 cm long. Flowers in many-flowered axillary or terminal panicles; calyx 3-5-lobed, lobes 3-5 mm long; stamens inserted on a short receptacle; ovary 3(-4)-loculed, style 3 mm long, slightly papillose at base, stigma peltate, 3-lobed. Capsule oblong or elliptic, 2 × 0.8 cm, prominently

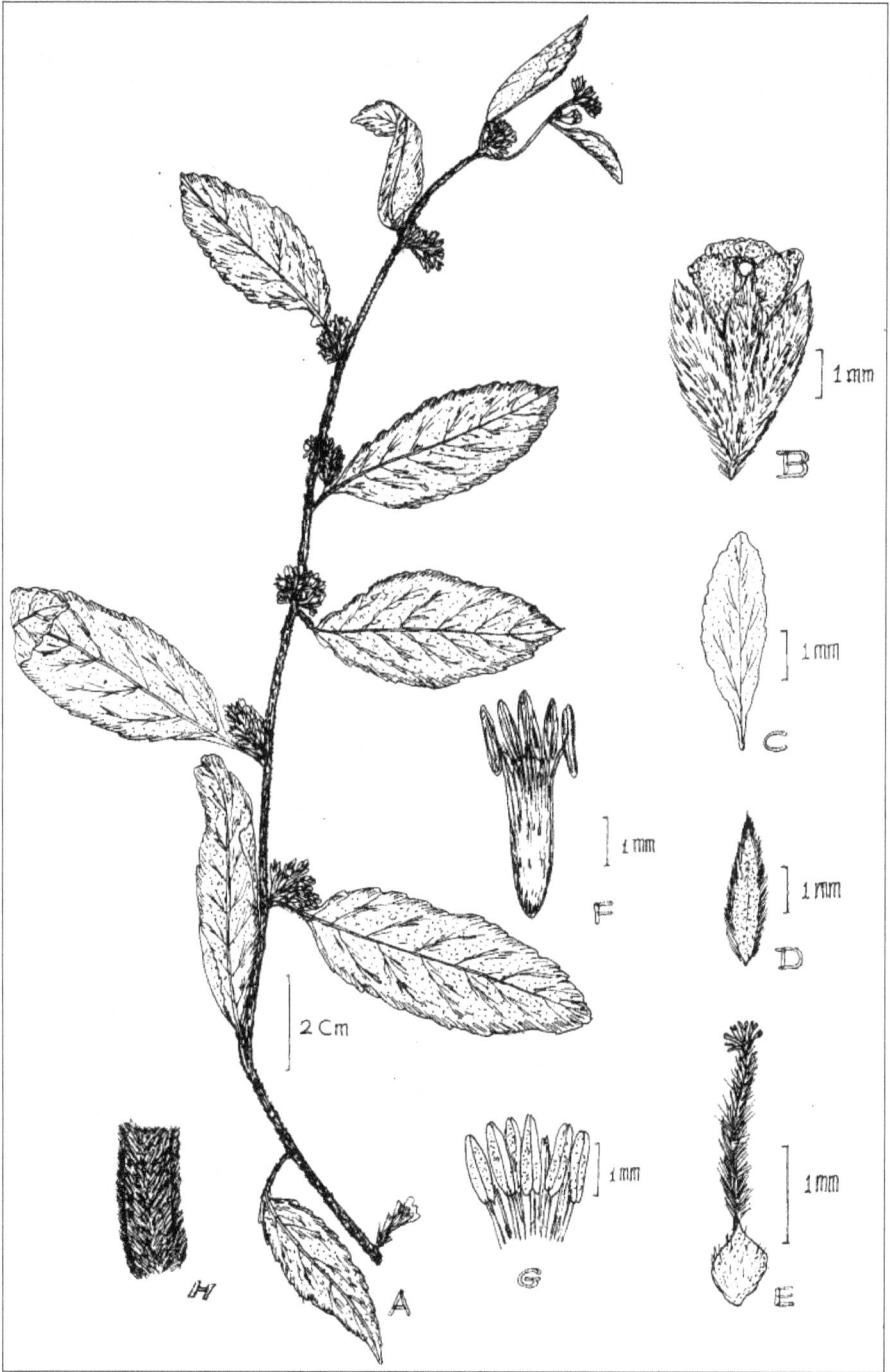

Figure 11: Waltheria indica L.

A. Twig, B. Flower, C. Petal, D. Bract, E. Pistil, F. Staminal column, G. Stamens enlarged, H. Part of stem enlarged.

nerved, 6-winged, villose; seeds 1-4 in each cell, covered with brown to yellow caducous bristles.

Cultivated species, and planted as avenue tree.

Fl.: December-January; Fr.: April.

CORCHORUS L.

1. Capsule depressed globose, without beak, valves woody .... **C. capsularis**
1. Capsule longer than broad, beaked, valves thin :
   2. Capsule angular :
      3. Capsule winged; beak trifid. .... **C. aestuans**
      3. Capsule not winged, beak entire :
         4. Capsules not falcate, 4-7 cm long, valves without partitions; seeds smooth. .... **C. trilocularis**
         4. Capsule falcate, to 3 cm long, valves with distinct transverse partitions; seeds wrinkled .... **C. urticifolius**
   2. Capsule cylindrical or linear-cylindrical :
      5. Beak entire :
         6. Capsule 4-7.5 cm long, 10-ribbed .... **C. olitorius**
         6. Capsule less than 4 cm long; not ribbed :
            7. Capsule glabrous .... **C. depressus**
            7. Capsule hairy .... **C. fascicularis**
      5. Beak divided .... **C. tridens**

Corchorus aestuans L., Syst. Nat. ed. 10. 1079. 1759; Daniel & Chandrabose in B.D.Sharma & Sanjappa, Fl. India 3: 485. 1993. *C. acutangulus* auct. non Forsskal 1755; Masters in Hook.f., Fl. Brit. India 1 : 398. 1874; Gamble, Fl. Madras 1: 121. 1915.

Annual, much branched, erect, hairy herbs, up to 50 cm high. Leaves ovate, elliptic or oblong, 2- 5 × 1-3 cm, sparsely hairy, base obtuse, serrate, apex acute or obtuse; petioles 0.5-4 cm long, grooved, purple; stipules 5-10 mm long, setaceous, purplish green. Flowers yellow, axillary, solitary or 2-3-flowered, in leaf-opposed cymes; pedicels ca 2 mm long, jointed near apex; bracts 4-6 mm long, filiform, purple; sepals linear-oblong, purple-dotted inside, green outside; petals 3-5 mm long, obovate, yellow; stamens 12-30; carpels 3, ovary cylindric, style 3-fid, stigma 2-lobed. Capsule elongate, 6-angled, septate, 3-winged, beak 3-fid, radiating; seeds numerous, dark brown, truncate.

Common weed of cultivated fields, wastelands and open forests, in most districts.

Fl. & Fr.: May-December

Sone (ADB), *TP* & *PVP* 4055; Sarvaipet (ADB), *GO* & *PVP* 4814; Manchippa RF (NZB), *TP* & *BR* 6032; Pocharam RF (MDK), *TP* & *CP* 12057; Central University campus (RR), *TP* & *MSM* 19247; Nagarjunasagar (NLG), *KMS* 9728 (CAL), 19084, 19085 (MH); Narsapur-way to Pagarikatta (MDK), *KMS* 6629 (MH); Kajipet (WGL), *RKP* 105206 (BSID); Guttakindipalle (MBNR), *SRS* 107490 (BSID); Mamillagudem (KMM), *RR* 105919 (BSID).

Corchorus capsularis L., Sp. Pl. 529. 1753; Masters in Hook.f., Fl. Brit. India 1: 397. 1874; Gamble, Fl. Madras 1: 122. 1915; Daniel & Chandrabose in B.D.Sharma & Sanjappa, Fl. India 3: 485. 1993.

Erect annual herbs or subshrubs, 1-2.5 m high, branchlets glabrous. Leaves ovate-lanceolate, 4-12 × 1.5-4 cm, glabrous or pubescent along nerves, base generally prolonged into tail-like appendages, margin serrate, apex acuminate; petioles up to 4 cm long; stipules linear. Flowers yellow, axillary or extra-axillary, solitary or 2-3-flowered fascicles; bracts linear-ovate; sepals 5, free, linear-oblong; petals 5, obovate; stamens 20-30; carpels 5, ovary subglobose, 5-loculed. Capsule depressed globose, 1 cm long, not beaked, ridged, 5-valved, muricate on ribs, green, black on drying; seeds smooth, ovoid, glabrous, deep brown or blackish brown.

Cultivated for the jute fibre obtained from this plant, occasionally found run wild.

Fl. & Fr.: August-December

Mudheli (NZB), *BR* & *CPR* 7217; Rechapalli (KRN), *GVS* 25694 (CAL); Pakhal RF (WGL), *KMS* 11637 (CAL); Rekapalli RF (KMM), *RCS* 98947 (BSID).

Corchorus depressus (L.) Vicary, J. Asiat. Soc. Bengal 16: 1160. 1847; Daniel & Chandrabose in B.D.Sharma & Sanjappa, Fl. India 3: 486. 1993. *Antichorus depressus* L., Mant. 1: 64. 1767. *Corchorus antichorus* Raeusch., Nomencl. Bot. ed. 3. 158. 1797; Masters in Hook.f., Fl. Brit. India 1: 389. 1874; Gamble, Fl. Madras 1 : 121. 1915.

Much branched, perennial, prostrate herbs, 15-35 cm long, glabrous, branches many from thick woody rootstock. Leaves 0.5-2 × 0.4-0.6 cm, oblong, glabrous, plicate, base 3-nerved, base rounded or cuneate, margin crenate- serrate, apex obtuse; petioles 1.5-2.5 cm long; stipules ca 3 mm long, subulate. Flowers yellow, numerous, in leaf-opposed fascicles, pedicels ca 1 mm long; bracts linear-lanceolate; sepals free, linear-oblong, reddish green; petals ovate-spathulate, yellow; stamens 8-10; carpels 4. Capsule 1.5-3 cm long, straight or often curved upwards, 4-valved; seeds trigonous, black.

Rare in Mahabubnagar district (Raghava Rao, 1990).

Fl. & Fr.: September – December.

Corchorus fascicularis Lam., Encycl. 2: 104. 1786; Masters in Hook.f., Fl. Brit. India 1: 398. 1874; Gamble, Fl. Madras 1: 122. 1915; Daniel & Chandrabose in B.D.Sharma & Sanjappa, Fl. India 3: 486. 1993.

Annual, erect, glabrous herbs or undershrubs, up to 60 cm high. Leaves oblong-lanceolate or elliptic, 2-5 × 0.5-1.5 cm, glabrous, base obtuse, margin serrate, apex obtuse or acute; petioles 3-10 mm long, hirsute; stipules 5 mm long, subulate. Flowers yellow, 3-5- flowered, in axillary, extra-axillary or leaf-opposed cymes; sepals linear-oblong; petals oblong to obovate; stamens 5-10; carpels 3, ovary oblong-ovoid to linear, stigma capitate. Capsule short, 1.5 cm, subcylindric, puberulous, shortly beaked, 3-loculed; seeds about 25, 3-gonous, black.

A weed in cultivated fields and in drying moist places of many districts.

Fl. & Fr.: September-February

Sone (ADB), *TP* & *PVP* 4054; Neelwai (ADB), *GO* & *DAM* 5163; Potchera (ADB), *GO* & *DAM* 5064; Yellareddy (NZB), *TP* & *BR* 6299; Patancheru (MDK), *BR* & *CP* 11308; Narsapur (MDK), *Ashrafunnisa* 7081; Rangammagudem (RR), *MSM* & *KH* 10574; Pakhal RF (WGL) *ANH* 15970 (CAL); Aklaspur (KRN), *GVS* 25629 (MH); Krishna river (NLG), *KMS* 9822 (MH); Bhadrachalam (KMM), *RCS* 104330 (BSID).

Corchorus olitorius L., Sp. Pl. 529. 1753; Masters in Hook.f., Fl. Brit. India 1: 397. 1874; Gamble, Fl. Madras 1: 122. 1915; Daniel & Chandrabose in B.D.Sharma & Sanjappa, Fl. India 3: 487. 1993.

Annual erect herbs, up to 75 cm high, branchlets glabrous. Leaves oblong, lanceolate or elliptic, 3-6 × 1.5-3.5 cm, glabrescent, base obtuse, serrulate, two lower serratures prolonged into long sharp points, margin serrate, apex acuminate; petioles 2-3 cm long, pubescent, stipules 8-12 cm long, subulate. Flowers yellow, solitary or 2-3 fascicled, axillary or extra axillary; sepals linear-oblong; petals yellow, oblong-spathulate; stamens many, somewhat united at base; carpels 5, ovary cylindric, 5-loculed, style short, stigma 5-lobed. Capsules cylindric, glabrous, 3-6-valved, 10-ribbed, greenish-brown; seeds trigonous, brown.

Common in moist places and hedges of the fields in many districts.

Fl. & Fr.: September-January.

Neredugunda (ADB), *GO* & *DAM* 5098; Mosra forest nursery (NZB), *TP* & *BR* 6060; Gandhari RF (NZB), *BR* & *CPR* 7246; Maddigunta (NZB), *BR* & *SPB* 9559; Patancheru (MDK), *BR* & *CP* 1309; Ramayampet (MDK), *BR* & *CP* 11501; Naskal (RR), *MSM* 15117, 15129; Pegarikutta-Narsapur (MDK), *KMS* 6645 (MH); Subedari (WGL), *RKP* 105255 (BSID); Krishnarajasagar (KMM), *RCS* 102461 (BSID).

Corchorus tridens L., Mant. Pl. 566. 1771; Masters in Hook.f., Fl. Brit. India 1: 398. 1874; Gamble, Fl. Madras 1: 122. 1915; Daniel & Chandrabose in B.D.Sharma & Sanjappa, Fl. India 3: 488. 1993.

Erect or suberct, slender herbs, up to 1 m high, branchlets glabrous. Leaves linear-oblong, elliptic or lanceolate, 4-10 × 1- 2 cm, basally 3- or 4-nerved, glabrous, base cuneate, margin serrate, basal serrature appendaged, apex acuminate; petioles up to 18 mm long; stipules 3-4 mm long. Flowers yellow, sessile or subsessile, in leaf-opposed, solitary or 2-4-flowered cymes; sepals linear-oblong; petals yellow, oblong; stamens 10-15; capels 3, ovary cylindric, stigma sparsely papillate. Capsule 2-6 cm, slender, linear-cylindric, terete, 3-valved, glabrous, beak 3-fid; seeds, truncate, brownish black.

Common on roadsides and waste places in Hyderabad, Mahabubnagar, Nalgonda, Karimnagar and Adilabad districts.

Fl. & Fr.: Through out the year.

Wankidi RF (ADB), *TP* & *GO* 5494; Aklaspur (KRN), *GVS* 25679; Wankidi RF (ADB), *TP* & *GO* 5494; Sandralagadda (NLG), *BVRS* 258 (NU).

Corchorus trilocularis L., Mant. Pl. 77. 1767; Masters in Hook.f., Fl. Brit. India 1 : 397. 1874; Gamble Fl. Madras 1: 122. 1915; Daniel & Chandrabose in B.D.Sharma & Sanjappa, Fl. India 3: 488. 1993.

Annual erect, much branched sparsely pubescent herbs, up to 80 cm high. Leaves oblong, lanceolate or ovate-oblong, 3-7 × 1.5-3 cm, base cuneate, margin serrate or crenate, basal serratures produced in filiform appendages, apex acute or obtuse, petioles 4-12 mm long, pilose; stipules lanceolate. Flowers yellow, solitary or 2-3-fascicled, leaf opposed, axillary, extra-axillary; pedicels ca 2.5 mm long; bracts ca 3 mm long, linear-lanceolate; sepals 4-5 mm long, linear-oblong; petals yellow; stamens 15-20; carpels 3, ovary cylindric, stigmas 3, capitate. Capsules straight or slightly curved, 3-4-gonous, stellate hairy; seeds trigonous, blackish-brown.

A weed of waste places and roadsides in all districts.

Fl. & Fr. : August-January.

Neredugonda (ADB), *GO* & *DAM* 5097; Tiryani RF (ADB), *PVP* 9814; Ghanapur (NZB), *BR* & *KH* 7107; Gangapur (MDK), *TP* & *CP* 12081; Kusumasamudram RF (RR), *MSM* & *NV* 10372; Hyderabad city, Moosi River Bank (HYD), *KMS* 5969 (MH); Wyra to Khammam (KMM), *RR* 108065 (BSID); Koriguttalu (KMM), *RR* 106033 (BSID).

Corchorus urticifolius Wight & Arn., Prodr. Fl. Ind. Orient. 73. 1834; "*urticaefolius*"; Masters in Hook.f., Fl. Brit. India 1 : 397. 1874; Gamble, Fl. Madras 1 : 122. 1915; Daniel & Chandrabose in B.D.Sharma & Sanjappa, Fl. India 3: 489. 1993.

Erect herbs or subshrubs, up to 1 m high, branchlets pubescent. Leaves ovate or lanceolate, 5-7 × 2-4 cm, basally 3 or 5-nerved, appressed hairy on both sides, base obtuse, sometimes oblique, margin dentate, basal serrature not appendaged, apex acute; petioles 0.8-2 cm long; stipules ca 5 mm long, linear. Flowers yellow, 2-4-flowered, in leaf-opposed cymes, or fascicles; bracts ca 3 mm long, oblong; petals yellow, ovate-spathulate; stamens ca 15; ovary 3-loculed. Capsule terete or slightly 3-angled, falcate, 2-3 cm long, pilose, beak short, entire, with internal transverse partitions; seeds 3-gonous, wrinkled.

Occasional in Hyderabad, Ranga Reddi, Medak and Nalgonda districts.

Fl. & Fr. : Through out the year.

ICRISAT (MDK) *LJGV* 3995 (CAL); Nagarjunasagar (NLG), *BVRS* 52 (NU).

GREWIA L.

1. Leaves 5- or more-nerved from the base of leaf blade :
    2. Drupes unlobed or obscurely lobed .... **G. rothii**
    2. Drupes 1-4-lobed;
        3. Plants trees :
            4. Stipules foliaceous-auricled at base .... **G. tiliifolia**
            4. Stipules lanceolate or linear-lanceolate or subulate .... **G. orbiculata**
        3. Plants shrubs :
            5. Flowers white, in leaf-opposed or axillary cymes .... **G. tenax**
            5. Flowers yellow, in extra-axillary or leaf-opposed cymes ..... **G. villosa**
1. Leaves 3-nerved from the base of the leaf blade :
    6. Shrubs or small trees :
        7. Flowers polygamous :
            8. Leaves oblique; fruits pilose at maturity;
               stamens more than 40 .... **G. hirsuta**
            8. Leaves not oblique; fruits glabrescent;
               stamens 16- 20 .... **G. helicterifolia**
        7. Flowers bisexual :
            9. Leaves glabrous beneath .... **G. rhamnifolia**
            9. Leaves pubescent beneath .... **G. damine**
    6. Plants stragglers :
        10. Drupes shallowly or deeply divided into 4 lobes; flowers white :
            11. Peduncle shorter than the petiole .... **G. abutilifolia**
            11. Peduncle twice the length of the petiole .... **G. orientalis**
        10. Drupes shallowly or deeply divided into 2 lobes;
            flowers yellow or green:
            12. Leaves glabrous beneath .... **G. bracteata**
            12. Leaves densely tomentose below .... **G. flavescens**

Grewia abutilifolia Vent. ex Juss., Ann. Mus. Natl. Hist. Nat. 4: 92. 1804; Masters in Hook.f., Fl. Brit. India 1: 390. 1874; Daniel & Chandrabose in B.D.Sharma & Sanjappa, Fl. India 3: 493. 1993. *G. aspera* Roxb., Fl. Ind. 2 : 591. 1832; Gamble, Fl. Madras 1: 119. 1915.

Vern.: *Guvuadada, Peddatadaki, Potucamanti.*

Scandent shrubs to small trees, up to 8 m high, branchlets densely tomentose, often velvety to woolly. Leaves rhomboid or ovate, at times obscurely lobed, 6-10 × 5-9 cm, chartaceous, 3-nerved, sparsely stellate-tomentose above, velvety below, base

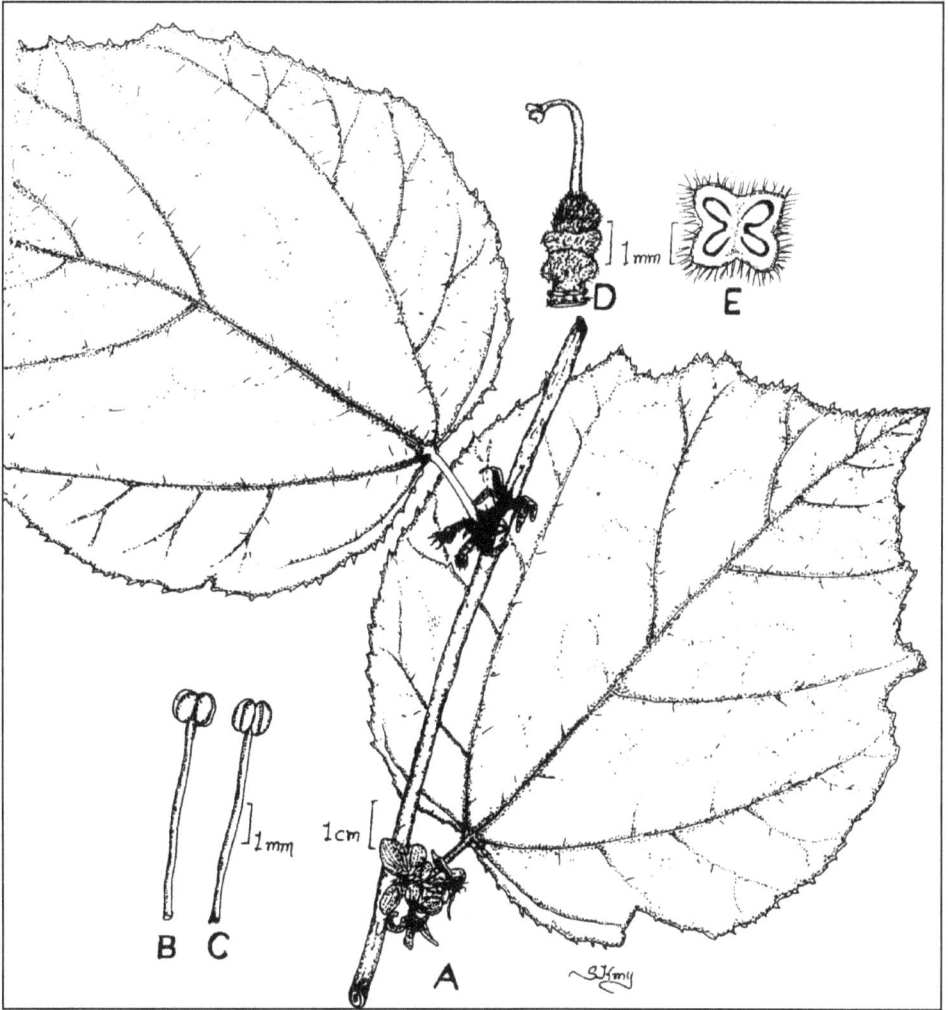

Figure 12: Grewia abutilifolia Vent. ex Juss.
A. Twig, B & C. Stamens, D. Pistil, Ovary c.s.

round, truncate or subcordate below, margin irregulary serrate or 2-dentate, apex acute-acuminate; petioles 0.5-4.5 cm long. Flowers white in axillary umbellate cymes; peducncles 1-3 together, up to 1 cm long; sepals 5, narrowly oblong or lanceolate, woolly outside; petals 5, white, oblong, clawed with a gland inside; stamens many; ovary subglobose, stigma laciniate. Drupe yellow, subglobose or turbinate obscurely 4-lobed, 8 mm across, stellate-tomentose, wrinkled.

Occasional in deciduous forests in Adialabad, Nizamabad and Ranga Reddi districts.

Fl. & Fr.: March-June.

Indervelli (ADB), *GO* & *PVP* 4538; Neredigonda (ADB), *PVP* 9819.

Grewia bracteata Heyne ex Roth, Nov. Pl. Sp. 243. 1821; Masters in Hook.f., Fl. Brit. India 1 : 389. 1874; Daniel & Chandrabose in B.D.Sharma & Sanjappa, Fl. India 3: 495. 1993. *G. obtusa* Wallich ex Gamble, Fl. Madras 2: 117. 1915. *G. wightiana* J.R.Drumm. ex Dunn in Gamble, Fl. Madras 2: 115. 1915.

Climbing shrubs, branches terete, younger ones rusty tomentose. Leaves elliptic-ovate, 11 × 3 cm, unequally rounded or acute at base, 3-nerved, margin crenate-serrulate, apex shortly acuminate; petioles 2-16 mm long; stipules ovate, sparsely pubescent. Flowers green in axillary, 3-flowered, umbellate cymes, peduncle up to 7 mm long; buds ovoid or ovoid-oblong, pedicels 4-10 mm long; sepals 5, free, ovate-lanceolate to lanceolate, pilose; petals 5, free, white or yellow, ovate or oblong, clawed with a gland inside; androgynophore ca 1 mm long, woolly; stamens numerous; ovary globose, stigma clavate. Drupes 2 cm across, shrivelled when dry, covered with yellow hairs, depressed at top, faintly 2-lobed, stones 2-3-celled.

Common in dry deciduous forests of Mahabubnagar district.

Fl. & Fr. : April-December

Rayalatipenta (MBNR), *TDCK* 15323.

Grewia damine Gaertn., Fruct. Sem. Pl. 2. 113.t.106.f.9. 1790; Gamble, Fl. Madras 1 : 118. 1915; Narayanaswamy & Seshagiri Rao, J. Indian Bot. Soc. 29 : 187. 1950; Daniel & Chandrabose in B.D.Sharma & Sanjappa, Fl. India 3: 496. 1993. *G. salvifolia* Heyne ex Roth, Nov. Pl. Sp. 329. 1821; Masters in Hook.f., Fl. Brit. India 1 : 386. 1874 *p.p.*

Shrubs or small trees, up to 5 m high, herbaceous portions covered with soft hoary pubescence. Leaves ovate, elliptic or lanceolate, 2.5-5 × 1-2 cm, 3-nerved, glabrous above, densely hoary-tomentose below, base obtuse proximally entire, distally serrulate, apex obtuse or acute; petioles 2-4 mm long. Flowers yellow, few in axillary, umbellate cymes; peduncles up to 1.5 cm long, buds ovoid to oblong, tomentose; pedicels up to 1.2 cm long, elliptic-oblong, clawed with a gland inside; stamens many, ovary subglobose, stigma 4-lobed. Drupe globose, deep reddish-yellow when ripe; 2- lobed; seeds stony.

Common in scrub jungles of Adilabad, Hyderabad and Mahabubnagar districts.

Fl.: July – September; Fr.: August - November.

Mallelatheertham (MBNR), *SKB* & *BSS* 30710, 30715.

Grewia flavescens Juss., Ann. Mus. Natl. Hist. Nat. 4: 91. 1804; Gamble, Fl. Madras 1: 119. 1915; Daniel & Chandrabose in B.D.Sharma & Sanjappa, Fl. India 3: 498. 1993. *G. carpinifolia* sensu Masters in Hook. f., Fl. Brit. India. 1: 387. 1874. non Juss. 1804.

Small trees, up to 4.6 m high, branchlets stellate-tomentose. Leaves oblong or oblanceolate, 3-8 × 1.5-4 cm, thin coriaceous, 3-nerved, scabrid, stellate-tomentose below, base slightly narrow, obtuse or truncate, margin crenate-serrate, apex slightly broad above middle, abruptly acute; petioles up to 7 mm long. Flowers yellow, in axillary 3- flowered cymes, peduncles exceeding the petiole; flower buds elliptic, dilated at base, pilose; sepals 5, oblong-lanceolate; petals 5, oblong, half the length of the sepals; stamens numerous, anthers smooth; ovary subglobose to ovoid, pilose; stigma bifid. Drupe deeply or obscurely 2-lobed, 0.8-1.5 cm across, wrinkled, densely stellate-tomentose, apiculate, rind crustaceous, mesocarp fibrous, yellowish brown.

Occasional in open forests and field hedges in Karimnagar, Nalgonda, Medak, Hyderabad, Ranga Reddi, and Nizamabad districts.

Fl. & Fr.: June-October.

Dharmaram (NZB), *BR* & *KH* 6461; Lingampet (NZB), *BR* 9298; Venkatapuram (MDK), *TP* & *CP* 14003; Edupayalu RF (MDK), *RG* 113418 (BSID); Mohammadabad RF (RR), *MSM* & *NV* 10427; Amrabad (MBNR), *SKB* & *BSS* 32530; Kondagutta (KRN), *GVS* 25685 (MH).

Grewia helicterifolia Wall. ex G. Don, Gen. Hist. 1: 548. 1831; Daniel & Chandrabose in B.D.Sharma & Sanjappa, Fl. India 3: 500. 1993. *G. hirsuta* Vahl var. *helicterifolia* (Wall. ex G. Don) Haines, For. Fl. Chota Nagpur 196. 1910. *G. polygama* auct. non Roxb. 1832; Masters in Hook.f. Fl. Brit. India 1: 391. 1874; Gamble, Fl. Madras 1 : 119. 1915.

Vern.: *Chinacipuru, Jibilika.*

Shrubs or small trees, branches spreading, branchlets pubescent. Leaves oblong, lanceolate, 5-10 × 1.5-2.5 cm, almost sessile, base rounded to subacute, margin irregularly serrate, apex acute, palmately 3-nerved, thinly stellate-pubescent above, velvety-tomentose beneath; petioles 3-6 mm long. Flowers white, polygamous in axillary cymes, peduncles more than twice the length of petiole, pedicels divergent; sepals 5, oblong-lanceolate, pilose; petals 5, white; stamens 16-20; ovary globose, pilose, stigma laciniate. Drupe obscurely 4-lobed, 1-2 cm across, glabrescent, brownish, stones 4, 1-seeded.

Rare in Medak and Mahabubnagar districts.

Fl.: May – October; Fr.: August – February.

Zaheerabad (MDK, *CP* 11589.

Grewia hirsuta Vahl, Symb. Bot. 1: 37. 1790; Masters in Hook.f., Fl. Brit. India 1: 391. 1874; Gamble, Fl. Madras 1: 119. 1915; Daniel & Chandrabose in B.D.Sharma & Sanjappa, Fl. India 3: 501. 1993.

Vern.: *Kadurupandlu.*

Erect shrubs, 1-2.5 m high, softly rufous-pubescent. Leaves distichous, oblong or lanceolate, 3-8 × 1-3 cm, scabrid, stellate-pubescent above, woolly below, base oblique, margin irregularly serrate, apex acuminate; petioles up to 7 mm long. Flowers polygamous (male + bisexual), pale to bright yellow, in axillary, extra-axillary, 2-4-flowered umbellate cymes; buds globsoe; pedicels 2-5 mm long; sepals 5, elliptic-lanceolate, hirsute; petals 5, oblong, glands ca half the length of petal; stamens more than 40, ovary globose, densely villous, stigma 5-lobed, lobes fringed. Drupe globose, obscurely 4-lobed, stiff- hirsute.

Common forest shrub in most districts.

Fl. & Fr. : May-October

Pulikunta RF (ADB), *PVP* 9471; Ankole – Banswada Road (NZB), *TP* & *BR* 6187; Manchippa (NZB), *TP* & *BR* 6053; Narsapur RF (MDK), *TP* & *CP* 11969; Pocharam RF (MDK), *CP* 11750; Lothuvagu (RR), *MSM* 10451 (BSID); Mannanur (MBNR), *TDCK* 13636; Aklaspur (KRN), *GVS* 20211 (MH); Rechapalli (KRN), *GVS* 22253 (MH); Westside of Pakhal (WGL), *KMS* 13115 (MH); Srinivaspur (NLG), *BVRS* 271 (NU); Tupakulagudem (WGL), *PVS* 76967 (BSID); Yenkur (KMM), *RR* 108540 (BSID).

Grewia orbiculata Rottl., Ges. Naturf. Freunde Berlin Neue Schr. 4: 205. 1803; Masters in Hook.f., Fl. Brit. India 1. 386. 1874; Daniel & Chandrabose in B.D.Sharma & Sanjappa, Fl. India 3: 504. 1993. *G. rotundifolia* Juss., Ann. Mus. Natl. Hist. Nat. 4: 92. 1804; Gamble, Fl. Madras 1: 118. 1915.

Vern.: *Jana, Nulithada.*

Small trees, up to 6 m high. Leaves orbicular, 1.5-2.5 × 0.5-1.5 cm, 5-nerved, pubescent above, glaucous and woolly below, base oblique, subcordate or rotund, margin serrate, dentate, apex obtuse; petioles 0.3-1.2 cm long. Flowers yellow in axillary umbellate 3-flowered cymes, buds oblong to elliptic-oblong, pedicels 7-10 mm long; sepals 5, lanceolate, woolly outside; petals 5, orange-yellow, oblong; stamens numerous; ovary obovoid, furrowed, pilose, stigma 4-lobed. Drupes orbicular, 2-lobed, greyish, pilose; seed stony.

Common in scrub forests and occasional in thick forests of most districts.

Fl. & Fr.: April-August

Vikarabad (RR), *TP* & *MSM* 10451; Manchippa (NZB), *BR* & *KH* 9031; Nagarjuna konda (NLG), *KTH* 9626 (CAL); Nagarjunasagar (NLG), *BVRS* 83, 130 (NU); Narsapur (MDK), *KMS* 8012 (CAL); Yenkur (KMM), *RR* 107944 (BSID); Hamsanpalli RF (MDK), *RG* 104072 (BSID); Pullajalamala (MBNR), *VBH* 84985 (BSID).

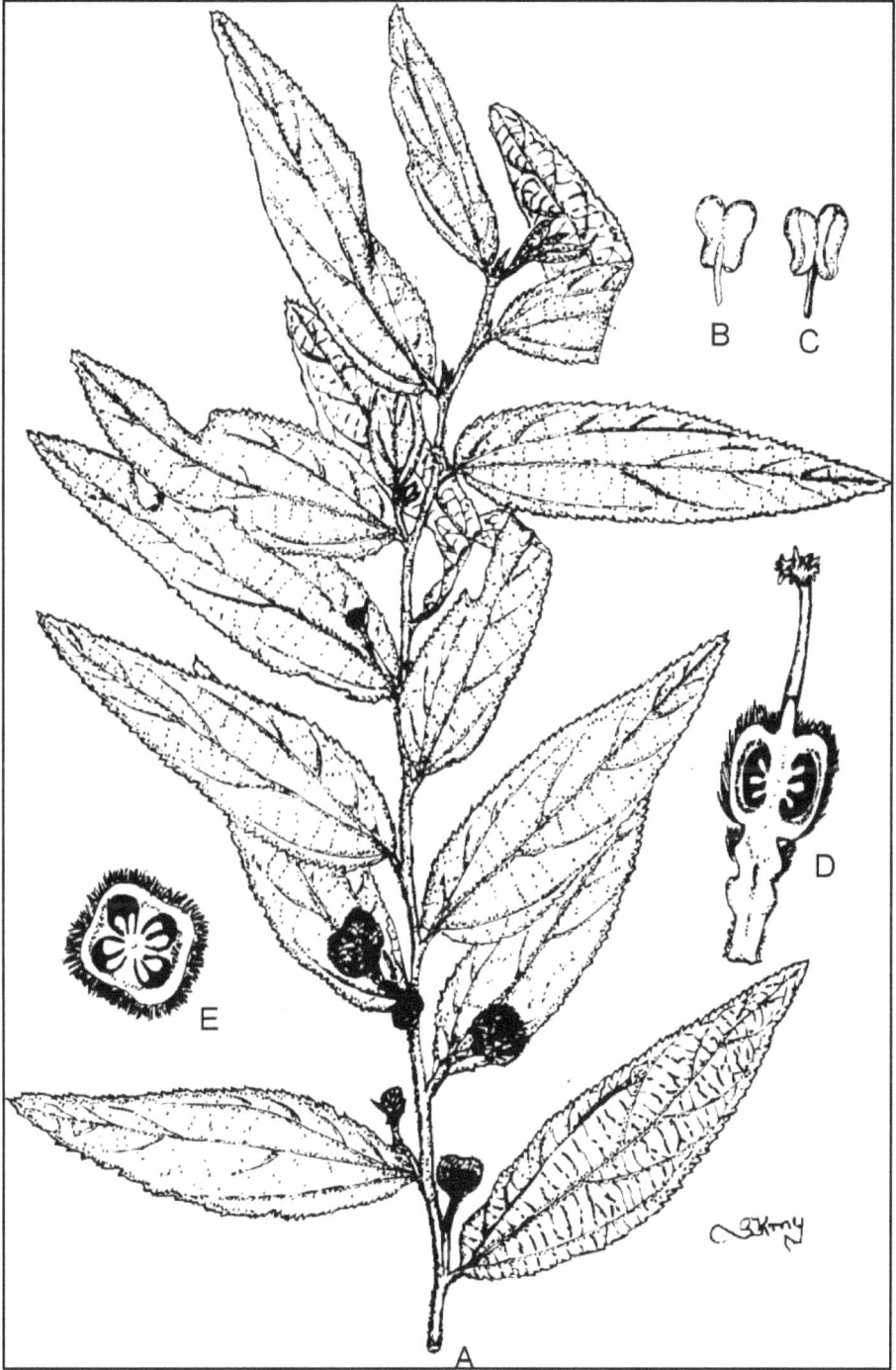

Figure 13: Grewia hirsuta Vahl.

A. Twig, B&C. Stamens, D. Ovary l.s., E. Ovary c.s.

Grewia orientalis L., Sp. Pl. 964. 1753; Gamble, Fl. Madras 1: 117. 1915; Daniel & Chandrabose in B.D.Sharma & Sanjappa, Fl. India 3: 506. 1993. *G. columnaris* J.E. Smith in Rees. Cycl. 17: no. 5. 1811; Masters in Hook.f., Fl. Brit. India 1: 383. 1874.

Shrubs or small trees, up to 2.5 m high, branches virgate, thinly stellate-hairy. Leaves elliptic-lanceolate, 4-10 × 2-4.5 cm, glabrescent, base slightly cordate, margin serrulate, apex acute or acuminate; petioles up to 7 mm long. Flowers white in 1-3-flowered, axillary umbellate cymes; sepals 5, oblong, pubescent, tomentose outside; petals 5, white, glands half the length of petals, ovate-lanceolate; ovary globose, pilose, stigma 5-lobed. Drupes globose, yellow, wrinkled, 2-4-lobed; seeds hairy.

Common in forests Adilabad, Nizamabad and Ranga Reddi districts.

Fl. & Fr.: July-December.

Ankusapuram (ADB), *GO* 4368; Mudheli RF (NZB), *BR* & *CPR* 9267; Central University campus (RR), *MSM* 15163.

Grewia rhamnifolia Heyne ex Roth, Nov. Sp. Pl. 244. 1824; Gamble, Fl. Madras 1 : 117. 1915; Daniel & Chandrabose in B.D.Sharma & Sanjappa, Fl. India 3: 507. 1993. *G. orientalis* auct. non L. 1753; Masters in Hook.f., Fl. Brit. India 1 : 384. 1874. *p.p.*

Shrubs, up to 3 m high, branchlets tomentose. Leaves elliptic or lanceolate, 4.5-8 × 2-4 cm, thin coriaceous, glabrous, base tapering, subcordate, obtuse, margin crenulate, apex acuminate or acute; petioles 4-10 mm long. Flowers white in leaf-opposed, axillary or terminal, 3-flowered umbellate cymes; peduncles twice the length of the petioles, flower buds ovoid, pilose, ribbed; sepals 5, lanceolate, woolly; petals 5, ovate-lanceolate, glands ca 3 mm long; androgynophore ca 2 mm long, grooved, woolly; stamens many, ovary globose, pilose, stigma 4-lobed. Drupe globose, 1.5 cm across, yellow, wrinkled, pilose, obscurely 4-lobed.

Common in Karimnagar (Naqvi, 2001), Hyderabad (Ramana, 2010) and Mahabubnagar districts.

Fl.: May –September; Fr.: August – February.

Grewia rothii DC., Prodr. 1 : 509. 1857; Gamble, Fl. Madras 1 : 117. 1915. *G. excelsa* auct. non Vahl 1790; Masters in Hook.f., Fl. Brit. India 1 : 385. 1874; *p.p.*; Daniel & Chandrabose in B.D.Sharma & Sanjappa, Fl. India 3: 507. 1993.

Tel.: *Cipuru, Jana, Jibilike, Peddcipuru, Putiki.*

Small trees, up to 3 m high; young branches tomentose. Leaves ovate or ovate-lanceolate, 4-10 × 1.5-3 cm, glabrous above, grey-pubescent beneath, 5-nerved, base obtuse, margin serrulate, apex acute or acuminate; petioles ca 5 mm long. Flowers yellow in axillary clusters, peduncles long, slender; sepals 5, ca 6 mm long, elliptic-oblong or elliptic-lanceolate; petals 5, ovate-lanceolate, glands ca 1 mm long, elliptic; stamens many; ovary globose, stigma 4-lobed. Drupe globose, unlobed, persistently hairy.

Common in open hilly areas of Adialabad, Hyderabad, Warangal, Karimnagar, Khammam and Mahabubnagar districts.

Fl.: April – October; Fr.: June – December.

Neelwai RF (ADB), *TP* & *GO* 5439; Satruguda (ADB), *GO* & *PVP* 4255; Pakhal (WGL), *KMS* 13116 (CAL); Bhadrachalam RF (KMM), *RCS* 102443 (BSID); north of Pasra (WGL), *RKP* 108204 (BSID).

Grewia tenax (Forssk.) Fiori, Agric. Colon. 5. Suppl. 23. 1912; Daniel & Chandrabose in B.D.Sharma & Sanjappa, Fl. India 3: 511. 1993. *Chadara tenax* Forssk., Fl. Aegypt.-Arab. 105. 1775. *Grewia populifolia* Vahl, Symb Bot. 1 : 33. 1790; Masters in Hook.f., Fl. Brit. India 1 : 385. 1874. *G. betulifolia* Juss., Ann. Mus. Natl. Hist. Nat. 4 : 92. t. 50. f. 1.1804; Gamble, Fl. Madras 1 : 117. 1915.

Vern.: *Gundukadira, Kadadarai, Kattekolupu.*

Erect, much branched shrubs, 2-3 m high, branchlets more or less stellate-hairy, ultimately glabrescent. Leaves mutiform, ovate- obovate, 1.5-4 × 1.5-2.5 cm, 3-5-nerved, glabrous, base cuneate or obtuse, margin dentate, apex acute; petioles up to 1 cm long. Flowers white in axillary or leaf-opposed cymes; sepals 5, linear-oblong, tomentose outside; petals 5, white, linear-oblong, glands ca 2 mm long; androgynophore ca 0.5 mm long; stamens many; ovary subglobose, 2-4-lobed, hirsute, stigma 4-5-lobed. Drupes yellowish, 1-4-lobed, glabrescent, edible.

Common in dry localities of Adilabad, Nalgonda and Mahabubnagar districts.

Fl. & Fr.: July-December.

Kudichintalabailu (MBNR), *TDCK* 15388.

Grewia tiliifolia Vahl, Symb. Bot. 1 : 35. 1790; Masters in Hook.f., Fl. Brit. India 1 : 386. 1874. '*tiliaefolia*'; Gamble, Fl. Madras 1 : 118. 1915; Daniel & Chandrabose in B.D.Sharma & Sanjappa, Fl. India 3: 511. 1993.

Vern. : *Charachi, Jana, Nulijana, Tada, Tada-jana.*

Moderate sized trees, up to 10 m high. Leaves broadly ovate, or oblong,1.7-36 × 1-24 cm, 3-5-nerved, hoary-tomentose beneath, base obliquely cordate, margin crenate, apex acute; petioles up to 4 cm long. Flowers yellow, 3-6 in axillary or extra-axillary cymes; buds subglobose or obovoid-oblong, tomentose; pedicels 4-13 mm long; sepals 5, elliptic-lanceolate, tomentose outside; petals 5, yellow, elliptic-oblong or spathulate, glands ca 0.5 mm long; stamens many; ovary globose, sparsely hirsute; stigma 4-lobed. Drupe globose, didymous, 2-lobed, sparsely hirsute, nuts 2.

Common in moist and dry deciduous forests of all districts.

Fl.: January-September; Fr.: May –October.

Jaipur RF (ADB), *GO* & *PVP* 4943; Lingaram (NZB), *BR* & *GO* 9060; Mustapur (NZB), *BR* & *GO* 9074; Narsapur RF (MDK), *TP* & *CP* 14032; Dharur RF (RR), *TP* & *MSM* 11094; Central University campus (RR), *MSM* 15164; Rangammagudem RF (RR), *MSM* & *KH* 10312; Farahabad (MBNR), *SKB* & *BSS* 34435; Mannanur RF (MBNR), *VBH* 86608 (BSID); way to Siddharam (KMM), *PVS* 84078 (BSID).

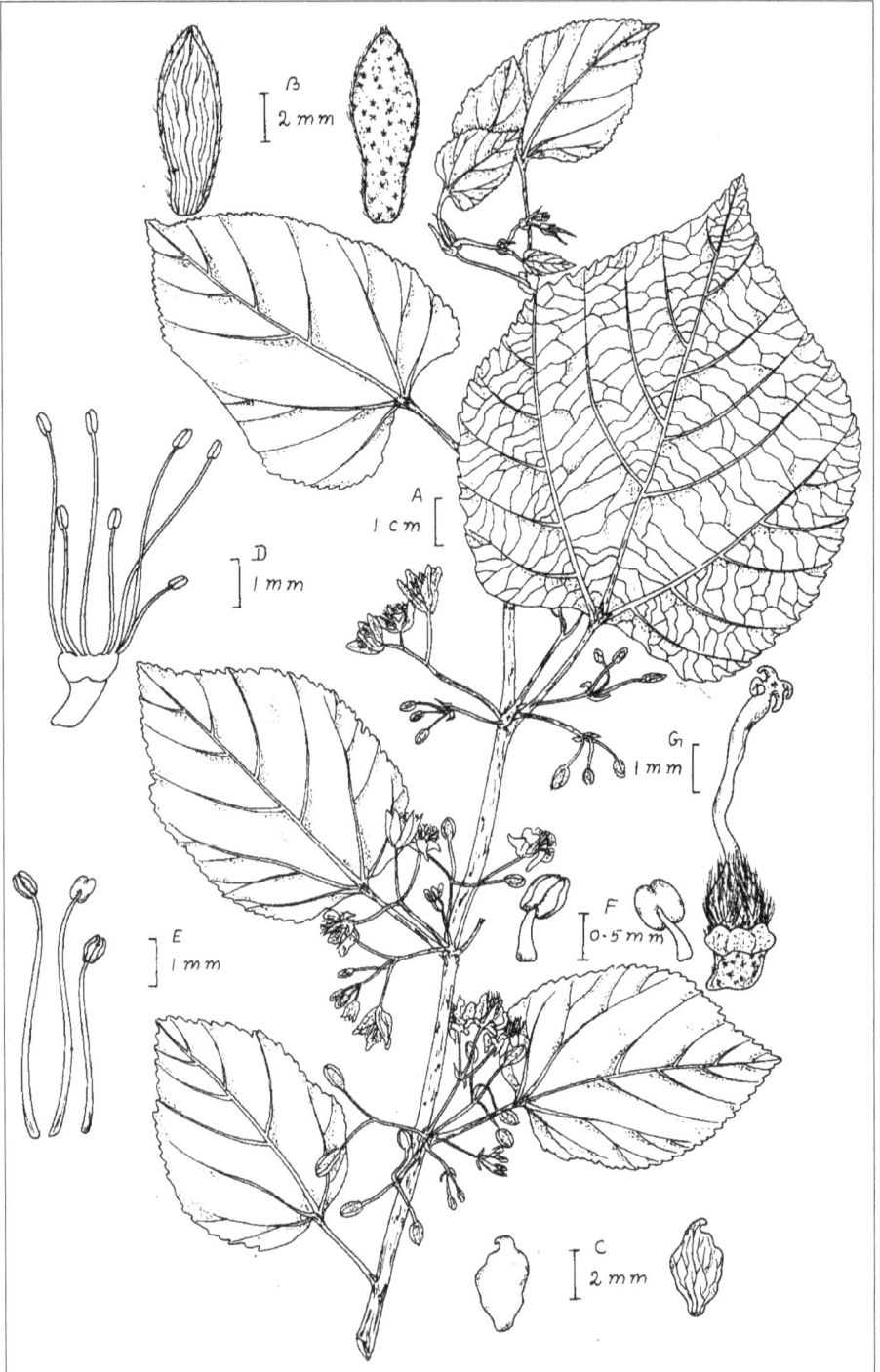

Figure 14: Grewia tiliifolia Vahl.

A. Twig, B. Sepals, C. Petals, D. Androecium, E. Stamens, F. Anther, G. Gynoecium.

Grewia villosa Willd. in Ges. Naturf. Freunde Berlin Neue Schriften 4 : 205. 1803; Masters in Hook.f., Fl. Brit. India 1 : 388. 1874; Gamble, Fl. Madras 1 : 119. 1915; Daniel & Chandrabose in B.D.Sharma & Sanjappa, Fl. India 3: 515. 1993.

Shrubs, up to 3 m high, herbaceous portions covered with long soft hairs. Leaves ovate, cordate or orbicular, 5-10 × 6-12 cm, rugose above, villous beneath, base 5-nerved, nerves prominent, villous, base cordate with a deep sinus, margin crenate or serrulate with villous tufts along serrations, apex round, obtuse or subacute; petioles up to 4 cm long. Flowers dull yellow in leaf-opposed or extra-axillary cymes; peduncles 1-5 mm long, buds ellipsoid, pilose; pedicels 2-5 mm long; sepals 5, ca 1 cm long, lanceolate, pilose; petals 5, dull yellow, spathulate, glands ca 1.5 mm long, obovoid; stamens many, ovary subglobose, densely villous, stigma laciniate. Drupe globose, stellate hairy, orange coloured, 4-stoned, rind crustaceous.

Common in dry scrub forests in Karimnagar, Hyderabad (Ramana, 2010), Ranga Reddi and Nalgonda districts.

Fl. & Fr.: April- September

Central University campus (RR), *MSM* 15161; Nagarjunakonda (NLG), *KTH* 9628 (CAL); Cherukupalli (NLG), *BVRS* 182 (NU).

TRIUMFETTA L.

1.   Leaves lobed :
   2.   Bristles on capsule hispid below .... **T. pentandra**
   2.   Bristles on capsule glabrous throughout .... **T. rhomboidea**
1.   Leaves not lobed :
   3.   Leaves ovate, younger branches densely hispid, woolly .... **T. pilosa**
   3.   Leaves orbiculr or rhomboid, younger branches
       stellate- tomentose .... **T. rotundifolia**

Triumfetta pentandra A. Rich in Guillemin & *al.*, Fl. Seneg. Tent. 93. t. 19. 1831; Gamble, Fl. Madras 1: 120. 1915; Daniel & Chandrabose in B.D.Sharma & Sanjappa, Fl. India 3: 519. 1993. *T. neglecta* Wight & Arn., Prodr. 75. 1834; Masters in Hook.f., Fl. Brit. India 1 : 396. 1874.

Annual, erect herbs or undershrubs, up to 2 m high, branchlets stellately-tomentose and simple hairs. Lower leaves palmately 3-lobed, upper ones elliptic or rhomboid, 4-7 × 1.6 cm, strigose above, stellate-tomentose below, base cuneate, margin crenate-serrate, apex acuminate; petioles up to 5.5 cm long, densely pubescent; stipules 5-6 mm long, subulate, ciliate and hispid-glandular along margins. Flowers bright yellow in terminal or extra- axillary clusters forming interrupted racemes; bracts 2-3 mm long, linear, hairy; sepals 5, lorate, cucullate, awned, stellate-pubescent; petals 5, yellow, as along as sepals, spathulate; stamens 5 (rarely up to 10); carpels 2; ovary ovoid, pubescent, stigma 2-fid. Capsule ovoid, ovoid-oblong, indehiscent, covered with hooked spines ciliated on the upper edge; seeds ovoid, slightly reddish-brown, glabrous.

Occasional in waste places and along roadsides in most districts.

Fl. & Fr.: August-January

Sirpur (ADB), *GO* 4392; Indervelli (ADB), *GO* 4532; Kalladi (NZB), *BR* & *PSPB* 9597; Narsapur – Hyderabad road (MDK), *BR* & *CP* 11214; Mohammadbad (RR), *MSM* 10439; Central University campus (RR), *MSM* 15156; Gacchibowli (RR), *MSM* 12772; Pegarikutta- Narsapur (MDK), *KMS* 6644 (CAL); Parnasala RF (KMM), *RCS* 102467 (BSID).

Triumfetta pilosa Roth, Nov. Pl. Sp. 223. 1821; Masters in Hook.f., Fl. Brit. India 1 : 394. 1874; Gamble, Fl. Madras 1 : 120. 1915; Daniel & Chandrabose in B.D.Sharma & Sanjappa, Fl. India 3: 519. 1993.

Vern.: *Tigebenda.*

Subshrubs, up to 2 m high, branchlets densely hispid and woolly. Leaves ovate, 7-12 × 3.5-7 cm, basally 3-5-nerved, thin coriaceous, stellate-tomentose above, woolly below, base obtuse or subcordate, margin unequally toothed, apex acuminate; petioles 1-3 cm long, pilose; stipules filiform. Flowers yellow, leaf-opposed, 5- flowered umbellate cymes, peduncle 1.5 cm, shorter than the petiole; sepals 5, lorate; petals 5, yellow, narrowly oblanceolate; stamens 10; carpels 4, ovary globose, 4-loculed, style subulate, stigma 4-lobed. Capsule globose, 8 mm across, covered with long hooked spines, apically glabrous, 4-celled, 8-seeded; seeds ovoid, dark brown to black, shiny.

Common in Adilabad and Ranga Reddi districts.

Fl. & Fr.: Through out the year.

Kadam dam (ADB), *GO* & *DAM* 5200.

Triumfetta rhomboidea Jacq., Enum. Syst. Pl. 22. 1760; Masters in Hook.f., Fl. Brit. India 1: 395. 1874; Gamble, Fl. Madras 1: 120. 1915; Daniel & Chandrabose in B.D.Sharma & Sanjappa, Fl. India 3: 520. 1993.

Erect, much branched, pubescent undershrubs, up to 2 m high. Leaves polymorphous, lower leaves palmately 3-lobed, upper ones rhomboid to narrow lanceolate or elliptic, 1-2.5 × 0.5-2 cm, basally 5-nerved, stellate-pubescent, base truncate or cuneate, margin serrate, apex acuminate to acute; petioles up to 5 cm long, pubescent; stipules 3-4 mm long, subulate, pubescent. Flowers yellow in terminal cymes; shortly pedicellate, buds oblong, club-shaped; sepals 5, oblong; petals 5, yellow, oblong-obovate; stamens 8-15; carpels 2-3, ovary subglobose, hairy, style subulate, stigma 2-3-lobed. Capsules globose, greenish-brown, clothed with smooth, hooked spines; seeds 1-2 in each locule.

A common weed of waste places, hedges and in undergrowth of forests of all districts.

Fl. & Fr.: August-January

Manchippa RF (NZB), *TP* & *BR* 6013; Gandivet RF (NZB), *BR* & *CPR* 7199; Sardhana (MDK), *BR* & *CP* 11627; Borapuram (MBNR), *BR* & *BSS* 32365; Pasra (WGL), *RKP* 108114 (BSID); Kambalpally (NLG), *BVRS* 207 (NU); Yellandu – Karepally (KMM), *RR* 105947 (BSID).

Triumfetta rotundifolia Lam., Encycl. 3. 421. 1792; Masters in Hook.f., Fl. Brit. India 1 : 395. 1874; Gamble, Fl. Madras 1: 120. 1915; Daniel & Chandrabose in B.D.Sharma & Sanjappa, Fl. India 3: 521. 1993.

Erect, suffruticose undershrubs, up to 2 m high, younger parts stellately hairy. Leaves orbicular-rhomboid, 1.5-3 × 1.5-2.5 cm, basally 3-5-nerved, scabrid above, woolly below, base cuneate, margin irregularly crenate or serrate, apex obtuse or round; petioles 0.5-3.5 cm long, pubescent; stipules ca 4 mm long, subulate, pubescent. Flowers yellow, in leaf-opposed cymes arranged in interrupted racemes; bracts ca 1.5 mm long, linear; sepals 5, lorate; petals 5, yellow, spathulate; stamens 10-25, as long as petals. Capsules ovoid or globose, indehiscent, densely pubescent, spines dialated at base; seeds ovoid, beaked.

Common in all districts in hedges and waste lands.

Fl. & Fr. : October-December.r

Jannaram (ADB), *TP* & *PVP* 4186; Sirpur (ADB), *GO* & *PVP* 4914; Manchippa RF (NZB), *TP* & *BR* 6054; Ghanapur (NZB), *BR* & *KH* 7106; Manjira Barrage (MDK), *BR* & *CP* 14097; Rechapalli (KRN), *GVS* 21896 (CAL); Narasampet (WGL), *KMS* 11590; Pegarikatta East-Narsapur (MDK), *KMS* 6679 (CAL); ICRISAT (MDK), *LJGV* 2630 (CAL); Kambalpally (NLG), *BVRS* 299 (NU); Poochavaram (KMM), *RCS* 102486 (BSID); Mannanur (MBNR), *SRS* 107457 (BSID).

ELAEOCARPACEAE

MUNTINGIA L.

Muntingia calabura L. Sp. Pl. 509. 1753; Murti in B.D.Sharma & Sanjappa, Fl. India 3: 570. 1993.

Large shrubs or small trees, to 6 m high; branchlets viscid, densely tawny-villous. Leaves inequilateral, oblong-lanceolate, villous above, woolly beneath; base oblique, obtuse, margin serrate, apex acuminate. Flowers axillary, solitary or in pairs; pedicels ca 1 cm long, sepals 5, free, lanceolate, acuminate; petals 5, free white, ovate-suborbicular; disc annular; glandular-pubescent; stamens numerous, inserted on the disc. Fruit globose berry, red when ripe, seeds numerous, minute.

Frequently cultivated as ornamental and avenue tree.

Fl. & Fr.: Throughout the year.

LINACEAE

1. Straggling shrubs; lower peduncles spirally hooked; stamens 10 .... **Hugonia**
1. Herbs or undershrubs; peduncles not hooked; stamens 5 .... **Linum**

HUGONIA L.

Hugonia mystax L., Sp. Pl. 675. 1753; Hook.f., Fl. Brit. India 1 : 413. 1874; Gamble, Fl. Madras 1 : 126. 1915; Hajra in B.D.Sharma & Sanjappa, Fl. India 3: 566. 1993.

Vern.: *Gatrinta.*

Rambling or climbing shrubs, branchlets tomentose. Leaves elliptic, obovate or obovate-oblong, 1-5 × 1-2 cm, coriaceous, glabrous, nerves pubescent below, base tapering, margin entire, apex obtuse or subacute; petioles ca 4 cm long, hairy. Flowers large, yellow in terminal and/or axillary cymes, lower peduncles spirally hooked; bracts ca 7 mm long, subulate; sepals 5, imbricate, ovate-lanceolate, fulvous pubescent; petals 5, unequal; twisted, alternate with sepals; stamens 10, alternately long and short, all fertile; ovary globular, glabrous, styles 4 mm long, stigmas lobed. Drupe globose, red, 1 cm across; seeds 5, compressed.

Occasional in forests of Adilabad district.

Fl. & Fr.: August - October.

Neelwai RF (ADB), *TP* & *GO* 5435.

LINUM L.

Linum usitatissimum L., Sp. Pl. 277. 1753; Hook.f., Fl. Brit. India 1: 410. 1874; Gamble, Fl. Madras 1 : 125. 1915; Hajra in B.D.Sharma & Sanjappa, Fl. India 3: 580. 1993.

Small slender, annual herbs or undershrubs, up to 50 cm high, simple or sparingly branched, branches glabrous. Leaves sessile, linear or lanceolate, 1-3.5 × 0.3-0.5 cm, base attenuate, margin entire, apex acute or acuminate, glabrous. Flowers blue, axillary, solitary and in terminal, few-flowered, corymbose panicles; sepals 5, ovate to elliptic; petals 5, blue or purple, rarely white, obovate or broadly rounded; stamens 5; ovary ovoid, styles free, stigmas linear-clavate, often cohering. Capsule 1 cm across, ovoid, subglobose, glabrous; seeds minute, compressed, dark-brown or nearly black.

Often cultivated for its oil seeds and occasionally found run wild.

Fl. & Fr. : September-March

Kadthala (ADB), *GO* & *DAM* 4994; Bhasar (ADB), *GO* & *PVP* 4642.

ERYTHROXYLACEAE

ERYTHROXYLUM P. Br.

Erythroxylum monogynum Roxb., Pl. Coromandel t. 88. 1798; Hook.f., Fl. Brit. India 1: 414. 1874; Gamble, Fl. Madras 1: 127. 1915; U. Chatterjee & B.D.Sharma in B.D.Sharma & Sanjappa, Fl. India 3: 590. 1993.

Large shrubs or small trees, 2-5 m high; bark brown, very rough, lenticellate. Leaves obovate, 2-5 × 1-2.5 cm, glabrous, base attenuate, margin entire, apex obtuse, shining above and pale glaucous brown beneath; petioles 3-8 mm long; stipules persistent, finely serrate. Flowers pale green with white disc, axillary, solitary or in clusters; calyx campanulate, persistent, glabrous, 5-lobed, lobes triangular; petals ligulate, 3-lobed, spreading; stamens 10, monadelphous, staminal tube 1-2 mm long; ovary oblong, stigmas 2-3. Drupes oblong-ellipsoid, reddish when ripe, 3-loculed, 1-seeded.

Common in deciduous forests in Nalgonda, Karimnagar, Warangal, Hyderabad and Mahabubnagar districts.

Fl. & Fr.: June- November

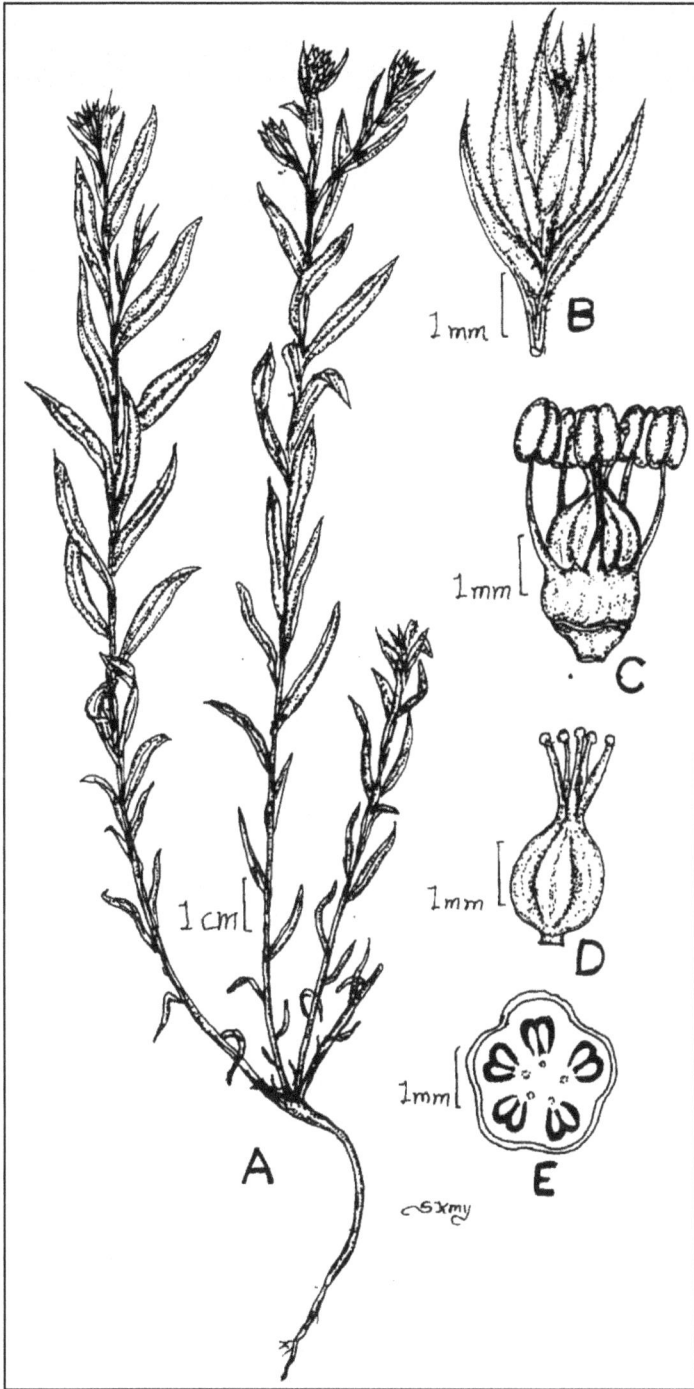

Figure 15: Linum usitatissimum L.

A. Habit, B. Flower-entire, C. Stamens & pistil, D. Pistil, E. Ovary c.s.

Chennoor RF (ADB), *GO* 4499; Jaipur RF (ADB), *GO* & *PVP* 4950; Mannanur (MBNR), *TDCK* 13640; Farahabad (MBNR), *MVS* & *MB* 38656; Pakhal (WGL), *ANH* 15950 (MH); way to Gangaram (WGL), *PVS* 76966 (BSID); Bhadrachalam RF (KMM), *RCS* 102427 (BSID); Pahapali RF (KMM), *RR* 112508 (BSID); Umamaheswaram (MBNR), *SRS* 109132 (BSID).

MALPIGHIACEAE

1. Flowers irregular; prominent gland present on one of the
    calyx lobes; samaras terminated by 1-3 lanceolate wings .... **Hiptage**
1. Flowers regular; gland absent on the calyx lobes; samaras
    surrounded by an orbicular or lanceolate wing .... **Aspidopterys**

ASPIDOPTERYS Juss.

1. Leaves pubescent beneath; samara wings more or less orbicular .... **A. cordata**
1. Leaves glabrous beneath; samara wings elliptic-oblong .... **A. indica**

Aspidopterys cordata (B. Heyne ex Wall.) A. Juss., Ann. Sci. Nat. Bot. ser. 2. 13. 267. 1840; Hook.f., Fl. Brit. India 1 : 421. 1874; Gamble, Fl. Madras 1 : 129. 1915; var. **cordata**; Srivastava in Hajra & *al.,* Fl. India 4: 5. 1997. *Hiraea cordata* B. Heyne ex Wall., Pl. Asiat. Rar. 1 : 13. 1829.

Climbing shrubs, whitish-tomentose in younger parts. Leaves broadly ovate or orbicular, 5-9 × 3-6 cm, appressed hairy to nearly glabrous above, softly whitish-pubescent beneath, base cordate, margin entire, apex acuminate; petioles 3.5-4 cm long. Flowers with yellow shade, mildly fragrant, in axillary and terminal tomentose panicles; peduncles decompound or compound; buds oblong, pedicels 3-6 mm long in flowers, 11-14 mm in fruits, slender, filiform; calyx 5-partite, lobes ovate or elliptic, glabrous; petals 5, broadly oblong or oblong-elliptic; stamens 10, filaments 2-5 mm long; ovary villous. Samaras orbicular, pale-brown, glabrous, finely reticulate, 3-winged; seeds globose.

Occasional in forests in Mahabubnagar, Adilabad, Nizamabad, Medak and Ranga Reddi districts.

Fl. & Fr.: September-February.

Kunthala waterfalls (ADB), *GO* & *DAM* 5120; Nasarullabad (NZB), *TP* & *BR* 6178; Annaram RF (NZB), *BR* & *CPR* 7178; Narsapur RF (MDK), *TP* & *CP* 14028; Mohammadabad RF (RR), *MSM* 10490, 10491; Pocharam RF (MDK), *TP* & *CP* 12063.

Aspidopterys indica (Roxb.) Hochr., Bull. Inst. Bot. Buitenzorg 19. 45. 1904; Srivastava in Hajra & *al.,* Fl. India 4: 8. 1997. *Triopteris indica* Roxb., Pl. Coromandel t. 160. 1798. *Aspidopteris roxburgiana* A. Juss., Ann. Sci. Nat. Bot. Ser. 2. 13 : 267. 1840; Hook.f., Fl. Brit. India 1: 420. 1874; Gamble, Fl. Madras 1 : 129. 1915.

Slender climbers, branches terete, younger rusty tomentose. Leaves ovate or elliptic-ovate, 7-10 × 4-6 cm, base cordate, margin entire, apex acuminate, glabrous; petioles rusty- tomentose, 8-15 mm long. Flowers greenish-white in axillary and terminal panicles; pedicels ca 16 mm long in fruits; calyx 5-partite, lobes ovate,

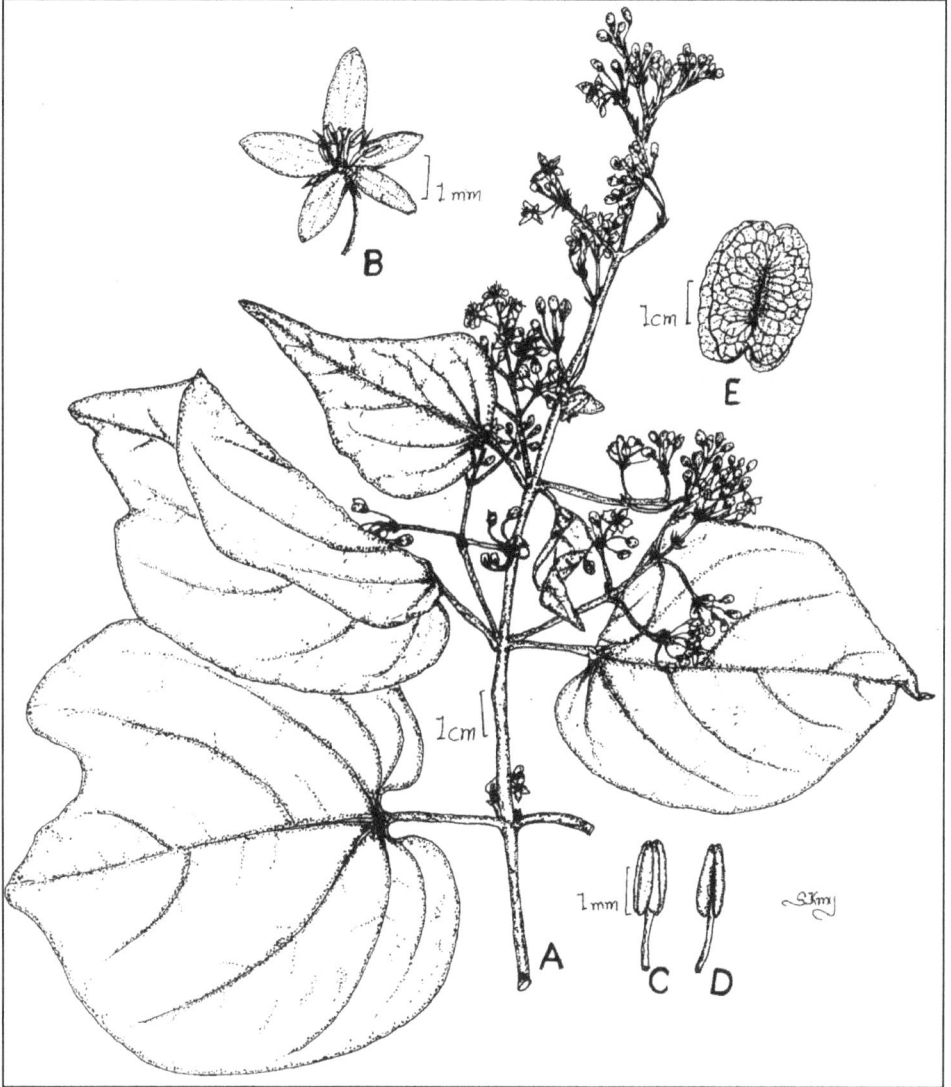

Figure 16: Aspidopterys cordata (Heyne ex Wall.) A. Juss.

A. Twig, B. Flower-entire, C&D. Stamens, E. Fruit.

tomentose; petals 5, ovate to obovate, concave; stamens 10, filaments connate at base; ovary pilose, styles 3, stigma capitate. Samaras ca 5 x 3 cm, elliptic-oblong, nerves on wings inconspicuous.

Common in forests of Karimnagar district (Naqvi, 2001).

Fl. & Fr. : September-January.

Panapally RF (KMM), *RR* 112531 (BSID); Murugampadu RF (KMM), *RCS* 102450 (BSID).

HIPTAGE Gaertn. *nom. cons.*

Hiptage benghalensis (L.) Kurz, J. Asiat. Soc. Bengal 43: 136. 1874. var. **benghalensis**; Srivastava in Hajra *& al.,* Fl. India 4: 14. 1997. *Banisteria bengalensis* L., Sp. Pl. 427. 1753. *Hiptage madablota* Gaertn., Fruct. 2: 169. 1791; Hook.f., Fl. Brit. India 1: 418. 1874; Gamble, Fl. Madras 1: 218. 1915.

Large climbing shrubs, branches brown, terete, lenticellate, branchlets densely woolly. Leaves ovate, elliptic-oblong, 6-13 × 4-5 cm, glaucous above, glabrous beneath, 2 glands at the base of leaf blade, base obtuse, margin entire, apex acute or acuminate; petioles ca 1 cm long, silky-pubescent. Flowers white in large axillary and terminal racemes; calyx 5-partite, lobes ovate, elliptic or oblong, fulvous sericeous outside with one large gland partly adnate to pedicel; petals 5, clawed, suborbicular to obovate, fringed on margins, fragrant; stamens 10; ovary pubescent. Samara with unequal wings, median one bigger than the laterals; seed solitary, globose, glabrous.

In hilly areas of the forests in Nizamabad and Mahabubnagar districts. Planted in Hyderabad as ornamental.

Fl. & Fr.: February-April.

Mudheli - Mondisadak road (NZB), *TP* & *BR* 6364; Pullaechelma (MBNR), *SRS* 109199 (BSID); Umamaheswaram (MBNR), *SRS* 109141.

ZYGOPHYLLACEAE

1. Prostrate silky herb; flowers yellow; fruit prickly .... **Tribulus**
1. Erect prickly herb; flowers pinkish purple; fruit without prickles .... **Fagonia**

FAGONIA L.

Fagonia indica Burm.f., Fl. Ind. 102. t 34. f.1. 1768; P. Singh & V. Singh in Hajra & *al.,* Fl. India 4: 44. 1997. *F. cretica* auct non L. 1753. *F. arabica* auct non L. 1753; Edgew. & Hook.f., Fl. Brit. India 1 : 425. 1874; Gamble, Fl. Madras 1: 130. 1915.

Vern.: *Chittigara, Hinuna.*

Small, green, spiny herbs with erect branches, more or less glandular, up to 50 cm high, nodes swollen. Leaves unifoliolate, sometimes a few basal 3-foliolate; stipular spines equal or shorter than leaves, occasionally reduced. Flowers small, pinkish-purple; pedicels 4-6 mm long; sepals 5, persistent, ovate, glandular outside; petals 5, spathulate; stamens 10, inserted on disc. Capsules ca 4 x 4 mm, softly hairy.

Occasional in wastelands in Mahabubnagar district.

Fl. & Fr.: Through the year.

TRIBULUS L.

1. Style inconspicuous, hardly 0.5 mm long; leaflets oblong .... **T. terrestris**

1. Style 1.5-2.5 cm long; leaflets ovate-oblong:

    2. Style glabrous; intrastaminal glands not ciliated .... **T. lanuginosus**

    2. Style puberulous; intrastaminal glands ciliated .... **T. subramanyamii**

Tribulus lanuginosus L., Sp. Pl. 1: 387. 1753; P. Singh & V.Singh in Hajra & *al.,* Fl. India 4: 51. 1997.

Annual or biennial, usually prostrate, silky-pubescent herbs. Leaves up to 5 cm long, leaflets 3-6 paris, 3-10 × 2-5 cm, ovate-oblong, base oblique, apex acute; stipules lanceolate or falcate. Flowers pale-yellow, solitary, axillary; sepals 5, lanceolate; petals 5, obovate-oblong; stamens 10, intrastaminal glands triangular, free, extra staminal glands thick; ovary ca 1 mm across, stigma faintly 5-rayed or sub-capitate. Fruits disc-shaped, mericarps ca 4 x 2.5 mm, 2-4-spined.

Rare in Warangal (C.S. Reddy, 2001) and Mahabubnagar districts (Raghava Rao, 1989).

Fl. & Fr.: Throughout the year.

Tribulus subramanyamii P. Singh & *al.,* Bull. Bot. Surv. India 25 : 197. 1983; P. Singh & V.Singh in Hajra & *al.,* Fl. India 4: 55. 1997.

Prostrate, spreading herbs, branches up to 40 cm long, solid, striate, clothed with silky white hairs, especially when young. Leaves usually opposite, one of each pair alternately smaller, aborting, sometimes lower ones alternate, paripinnate, leaflets usually 5 pairs, rarely 3,4 or 6 pairs, terminal pair always directed upward, subsessile, ovate or ovate-oblong, 6-17 × 6-10 mm, unequal-sided, apex apiculate or mucronate, brownish pubescent above, densely silky villous beneath. Flowers yellow, axillary, solitary, alternating from the axils of smaller leaves; sepals 5, linear-lanceolate; petals 5, obovate; stamens 10, intrastaminal glands free, extrastaminal glands thick, orbicular to oblong; ovary ca 2.5 mm across, stigma pyramidal, 5-rayed. Fruit a 5-lobed schizocarp, horizontally depressed with 5 mericarps; seeds small, 3-5 in each mericarp.

Occasional in wastelands in Warangal district.

Fl. & Fr. : June-December.

Subedari (WGL), *RKP* 105247 (BSID).

Tribulus terrestris L., Sp. Pl. 387. 1753; Edgew. & Hook.f., Fl. Brit. India 1: 423. 1874; Gamble, Fl. Madras 1: 130. 1915; P. Singh & V.Singh in Hajra & *al.,* Fl. India 4: 55. 1997.

Vern. : *Palleru.*

Prostrate, much branched, hirsute or silky hairy, annual herbs. Leaves 2-6 cm, opposite or alternate, paripinnate, leaflets 4-7 pairs, oblong, 0.5-2 × 0.2-1 cm, villous-hairy, base oblique, margin entire, apex mucronate. Flowers yellow, axillary or leaf-opposed, solitary; sepals 5, ovate-lanceolate; petals 5; stamens 10; ovary ca 1.5 mm across, with stiff upward spreading bulbous-based hairs, style ca 1 mm long, stigma 5-rayed, slightly asymmetrical. Fruit globose, spiny, of 5 cocci, each coccus hairy, each with two long, hard, sharp divaricate spines; seeds minute, oblong, numerous.

Common weed in wastelands, road sides and open forests during rainy season in all districts.

Fl. & Fr.: May-December

Bijjur RF (ADB), *GO* & *PVP* 4853; Kammarpally (NZB), *TP* & *BR* 6193; Manjira barrage (MDK), *BR* & *CP* 14093; Thandur (RR), *TP* & *MSM* 12702; Rachakonda (NLG), *BVRS* 361 (NU); Vijayapuri (NLG), *KMS* 9838 (MH); Moosi River bank (HYD), *KMS* 5963 (MH); Kodimial (KRN), *GVS* 20045 (MH); Appapur (MBNR), *BSS* & *SKB* 34454; Amangal to Shadnagar (MBNR), *SRS* 104581 (BSID); Yenkur (KMM), *RR* 108542 (BSID).

Note: This is an exceedingy variable species in both vegetative and reproductive characters. Hadidi in Rech.f., Fl. Iran. No. 98: 15-18. 1972 recognised 4 varities under this species *viz.*, i) var. *orientalis* (Kerner) G. Beck., ii) var. *robustus* (Boiss. & Boe) Boiss., iii) var. *bicornutus* (Fischer & Mey) Hadidi and iv) var. *inermis* Boiss; but the characters used for separating them are overlapping and variable. Hence not recognized here.

OXALIDACEAE

1. Diffuse herbs; leaves digitate, 3-foliolate; flowers 1-4 on long axillary peduncles; fruit linear-cylindric, far exceeding the persistent calyx .... **Oxalis**
1. Erect herbs; leaves even pinnate, leaflets 6-15 pairs; flowers many in terminal, crowded long peduncled umbels; fruit ellipsoid, not exceeding the persistent calyx .... **Biophytum**

BIOPHYTUM DC.

1. Seeds with two longitudinal ridges with transverse rows of tubercles in between .... **B. umbraculum**
1. Seeds spirally, obliquely or transversely ridged or tuberculate:
   2. Leaflets conspicuously nerved beneath .... **B. nervifolium**
   2. Leaflets not conspicuously nerved beneath:
      3. Sepals almost as long as petals and longer than the capsule .... **B. sensitivum**
      3. Sepals about half as long as petals and as long as or shorter than the capsules .... **B. reinwardtii**

Biophytum nervifolium Thwaites, Enum. Pl. Zeyl. 64. 1858; Manna in Hajra & *al.,* Fl. India 4: 234. 1997. *B. sensitivum* DC. var. *nervifolium* (Thwaites) Edgew. & Hook. f. in Hook. f., Fl. Brit. India 1: 437. 1874.

Annual, unbranched herbs, 6-25 cm high. Leaves radical, 5-10 cm long, paripinnate, leaflets oblong, 0.5-1.5 × 0.4-0.7 cm, glabrous, base truncate, margin entire, apex apiculate. Flowers yellow, often tinged red, in terminal umbels; pedicels shorter than sepals; bracts hairy; sepals 5, linear-lanceolate; petals 5, much exceeding sepals, oblanceolate, attenuate at base, entire at margin, rounded at apex, with a distinct red midnerve; ovary 2.5-3 mm long, styles hairy, stigma papillose. Capsule oblong-ovoid, glabrous, minutely apiculate; seeds ovoid, 2-4 per cell, light- brown, transversely ridged.

Common weed in wastelands, cultivated fields, rare in the forests in Adilabad, Khammam and Warangal districts.

Fl. & Fr.: September-December.

Sone (ADB), *TP* & *PVP* 4058; Dabba (ADB), *GO* & *PVP* 4836; Pakhal (WGL), *KMS* 11686 (MH); Koriguttalu (KMM), *RR* 106048 (BSID); Veerapuram (WGL), *PVS* 76962 (BSID).

Biophytum reinwardtii (Zucc.) Klotzsch, Naturw., Reise Mossamb. Bot. 1: 85, 1862; Edgew. & Hook.f. in Hook.f., Fl. Brit. India 1: 437. 1874; Gamble, Fl. Madras 1: 133. 1915; Manna in Hajra & *al.,* Fl. India 4: 236. 1997. *Oxalis reinwardtii* Zucc., Abh. Math. Phys. Cl. Koenigl. Bayer Akad. Wiss. 1 : 274. 1829. 1830.

Herbs, up to 10 cm high, branches reddish, sparsely retrorsely hairy. Leaves 8 cm long, crowded at the top of stem, leaflets 6-11 pairs, ovate or obovate, base rounded or truncate, about 10 pairs, increasing in size towards ends of branches, pale beneath, lateral nerves faint, parallel. Flowers yellow with red streak near the base of each lobe, in terminal umbels, peduncles hairy; sepals 5, ovate-lanceolate; petals 5, elliptic to oblanceolate; filament 1-1.5 mm long, glabrous; ovary glabrous, stigma flattened. Capsule 0.3 cm long, ovoid; seeds grooved, muricate.

Occasional in forests of Mahabubnagar district.

Fl. & Fr.: July-January.

Mannanur (MBNR), *BSS* & *SKB* 30653 (BSID).

Biophytum sensitivum (L.) DC., Prodr. 690. 1824; Edgew. & Hook. f. in Hook. f., Fl. Brit. India 1: 436. 1874. *p.p.*; Gamble, Fl. Madras 1: 133. 1915; var. **sensitivum;** Manna in Hajra & *al.,* Fl. India 4: 238. 1997. *Oxalis sensitiva* L., Sp. Pl. 434. 1753; *p.p.*

Erect herbs, about 8 cm high, sensitive to touch, stems reddish pink, downy pubescent. Leaves about 2-6 cm long, several, crowded at top of stem, rachises sparsely hairy; leaflets 7-15 pairs, increasing in size towards ends of rachises, oblong-obovate, about 1 × 0.5 cm, base truncate, equal or unequal, nerves irregular, prominent beneath, marginally jointed. Flowers deep yellow, often tinged red, shortly pedicellate, in terminal umbels; sepals 5, lanceolate, to 5 mm long, glandular-pubescent, 5-nerved;

petals 5, oblong, to 8 mm long; stamens 10, short ones to 1 mm long, long ones to 3 mm long, staminal tube to 1.5 mm long, sparsely puberulous. Capsules 0.5 cm long, ovoid; seeds 2-4 in each locule, concentrically 9-ridged.

In open places, on roadsides, and as a weed in cultivated fields in almost all districts.

Fl. & Fr.: June-December.

Manchippa RF (NZB), *TP* & *BR* 6027; Ghanapur RF (NZB), *BR* & *KH* 6499; Ganpur RF (MDK), *TP* & *CP* 12073; Anantagiri RF (RR), *TP* & *MSM* 11086; Pargi (RR), *MSM* 15106; Madimalakala (MBNR), *TDCK* 16836; Bhongir (NLG), *BVRS* 429 (NU); Murugampadu RF (KMM), *RCS* 102433 (MH, BSID).

Key to Varieties
1. Leaflets more or less glabrous; seeds prominently ridged,
   not tubercled .... var. **sensitivum**
1. Leaflets strigosely hirsute; seeds ridged and tubercled .... var. **candolleanum**

Biophytum sensitivum (L.) DC. var. **candolleanum** (Wight) Edgew. & Hook. f. in Hook.f., Fl. Brit. India 1: 437. 1874. *B. candolleanum* Wight, Ill. 1: 161. t. 62. 1840; Gamble, Fl. Madras 1: 133. 1915.

Annual herbs, up to 35 cm high, hispidly pubescent. Leaves sensitive, crowded on top of stem; rachis 8-15 cm long, petioles short, leaflets opposite, 6-15 pairs. Peduncles of various lengths, up to 14 cm long, appressed-strigose and patently glandular-hairy, bracts lanceolate, small, crowded beneath short pedicels, sepals ovate-lanceolate, acute, 4-7 × 0.5-1.5 mm, strigose and glandular hairy, longer than pedicels, petals lanceolate, 5-7 × 1-2 mm, yellow, anthers 1.5-2.5 mm long, ovary 0.5-0.75 mm, apically ciliate, ovules 2-5 per locule, style 0.5-1 mm, stigma flattened.

Rare, occurs in forest in Khammam district.

Fl. & Fr.: September- February.

Mothugudem (KMM), *VSR* 3476 (KU).

Biophytum umbraculum Welw., Apont. 590. 1859. *B. petersianum* Klotzch in Peters, Reise Mossamb. Bot. 1: 81, t. 15. 1862; Veldk., Fl. Thailand 2: 18. 1970 & in Steenis Fl. Males. Ser. 1. 7(1): 161. 1971; Manna in Hajra & *al.,* Fl. India 4: 235. 1997. *Oxalis apodiscias* Turcz., Bull. Soc. Imp. Naturalistes Moscou 36: 595. 1863. *Biophytum apodiscias* (Turcz.) Edgew. & Hook.f. in Hook.f., Fl. Brit. India 1: 437. 1874; Gamble, Fl. Madras 1: 133. 1915.

Annual herbs, up to 15 cm tall. Leaves 3-9 pairs, rachis 3-5 mm long, glabrous to appressed-pubescent mainly at nodes; leaflets overlapping, terminal ones 2-8 × 2-5 mm, obovate, midnerve eccentric, others triangular, orbicular or elliptic, apex rounded to obtuse, glabrous or sparsely ciliate along margins, midnerve median. Pedicels 1-3 mm long with few bristly hairs, sepals ovate-lanceolate, acute, 3-5 × 1-2 mm, sparsely hairy, petals lanceolate, 5-6 × 1-2 mm, yellow, ovary 1-2 × 1 mm, ovules 4-5 per locule; styles 1-2 mm long, stigma flattened, bifid. Capsules 3-4 × 2-2.5 mm.

Occasionally met with in cultivated fields and forest fringes in Medak district.

Fl. & Fr.: September – February.

Sivampet RF (MDK), *RG* 116474.

OXALIS L.

1. Plants acaulescent; leaves radical .... **O. dehradunensis**
1. Plants caulescent, creeping; leaves cauline:
    2. Flowers 1-2 or in simple umbels .... **O. corniculata**
    2. Flowers in broad umbelliform false racemes .... **O. spiralis**

Oxalis corniculata L., Sp. Pl. 435. 1753; Edgew. & Hook.f. in Hook.f., Fl. Brit. India 1 : 436. 1874. Gamble, Fl. Madras 1 : 132. 1915; Manna in Hajra & *al.,* Fl. India 4: 242. 1997.

Annual or perennial herbs, branchlets creeping, rooting at nodes. Leaves digitately 3-foliolate, leaflets obcordate, 0.4-1.5 × 0.5-1.6 cm, chartaceous, pilose, base cuneate, entire, tip emarginate. Flowers yellow, subumbellate, axillary, 1-6-flowered; peduncles up to 20 cm high, usually much shorter; pedicels up to 20 mm long; sepals 5, lanceolate, obliquely retuse with pale narrow margins; petals 5, oblong, yellow, obtuse to emarginate, puberulous outside, 6-nerved; stamens 10, united, long ones 5 mm long, short ones 4 mm long; ovary ellipsoid to cylindric, stigma small, sometimes flattened and minutely bifid. Capsule oblong, beaked, glabrescent; seeds brown or reddish brown.

Common in shady, moist localities in most districts.

Fl. & Fr.: July-December.

Alisagar (NZB), *BR* 7285; Medak (MDK), *BR* & *CP* 11469; Central university campus (RR), *TP* & *MSM* 15134; Mannanur (MBNR), *TDCK* 16801, *AJR* 20485; Srinivasaur (NLG), *BVRS* 267, 401 (NU).

Oxalis dehradunensis Raizada, Suppl. Fl. Upp. Gang. Plain 37-38. 1976; Manna in Hajra *& al.,* Fl. India 4: 246. 1997. *O. intermedia* A. Rich., Ess. Fl. Cuba 315. 1845, non Steudel. 1841. *O. richardiana* Babu, Herb. Fl. Dehra Dun 104. 1977. *O. latifolia* auct. plur non Kunth 1822; Pullaiah & Chennaiah, Fl. Andhra Pradesh 1: 167. 1997

An annual, acaulescent herb, 15-25 cm high, with conical bulbs, with numerous basal erect stolons with a few small scales. Leaves 2-6, radical, 3-foliolate, leaflets subsessile, fish tail shaped, broadly deltoid, sub-bilobed with lobes divergent, apex acute, base broadly but distinctly cuneate, ca 1.5 × 3 cm at broadest part. Flowers rose, in 5-6-flowered umbels, pedicellate, sepals 5, elliptic-oblong, equal, ca 4 mm × 1.5 mm, corolla infundibuliform, 1-1.5 cm long; petals 5; stamens 10, alternately long and short, longer filaments inappendiculate, pubescent, ca 4 mm long, short ones hairy, ca 2 mm long; styles 5, distinct, glabrous, stigmas 5, discoid.

Native of central and tropical south America. Weed in Ranga Reddi district.

Fl. & Fr.: May – November.

Central University campus (RR), *TP* & *MSM* 15135.

Oxalis spiralis G.Don, Gen. Hist. 1: 755. 1831. *O. pubescens* Kunth, Nov. Gen. Sp. 5: 240. 1820; Calder, Rec. Bot. Surv. India 6(8): 333. 1919; Fyson, Fl. S. Indian Hill Stat. 74, t. 48. 1932; Manna in Hajra & *al.,* Fl. India 4: 253. 1997.

Procumbent herbs, rooting at nodes. Leaves 3-foliolate, petioles 4-7 cm long, sparsely hairy; stipules oblong; leaflets unequal, median leaflet obcordate, ca 2.5 × 1.5 cm, membranous, very sparsely hairy above, appressed-hairy beneath. Cymes umbelliform with irregular branches with flowers arranged in a pseudo-raceme, peduncles axillary, solitary, 15-17 cm long, arising chiefly towards apices of stem, pedicels hairy, 5-7 mm long; bracts subulate, minute, 2-3 mm long, pubescent; sepals 5, lanceolate, acute, ca 4 × 1 mm, pubescent, membranous, persistent; petals 5, twice as long as sepals, ca 8 mm long, connate in middle, bases and apices free, orange-yellow; stamens 10, 5 long and 5 short, glabrous; styles 5, erect, capillary, pubescent, stigmas 5.

Occasional, occurs in moist shady places. Shiva Kumar and Prabhakar (2002) recorded first time in Andhra Pradesh.

Fl. & Fr.: August – January.

Hyderabad, *Shiva Kumar* 537 (OU).

BALSAMINACEAE

IMPATIENS L.

Impatiens balsamina L., Sp. Pl. 938. 1753; Hook.f., Fl. Brit. India 1 : 453. 1874; Gamble, Fl. Madras 1: 142. 1915; Vivekananthan & *al.* in Hajra *&al.,* Fl. India 4: 123. 1997.

1. Spur long, not incurved; flowers medium sized; leaves lanceolate or elliptic-lanceolate .... var. **coccinea**

1. Spur short, incurved; flowers small; leaves linear-lanceolate ...... var. **balsamina**

Impatiens balsamina L. var. **balsamina**; Hook. f., Fl. Brit. India 1 : 453. 1874; Gamble, Fl. Madras 1: 142. 1915.

Erect, annual, glabrescent herbs, up to 1 m high. Leaves alternate, linear-lanceolate, 2.5-5 × 1.5-3 cm, sparsely pubescent, base cuneate, margin serrate, apex acute. Flowers pale pink, solitary or in clusters of 2-3, purple or white; pedicels 1-2 cm long; lateral sepals ovate-lanceolate, lip cymbiform, petaloid with yellowish blotch, spur filiform, incurved, standard with a crest terminating into a mucro, wings obovate, deeply notched, auricles ovate-rounded. Capsules oblong, tomentose, seeds ovoid, smooth.

Occasional in most districts in hilly regions at low levels.

Fl. & Fr.: September-March.

Bothampalli (ADB), *GO* & *PVP* 4808; Tiryani RF (ADB), *PVP* 9810; Mudheli (NZB), *BR* & *CPR* 7214; Medak (MDK), *CP* 11729; Anantagiri (RR), *TP* & *MSM* 11081.

Impatiens balsamina L. var. **coccinea** Hook. f., Fl. Brit. India. 1 : 454. 1874 & Rec. Bot. Surv. India. 4: 7. 1904 & 4 : 32. 1905.

Erect, 30-60 cm high, simple or sparingly branched, nearly glabrous herbs. Leaves lanceolate or elliptic-lanceolate, 2.5-16 × 1-3 cm, glabrous, base attenuate, margin very deeply serrate, apex acuminate. Flowers light to deep-purple or bright reddish-purple, axillary, solitary or 2-6 fascicled. Capsules 0.5-1.2 cm long, ellipsoidal or ovoid, apiculate, tomentose.

Common in Karimnagar district.

Fl. & Fr.: August-December

Aklaspur (KRN), *GVS* 25660 (MH).

RUTACEAE

1. Plants armed :
   2. Leaves uni-foliate
      3. Petioles usually winged .... **Citrus**
      3. Petioles not winged .... **Atalantia**
   2. Leaves 3-many-foliolate :
      4. Stragglers or lianas .... **Toddalia**
      4. Shrubs or trees :
         5. Petiole not winged. Berries with a woody rind .... **Aegle**
         5. Petiole winged:
            6. Disc annular, stamens 10-12; seeds numerous .... **Limonia**
            6. Disc columnar, stamens 8-10; seeds 3-4 .... **Naringi**
1. Plants unarmed :
   7. Plants with pinnately compound leaves .... **Murraya**
   7. Plants with 1 or 3-foliolate leaves .... **Glycosmis**

AEGLE Correa *nom. cons.*

Aegle marmelos (L.) Correa, Trans. Linn. Soc. London 5: 223. 1800; Hook.f., Fl. Brit. India 1: 516. 1875; Gamble, Fl. Madras 1 : 161. 1915; K.N.Nair & M.P.Nayar in Hajra & *al.*, Fl. India 4: 264. 1997. *Crataeva marmelos* L., Sp. Pl. 444. 1753.

Vern.: *Maredu, Bilvamu, Sriphalamu.*

Armed, densely foliaceous deciduous trees, up to 12 m high, branchlets pubescent, spines axillary, straight, single or paired, reddish-brown in colour, bark grey-white with longitudinal wrinkles. Leaves 3-foliolate (rarely 5), leaflets ovate, elliptic or lanceolate, 3-7 × 2-2.5 cm, glabrous, base cuneate or rounded, margin subcrenulate,

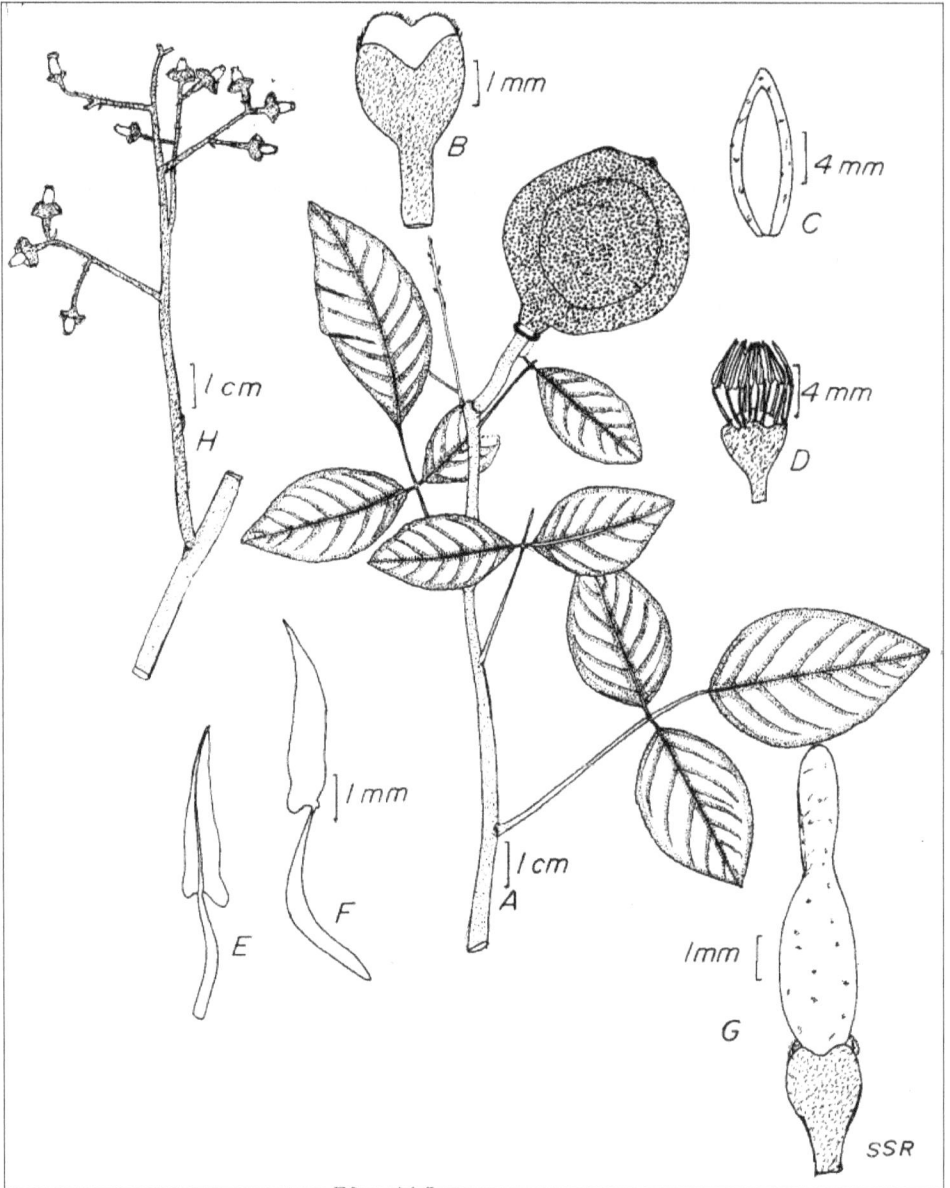

Figure 17: Aegle marmelos (L.) Correa
A. Twig, B. Calyx, C. Petals, D. Androecium, E & F & Stamens, G. Pistil, H. Flowering tiwg.

apex obtuse. Flowers greenish white in axillary panicles; pedicels 2-4 mm long, densely puberulent; calyx cupular with 5 small deltate or suborbicular teeth, caducous; petals 4 or 5, ovate-oblong; stamens 30-40 in irregular 2 or 3 series, free or irregularly coherent at base, unequal; ovary ovoid-oblong, style very short, stigma cylindric or bluntly conical. Fruit large, ovoid or globose berry with rough woody rind; seeds numerous, embedded in aromatic pulp, oblong, compressed.

Occasional in forests and grown in backyards and in temples in almost all districts.

Fl.: March – April; Fr.: September- December.

Mahaboobghat (ADB), *GO* & *PVP* 4260; Vempalli RF (ADB), *GO* 4434; Mudheli RF (NZB), *BR* & *KH* 7130; Pathur (MDK), *TP* & *CP* 12129; Pocharam sanctuary (MDK), *BR* & *CP* 11608; D. Thanda (RR), *MSM* & *KH* 10253; Mannanur (MBNR), *TDCK* 13655; Pakhal (WGL), *ANH* 15919 (MH); Rechapalli (KRN), *GVS* 22257 (MH); Waddipatla (NLG), *BVRS* 422 (NU); Umamaheswaram (MBNR), *SRS* 109124 (BSID).

ATALANTIA Correa ex Koenig *nom. cons.*

Atalantia monophylla (L.) DC., Prodr. 1: 535. 1824; Hook.f., Fl. Brit. India 1 : 511. 1875; Gamble, Fl. Madras 1: 159. 1915; var. **monophylla**; K.N.Nair & M.P.Nayar in Hajra & *al.,* Fl. India 4: 266. 1997. *Limonia monophylla* L., Mant. Pl. 237. 1771.

Vern.: *Adavi-nimma, Yerramunuludu.*

Armed trees, up to 8 m high, branchlets velutinous, spines 1 or 2, axillary on branches. Leaves simple, lanceolate, 3-7 x 2-2.5 cm, glabrous, base obtuse or cuneate, margin entire, apex retuse; petioles up to 1 cm long. Flowers white, fragrant, in axillary clusters; pedicels slender, filiform; calyx cupular, splitting irregularly into 2, or sometimes 3 or 4 irregular lobes at anthesis, persistent; petals 4 or 5, oblong-elliptic, to 7 mm long, clawed below, sometimes adnate to staminal tube at base, glabrous, white; staminal tube long, up to 6 mm, stamens 8 or 10, disc annular, obscurely 8- or 10-lobed; ovary ovoid, styles as long as ovary, stigma capitate. Berry globose, greenish yellow; seeds 4, ovoid, ca 7, 5x 4 mm.

Common in dry deciduous and thorny scrub forests in Khammam, Mahabubnagar and Nalgonda districts.

Fl. & Fr.: September-May.

Bhadrachalam (KMM), *JSG* 15871 (CAL); Rachakonda (NLG), *BVRS* 436 (NU).

CITRUS L.

1. Petioles not winged; fruits longer than broad .... **C. medica**

1. Petioles margined or narrowly to broadly winged; fruits not as above:

   2. Petioles narrowly margined or winged .... **C. limon**

   2. Petioles narrowly to broadly winged:

      3. Flowers 2-2.5 cm across; fruits to 5 cm across;
         pericarp thin, leathery .... **C. aurantifolia**

3. Flowers 4 – 4.5 cm across; fruits more than 6 cm across; pericarp rather thick:

    4. Fruits with loosely attached pericarp.

        5. Leaf base subcordate, sparsely pubescent on twigs and undersurface of leaves; fruits 10-20 cm diam ..... **C. grandis**

        5. Leaf base other than subcordate; glabrous on twigs and undersurface of leaves; fruits 6 cm in diam .... **C. aurantium**

    4. Fruits with adherant pericarp .... **C. sinensis**

Citrus aurantifolia (Christm.) Swingle, J. Wash. Acad. Sci. 3: 465. 1913. K.N. Nair & M.P. Nayar in Hajra & al., Fl. India 4: 278.1997. *Limonia aurantifolia* Christm. in L., Pflanzensyst. 1: 618. 1777. *Citrus medica* L. var. *acida* (Roxb.) Hook.f., Fl. Brit. India 1: 515. 1875; Gamble, Fl. Madras 1: 115. 1915. *Citrus acida* Roxb., Fl. Ind. 3: 391. 1832.

Vern.: *Nimma*

A shrub or small tree, up to 3 m high, branchlets slender with sharp spines. Leaves 1-foliate, petiole narrowly winged, leaflets ovate-lanceolate, coriaceous, glabrous. Flowers in axillary cymes; calyx cupular, lobes 4, deltoid, glandular; petals 4, oblong, glandular, white; stamens 20-28, filaments polyadelphous, white; ovary globose-depressed, style cylindric, stigma capitate. Fruit oblong, pitted, glandular, green; seeds ovoid.

Cultivated for its fruits.

Fl. & Fr.: April – January.

Citrus aurantium L., Sp. Pl. 782. 1753; Hook.f., Fl. Brit. India 1: 515. 1875. K.N. Nair & M.P. Nayar in Hajra & al., Fl. India 4: 279.1997.

Vern.: *Mallika narangi.*

Shrubs or small trees, 3-5 m high; branchlets angular. Leaves elliptic or obovate, 6-9 × 3-4 cm. Flowers white; sepals 5 (4), gamosepalous; petals 5 (4), free, gland dotted; stamens as many as petals. Fruits depressed globose, rind loose, orange-yellow when ripe.

Cultivated occasionally for its edible fruits.

Fl. & Fr.: May – October.

Citrus grandis (L.) Osbeck, Degbok Ostind Res. 98. 1757. *C. aurantium* L. var. *grandis* L., Sp. Pl. 793. 1753. *C. decumana* Murr., Syst. ed. 13. 508. 1774; Hook.f., Fl. Brit. India 1: 515. 1875.

Vern.: *Pampala masam.*

Small trees, young shoots pubescent. Leaves 15-20 cm long, ovate oblong, apex emarginated; pubescent beneath; petioles broadly winged. Flowers white, axillary, solitary or clustered; sepals 5(4), gamosepalous; petals 5 (4), free, gland dotted;

stamens as many as petals. Fruits globose or obovate, pale yellow, rind thick, pulp pinkish, acid-sweet.

Cultivated occasionally.

Fl. & Fr.: May – October.

Citrus limon (L.) Burm.f., Fl. Ind. 173. 1768. K.N. Nair & M.P. Nayar in Hajra & al., Fl. India 4: 282. *Citrus medica* L. var. *limon* L., Sp. Pl. 782. 1753. *Citrus medica* L. var. *limonum* (Risso) Hook.f., Fl. Brit. India 1: 515. 1875; Gamble, Fl. Madras 1: 115. 1915. *Citrus limonum* Risso, Ann. Mus. Natl. Hist. Nat. 20: 201. 1813.

Vern.: *Dabba.*

Tree, 3-4 m high, spines strong, axillary. Leaves 1-foliate, petiole winged, leaflets elliptic or ovate-lanceolate, base rounded, margin serrate, apex obtuse, coriaceous, glabrous. Flowers white, in axillary cymes; sepals orbicular, minute; petals ovate-oblong; stamens 20-30, filaments connate at base, free above, glabrous; disc annular; ovary cylindric, styles ca 3.5 mm long, stigma globose. Fruit oblong, 8-10 cm in diam., yellow when ripe; seeds ovoid.

Cultivated for its fruits.

Fl. & Fr.: May – November.

Citrus medica L., Sp. Pl. 752. 1753; Hook.f., Fl. Brit. India 1: 514. 1875; Gamble, Fl. Madras 1: 161. 1915; K.N.Nair & M.P.Nayar in Hajra & *al.,* Fl. India 4: 284. 1997.

Vern.: *Madiphalamu.*

Shrubs or small trees, up to 3 m high, young shoot glabrous, purple, with sharp, stout, axillary spines (ca 4 cm long). Leaves 1- foliolate, leaflets oblong, up to 15 cm, base acute, margin crenate-serrate, apex usually obtuse, petioles ca 10 mm long, wingless or slightly marginate. Flowers white, usually tinged with red, small, 5-10 in a raceme; calyx urceolate, 4- or 5-lobed; petals 4 or 5, oblong or oblanceolate; stamens 30-40, polyadelphous; ovary cylindric, stigma globose. Fruit large, oblong or obovoid, mamilla obtuse, rind usually warted, thick, tender, aromatic, pulp scanty, subacid.

Cultivated for its fruits.

Fl.: September – December; Fr.: November – March.

Citrus sinensis (L.) Osbeck, Deg. Book. Osting. Res. 98. 1757. *C. aurantium* L. var. *sinensis* L., Sp. Pl. 783, 1753.

Vern.: *Narinja.*

Small trees, to 4 m high; branchlets glabrous. Leaves ovate-oblong, shortly acuminate, margin crenate or sinuate. Flowers white; sepals 5 (4), gamosepalous; petals 5 (4), free, gland dotted; stamens as many as petals. Fruit globose, rind green turning to yellow, pulp sweet.

Cultivated on a large scale especially in Nalgonda district.

Fl. & Fr.: Through out the year.

Sattenapally (ADB), *TP* & *PVP* 4150.

GLYCOSMIS Correa, *nom. cons.*

Glycosmis pentaphylla (Retz.) DC., Prodr. 1 : 538. 1824; quoad basionym; Tanaka, J. Indian Bot. Soc. 16 : 229. 1937; K.N.Nair & M.P.Nayar in Hajra & *al.,* Fl. India 4: 343. 1997.; var. **pentaphylla**., *Limonia pentaphylla* Retz., Observ. Bot. 5 : 24. 1789. *Glycosmis cochinchinensis* auct. non (Lour.) Pierre ex Engl. 1896; Gamble, Fl. Madras 1 : 153. 1915, *p.p. G. pentaphylla* auct. non DC. 1824; Hook. f., Fl. Brit. India 4 : 499. 1785; *p.p.*

Shrubs or small trees, up to 4 m high. Leaves usually with 3-7 leaflets, petiole 5.5 cm long, leaflets up to 24 × 7 cm, usually paler (whitish) below, base cuneate or obtuse, margin entire to subdentate, subcrenate, apex acute, obtuse, rounded or acuminate, lateral nerves up to 12 pairs. Flowers in dense axillary racemes, peduncle up to 8 cm; sepals 5, deltate to suborbicular; petals 5, elliptic-obovate, glandular, creamy white; stamens 10, filaments tapering above, anthers usually with one prominent dorsal gland on the connective; disc annular, ovary ovoid-cylindric, 5-celled. Berry subglobose, 1 cm across, white to pink mostly 1-(2-3) seeded, seeds green.

Occasional in Mahabubnagar district, as undergrowth of forests.

Fl. & Fr. : June-August.

Mallelatheertham (MBNR), *BR* & *SKB* 30634.

LIMONIA L.

Note: Swingle (Journ. Wash. Acad. Sci. 4: 328. 1914) is of the opinion that the generic name *Limonia* L. (1764) is illegitimate "being a mere variant of *Limonium* Miller (1764)" of Plumbaginaceae. He therefore took the next available name *Feronia* Cor. for the genus which has been subsequently followed by many. Airy Shaw (Kew Bull. 1939: 293. 1939) states that *Limonia* is not an orthographic variant of *Limonium* of Plumbaginaceae and the legitimate name *Limonia* L. must be retained.

Panigrahi (Taxon 26: 576. 1977) has argued for the rejection of the name *Limonia acidissima* L. as *nomen ambiguum.* However Stone and Nicolson (Taxon 27: 551-552. 1978*)* have argued for retaining the original name. The concept of Stone and Nicolson is followed in the present work.

Limonia acidissima L., Sp. Pl. ed. 2. 1: 554. 1762; B.C. Stone & D.H. Nicolson, Taxon 27: 551-552. 1978; K.N.Nair & M.P.Nayar in Hajra & *al.,* Fl. India 4: 294. 1997. *Feronia limonia* (L.) Swingle, Journ. Wash. Acad. Sci. 4: 328. 1914. *Schinus limonia* L., Sp. Pl. 389. 1753. *Limonia elephantum* (Correa) Panigrahi, Taxon 26: 577. 1977. *Feronia elephantum* Correa, Trans. Linn. Soc. London 5: 224. 1800; Hook.f., Fl. Brit. India 1: 516. 1875; Gamble, Fl. Madras 1: 160. 1915.

Vern.: *Velaga.*

Armed, deciduous trees, up to 12 m high, spines axillary, strong, straight, branchlets densely tomentose, bark deeply fissured, black. Leaves 1-3 in a cluster, 3-9-foliolate, leaflets opposite, oblong or obovate, 2-4.5 × 1-2 cm, thin coriaceous,

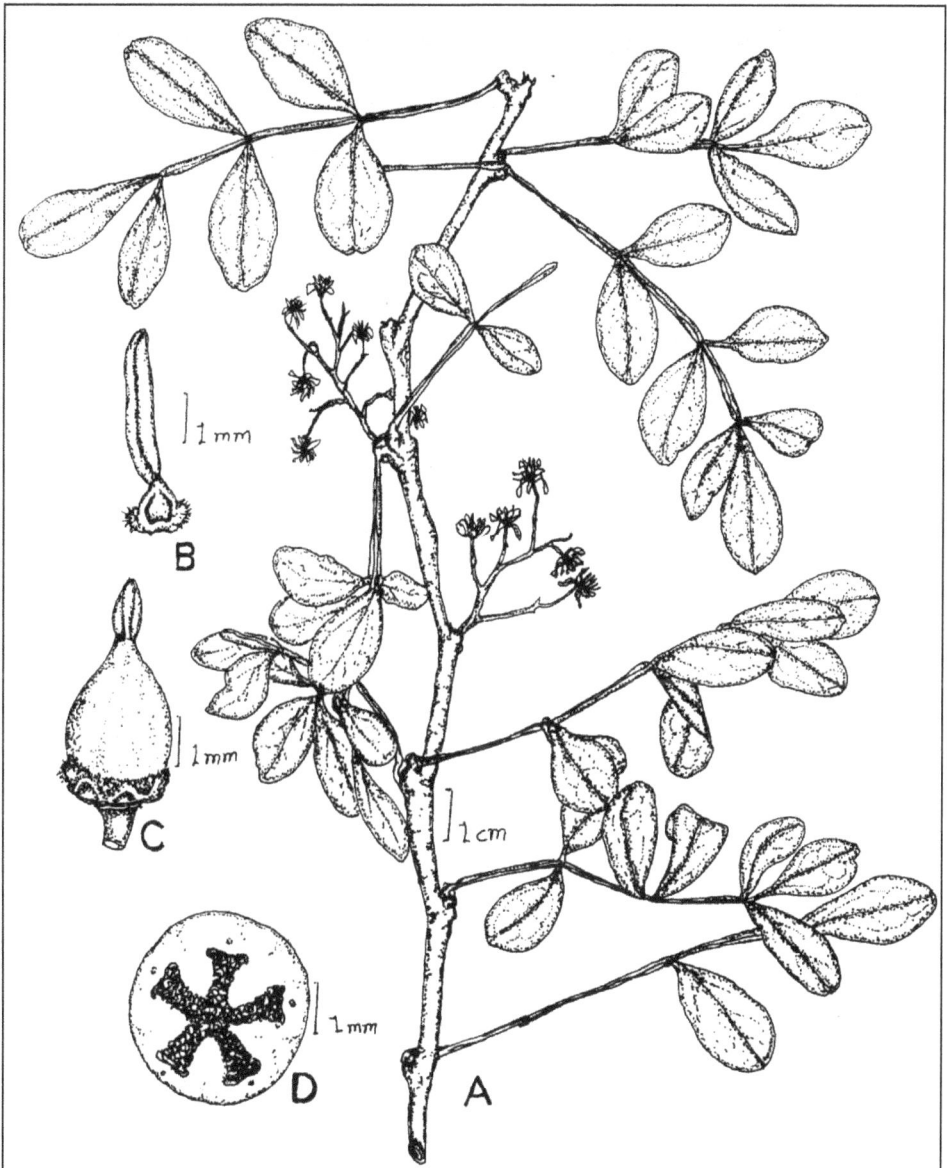

Figure 18: Limonia acidissima L.

A. Twig, B. Petal, C. Pistil, D. Ovary c.s.

glabrous, base cuneate, margin entire, apex obtuse; petiole and rachis flattened, narrowly winged. Flowers creamish in axillary and terminal panicles; calyx 3-toothed, teeth lobes deltoid; petals 5, ovate-oblong, glandular, glabrous, white or pale yellowish; stamens 10 or 12, free, inserted around disc, filaments subequal, to 3.5 mm long, pubescent inner face, glabrous above, anthers oblong; disc minute; ovary globose, 4-loculed, ovules many, many-seriate, style to 2 mm long, thick, fleshy, stigma capitate. Berry globose, woody, indehiscent; seeds many, compressed, embedded in mucilaginous pulp.

Widespread deciduous tree in open dry forests and also in open fields, often cultivated along the road sides.

Fl.: March- April; Fr.: Ripens during August -December

Gowraram RF (NZB), *TP* & *BR* 6354; Mamidipally RF (NZB), *BR* & *KH* 9022; Manchippa RF (NZB), *BR* & *KH* 9034; Narsapur RF (MDK), *TP* & *CP* 14062; Central University (RR), *MSM* 19215; Appapur (MBNR), *TDCK* 15366; Umamaheswaram (MBNR), *BR* & *TSS* 28297; Eturunagaram (WGL), *SSR* 18536; Poosugundi (KMM), *PVS* 84045 (BSID); Aklaspur (KRN), *GVS* 25655 (MH).

MURRAYA L. *nom. cons.*

1. Leaflets 10-25; flowers many in panicle; fruit 2-celled, purplish black when ripe .... **M. koenigii**

1. Leaflets 3-7; flowers a few in simple cymes; fruit 1-celled, red when ripe .... **M. paniculata**

Murraya koenigii (L.) Spreng., Syst. Veg. 2 : 315. 1826; Hook.f., Fl. Brit. India 1 : 503. 1875; Gamble, Fl. Madras 1 : 156. 1915; K.N.Nair & M.P.Nayar in Hajra & *al.,* Fl. India 4: 351. 1997. *Bergera koenigii* L., Mant. Pl. 565. 1771.

Vern.: *Karepaku, Karivepaku.*

Shrubs or small trees, up to 6 m high; bark green or greyish. Leaves pinnate, leaflets alternate, 10-25, somewhat asymmetrical, oblong-lanceolate, 2-4 × 1-2 cm, glabrous, aromatic, base oblique, margin entire and serrate, apex acute. Flowers small, greenish white, fragrant in terminal corymbose panicles; calyx saucer-shaped, sepals 5, fused at base, caducous; petals 5, linear; glandular, glabrous, greenish white; stamens 10; disc annular, slightly conical, 5-lobed; ovary oblong-ovoid. Berry subglobose, purplish black when ripe, pulp whitish; seeds 1 or 2, green, globose.

Common on the hill slopes and in dense forests and also grown in back yards of houses, as the aromatic leaves are used for curry flavouring.

Fl. & Fr.: March-August.

Mohammadabad (RR), *MSM* 19229; Kudichintalabailu (MBNR), *TDCK* 15370; Kambalpally (NLG), *BVRS* 426 (NU).

Murraya paniculata (L.) Jack., Malayan Misc. 1(5): 31. 1820; K.N.Nair & M.P.Nayar in Hajra & *al.,* Fl. India 4: 352. 1997. *Chalcas paniculata* L., Mant. 1: 68. 1767. *Murraya exotica* L., Mant. Pl. 563. 1771. '*Murraea*'; Hook.f., Fl. Brit. India 1: 502. 1875; Gamble, Fl. Madras 1: 156. 1915.

Vern. : *Naga golunga.*

Evergreen shrubs or small trees, up to 10 m high; bark pale white, lenticellate. Leaves pinnate, up to 20 cm long, leaflets 3-7, rarely 1-foliate, leaflets ovate or elliptic-ovate, base oblique, margin entire or at most obscurely crenate, apex acuminate. Flowers white, solitary or in 1-3-flowered axillary or terminal cymes respectively; calyx lobes 5, deltate; glabrous; petals 5, oblong-elliptic or obovate, narrow at base, obtuse at apex, glabrous, white; stamens 10, filaments linear, flat, tapering, to 10 mm long, white, anthers ellipsoid, yellowish; disc annular, lobulate, ovary ovoid-ellipsoid, style cylindric, stigma capitate, 2-3-lobed, broader than style. Berry globose, oblong or ovoid, pointed, reddish when ripe; seeds solitary, woolly.

Often cultivated in garden.

Fl.: February – April; Fr.: June – July.

NARINGI Adans.

Naringi crenulata (Roxb.) Nicolson in C.J. Saldhana & Nicolson, Fl. Hassan Dist. 387. 1976; K.N.Nair & M.P.Nayar in Hajra & *al.,* Fl. India 4: 294. 1997. *Limonia crenulata* Roxb., Pl. Coromandel t. 86. 1798; Gamble, Fl. Madras 1: 157. 1915. *L. acidissima* auct. non. L. 1762; Hook.f., Fl. Brit. India 1: 507. 1875.

Vern.: *Tor-elaga.*

Armed, deciduous trees, up to 15 m high, spines solitary or geminate, axillary, straight, branchlets glabrous; bark greyish, rough. Leaves 3-5-foliolate, petiole and rachis jointed, winged; leaflets ovate or obliquely oblong, 2-6 × 1-2.5 cm, glabrescent, base obtuse, margin crenulate, apex acute. Flowers white in axillary racemes; peduncles to 2 cm long, pedicels filiform, 8-12 mm long; sepals 4, ovate-orbicular or deltate, glabrous or puberulous; petals 4 or 5, elliptic or oblong, concave, glabrous, white; stamens 8, subequal or alternately shorter and longer, free, inserted around disc, anthers oblong, yellowish; disc annular or cupular, ovary subobovoid, 4-loculed, with 1 ovule in each, stigma subglobose. Berry globose, black; seeds shining, flattened.

Common in dry forests of most districts.

Fl.: March – April; Fr.: August onwards.

Bheemaram (ADB), *GO* & *DAM* 5181; Narsapur RF (MDK), *BR* & *CP* 14106; Anantagiri RF (RR), *MSM* & *TP* 11056 Bourapur (MBNR), *TDCK* 13637; Pullayapally (MBNR), *MVS* & *DV* 38648; Farahabad (MBNR), *SKB* & *BSS* 34438; way to Parantapalli (KMM), *PVS* 84059 (BSID).

TODDALIA Juss. *nom. cons.*

Toddalia asiatica (L.) Lam., Tab. Encycl. Meth. 2 : 116. 1797; Gamble, Fl. Madras 1. 150. 1915; K.N.Nair & M.P.Nayar in Hajra & *al.* Fl. India 4: 403. 1997. *Paullinia asiatica* L., Sp. Pl. 365. 1753. *Toddalia aculeata* (Smith) Pers., Syn. 1 : 249. 1805, *nom.superfl.*; Hook.f., Fl. Brit. India 1 : 479. 1875. *Scopalia aculeata* Smith, Pl. Ic. Hact. ed. 1, 2: t.34. 1790. *nom.superfl.*

Vern.: *Konda kashinda*

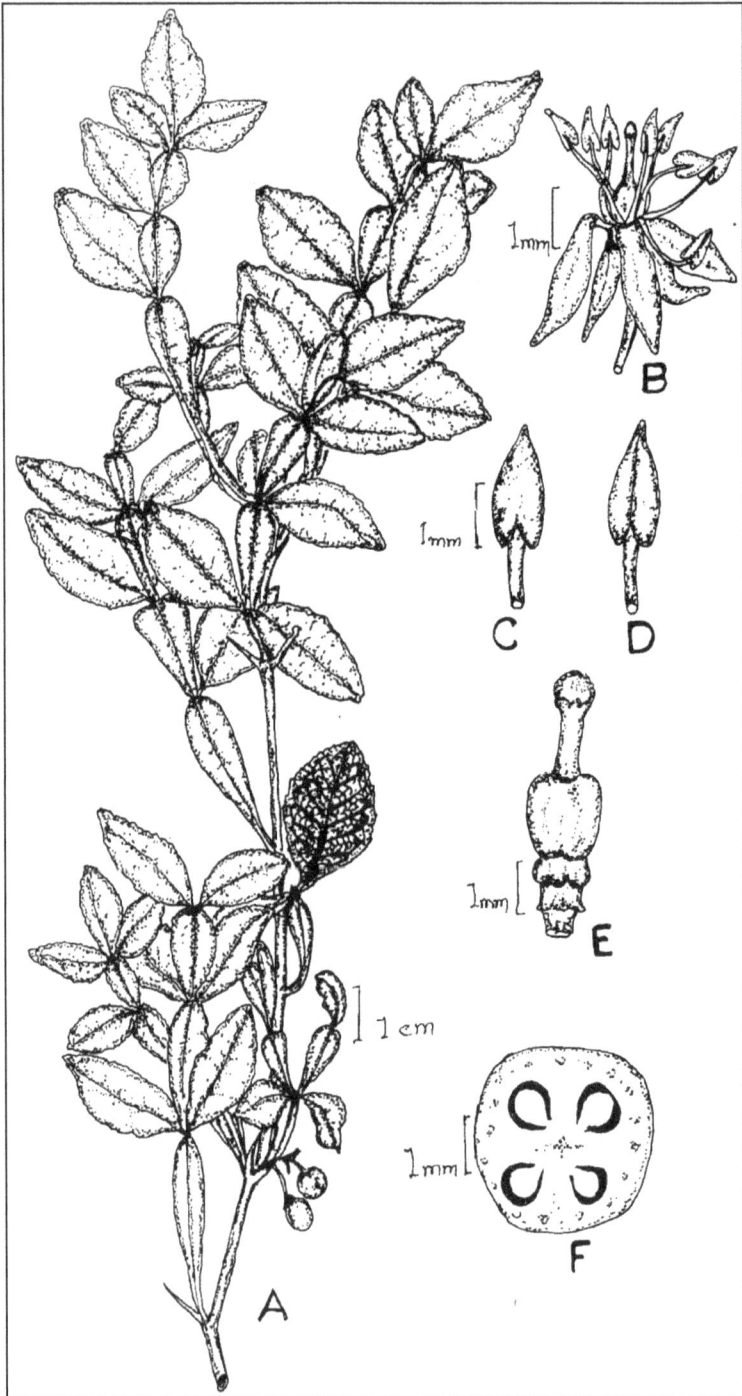

Figure 19: Naringi crenulata (Roxb.) Nicolson

A. Twig, B. Flower-entire, C&D. Stamens, E. Pistil, F. Ovary c.s.

Armed straggler, 3-15 m long, branchlets glandular-tomentose, prickles recurved. Leaves digitately 3-foliolate, sessile; petioles 1-3.5 cm long, slender; leaflets sessile, variable, oblong-elliptic, elliptic-obovate, oblanceolate, base narrow to cuneate or acute, apex obtuse or obtusely acute or acuminate. Flowers creamy white in axillary panicles, pentamerous, unisexual; Male flowers: calyx campanulate, lobes 5, deltoid, glandular, short-pubescent outside; petals 5, narrow to broadly oblong, hooded at apex, glabrous, glandular; stamens 5, slightly exserted, filaments 2-5 mm long, white, anthers oblong, ca 1 mm long with pellucid gland on dorsal side, yellow; disc 5- or more-lobed, pistillodes 1-2 mm high; Female flowers: sepals broadly triangular, otherwise same as in male flowers; petals narrowly oblong; staminodes ca 1.5 mm long; disc broader than in male flowers; ovary subglobose or oblong, 5-loculed, style short or absent, stigma 5-lobed, sessile. Drupe 1-5-seeded; seed brownish, shiny.

Common in forests of Mahabubnagar district.

Fl.: Twice in a year, in November-January and again in May-June; Fr.: Through the year.

Mannanur (MBNR), *TDCK* 13642; Vatavarlapalli (MBNR), *BR* & *SKB* 32901.

SIMAROUBACEAE

AILANTHUS Desf. *nom. cons.*

Ailanthus excelsa Roxb., Pl. Coromandel t. 23. 1795; Bennett in Hook.f., Fl. Brit. India 1: 518. 1875; Gamble, Fl. Madras 1: 163. 1915; Basak in Hajra & *al.,* Fl. India 4: 410. 1997.

Vern. : *Peddamanu.*

Large, deciduous trees, up to 25 m high; bark rough, greyish brown; branchlets with persistent leaf-scars. Leaves imparipinnate, 30 × 15 cm; leaflets 8-14 pairs, subopposite, oblong-lanceolate, up to 9 x 0.5 cm, inequilateral at base, margin irregularly toothed to sublobate, apex acuminate, petiole 10 cm long. Flowers large, polygamous, in axillary or terminal panicles; calyx 5-lobed, lobes ovate-triangular; petals ovate-lanceolate, reflexed, glabrous; stamens 10 in male flowers, filaments glabrous, anthers ca 1 mm long; ovary sparsely hairy, styles free or connate, stigma curling. Samara oblong, 2.5 × 1 cm, strongly nerved, blunt or pointed at both ends, copper red, always once or twice twisted at the base.

Occasional in forests of Hyderabad, Ranga Reddi, Medak, Nizamabad, Adilabad, Mahabubnagar, Warangal and Nalgonda districts.

Fl.: January-March; Fr. : March onwards.

Kadam RF (ADB), *GO* & *DAM* 5195; Domakonda (NZB), *BR* 9688; Domadugu (MDK), *BR* & *CP* 14123; Central University campus (RR), *MSM* 19230; Rayagir (NLG), *BVRS* 430 (NU).

BALANITACEAE

BALANITES Delile

Balanites roxburghii Planch., Ann. Sci. Nat. Bot. ser. 4. 4: 258. t. 2. 1854; Bennett in Hook.f., Fl. Brit. India 1: 522. 1875; Gamble, Fl. Madras 1: 164. 1915; P.Singh & V.Singh in Hajra & *al.,* Fl. India 4: 40. 1997. *B. aegyptiaca* auct non (L.) Delile 1813.

Small trees, up to 4 m high, branches terete, striate, puberulous or glabrous, thorns up to 2 cm long, bark grey. Leaves 2-foliolate; leaflets up to 4 × 2 cm, elliptic-ovate or oblong, base rounded or cuneate, apex obtuse or acute, coriaceous, finely nerved, glaucous green, pubescent; petiolules 3-5 mm long. Flowers whitish, 7-8 in axillary or supra axillary, pseudo-umbels; pedicels 5-15 mm long; sepals 5, imbricate, elliptic-ovate, densely pubescent outside, long silky hairy inside; petals 5, elliptic-oblong, glabrous outside, silky hairy inside; stamens 10, 2-3 mm long, filaments 1.5-2 mm long; ovary globose. Drupes ashy green, elliptic-oblong or ovoid, slightly 5-grooved, about 4 × 2.5 cm; seed solitary, embedded in pulp.

Common in dry forests of Adilabad, Nizamabad, Medak, Hyderabad, Nalgonda and Mahabubnagar districts.

Fl. & Fr.: September-July.

Bhoopalapatnam RF (ADB), *GO* 4401; Sirpur (ADB), *GO* & *PVP* 4932; Nasarullabad (NZB), *TP* & *BR* 6149; Sriramsagar project (NZB), *BR* 7299; Mamillagudem (KMM), *RR* 105932 (BSID); Tadwai (WGL), *SSR* & *KSM* 18547.

Note: Planchon (*loc.cit.)* bases his name *B. roxburghii* on *Ximenia aegyptiaca* sensu Roxb. (1832) and considers it distinct from *X. aegyptiaca* L. (1753). He points out that *B. roxburghii* differs from African *B. aegyptiaca* (L.) Del. in petals being villous on inner surface (Sprague, Bull. Misc. Inform. 1913: 135. 1913). The former also differs from latter in having shorter petiolule. Further, the ovary does not lengthen out after flowering. Hence treated as a distinct species here.

OCHNACEAE

OCHNA L.

1. Inflorescence of 2 or 3 flowers or flowers solitary; petals 5-10 × 2-5 mm, as long as sepals .... **O. lanceolata**

1. Inflorescence of dense racemes or panicles; petals 15-25 × 7-15 mm, longer than sepals:

    2. Leaves waxy, glaucous, rounded at apex .... **O. gamblei**

    2. Leaves not waxy or glaucous, acute or acuminate at apex .... **O. obtusata**

Ochna gamblei King ex Brandis, Indian Trees 128. 1906; Gamble, Fl. Madras 1: 166. 1915; Safui & M.P.Nayar in Hajra & *al.,* Fl. India 4: 426. 1997. *O. obtusata* DC. var. *gamblei* (King ex Brandis) Kanis, Blumea 16(1): 1-82. 1968. *O. beddomei* Gamble *loc.cit.* & in Bull. Misc. Inform. 1916: 34. 1916.

Vern.: *Kuka-moi.*

Large shrubs or small trees, bark thick, whitish. Leaves crowded at ends of branchlets, almost sessile, obovate, oblong or broadly elliptic, 6-14 × 3.5 – 7 cm, apex obtuse or rounded, margin serrate or subentire, coriaceous, waxy, glaucous. Flowers in many-flowered compound cymes; sepals 5, ovate or ovate-oblong; petals 5-8, obovate, yellow; stamens 35-70; ovary 5-10-lobed. Fruits 3-5.

Common in most districts.

Fl. & Fr.: March – October

Eturnagaram (WGL), *SSR* & *KSM* 18531.

Ochna lanceolata Spreng., Syst. Veg. 2. 592. 1825; Kanis, Blumea 16: 24. 1968; Safui & M.P.Nayar in Hajra & *al.,* Fl. India 4: 428. 1997. *O. wightiana* Wall. ex Wight & Arn., Prodr. Fl. Ind. Orient 1: 152. 1834; Bennett in Hook. f., Fl. Brit. India 1: 524. 1875; Gamble, Fl. Madras 1: 166. 1915. *O. heyneana* Wight & Arn., Prodr. Fl. Ind. Orient. 1: 152. 1834; Gamble, Fl. Madras 1: 166. 1915.

Large shrubs or small trees, up to 8 m high. Leaves ovate, lanceolate or ellptic, 4-5 × 1-2 cm, thin coriaceous, glabrous, base obtuse, margin serrate, apex apiculate; petioles 1-3 mm; stipules slender. Flowers yellow in axillary, 2-3-flowered cymes or rarely solitary; sepals 5, oblong, reflexed; petals 5, oblong, caducous; stamens 25-50, filaments 2-4 mm long, anthers 2-5 mm long; ovary 5-lobed, one ovule per lobe, stigma shortly branched. Drupe 1-3-lobed, ca 1 × 0.7 cm, with persistent style; seeds 5 mm.

Occasional in Mahabubnagar and Nalgonda districts.

Fl. & Fr.: February-June.

Nagarjunakonda (NLG), *KTH* 9731 (CAL); Achampet RF (MBNR), *SJRC* 503 (MH).

Ochna obtusata DC., Ann. Mus. Natl. Hist. Nat. 17. 481. 1811; Safui & M.P.Nayar in Hajra & *al.,* Fl. India 4: 429. 1997. *O. squarrosa* auct. non L. 1753; Bennett in Hook.f., Fl. Brit. India 1 : 523. 1875; Gamble, Fl. Madras 1: 165. 1915.

Deciduous shrubs or trees, up to 10 m high, branchlets pubescent. Leaves alternate or apically clustered, obovate or elliptic, 5-16 × 3-7 cm, glabrous, base and apex obtuse or cuneate, margin crenate or serrate; petioles up to 5 mm long; stipules 3-8 mm long. Flowers golden yellow, bracteate, in terminal and/or axillary racemes or panicles, sometimes umbellate or corymbose; sepals 5, ovate-oblong, green turning pale pinkish in ripe fruit; petals 5-10, deciduous, obovate, cuneate at base, rounded at apex, yellow; stamens 30-75, filaments to 4 mm long, anthers 4-20 mm long; ovaries 5-10, 1 ovule per lobe, styles twisted. Drupe 3-10-lobed, 1 cm across, with a solitary seed per lobe.

Common in Karimnagar, Mahabubnagar, Nalgonda and Hyderabad districts.

Fl. & Fr.: February-May.

Pedduti (MBNR), *TDCK* 15375; Mahadevpur RF (KRN), *SLK* 70849 (NBG); Nagarjunakonda (NLG), *BVRS* 492 (NU); Kudichintalabailu (MBNR), *TDCK* 15375; Katapur (WGL), *RKP* 110846 (BSID); Vatavarlapally (MBNR), *VBH* 84954 (BSID); Tekupalli (MM), *PVS* 84053 (BSID).

BURSERACEAE

1. Leaf rachis winged; cultivated .... **Bursera**

1. Leaf rachis not winged:
    2. Leaves present during flowering; drupes ovoid, globose or
        oblong, indehiscent :
        3. Disc adnate to calyx; pyrenes free, bony pitted, all seed bearing ... **Garuga**
        3. Disc free; pyrenes combined, one only seed bearing,
            the rest barren .... **Commiphora**
    2. Leaves absent during flowering; drupes trigonous, dehiscent .... **Boswellia**

BOSWELLIA Roxb. ex Colebr.

Boswellia serrata Roxb. ex Colebr. in Asiat. Res. 9. 379. t. 5. 1807; Bennet in Hook.f.,
Fl. Brit. India 1 : 528. 1875; Gamble, Fl. Madras 1: 168. 1915; Chitra & Henry in
Hajra & *al.,* Fl. India 4: 432. 1997. *B. glabra* Roxb., Pl. Coromandel t. 207. 1811;
Gamble, Fl. Madras 1: 168. 1915. *B. serrata* Roxb. var. *glabra* (Roxb.) Bennet in
Hook. f., Fl. Brit. India 1: 528. 1875.

Vern. : *Anduga.*

Large trees, up to 15 m high, bark papery. Leaves apically clustered,
imparipinnate, leaflets 8-23 pairs, variable in size, shape and degree of pubescence,
oblong to lanceolate, 3-8 × 1-3 cm, glabrous, rusty below, base cuneate, oblique, margin
dentate-serrate, apex acute. Flowers pale pink or white in axillary panicles; calyx
tube broadly campanulate, calyx-lobes triangular-ovate, pubescent, persistent; petals
5, ovate-oblong, shortly clawed, inflexed at apex, pink or white, pubescent outside,
except margin; stamens 10, alternately long and short. Drupes trigonous, greenish-
yellow, pyrenes 3, each 1-seeded; seeds heart-shaped.

Common in rocky areas of dry forests of Adilabad, Nizamabad, Medak, Ranga
Reddi, Karimnagar, Mahabubnagar and Nalgonda districts.

Fl. & Fr.: March-August.

Wankidi RF (ADB), *PVP* 9485; Manchippa RF (NZB), *SSR* 19339, *BR* & *KH* 9023;
Narsapur RF (MDK), *TP* & *CP* 14061; Anantagiri RF (RR), *TP* & *MSM* 11071;
Mannanur (MBNR), *TDCK* 15369; Mallelatheertham (MBNR), *BR* & *SKB* 30712;
Rekularam (NLG), *BVRS* 501; Bhupatipur (KRN), *GVS* 22296 (MH); Pakhal (WGL),
ANH 15951 (MH).

BURSERA Jacq. ex L.

Bursera penicillata (DC.) Engl., Bot. Jahrb. Syst. 1 : 44. 1881; Mcvaugh & Rzed., Kew
Bull. 18: 334. 1965; Chitra & Henry in Hajra & *al.*, Fl. India 4: 434. 1997. *Elaphrium
penicillatum* Sesse & Mocino ex DC., Prodr. 1 : 724. 1824.

Medium sized deciduous tree; stems soft, young branches hairy, reddish brown.
Leaves clustered, odd pinnate, 12 cm long; petioles to 3 cm long, rachis winged, wing
2-3 mm long; leaflets 3-4 pairs, opposite, obovate or elliptic, 3-3.5 x 1.5 cm. thin

coriaceous, densely villous above, pubescent below, base sub acute or cuneate, margins dentate, apex acute or obtuse. Flowers in axillary racemes, 2-4-flowered on 3-7 cm long peduncles; pedicels 1 cm long; calyx campanulate; lobes ovate 1mm long. Petals cream colored, oblong, 3 mm long. Drupe ovoid, greenish, single seeded, 1 x 0.7 cm.

Planted in gardens and national parks as the source of Linaloe oil.

Fl. & Fr.: March-May.

Mohammadabad F.O. (RR), *MSM* & *NV* 10416.

COMMIPHORA Jacq. *nom. cons.*

Commiphora caudata (Wight & Arn.) Engler in A. & C.DC. Monog. Phan. 4 : 27. 1833; Gamble, Fl. Madras 1: 171. 1915; Chitra & Henry in Hajra & *al.,* Fl. India 4: 443. 1997. *Protium caudatum* Wight & Arn., Prodr. Fl. Ind. Orient. 176. 1834; Bennet in Hook.f., Fl. Brit. India 1: 530. 1875.

Vern. : *Kondamamidi, Konda ragi, Gugilam.*

Deciduous unarmed trees, up to 18 m high, branchlets glabrous. Leaves alternate, 3-7-foliolate, leaflets opposite, the terminal leaflet with a long petiolule, ovate or lanceolate, 3-10 × 1.5-6 cm, glabrescent, base alternate, margin entire, apex caudate. Flowers red, scented, polygamous (male and bisexual) in axillary long peduncled dichasial cymes; calyx campanulate or cupular, 4-lobed, lobes ovate or deltoid; petals 4, linear-oblong or oblanceolate, disc cupular, intrastaminal; stamens 8, free, alternately long and short, pistillode short in male flowers; ovary in bisexual and female flowers ovoid or oblong, stigma 2-lobed. Drupes globose, black, pyrenes 1 or 2; seed solitary, ca 2 mm long with radiating wings.

Common in dry forests of Nalgonda and Mahabubnagar districts.

Fl. & Fr.: March-October

Bourapur (MBNR), *TDCK* 13652; Nagarjunakonda (NLG), *BVRS* 135 (NU).

GARUGA Roxb.

Garuga pinnata Roxb., Pl. Coromandel t. 208. 1819; Bennet in Hook.f., Fl. Brit. India 1: 528. 1875; Gamble, Fl. Madras 1: 169. 1915; Chitra & Henry in Hajra & *al.,* Fl. India 4: 448. 1997.

Vern. : *Garuga.*

Large trees, up to 25 m high, bark thick and reddish brown, peeling off in flakes, branchlets pubescent. Leaves crowded towards the ends of the branches, alternate, imparipinnate, leaflets 11-17, variables in size, opposite, oblong, ovate- lanceolate, 8-14 × 1.5-4 cm, puberulous with characteristic fruit like reddish galls, base cuneate or oblique, margin crenate, apex acuminate; petiole to 10 cm long. Flowers cream or yellow in large terminal panicles; calyx 3-8 mm long, lobes 5, ovate-oblong, densely tomentose; petals 5, oblong-lanceolate, 5-8 × 1.5-2 mm, tomentose, cream-coloured or pale yellow; stamens 3-3.5 mm long, inserted on calyx, filaments pubescent; disc crenate, yellow; ovary globose, pilose, style pilose, stigma 5-lobed. Drupes oblong, horned, pyrenes 2 or 3, seed solitary.

Common in deciduous forests of Karimnagar, Adilabad, Nizamabad, Ranga Reddi, Warangal and Mahabubnagar districts.

Fl.: March – April; Fr.: June - August

Jaipur RF (ADB), *GO* 4491; Rampur (ADB), *SSR* 19314; Mudheli RF (NZB), *TP* & *BR* 6360, *BR* & *KH* 7132; Bourapur (MBNR), *BR* & *SKB* 32394; Pakhal (WGL), *ANH* 15952 (MH); Mahadevpur (KRN), *SLK* 70847 (NBG).

MELIACEAE

1. Seeds winged; ovules numerous in each cell .... **Soymida**

1. Seeds not winged; ovules 1-2 in each cell :

   2. Staminal tube not appendaged .... **Aglaia**

   2. Staminal tube appendaged:

      3. Shrubs or small trees; filaments connate below only .... **Cipadessa**

      3. Medium sized trees; filaments connate throughout:

         4. Leaves unipinnate; stigma 3-lobed; seeds exalbuminous; cotyledons fleshy .... **Azadirachta**

         4. Leaves 2 or 3-pinnate; stigma 4-6-lobed; seed albuminous; cotyledons thin .... **Melia**

AGLAIA Lour. *nom. cons.*

Aglaia elaeagnoidea (A. Juss.) Benth., Fl. Austral. 1: 383. 1863; Ramamoorthy in C.J.Saldanha & Nicolson, Fl. Hassan 392. 1976; var. **beddomei** (Gamble) K.K.N. Nair, J. Bombay Nat. Hist. Soc. 78: 426. 1981; S.S.Jain & S.S.R.Bennet in Hajra & *al.,* Fl. India 4: 462. 1997. *A. roxburghiana* (Wight & Arn.) Miq. var. *beddomei* Gamble, Fl. Madras 1 : 180. 1915. *A. roxburghiana* sensu Bedd., Fl. Sylv. t. 130. A. 1871, non (Wight & Arn.) Miq. 1868.

Vern. : *Yerra aduga.*

Trees, up to 15 m high, bark light brown, smooth, branchlets stellate-tomentose. Leaves odd-pinnate, 10-15 cm long, leaflets 7, narrow-lanceolate, rarely oblanceolate, 8-10 × 2- 2.5 cm, base rounded, margin entire, apex subacute or obtuse, glabrous above, pale beneath and drying grey. Flowers 5-merous, polygamodioecious, pedicelled; calyx cup-shaped, 5-lobed, scaly; petals 5, elliptic-oblong, longer than calyx, staminal tube shorter than petals, stamens 5; disc obscure; ovary depressed, stigma ovoid or obscurely lobed. Berry globose, 3.8 cm in diam, ferruginous, 1-seeded.

Common in Mahabubnagar district.

Fl. & Fr.: January-March.

Saleswaram (MBNR), *SKB* & *BR* 30738; Umamaheswaram (MBNR), *SRS* 109154 (BSID).

AZADIRACHTA A. Juss.

Azadirachta indica A. Juss., Mem. Mus. Hist. Nat. 19: 221. 1830; Gamble, Fl. Madras 1: 177. 1915; S.S.Jain & S.S.R.Bennet in Hajra & *al.,* Fl. India 4: 478. 1997. *Melia azadirachta* L., Sp. Pl. 385. 1753; Hiern in Hook.f., Fl. Brit. India 1: 544. 1875.

Semievergreen trees, up to 15 m high, bark brown with vertical furrows, branchlets glabrous and lenticellate. Leaves alternate, pari or imparipinnate, 10-45 cm long, leaflets ca 5-6 pairs, obliquely ovate-lanceolate, falcate, 3-5.5 × 1-2 cm, glaucous, base oblique, margin crenate-serrate, apex acute; petiolules 2-3 mm long, that of terminal leaflet 1-1.5 cm long. Flowers white in axillary panicles, panicles as long as or shorter than leaves; sepals 5, free, connate at base, ciliate; petals 5, free, oblong spathulate, tomentose; staminal tube cylindrical, mouth expanded with 10 appendages, anthers 10; disc annular, adnate to base of ovary; ovary 3-locular, style slender, stigma 3-lobed. Drupes oblong, yellow when ripe; seed solitary, ellipsoid, glabrous, fleshy.

Common in plains, roadsides and rare in forests in all districts.

Fl. & Fr.: January-May.

Sirpur (ADB), *GO* & *PVP*4809; Mudheli forest guest house (NZB), *TP* & *BR*6361; Sirpur RF (ADB), *GO* & *PVP*4809; Narsingi (MDK), *BR* & *CP*11527; Central University campus (RR), *MSM*19231; Rachakonda (NLG), *BVRS*437 (NU).

CIPADESSA Blume

Cipadessa baccifera (Roth) Miq., Ann. Mus. Bot. Lugduno-Batavi 4: 6. 1868; Gamble, Fl. Madras 1: 176. 1915; S.S. Jain & S.S.R. Bennet in Hajra & *al.,* Fl. India 4: 482. 1997. *Melia baccifera* Roth, Nov. Pl. Sp. 215. 1821. *Cipadessa fruticosa* Blume, Bijdr. 162. 1825; Hiern in Hook.f., Fl. Brit. India 1: 545. 1875. *Mallea rothii* A. Juss., Mem. Mus. Hist. Nat. 19: 221. 1830.

Vern. : *Chandbera, Purudona.*

Small trees or shrubs, up to 7 m high, branchlets tomentose. Leaves alternate or subopposite, imparipinnate, petiole to 3 cm, leaflets 4 pairs, opposite, oblong, lanceolate, 3.5-10 × 2-4 cm, glabrous above, pilose below, base cuneate, margin proximally entire, distally serrate, apex acuminate; petiolule to 4 mm. Flowers greenish-white in axillary corymbose panicles; calyx with 5, short, 3-angular teeth, pubescent outside; petals oblong, ca 4 ×1.5 mm, pubescent outside; stamens slightly shorter than petals; ovary glabrous, stigma 5-lobed. Drupes globose, 5-lobed, red when ripe; seeds 5, ovoid.

Common on laterite hills, near villages and in dry forests in Mahabubnagar district.

Fl. & Fr.: October-April.

Pullajalamala (MBNR), *VBH*84990 (BSID).

MELIA L.

Melia azedarach L., Sp. Pl. 384. 1753; Hiern in Hook.f., Fl. Brit. India 1: 544. 1875; Gamble, Fl. Madras 1: 176. 1915; S.S.Jain & S.S.R.Bennet in Hajra & *al.,* Fl. India 4: 494. 1997.

Vern.: *Thuraka vepa.*

Medium sized trees, up to 10 m high; bark thick and deeply fluted, smooth in young; young branches sparsely clothed with deciduous stellate hairs. Leaves 2- to 3-pinnate, 30 cm long, petiole to 15.2 cm long, pinnae 3-5-paired; leaflets 1-5 pairs, opposite, ovate-obovate or lanceolate, 3-5.5 × 1-2 cm, glabrous, base oblique, cuneate, margin serrate, apex cuneate. Flowers lilac in long panicles; calyx 5-lobed to base, pubescent outside; petals 5, linear-oblong or oblanceolate; staminal tube cylindric, 20-30-toothed, purple, anthers at mouth of tube; ovary glabrous, style clavate at apex, stigma 4-6-lobed. Drupes ellipsoid, fleshy, yellow when ripe; seed solitary, elliptic.

Cultivated in most districts and occasionally found run wild.

Fl. & Fr.: April-December.

Bodh (ADB), *GO* & *DAM* 5038; Bellampalli (ADB), *PVP* 9444; Mamidipally (NZB), *BR* & *KH* 9007; Naskal (RR), *MSM* & *PRSR* 15113; Mombojipally (MDK), *BR* & *CP* 11630; Kodimial (KRN), *GVS* 20093 (MH).

SOYMIDA A.Juss.

Soymida febrifuga (Roxb.) A. Juss., Mem. Mus. Natl. Hist. Nat. 19: 251. 1830; Hiern in Hook.f., Fl. Brit. India 1 : 567. 1875; Gamble, Fl. Madras 1: 185. 1915; S.S.Jain & S.S.R.Bennet in Hajra & *al.,* Fl. India 4: 501. 1997. *Swietenia febrifuga* Roxb., Pl. Coromandel 1: 18. t. 17. 1795.

Vern. : *Somi, Somida.*

Lofty glabrous trees, up to 10 m high; bark dark brown, rough with oblong flakes; branchlets with persistent leaf-scars. Leaves alternate, paripinnate, petiole to 7 cm, red, leaflets 6 pairs, opposite, ovate-oblong, 5-13 × 6-7.5 cm, thick-coriaceous, glabrous, base oblique, margin entire, apex obtuse. Flowers greenish white in axillary or terminal panicles; calyx 5-lobed, lobes ovate, ca 2 mm long, pubescent; petals 5, obovate, ca 5 mm long, pubescent out side, greenish white; staminal tube cup-shaped, anthers 10; ovary glabrous, ovoid, ovules 2-seriate, stigma sessile, 5-angled. Capsule woody, 5-celled, septifragally 5-valved, the valves 2-lamellate, separating from the thick, 5-winged axis; seeds oblong, compressed, winged at both ends.

Common in dry forests of most districts.

Fl.: February – April; Fr.: May – October.

Ichampalli RF (ADB), *GO* & *DAM 5155;* Gouraram RF (NZB), *TP* & *BR* 6396; Yerlapenta (MBNR), *TDCK* 13689; Central University campus (RR), *MSM* 19232; Kambalpally (NLG), *BVRS* 427 (NU); way to Penuloddi (KMM), *PVS* 84094 (BSID); way to Dabbachelama (MBNR), *SRS* 107443 (BSID); Sivampet RF (MDK), *RG* 116466 (BSID); Pocharam WLS (MDK), *RG* 113596 (BSID).

FLINDERSIACEAE

CHLOROXYLON DC.

Chloroxylon swietenia DC., Prodr. 1: 625. 1824; Hiern in Hook.f., Fl. Brit. India 1 : 569. 1875; Gamble, Fl. Madras 1 : 152. 1975; K.N.Nair & M.P.Nayar in Hajra & *al.,* Fl. India 4: 355. 1997.

Vern.: *Billu, Billudu.*

Moderate-sized, deciduous trees, up to 10 m high, with rough corky bark and yellow close grained wood. Leaves pinnate, up to 30 cm long, petiole slender; leaflets 10-20 pairs, alternate and opposite, oblong-lanceolate, 2-3.5 × 0.5-1.5 cm, gland-dotted, base oblique, margin entire, apex obtuse. Flowers creamy white in terminal or axillary panicles; sepals 5, minute, deltoid, puberulent; petals 5, imbricate, spathulate, clawed; stamens 10, shorter than petals; disc pulvinate, 10-lobed, ovary immersed in disc, greenish, 3-lobed, style slender, stigma obliquely-capitate, obscurely 3-lobed. Capsule glabrous, oblong, dark-brown, 3-celled, loculicidally 3-valved; seeds compressed, margins angular, winged above.

Common in forests of most districts.

Fl.: March-April; Fr.: July – October.

Devapur RF (ADB), *GO* & *PVP* 4239; Manchippa (NZB), *SSR* 19336; Laxmapur RF (NZB), *TP* & *BR* 6258; Gadirayal RF (RR), *MSM* & *KH* 10257; Sarlapally (MBNR), *TDCK* 15379; Mahadevpur (KRN), *SLK* 70823 (NBG); Srinivaspur (NLG), *BVRS* 397 (NU).

OLACACEAE

1. Staminodes present; calyx accrescent and enveloping the drupe to half or more the length .... **Olax**

1. Staminodes absent; calyx not accrescent, enveloping the drupe only at base .... **Ximenia**

OLAX L.

Olax scandens Roxb., Pl. Coromandel t. 102. 1799; Masters in Hook.f., Fl. Brit. India 1 : 575. 1875; Gamble Fl. Madras 1: 190. 1915; Uniyal in N.P.Singh & *al.*, Fl. India 5: 9. 2000.

Vern.: *Kurpodur.*

Armed stragglers, up to 5 m high, branchlets sparsely fulvous pubescent, thorns blunt. Leaves oblong or ovate-lanceolate, 3-7 × 2-3.5 cm, thin coriaceous, glabrous above, glabrous or puberulous beneath, base cuneate, margin ciliate, apex obtuse-acute; petiole ca 0.8 cm long, pubescent. Flowers white in axillary racemes, sweet scented, bracts ovate-oblong, caducous; calyx cup-shaped, finely ciliate; petals usually 5, rarely 6, more or less connate, linear, cleft; stamens 3, staminodes 2-cleft at apex; ovary ovoid, glabrous, style linear, half as long as the petals, stigma 3-lobed. Drupe globose, oblong, basal half enclosed by accrescent calyx except at tip, orange in colour, 1-seeded.

Common near streams in dense forests and in ravines in most districts.

Fl. & Fr.: April-July.

Bijjur RF (ADB), *GO* & *PVP* 4888; Manchippa (NZB), *TP* & *BR* 6007; Laxmapur RF (NZB), *TP* & *BR* 6261; Gandhari RF (NZB), *TP* & *BR* 6331; Gangapur RF (MDK), *BR* & *CP* 11446; Burugupally (MDK), *CP* 11766; Kusumasamudram RF (RR), *MSM* &

NV 10353, 10369; Anantagiri RF (RR), *MSM* 11067; Sarlapally (MBNR), *TDCK* 15383; Rechapalli (KRN), *GVS* 20172 (MH); Pegarikutta (MDK), *KMS* 6652 (MH); Pakhal (WGL), *KMS* 13142 (MH); Pasra (WGL), *RKP* 108247 (BSID); Gowraram (KMM), *PVS* 76951 (BSID).

XIMENIA L.

Ximenia americana L., Sp. Pl. 1193. 1753; Masters in Hook.f., Fl. Brit. India 1: 574. 1875; Gamble, Fl. Madras 1: 189. 1915; Uniyal in N.P.Singh & *al.*, Fl. India 5: 15. 2000.

Vern.: *Uranechra.*

Small trees, up to 3 m high; branches terete, thorns axillary. Leaves broadly elliptic-ovate, obovate, up to 7 × 5 cm, base broadly acute, margin entire or shortly crenulate, apex retuse; petiole 3-5 mm long. Flowers pale green in umbellate or terminal racemes, fragrant, bracts minute; calyx cupular, glabrous, deeply divided, lobes 5, ovate-triangular, reflexed at length, ciliate, persistent; petals linear, much exceeding the calyx, frilled, hairy within; stamens as long as the petals, filaments stigmoid near apex. Drupes globose or ovoid, 3 x 3 cm, orange when ripe, pulp edible; seed solitary.

Common in most of the districts, in dry forests on stony ground.

Fl. & Fr.: March-July.

Gowraram RF (NZB), *TP* & *BR* 6382; Narsapur RF (MDK), *TP* & *CP* 11965; Gangapur RF (MDK), *BR* & *CP* 11440; Mohammadabad RF (RR), *MSM* & *KH* 10217, 10256; Molachintapally (MBNR), *TDCK* 15367; Uskavagu (MBNR), *SRS* 108908 (BSID); Akkannapet (MDK), *RG* 106729 (BSID).

OPILIACEAE

1. Tepals connate, urceolate; flowers tetramerous in spikes; stamens shorter than tepals; stigma on a short style, shallowly 4-lobed .... **Cansjera**

1. Tepals free; flowers pentamerous in racemose cymes; stamens longer than sepals; stigma sessile, simple .... **Opilia**

CANSJERA A.L. Juss. *nom. cons.*

Cansjera rheedii J.F. Gmel., Syst. Nat. 1: 280. 1791; Masters in Hook.f., Fl. Brit. India 1: 582. 1875; Gamble, Fl. Madras 1: 193. 1915; Mathur in N.P.Singh & *al.*, Fl. India 5: 40. 2000.

Twining shrubs, up to 5 m long, branches sometimes spiny, younger ones as well as the inflorescence, petioles and tube of the flower pubescent. Leaves ovate or oblong-lanceolate, 3-5 × 2-3 cm, thin coriaceous, glabrous, base obtuse, sometimes oblique, margin entire, apex acuminate to caudate, petiole 5 mm. Flowers green or yellowish- green, 0.2-0.3 cm long, axillary, solitary or 2 or 3 in spikes; corolla urceolate, greenish yellow, tube 2.5-3 mm long, lobes recurved, ca 0.5 mm; filaments ca 2 mm long, anthers broadly ovate, reaching as far as throat of perianth tube; disc scales ovate, acute, irregularly toothed, slightly fleshy. Drupe ovoid or ellipsoidal, reddish-orange when ripe.

Common in Adilabad, Nizamabad and Mahabubnagar districts chiefly in hilly areas.

Fl. & Fr.: November-March.

Mahaboobghat (ADB), *GO* & *PVP* 4261; Mudheli RF (NZB), *BR* & *KH* 9043.

OPILIA Roxb.

Opilia amentacea Roxb., Pl. Coromandel 2 : 31 t. 158. 1802; Masters in Hook.f., Fl. Brit. India 1: 583. 1875; Gamble, Fl. Madras 1: 192. 1915; Mathur in N.P.Singh & *al.*, Fl. India 5: 44. 2000.

Scandent shrubs or low trees, up to 10 m high, branchlets pubescent. Leaves elliptic or oblong-lanceolate, 5-12 × 3-5 cm, inequilateral at times, thin coriaceous, glabrous, base obtuse, attenuate or subacute, margin entire, apex acuminate; petiole 8-9 mm long. Flowers greenish yellow, numerous in axillary racemose cymes, before flowering resembling cones; bracts peltate, 2-3 mm wide, pedicels 1-2 mm long; perianth 1-2 mm long, ovate, shortly pubescent outside; filaments filiform, ca 1.5 mm long, anthers ovoid; disc lobes club-shaped to ovoid; stamens 5, antipetalous; ovary ellipsoid. Drupe ovoid-globose, 1 × 1 cm, apiculate, yellow-red when ripe; seed 1, ovoid, blackish.

Occasional in Karimnagar, Warangal, Nizamabad, Ranga Reddi, Khammam and Mahabubnagar districts.

Fl. & Fr.: April-July.

Gowraram RF (NZB), *TP* & *BR* 6353; Mohammadabad RF (RR), *MSM* & *NV* 10412; Kudurpalli (KRN), *SLK* 70731 (NBG); Nimmagudem (KRN), *SLK* 70915 (NBG); Katapur (WGL), *RKP* 110849 (BSID); way to Tekuloddi (KMM), *PVS* 84073 (BSID); Mannanur (MBNR), *SRS* 108910 (BSID).

CELASTRACEAE

1. Leaves opposite:
    2. Leaves entire; disc copular; fruits with persistent lateral style .... **Pleurostylia**
    2. Leaves crenate-serrate; disc cupular; fruits with persistent terminal style or its scar .... **Cassine**
1. Leaves alternate:
    3. Plants armed with spines; erect herbs or small trees; seeds partly enveloped by aril .... **Maytenus**
    3. Plants unarmed; scandent shrubs; seeds completely enveloped by aril .... **Celastrus**

CASSINE L.

Cassine paniculata (Wight & Arn.) Lobr.-Callen in Adansonia Ser. 2: 15: 220. 1975; Ramam. in N.P.Singh & *al.*, Fl. India 5: 83. 2000. *Elaeodendron paniculatum* Wight & Arn., Prodr. Fl. Ind. Orient. 157. 1834. *E. glaucum sensu* Lawson in Hook.f., Fl. Brit. India 1 : 623. 1875 (non Pers.) *p.p.*; Gamble, Fl. Madras 1: 211. 1918. *Cassine* *glauca* (Rottb.) Kuntze, Revis. Gen. Pl. 1: 114. 1891; Pullaiah & Chennaiah, Fl. Andhra Pradesh 1: 197. 1997.

Vern. : *Butankush.*

Medium sized trees, up to 10 m high, branchlets reddish, glabrous. Leaves opposite decussate, ovate-oblong, 5-10 × 4.6 cm, coriaceous, base obtuse or acute, margin crenate-serrate, apex acute. Flowers green in axillary, divergently branched, loose corymbose cymes; sepals orbicular; petals spathulate, ca 5 × 3 mm, membranous margin; stamens arising from lobes of disc; disc thick, fleshy. Drupes 1.5 × 0.6 cm, oblong, 1-seeded.

Common in most districts.

Fl. & Fr.: May-December.

Ichampalli (ADB), *GO* & *DAM* 5153; Manchippa RF (NZB), *TP* & *BR* 6009; Narsapur RF (MDK), *BR* & *CP* 14105; Thoniganla RF (MDK), *RG* 112473 (BSID); Srinivasapur (NLG), *BVRS* 431 (NU); Thattilanga RF (KMM), *RC* 102420 (BSID).

CELASTRUS L.

Celastrus paniculatus Willd., Sp. Pl. 1 : 1125. 1797; Lawson in Hook.f., Fl. Brit. India 1 : 617. 1875; Gamble, Fl. Madras 1: 208. 1918; Ramam. in N.P.Singh & *al.*, Fl. India 5: 87. 2000.

Vern. : *Maner tiga.*

Climbing shrubs, stem dark pink, angled or grooved, prominently lenticellate. Leaves alternate, ovate, obovate, orbicular, 4-7.5 × 3-6 cm, thin-coriaceous, base attenuate, margin crenate- serrate, apex acuminate to caudate, pubescent beneath, stipules laciniate; petioles up to 2 cm long. Flowers pale yellow, in terminal paniculate cymes, polygamous; Male flowers: sepals 5-lobed, lobes semi-orbicular; petals oblong to obovate-oblong; stamens 3 mm long; ovary sessile, columnar, disc cupular; Female flowers: sepals, petals and disc as in male flowers, stamens sterile; ovary globose, style columnar, stigma 3-lobed. Capsule 1 cm across, globose, orange when ripe; seeds 2, brownish, enclosed in arils.

Common in most districts of the State.

Fl. & Fr.: April-December.

Kundaram RF (ADB), *TP* & *GO* 5461; Mavala (ADB), *GO* & *PVP* 4247; Gandivet RF (NZB), *BR* & *CPR* 7193; Mudheli RF (NZB), *BR* & *KH* 9042; Mohammadabad RF (RR), *MSM* & *KH* 10245, 10246; Dharmapur (RR), *MSM* & *KH* 10293; Narsapur (MDK), *TP* & *CP* 14036, *KMS* 7955 (MH); Thattilanga RF (KMM), *RC* 102414 (BSID); Hansanpalli (MDK), *RG* 104041(BSID); Bureddipalli (MBNR), *VBH* 86603(BSID).

MAYTENUS Molina emend Bosch

1. Seed aril thin, membranous, aril covers half or nearly the whole seed .... **M. rufa**

1. Seed aril fleshy, covers only the base of seed:

    2. Leaves obovate with entire margin and emarginated tip .... **M. emarginata**

    2. Leaves otherwise .... **M**. **heyneana**

Maytenus emarginata (Willd.) Ding Hou in Steenis. Fl. Males Ser. 1. (62): 241. 1962; Ramam. in N.P.Singh & *al.*, Fl. India 5: 120. 2000. *Celastrus emarginatus* Willd., Sp. Pl. 1: 2. 1798. *Gymnosporia emarginata* (Willd.) Thwaites, Enum. Pl. Zeyl. 409. 1864; Lawson in Hook.f., Fl. Brit. India 1: 621. 1875; Gamble, Fl. Madras 1: 210. 1918. *Gymnosporia montana* (Roth) Benth., Fl. Austral. 1: 400. 1863; Lawson in Hook.f., Fl. Brit. India 1 : 621. 1875; Gamble, Fl. Madras 1: 209. 1918.

Vern.: *Danti.*

Shrubs to small trees, up to 3 m high, branchlets pink, terete, spines either from main branch or apicllay on short shoots, 2-4 cm long. Leaves ovate, obovate, 2-8 × 1-4.5 cm, coriaceous, glabrous, base attenuate, gradually or abruptly tapering, margin crenate-serrulate, apex usually emarginate, sometimes obtuse; petioles 2-12 mm long. Flowers white or pale green, often with a pink tinge, in axillary, dichotomous cymes; bracts deltoid, sepals lobed, petals oblong or ovate-oblong; stamens 2-3 mm long, inserted beneath the margin of the disc; ovary partially immersed, 3-loculed, stigmas 3, distinct, disc fleshy, rounded. Capsules subglobose, 3-loculed, seeds 2 per each locule, ellipsoid, arillate.

Common in dry forests of all districts.

Fl. & Fr.: November-June.

Sattenapalli RF (ADB), *TP* & *PVP* 4077; Sirnapally RF (NZB), *BR* & *GO* 9092; Ramareddy (NZB), *BR* & *PSPB* 9553; Medak (MDK), *BR* & *CP* 11415; Gadirayal RF (RR), *MSM* & *KH* 10255; Nagarjunasagar (NLG), *KMS* 9769 (MH); Narsapur (MDK), *KMS* 6751 (MH); Rechapalli (KRN), *GVS* 22250 (MH); Cherukupalli (NLG), *BVRS* 171 (NU);   Manillagudem (KMM), *RR* 105913 (BSID); Yenkur (KMM), *RR* 108554 (BSID).

Note: Ding Hou (*loc. cit.*) considered *M. emarginata* (*Gymnosporia montana*) as distinct from the African *M. senagalensis.* Ramamurthy (*loc. cit*) reported that *M. senegalensis* occurs in Andhra Pradesh.

Maytenus heyneana (Roth) Raju & Babu, Bull. Bot. Surv. India 10: 348. 1968; Ramamurthy in N.P.Singh & *al.*, Fl. India 5: 122. 2000. *Celastrus heyneanus* Roth in Roem. & Schult., Syst. 5. 421. 1819. *p.p. Gymnosporia heyneana* (Roth) Lawson in Hook. f., Fl. Brit. India 1: 620. 1875; Gamble, Fl. Madras 1: 209. 1915.

Armed shrubs, 2-4 m high, spines straight, terminating short shoots and axillary. Leaves ovate, obovate, 3-8 × 2-5 cm, base obtuse or subacute, margin serrate, apex acute. Flowers white in axillary dichotomous cymes; sepals 5, margins ciliate; petals broadly ovate, yellowish; stamens 5, arising from in between the petals below the margins of the disc; ovary immersed in the disc, stigma trilobed. Capsule to 1 cm, obovoid, 3-celled, cells 2-seeded each, on drying blackish, aril fleshy.

Rare in deciduous forests in Medak district.

Fl. & Fr.: July - February.

Thanigonla RF (MDK), *RG* 112489 (BSID); Akkannapet to Lakshmipuram (MDK), *RG* 106720 (BSID).

Maytenus rufa (Wall.) Hara, J. Jap. Bot. 40. 327. 1965. var. **rufa**; Ramamurthy in N.P.Singh & *al.*, Fl. India 5: 126. 2000. *Celastrus rufa* Wall. in Roxb., Fl. Ind. 2. 397. 1824. *Gymnosporia rufa* (Wall.) Lawson in Hook. f., Fl. Brit. India 1 : 271. 1875; Gamble, Fl. Madras 1: 209. 1918.

Small trees, with slender sparingly armed branches, spines slender. Leaves ovate-lanceolate or lanceolate, 5-13 × 1-4 cm, base narrowed, margin minutely serrulate, apex obtusely acute, subcoriaceous, glabrous, green above, pale beneath, reddish when dry. Flowers white, small, in capillary cymes; peduncles up to 2 cm long; calyx 5-toothed, triangular; petals suborbicular. Capsule obovoid, 0.8 cm in diam., pale yellow inside; seeds black.

Common in Medak district up to 1500 m.

Fl. & Fr.: April- August.

Edupayalu RF (MDK), *RG* 113412 (BSID).

PLEUROSTYLIA Wight & Arn.

Pleurostylia opposita (Wall.) Alston in Trimen, Handb. Fl. Ceylon 6 (Suppl.): 48. 1931; Ramamurthy in N.P.Singh & *al.*, Fl. India 5: 134. 2000. *Celastrus opposita* Wall. in Roxb., Fl. Ind. 2: 398. 1824. *Pleurostylia wightii* Wight & Arn., Prodr. 157. 1834; Lawson in Hook.f., Fl. Brit. India 1 : 617. 1875; Gamble, Fl. Madras 1: 211. 1918.

Unarmed shrubs or small trees. Leaves opposite decussate, oblong- obovate or lanceolate, 3-8 × 1.5-5.5cm, thin coriaceous, base attenuate, margin entire, apex obtuse, 2-3 cm long. Flowers green in axillary few-flowered cymes; peduncle 2-3 mm long; pedicel up to 2 mm long; calyx lobes rounded or subreniform; petals elliptic to broadly ovate; stamens 2 mm long; filaments subulate, attached just beneath margin of disc; disc cupular, to 2 mm long, fleshy; ovary flask-shaped, style short, stigma capitate. Drupe 6 x 4 mm, 1-2-celled, ellipsoid, indehiscent; seed 1, rarely 2.

Occasional in deciduous forests in Warangal district.

Fl. & Fr.: July-December.

Kawal RF (WGL), *RKP* 108172 (BSID).

RHAMNACEAE

1. Plants unarmed; flowers in panicles; fruit samara; leaves penninerved .... **Ventilago**
1. Plants armed (except *Ziziphus glabrata*); flowers in axillary clusters or cymes; fruit drupe; leaves palminerved .... **Ziziphus**

VENTILAGO Gaertn.

1. Calyx tube adherent up to middle of ovary; disc pubescent .... **V. denticulata**
1. Calyx tube adherent at base of ovary; disc glabrous except near its centre .... **V. maderaspatana**

Ventilago denticulata Willd., Ges. Naturf. Freunde Berlin Neue Schriften. 3: 417. 1801; var. **denticulata**; Bhandari & Bhansali in N.P.Singh & *al.*, Fl. India 5: 217. 2000. *V. calyculata* Tul. in Ann. Sci. Nat. Bot. Ser. 4.8: 124. 1857; Lawson in Hook.f., Fl. Brit. India 1 : 631. 1875; Gamble, Fl. Madras 1: 218. 1918.

Climbing shrubs; branches faintly angled, striate, young branches fulvous-pubescent. Leaves 4- 15 × 2.3-7 cm, elliptic-ovate, base unequally rounded, margin crenate, apex acute; petioles 3-14 mm long, pubescent, channelled. Flowers pale yellow in terminal, sometimes axillary, dense, 12 cm long panicles; calyx lobes deltoid, hairy within; petals spathulate, truncate; stamens 1.5 mm long, connectives prolonged; disc 5-lobed, rarely 10-lobed, ovary bicarpellary, stigmatic arms divergent. Samaras about 4 cm across, with median line single or often double from middle, enclosed at base by adherent calyx tube to more than half the nut, tawny pubescent, 1-seeded.

Common in dry forests of most districts.

Fl. & Fr.: September-February.

Bersaipet (ADB), *GO* & *PVP* 4609; Karepally (NZB), *TP* & *BR* 6224; Pocharam RF (MDK), *BR* & *CP* 11607; Anantagiri RF (RR), *MSM* & *PRSR* 12779; Srinivasapur (NLG), *BVRS* 268 (NU); Mannanur to Farhabad (MBNR), *SRS* 107449 (BSID); Umamaheswaram (MBNR), *SRS* 109143 (BSID); Rollapadu (KMM), *RR* 108023 (BSID).

Key to Varieties

1. Apex of the fruit wing bifid .... var. **bifida**
1. Apex of the fruit wing not bifid .... var. **denticulata**

Ventilago denticulata Willd. var. **bifida** Bhandari & Bhansali in Nayar & *al.*, Fasc. Fl. India 20: 84. 1990 & in Singh & *al.*, Fl. India 5: 218. 2000

Rare in Medak district (Gopalan & *al.*, 2006).

Edupayalu RF (MDK), *RG* 113429 (MH).

Ventilago madraspatana Gaertn. Fruct. Sem. Pl. 1: 223. t. 47. f. 2. 1788; Lawson in Hook.f., Fl. Brit. India 1 : 631. 1875; Gamble, Fl. Madras 1: 218. 1918; var. **madraspatana**; Bhandari & Bhansali in N.P.Singh & *al.*, Fl. India 5: 220. 2000.

Climbing shrubs; branches terete, striate, sparingly puberulous or glabrous. Leaves elliptic-ovate or lanceolate, ca 8 × 3 cm, thin- coriaceous, base acute, margin entire, apex acute-acuminate. Flowers pale green in terminal and axillary branched panicles; calyx lobes spreading; petals emarginated with a tooth in the middle; stamens 1.5 mm long; disc glabrous, ovary villous, style arms divergent or curved. Fruits of winged nuts, about 3 cm across, enclosed at base by adherent calyx tube near base, wings glabrous, 1- seeded; seeds globose, brown.

Common in most districts.

Fl. & Fr.: December-April.

Mannanur (MBNR), *TDCK* 13667, 15394; Veerapuram (WGL), *PVS* 76902 (BSID).

Note: The spelling of specific epithet, *madraspatana,* as published by Gaertner (*op.cit.*) has been reinstated by Oza (in Taxon 19: 822. 1970) which had been changed to *maderaspatana* by Gamble.

ZIZIPHUS Tourn. ex Mill.

1. Plants unarmed .... **Z. glabrata**

1. Plants armed :

   2. Flowers in long peduncled cymes or panicles; petals 0 .... **Z. rugosa**

   2. Flowers in sessile or subsessile umbels or fascicles; petals 5 :

      3. Flowers in fascicles or short peduncled cymes; styles 2, connate to the middle :

         4. Fruit more than 1.5 cm in diameter; leaves tomentose beneath .... **Z. mauritiana**

         4. Fruit less than 1.5 cm in diameter; leaves tomentose on both sides or silky hairs below :

            5. Spines in pairs, one straight and sharp, the other shorter and hooked; leaves tomentose on both sides .... **Z. nummularia**

            5. Spines solitary, recurved; leaves clothed with silky hairs beneath .... **Z. oenoplia**

      3. Flowers in pedunculate cymes; styles 3, distinct or nearly so .... **Z. xylopyra**

Ziziphus glabrata Heyne ex Roth, Nov. Pl. Sp. 159. 1821; Lawson in Hook.f., Fl. Brit. India 1: 633. 1875; Bhandari & Bhansali in N.P.Singh & *al.*, Fl. India 5: 229. 2000. *Z. trinervia* Roxb., Fl. Ind. 1. 606. 1832; Gamble, Fl. Madras 1: 220. 1918.

Unarmed trees, up to 12 m high, branchlets glabrous. Leaves ovate or oblong-elliptic, 3.5-7.5 × 2-5 cm, asymmetrical, 3-nerved, convergent, pubescent above, tomentose below, base obtuse-subcordate, margin crenate, apex obtuse; petioles 3-9 mm long. Flowers yellow, axillary, in short peduncled cymes; calyx lobes glabrous within, keeled inside up to middle; petals obtriangular; stamens ca 3 mm long, filaments flattened; disc faintly 10-lobed, fleshy, ovary glabrous, style 2-cleft, connate up to middle. Drupe yellow, globose- obovoid, rugose, with a sweet gelatinous pulp, 1-2-celled; seeds 1 or 2, brownish.

Rare, along hilly forests in Adilabad, Nizamabad and Ranga Reddi districts.

Fl. & Fr.: March-November.

Ankusapuram RF (ADB), *GO* 4370; Nasarullabad (NZB), *TP* & *BR* 6165; Rudraram RF (RR), *TP* & *MSM* 12714.

Ziziphus mauritiana Lam., Encycl. 3: 319. 1879; Bhandari & Bhansali in N.P.Singh & *al.*, Flora of India 5: 233. 2000. *Z. jujuba* (L.) Gaertner, Fruct. 1: 203. 1788. non

Miller 1768; Lawson in Hook.f., Fl. Brit. India 1 : 632. 1875; Gamble, Fl. Madras 1: 219. 1918. *Rhamnus jujuba* L., Sp. Pl. 184. 1753.

Key to Varieties

1. Large much branched tree or large shrub; leaves fulvous tawny; drupe 1.5 cm in diameter .... var. **mauritiana**

1. Low shrub; leaves smooth; drupe 0.8 cm in diameter .... var. **fruticosa**

Ziziphus mauritiana Lam. var. **mauritiana**

Vern.: *Regu, Regi, Pedda regu.*

Much branched thorny trees, up to 10 m. high, bark rough grey, younger parts rusty tomentose, spines paired, one straight, another recurved. Leaves sub-orbicular, 2-3 × 1.5-3 cm, nearly symmetrical, basally 3-nerved, convergent, grey and glabrous above, rusty tomentose below, base subcordate, margin crenate, apex round or retuse. Flowers pale green in axillary pedicillate clusters; calyx lobes glabrous within, tomentose without, tube campanulate; petals spathulate; stamens equal to petals; distinctly 10-grooved, fleshy; ovary bicarpellary, bilocular, glabrous, style short, 2-cleft, united to the middle, stigmatic lobes curved. Drupes globose, yellow or red when ripe, 1-2-celled, rugose; seeds 1 or 2, compressed.

Common in waste places, roadsides and scrub jungles in all districts.

Fl. & Fr.: October-January

Sone (ADB), *TP* & *PVP* 4066; Chandoor (NZB), *BR* & *GO* 9206; Pochammaralu (MDK), *BR* & *CP* 11482; Mancherla (RR), *MSM* & *KH* 10270; Rudraram (RR), *MSM* & *TP* 12713; Rusthampet (MDK), *KMS* 6792 (CAL); Suryapet (NLG), *BVRS* 49 (NU); Bhadrachalam (KMM), *RC* 98971 (BSID).

Note: The correct name of this taxon has been discussed by Santapau in J. Bombay Nat Hist. Soc. 51: 807. 1955. *Z. jujuba* Gaertn. is a latter homonym of *Z. jujuba* Mill. and hence invalid. *Z. jujuba* was first employed by Miller for a different tree.

Ziziphus mauritiana Lam. var. **fruticosa** (Haines) Sebastine & Balakr., Indian Forester 89: 525. 1963. *Z. jujuba* (L.) Gaertn. var. *fruticosa* Haines, For. Fl. Chota Nagpur 270. 1910; Gamble, Fl. Madras 1: 219. 1918.

Vern.: *Regu, Regi,*

Armed, much-branched shrubs. Leaves suborbicular, 2-3 × 1-1.5 cm, basally 3-nerved, convergent, grey and glabrous above, fuscous-tomentose beneath, base subcordate, margin crenate, apex round or retuse. Flowers pale green in axillary, pedicellate clusters. Drupes 0.8 cm in diam., disc grooved.

Common on stony wastelands in all districts. Much used for fencing purposes.

Fl. & Fr.: September-May.

Rusthumpet (MDK), *KMS* 6803 (CAL); Eturnagaram (WGL), *RKP* 110828 (BSID); Dhaniyanigudem (KMM), *RR* 108090 (BSID).

Ziziphus nummularia (Burm. f.) Wight & Arn., Prodr. Fl. Ind. Orient. 162. 1834; Lawson in Hook.f., Fl. Brit. India 1: 633. 1875; Gamble, Fl. Madras 1: 219. 1918; var. **glabrescens** Bhandari & Bhansali in M.P.Nayar & *al.*, Fasc. Fl. India 20: 103. 1990 & in N.P.Singh & *al.*, Fl. India 5: 236. 2000.

Straggling shrubs, up to 3.5 m high, branches shortly zig-zag, tomentose. Leaves alternate, orbicular or elliptic-oblong or broadly ovate, serrulate, densely tomentose on both surfaces, nerves 3 from the base, stipular prickles of 2 types, one longer and straight, other short, curved. Flowers pale-yellow in axillary cymes, about 2 cm long; calyx lobes deltoid, keeled nearly to the base, campanulate; petals obovate-spathulate, rounded or truncated at apex; stamens 0.8-.2 mm long, disc slightly 10-lobed; ovary bicarpellary, bilocular, style 2-cleft, united to above the middle. Drupes 1 × 0.5 cm, obovid, stony, smooth, deep red when ripe; seeds 2, compressed, black.

Common in Karimnagar, Warangal, Khammam, Mahabubnagar and Nalgonda districts.

Fl. & Fr.: November-January.

Kodimial (KRN), *GVS* 21869 (MH & CAL); Rechapalli (KRN), *GVS* 22269 (CAL); Narasampet (WGL), *KMS* 11706 (CAL); Suryapet (NLG), *BVRS* 499 (NU); Ramavaram (KMM), *RR* 112662 (BSID).

Ziziphus oenoplia (L.) Mill., Gard. Dict. ed. 8. n. 3. 1768; Lawson in Hook.f., Fl. Brit. India 1 : 634. 1875; Gamble, Fl. Madras 1: 220. 1918; var. **oenoplia**; Bhandari & Bhansali in N.P.Singh & *al.*, Fl. India 5: 236. 2000. *Rhamnus oenoplia* L., Sp. Pl. 194. 1753.

Straggling, prickly, tomentose shrubs, up to 6 m high, prickle solitary, recurved. Leaves ovate-lanceolate, 2-5 × 1-3 cm, 3-nerved, divergent, glabrescent or pubescent above, densely sericeous below, base oblique, subacute-obtuse, margin denticulate, apex acute-acuminate. Flowers pale green, in axillary, peduncled cymes; calyx lobes 1.5-2 mm long, brownish; petals 0.8-1 mm long; stamens 0.7-0.9 mm long; disc 10-lobed, pitted or grooved, lobes opposite each calyx lobe; ovary glabrous, style 2-cleft, united above the middle, stigma obtuse. Drupes globose, rugose, woody, dark- pink, edible; seed solitary, ovoid.

Common in all districts, especially in dry forests and open bushy places.

Fl. & Fr.: October-January

Sattenapalli (ADB), *TP* & *PVP* 4072; Boath (ADB), *GO* & *PVP* 4636; Manchippa (NZB), *TP* & *BR* 6046; Mamidipally (NZB), *BR* & *KH* 9013; Marrikunta (MDK), *CP* 11384; Ramayampet (MDK), *BR* & *CP* 11508; Anantasagar RF (RR), *MSM* & *NV* 10401; Kodimial (KRN), *GVS* 20062; Suryapet (NLG), *BVRS* 22 (NU); Lakshmipuram RF (KMM), *RC* 99008 (BSID).

Ziziphus rugosa Lam., Encycl. 3: 319. 1789; Lawson in Hook.f., Fl. Brit. India 1 : 636. 1875; Gamble, Fl. Madras 1: 221. 1918; var. **rugosa**; Bhandari & Bhansali in N.P.Singh & *al.*, Fl. India 5: 240. 2000.

Straggling, evergreen shrubs or small trees, 3-6 m high, branches brown-pubescent, becoming glabrous, strongly prickly, prickles solitary, recurved, broad based. Leaves elliptic, ovate- oblong, up to 15 × 8 cm, base unequally cordate, margin serrulate, apex slightly emarginated or mucronate, 3-nerved, coriaceous, glabrous above, rufous pubescent beneath, petioles up to 1 cm long. Flowers pale-yellow or greenish-yellow in 12 cm long, pubescent or tomentose, cymose panicles; calyx lobes 5, pubescent without; stamens 1-2 mm long; disc 5-lobed; ovary bicarpellary, style 20-cleft, united below the middle, curved. Drupes globose or pyriform, 1 × 0.8 cm, whitish when ripe; seeds compressed, black.

Common in Mahabubnagar district in forests.

Fl. & Fr.: March-May.

Ziziphus xylopyra (Retz.) Willd., Sp. Pl. 1 : 1104. 1798; Lawson in Hook.f., Fl. Brit. India 1: 634. 1875 ('*xylopyrus*'); Gamble, Fl. Madras 1: 220. 1918; Bhandari & Bhansali in N.P.Singh & *al.*, Fl. India 5: 243. 2000. *Rhamnus xylopyrus* Retz. Observ. Bot. 2: 11. 1781.

Vern.: *Gotti, Gotika.*

Straggling prickly tree-like shrubs, up to 5 m high, stipular prickles not frequent, but when present one straight, other recurved. Leaves ovate, ovate-oblong, 5-8 × 3-4 cm, asymetrical, 3-nerved, glabrous above, tomentose below, base subcordate or obtuse, margin serrulate, apex rounded; petioles 2-7 mm long, fulvous tomentose. Flowers yellow, in axillary and extra-axillary cymes; calyx lobes keeled up to the middle, thickened at apex, glabrous within, pubescent without; petals 1.5-2 mm long; stamens equal to petals; disc 10-lobed, pitted or grooved, rarely 5-lobed, glabrous; ovary tricarpellary, globose, nearly hidden by the disc, style cleft up to half to two-third of the length. Drupes globose, 2 cm across, brown tomentose, 2 or 3-celled; seeds oblong, black.

Common in dry deciduous forests of all districts.

Fl. & Fr.: May-September.

Bheemaram RF (ADB), *GO* 4497; Ghanapur RF (NZB), *BR* & *KH* 6497; Pocharam RF (MDK), *BR* & *CP* 11606; Narsapur RF (MDK), *TP* & *CP* 11981; Kothapalli RF (RR), *MSM* & *KH* 10301; Kusumasamudram RF (RR), *MSM* & *NV* 10336; Bourapur (MBNR), *TDCK* 16813; Naraspur (MDK), *KMS* 7951 (CAL); Aklaspur (KRN), *GVS* 46879 (MH); Pakhal (WGL), *KMS* 13127 (MH); Nagarjunakonda (NLG), *BVRS* 82 (NU); Bhadrachalam RF (KMM), *RC* 102442 (BSID).

VITACEAE

1. Petals united at apex and falling off as a cup at anthesis; leaves always simple .... **Vitis**

1. Petals free, expanding at anthesis; leaves simple or compound:

    2. Flowers 5-merous .... **Ampelocissus**

    2. Flowers 4-merous:

3. Flower buds flask-shaped, nectariferous; disc of
    4 free glands .... **Cyphostemma**

3. Flower buds globose or oblong, nectariferous disc entire:

    4. Inflorescence leaf-opposed; berries 1 or 2-seeded .... **Cissus**

    4. Inflorescence axillary; berries 2-4-seeded .... **Cayratia**

Note: Planchon in DC. Monogr. Phan. divided *Vitis* into number of genera, the species
here noticed belong to the following : 1. *Vitis* petals 5, united at apex 2.
Ampelocissus petals 4, mostly 5, free. 3. *Cissus* petals 4, style long. 4. *Tetrastigma,*
petals 4, stigma nearly sessile, 4-lobed, inflorescence axillary, flowers
unisexual, 5. *Cyphostemma,* branchlets bristly setose, 6. *Cayratia,* cymes axillary.

AMPELOCISSUS Planch. *nom. cons.*

1. Leaves angled or shallowly lobed, glabrous; flowers in small
    compact, puberleous thyrses; disc 5-furrowed .... **A. latifolia**

1. Leaves deeply lobed, pubescent or woolly tomentoses, flowers
    in shortly peduncled, compact, dense woolly cymes;
    disc not furrowed .... **A. tomentosa**

Ampelocissus latifolia (Roxb.) Planch. in Vigne Amer. Vitic. Europe 8: 375. 1884;
Gamble, Fl. Madras 1: 230. 1918; Shetty & P.Singh in N.P.Singh & *al.* Flora of
India 5: 256. 2000. *Vitis latifolia* Roxb., Fl. Ind. 1: 661. 1820; Lawson in Hook.f., Fl.
Brit. India 1: 652. 1875.

Climbing shrubs, branches terete, striate, hollow, glabrous, tendrils forked, only
young shoots and inflorescence pubescent. Leaves orbicular, 5-17 × 4-16.5 cm,
shallowly 3-5-lobed, base cordate, margin dentate-serrate or serrulate, apex acuminate,
5-nerved from base; petioles 5-15 cm long. Flowers reddish-green or reddish-brown
in leaf-opposed thyrsoid cymes, infertile branches tendrillar; calyx saucer-shaped,
entire, glabrous; petals oblong, slightly incurved at apex, glabrous; stamens ca 1.2
mm long; disc enclosing about half of ovary, 5-grooved, ovary glabrous, stigma
subsessile. Berries globose, black when ripe, edible; seeds ellipsoid, usually 2, smooth.

Occasional in hilly forests of all districts.

Fl. & Fr. : June- September.

Kotapalli RF (ADB), *TP* & *GO* 5419; Mamidipally RF (NZB), *BR* & *KH* 6430;
Sirnapally RF (NZB), *BR* & *GO* 9094; Mudheli RF (NZB), *BR* 9538; Narsapur RF
(MDK), *TP* & *CP* 14032; Burugupally (MDK), *CP* 11767; Mohammadabad RF (RR),
MSM & *NV* 10402; Pakhal (WGL), *KMS* 13123 (MH); Durachipalli (NLG), *BVRS* 26
(NU).

Ampelocissus tomentosa (Heyne ex Roth) Planch., J. Vigne Amer. Vitic. Europe 8:
375. 1884 & in DC., Monogr. Phan. 5: 376. 1887; Gamble, Fl. Madras 1: 230. 1918;
Shetty & P.Singh in N.P.Singh & *al.*, Fl. India 5: 261. 2000. *Vitis tomentosa* Heyne
ex Roth, Nov. Pl. Sp. 157. 1821; Lawson in Hook.f., Fl. Brit. India 1 : 650. 1875.

Shruby climbers, up to 6 m, branches striate, white to brownish-red, woolly tomentose. Leaves palmately 3-5-deeply lobed, 8-20 × 7-22 cm, pubescent above, tomentose below, base cordate, margin crenate, apex acuminate; petioles up to 15 cm long. Flowers brick red on leaf-opposed thyrsoid cymes, dense woolly tomentose; peduncles ca 10 cm long, tendrils branched; calyx cupular, 4-5-lobed, woolly; petals oblong-ovate, incurved at apex; stamens ca 1 mm long, disc annular, restricted to lower half of ovary, ovary 10-furrowed, stigma sessile, concave. Berries ovoid, black when ripe, 1-4-seeded; seeds ovate-oblong, rugose, furrowed along outer edge.

Common in hill forests of Warangal and Mahabubnagar districts.

Fl. & Fr.: July-November

Lawal beat, Pasra range (WGL), *RKP* 108175 (BSID); Pakhal RF (WGL), *KMS* 13122 (MH); Mannanur core range (MBNR), *VBH* 86615 (BSID).

CAYRATIA Juss. *nom. cons.*

1. Leaves pedately 3-foliolate, leaflets ovate-orbicular; seed trigonous ..... **C. trifolia**
1. Leaves pedately 5-7-foliolate; leaflets oblong; seed hemispherical .... **C. pedata**

Cayratia pedata (Lam.) Juss. ex Gagnep. in Lecomte, Notul. Syst. (Paris) 1 : 346. 1911; Gamble, Fl. Madras 1: 236. 1918; var. **pedata**; Shetty & P.Singh in N.P.Singh & *al.*, Fl. India 5: 271. 2000. *Cissus pedata* Lam., Encycl. 1 : 31. 1783. *Vitis pedata* (Lam.) Wall. ex Wight & Arn., Prodr. Fl. Ind. Orient. 128. 1834; Lawson in Hook.f., Fl. Brit. India 1 : 661. 1875.

Large, climbing shrubs, branches terete, striate, puberulous. Leaves pedately 5-7-foliolate, 8-15 cm long, leaflets 1-12 × 3-6 cm, terminal one oblong, laterals pedately lobed, inequilateral, membranous, pubescent, base cuneate, margin serrate, apex acuminate, tendrils leaf-opposed, branched, pubescent; Flowers pale green in axillary branched cymes; calyx cupular, truncate, pubescent; petals incurved at apex; stamens ca 2 mm long; disc cupular, 4-lobed, covering ovary; ovary ca 1.5 mm across, style subulate, stigma subcapitate. Berries subglobose, pale yellow when ripe; seeds 2 or 4, hemispherical, deeply pitted towards the centre.

Common in the hills of Mahabubnagar, Medak and Adilabad districts.

Fl. & Fr.: January-November

Wankidi (ADB), *GO* 4471; Madharam RF (ADB), *TP* & *GO* 5475; Narsapur RF (MDK), *TP* & *CP* 11991.

Cayratia trifolia (L.) Domin., Biblioth. Bot. 89: 370. 1917; Shetty & P.Singh in N.P.Singh & *al.*, Fl. India 5: 275. 2000. *Vitis trifolia* L., Sp. Pl. 203. 1753. *Cissus carnosa* Lam., Encycl. 1: 31. 1783. *Cayratia carnosa* (Lam.) Gagnepain in Lecomte, Notul. Syst. (Paris) 1: 347. 1911; Gamble, Fl. Madras 1: 237. 1918. *Vitis carnosa* (Lam.) Wall. ex Lawson in Hook.f., Fl. Brit. India 1: 654. 1875.

Climbing shrubs, rather succulent, branches striate, glandular, crisped-hirtellous when young. Leaves 3-foliolate, 7-9 cm, leaflets 7 × 3.5 cm, ovate-orbicular, unequally cordate or acute at base, crenate-dentate, tendrils leaf-opposed, stipules foliaceous, ovate, fimbriate along margins. Flowers pale green in a greenish-white dichotomously branched umbellate or corymbose cymes. Berry obovid, seeds 4, triangular rugose.

Common in most of the districts.

Fl. & Fr.: July-October.

Servaipet (ADB), *GO* & *PVP* 4819; Mosra-Malkapur road (NZB), *TP* & *BR* 6089; Nagulabanda (MDK), *CP* 11338; Anantagiri RF (RR), *TP* & *MSM* 11075; Kodimial (KRN), *GVS* 24485 (MH); Kambalpally (NLG), *BVRS* 234 (NU).

Note: Though some earlier authors treated *Cissus carnosa* Lam. as a distinct taxon from L. *Vitis trifolia,* it is now evident from the critical studies by Merrill (Inter. Herb. Amb.) and subsequently by Backer & Bakhuizen f. (Fl. Jav. 1965) that both are conspecific and hence its correct name under the genus *Cayratia* should be *C. trifolia* (L.) Domin.

CISSUS L.

1. Stems and branches acutely angled or winged .... **C. quadrangularis**
1. Stems terete or obscurely angled :
    2. Leaves quite glabrous .... **C. arnottiana**
    2. Leaves pubescent or tomentose beneath :
        3. Tendril deciduous .... **C. vitiginea**
        3. Tendrils not deciduous :
            4. Leaves when young densely tomentose beneath; flowers pink .... **C. repanda**
            4. Leaves when young orange-red pubescent beneath; flowers greenish-yellow .... **C. adnata**

Cissus adnata Roxb., Fl. Ind. 1: 423. 1820; Gamble, Fl. Madras 1: 234. 1918; Shetty & P.Singh in N.P.Singh & *al.,* Fl. India 5: 279. 2000. *Vitis adnata* (Roxb.) Wall. ex Lawson in Hook.f., Fl. Brit. India 1 : 649. 1875.

Hairy climbing shrubs, stems and inflorescence clothed with orange red pubescence, at length glabrate, tendrils forked. Leaves broadly-ovate, 7.5-12.5 × 5-8 cm, base cordate, margin bristle-serrate, apex shortly acuminate or cuspidate, densely clothed with orange-red pubescence beneath, more or less glabrous above, main nerves 4-5 pairs, prominent beneath. Flowers greenish yellow, in much branched, peduncled, compound umbellate cymes; calyx cupular, truncate, hairy; petals oblong, acute, hooded, hairy; stamens ca 1 mm long; disc 4-notched, conspicuous, covering ovary, ovary ca 1 mm across, pubescent at summit, style stout, stigma minute. Berry 0.6 cm in diameter, obovoid or subglobose, glabrous, apiculate, 1-seeded, rarely 2-seeded, black when ripe.

Occasional in Medak district.

Fl. & Fr.: October-December.

Pagarikutta (MDK), *KMS* 6685 (CAL, BSID).

Cissus arnottiana Shetty & P. Singh, Kew Bull. 44: 473. 1989 & in N.P.Singh & *al.*, Fl.
India 5: 281. 2000. *Vitis pallida* sensu Lawson in Hook.f., Fl. Brit. India 1: 645.
1875, *p.p.* (non Wight & Arn., 1883). *Cissus pallida* sensu Planch. in DC., Monogr.
Phan. 5: 477. 1887, *p.p.* (non Salisb., 1796); Gamble, Fl. Madras 1 : 234. 1918.

Erect shrubs, up to 1.5 m high, stems reddish, terete, striate, woody, branchlets
glabrescent, tendrils leaf-opposed, simple. Leaves simple, orbicular, 7-18 × 8.5-20 cm,
thin coriaceous, cordate at base with wide sinus, nerves 5 from base, margin serrate,
sometimes crenate, apex abruptly acuminate, petiole up to 20 cm. Flowers pale yellow
in compound umbellate cymes; peduncles thick, 1.5-5 cm long, calyx cupular, truncate
to obscurely 4-lobed; petals ovate, glabrous; stamens ca 2 mm long, disc 4-lobed,
covering ovary, fleshy, white; ovary glabrous, style slender. Berries globose,
mucronate, 1-2-seeded; seeds solitary, pyriform.

Common in forests of Adialabad, Medak and Mahabubnagar districts.

Fl. & Fr.: January-December

Jaipur RF (ADB), *GO* & *PVP* 4825; Killah (MDK), *RG* 104004 (BSID); near
Domalapenta (MBNR), *VBH* 84948 (BSID).

Cissus quadrangularis L., Mant. Pl. 39. 1767; Gamble, Fl. Madras 1: 234. 1918; Shetty
& P.Singh in N.P.Singh & *al.*, Fl. India 5: 288. 2000. *Vitis quadrangularis* (L.) Wall.
ex Wight & Arn., Prodr. Fl. Ind. Orient. 125. 1834; Lawson in Hook.f.,Fl. Brit.
India 1: 645. 1875.

Vern.: *Nalleru.*

Rambling shrubs, stem fleshy, 4-angular, glabrous, winged or margined,
contracted at nodes. Leaves simple, ovate or reniform, 2.5 × 1.5-4.5, thick coriaceous,
apex and base round, margin serrate, tendril leaf-opposed, highly coiled. Flowers
greenish- red in leaf-opposed, short-peduncled umbellate cymes; calyx cupular,
obscurely lobed, glabrous; petals ovate-oblong, acute, hooded; stamens ca 1.5 cm
long; disc 4-lobed, covering ovary, ovary ca 1.5 mm across, style short. Berry globose,
reddish brown; seed solitary, obovate, smooth.

Common in scrub jungles, roadsides and wastelands in all districts.

Fl. & Fr.: June-December.

Mamidipally (NZB), *BR* & *KH* 9003; Maddigunta (NZB), *BR* & *PSPB* 9558; Central
University campus (RR), *MSM* 19209; Banala (MBNR), *TDCK* 13699; Bhongir fort
(NLG), *BVRS* 18 (NU); Khammam to Dhanaiyaigudem (KMM), *RR* 108515 (BSID);
Nadisisagutta (KMM), *RR* 106051 (BSID); Vatavarlapalli (MBNR), *VBH* 84955 (BSID).

Cissus repanda Vahl, Symb. Bot. 3: 18. 1794; Gamble, Fl. Madras 1: 234. 1918. Shetty
& P.Singh in N.P.Singh & *al.*, Fl. India 5: 290. 2000. *Vitis repanda* (Vahl) Wight &
Arn., Prodr. Fl. Ind. Orient. 125. 1834; Lawson in Hook.f., Fl. Brit. India 1: 648.
1875.

Perennial, climbing shrubs with tendrils, tendrils with flattened discs, young
parts reddish, adpressed woolly. Leaves broadly ovate or orbicular, base cordate,
margin crenate-serrate, apex acuminate, pubescent above, silky-tomentose beneath;

petioles 3-7 cm long. Flowers pink in umbellately branched cymes; calyx cupular, obscurely 4-lobed, hairy; petals ovate, hooded; stamens ca 1.5 mm long; disc 4-lobed, ovary ca 1 mm across, style stout, stigma capitate. Berry pyriform, 1-seeded; seed obovoid.

Occasional in forests in Hyderabad and Mahabubnagar districts.

Fl. & Fr.: March- June.

Appapur (MBNR), *BSS* & *BR* 32523.

Cissus vitiginea L., Sp. Pl. 117. 1753; Gamble, Fl. Madras 1: 234. 1918; Shetty & P.Singh in N.P.Singh & *al.*, Fl. India 5: 294. 2000. *Vitis linnaei* Wall. ex Wight & Arn., Prodr. Fl. Ind. Orient. 126. 1834; Lawson in Hook.f., Fl. Brit. India 1 : 649. 1875.

Vern.: *Godhuma theega, Gummadi, Nela gummadi, Vorupaku.*

Woody straggling or climbing shrubs, branchlets densely pubescent. Leaves simple or broadly cordate, 5-12 × 4-11 cm, 5-angular or deeply lobed, thick coriaceous, basally 3-nerved, pubescent, base cordate, margin dentate, apex, acuminate, tendril leaf-opposed, highly coiled. Flowers pale yellow in decompound umbellate cymes; calyx cupular, 4-lobed, pubescent; petals oblong-ovate, hooded, pubescent; stamens ca 1 mm long; disc 4-lobed, fleshy, covering ovary, ovary ca 1 mm across, style short, stigma minute. Berries ovoid, reddish-brown, rugose, apiculate, 1-seeded; seeds tessellate.

Common in scrub jungles and hedges in plains in most districts.

Fl. & Fr.: July-May.

Ankusapuram RF (ADB), *GO* 4377; Vempalli RF (ADB), *GO* 4431; Mosra-Malkapur Road (NZB), *TP* & *BR* 6094; Dichpally (NZB), *BR* & *KH* 6444; Nadipally (NZB), *BR* & *KH* 6454; Medak (MDK), *CP* 11726; Kusumasamudram RF (RR), *MSM* & *NV* 10366; Pakhal (WGL), *KMS* 13138 (MH); Aklaspur (KRN), *GVS* 20224 (MH); Thunnikkicheruvu RF (KMM), *RC* 104274 (BSID); Thallada – Kothagudem (KMM), *RR* 106098 (BSID).

CYPHOSTEMMA (Planch.) A.H.G. Alston

1. Leaves 5-foliolate; petioles 7-15 cm long; berries glabrous .... **C. auriculatum**
1. Leaves 3-foliolate, subsessile; berries glandular-hispid .... **C. setosum**

Cyphostemma auriculatum (Roxb.) P. Singh & Shetty, Taxon 35: 596. 1986; Shetty & P.Singh in N.P.Singh & *al.*, Fl. India 5: 298. 2000. *Cissus auriculata* Roxb., Fl. Ind. 1: 430. 1820. *Cayratia auriculata* (Roxb.) Gamble, Fl. Madras 1: 230. 1918. *Vitis auriculata* (Wall.) Wight & Arn., Prodr. Fl. Ind. Orient. 129. 1834; Lawson in Hook.f., Fl. Brit. India 1: 658. 1875.

Vern.: *Kurapalleru, Kutamu, Pulusa.*

Large climbers, branchlets cylindric, succulent, the young parts softly pubescent. Leaves 5-foliolate, leaflets obovate, 6-12(20) × 6.5-15(25) cm, stipules ear-shaped, tendrils 2-3-cleft, equilateral or not, pubescent along nerves above, densely pubescent below, base cuneate, margin crenate-serrate, apex abruptly acuminate to shortly caudate. Flowers small, greenish to cream in dichasial cymes on long, thick, succulent peduncles, divaricating, longer than the petiole; calyx cupular, truncate, pubescent; petals oblong, hooded; stamens ca 2 mm long; disc of 4 glands, almost covering the ovary; ovary pyramidal, hairy. Berry globose, red, 1.2 cm across, 1- seeded; seeds oblong-ovoid to subglobose.

Occasional at higher elevation in Karimnagar, Nizamabad, Medak, Mahabubnagar and Ranga Reddi districts.

Fl. & Fr.: June-October.

Ghanpur RF (NZB), *BR* & *KH* 6500; Mohammadabad RF (RR), *MSM* & *KH* 10300; Mallelatheertham (MBNR), *TDCK* 15389.

Cyphostemma setosum (Roxb.) Alston in Trimen, Handb. Fl. Ceylon 6 (Suppl.): 53. 1931; Shetty & P.Singh in N.P.Singh & *al.*, Fl. India 5: 300. 2000. *Cissus setosa* Roxb., Fl. Ind. 1: 410. 1820; Gamble, Fl. Madras 1: 235. 1918. *Vitis setosa* (Roxb.) Wall. ex Wight & Arn., Prodr. Fl. Ind. Orient. 127. 1834; Lawson in Hook.f., Fl. Brit. India 1: 654. 1875.

Vern.: *Barrebachali, Pullabachali.*

Straggling shrubs with bristly setose branches. Leaves 3- foliolate, leaflets oblong, ovate, 4-7.5 × 2-6 cm, terminal one larger, laterals inequilateral, subsucculent, glandular- puberulous, apex and base rounded, margin dentate-serrate, stipules ovate, tendrils forked. Flowers green in apparently terminal cymes; calyx cupular, subtruncate, glabrous; petals oblong, hooded; stamens ca 2 mm long; disc of 4 truncate glands almost covering the ovary; ovary pyramidal, hairy, style subulate. Berries ovoid, red when ripe; seeds solitary, subglobose, pitted and crenate on margins.

In dry localities of the forests, in Mahabubnagar district.

Fl. & Fr.: September-January.

VITIS L.

Vitis vinifera L., Sp. Pl. 202. 1753; Lawson in Hook.f., Fl. Brit. India 1: 652. 1875; Gamble, Fl. Madras 1: 163. 1918; Khan, For. Fl. Hyd. State 108. 1953. Shetty & P.Singh in N.P.Singh & *al.*, Fl. India 5: 324. 2000.

Vern.: *Draksha.* English: Grapes.

A large woody climber. Leaves simple, orbicular-ovate, dentate, acuminate, glabrous above, slightly tomentose beneath; tendrils bifid, leaf-opposed. Flowers in cymes, fragrant, green. Berry edible.

Cultivated for its fruits in many districts.

Fl. & Fr.: August – December.

Suryapet (NLG), *BVRS* 48 (NU).

LEEACEAE

LEEA L. *nom. cons.*

1. Lobes of staminal tube rounded or truncate; leaflets serrulate ..... **L. macrophylla**

1. Lobes of staminal tube notched or bifid; leaflets crenate- serrate or dentate;

    2. Lateral nerves 15-20, close, prominent; leaflets asperous above ..... **L. asiatica**

    2. Lateral nerves 10-15, rather distant, not prominent above,
        leaflets glabrous or with only a few hirsute haris .... **L. indica**

Leea asiatica (L.) Ridsdale in Manilal, Bot. Hist. Hort. Malab. 189. 1980; Naithani in N.P.Singh & *al.*, Fl. India 5: 330. 2000. *Phytolacca asiatica* L., Sp. Pl. 474. 1753. *Leea crispa* L., Syst. Nat. ed. 12. 2. 627. 1767; Lawson in Hook.f., Fl. Brit. India 1: 665. 1875; Gamble, Fl. Madras 1: 240. 1918. *L. aspera* Edgew., Trans. Linn. Soc. London 20: 36. 1846. non Wall. ex G. Don 1831; Lawson in Hook.f., Fl. Brit. India 1 : 665. 1875; Gamble, Fl. Madras 1: 240. 1918. *L. herbacea* Buch-Ham., Trans. Linn Soc. 20 : 36. 1846; Lawson in Hook.f., Fl. Brit. India 1: 665. 1875; Gamble, Fl. Madras 1: 240. 1918. *L. edgeworthii* Sant., Rec. Bot. Surv. India 16: 54. 1953.

Shrubs, up to 2 m high, branches shallowly 8-10-angled, grooved, glabrous. Leaves about 32 cm long, leaflets 5-7, elliptic-ovate, or oblong, caudate, or rhomboid, 10-15 × 3-6 cm, base rounded or cordate, margin crenate-serrate, apex acuminate to caudate, lateral nerves 15-20 pairs, transverse nerves prominent, terminal leaflets petiolule up to 5.5 cm long. Flowers pale green in terminal and axillary branched cymes; corolla tube along with staminal lobes 3-4 mm long; staminal lobes deeply bifid; stamens free; ovary 4-8-loculed. Berries 6-8 mm, depressed, 6-lobed, purple black; seeds wedge-shaped.

Rare in Adilabad, Karimnagar, Ranga Reddi and Mahabubnagar districts.

Fl. & Fr.: August-November

Bijjur RF (ADB), *GO* & *PVP* 4864; Bhadrachalam RF (KMM), *RC* 104301 (BSID).

Leea indica (Burm.f.) Merr., Philipp. J. Sci. 14: 245. 1919; Ridsdale, Blumea 22: 95. 1974; Naithani in N.P.Singh & *al.* Flora of India 5: 337. 2000. *Staphylea indica* Burm. f., Fl. Ind. 75. t. 23. f. q. 1768. *Leea sambucina* (L.) Willd., Sp. Pl. 1 : 1177. 1798; Lawson in Hook.f., Fl. Brit. India 1: 666. 1875; Gamble, Fl. Madras 1: 240. 1918. *Aquilicia sambucina* L., Mant. Pl. 2: 211. 1771.

A large shrub or small tree, up to 6 m high, branchlets puberulous. Leaves 2-3-pinnate, pinnae 3-paired, leaflets odd-pinnate, 2-5-paired, opposite, oblong-lanceolate, sometimes ovate 5-10 × 2-3.5 cm, thin coriaceous, nerves ca 15-paired, midnerve prominent below, glabrous above, pubescent along nerves below, base obtuse, margin 2-dentate, serrate, apex acuminate. Flowers greenish-white in lateral or leaf-oppsed corymbose cymes, divaricate; calyx lobed up to middle, corolla tube and staminodal lobes 2.5-3.5 mm long, staminal lobes cleft or notched; ovary 6-loculed, style 1-2 mm long. Berry globose, depressed, black-purple, pyrenes 5 or 6, each 6 mm across, 1-seeded.

Rare in forests in Warangal district.

Fl. & Fr.: Through the year.

Pasra (WGL), *RKP* 108228 (BSID).

Leea macrophylla Roxb. ex Hornem. Hort. Hafin 1: 231. 1813; Lawson in Hook.f., Fl. Brit. India 1: 664. 1875; *p.p.* Gamble, Fl. Madras 1: 240. 1918; Ridsdale, Blumea 22: 85. 1974; Naithani in N.P.Singh & *al.*, Fl. India 5: 339. 2000. *L. integrifolia* Roxb., Fl. Ind. 2: 472. 1824; Lawson in Hook.f., Fl. Brit. India 1: 667. 1875. *L. robusta* Roxb., Fl. Ind. 1: 665. 1820; Lawson in Hook.f., Fl. Brit. India 1 : 667. 1875 *p.p.*; Gamble, Fl. Madras 1: 240. 1918. *L. diffusa* Lawson in Hook. f., Fl. Brit. India 1 : 667. 1875. *L. latifolia* Wall. ex Kurz, J. Asiat. Soc. Bengal 44: 178. 1875; Gamble, Fl. Madras 1: 240. 1918. *L. cinerea* Lawson in Hook. f., Fl. Brit. India 1: 665. 1875. *L. coriacea* Lawson in Hook. f., Fl. Brit. India 1: 665. 1875. *L. venkobarowii* Gamble, Kew Bull. 1977. 26. 1917 & Fl. Madras 1: 240. 1918.

Vern.: *Booradipokai.*

Shrubs, about 1.5 m high, branches terete, striate, young branches, petioles, rachis, petiolules and inflorescence faintly puberulous. Leaves about 35 cm long, 3-5-foliolate, leaflets elliptic-ovate, 9-25 × 4-15 cm, base unequally rounded, margin serrulate, apex acute or obtuse, sparingly strigose, lateral nerves about 12 pairs, parallel. Flowers pale green, in 10 cm long branched corymbose cymes, rachises rusty-puberulous; calyx mealy pubescent, corolla tube with staminodal lobes 3-4 mm long, corolla lobes greyish pubescent to papillose, staminodal lobes slightly retuse or shallowly cleft; ovary 6-loculed. Berries 0.9 x 0.5 cm, depressed, prominently 6-lobed, seeds black with white-scrufy surface, wedge-shaped, rugose.

Rare in Warangal, Mahabubnagar, Hyderabad and Medak districts.

Fl. & Fr.: April-November.

Way to Gangaram (WGL), *PVS* 84011 (BSID).

SAPINDACEAE

1. Stamens inserted inside the disc :
   2. Flowers irregular; disc unilateral :
      3. Climbing herbaceous plants with tendrils; fruit a membranous inflated capsule .... **Cardiospermum**
      3. Erect shrubs or trees, tendrils absent, fruit not inflated, indehiscent :
         4. Leaves imparipinnate; flowers not very small .... **Lepisanthes (rubiginosa)**
         4. Leaves digitately 3-foliolate; flowers very small .... **Allophylus**
   2. Flowers regular; disc annular :
      5. Petals O, ovary entire; fruit tipped with persistent, rigid style .... **Schleichera**

5. Petals present, ovary lobed or angled, style deciduous :

   6. Inflorescence terminal; leaflets emarginate,
generally more than 2 pairs .... **Sapindus**

   6. Inflorescence axillary; leaflets not emarginate,
only two pairs .... **Lepisanthes (tetrasperma)**

1. Stamens inserted outside the disc; petals O; fruit winged capsule ...... **Dodonaea**

ALLOPHYLUS L.

1. Berry black; leaflets sessile; thyrses longer than leaves .... **A. cobbe**

1. Berry red; leaflets petiolulate; thyrses shorter than leaves .... **A. serratus**

Allophylus cobbe (L.) Raeusch., Nomencl. Bot. ed. 3. 108. 1797; Hiern in Hook.f., Fl. Brit. India 1 : 673. 1875; Leenhouts, Blumea 15: 322. 1967; Pant in N.P.Singh & *al.*, Fl. India 5: 330. 2000. *Rhus cobbe* L., Sp. Pl. 267.1753. *Allophylus rheedii* (Wight) Radlk. in Engler & Prantl, Pflazenf. 3(5): 313. 1895; Gamble, Fl. Madras 2: 246. 1918.

Small trees or large shrubs, 8-10 m high. Leaves about 25 cm long, 3-foliolate; leaflets rhomboid, ovate, elliptic-ovate, 6-12 × 3-6 cm, acute or rounded at base, margin crenate and dentate-serrate, apex obtuse or tapering. Flowers pale green or white in terminal and axillary, long, linear racemes, about 10 cm long; sepals 4, in two whorls, outer pair elliptic, ciliate, inner obovate; petals 4, unilateral, oblong-obovate, clawed, white; male flowers ca 2 mm long, filaments unequal, glabrous, anthers ovoid; female flowers ovary 2-lobed, 2-loculed, style deeply bifid. Berry 0.8 x 0.4 cm in diam., globose, smooth, black, seed solitary.

Common in dry deciduous forests and scrub jungles of most districts.

Fl. & Fr.: August – February.

Parnasala RF (KMM), *RC* 104243 (BSID).

Allophylus serratus (Roxb.) Kurz, J. Asiat. Soc. 44(2): 185. 1876; Radlk. in Engler, Pflanzenr. 98b: 562. 1932; Leenhouts, Blumea 15: 351. 1967; S.K. Mukerjee, Indian Forest. 98: 494. 1972; Pant in N.P.Singh & *al.*, Fl. India 5: 330. 2000. *Ornitrophe serrata* Roxb., Pl. Coromandel 1: 44. t. 61. 1796. *Allophylus cobbe* (L.) Raeusch. forma *serratus* (Roxb.) Heirn in Hook. f., Fl. Brit. India 1: 673. 1875.

Climbing shrubs, 2 m long, branchlets whitish. Leaves 3-foliolate; leaflets elliptic or obovate, 7-10 × 5-6 cm, apex shortly acuminate, base cuneate, margin serrate-denticulate, petiolule up to 2.5 cm long. Flowers yellowish or white, in simple axillary racemes, peduncle to 3 cm long. Fruit globose to subglobose, orange red on ripening.

In deciduous forests in Ranga Reddi district.

Fl. & Fr.: June – September.

Puttapahad – Anantasagar (RR), *MSM* & *NV* 10398.

CARDIOSPERMUM L.

1. Capsules not winged; seeds with a small orbicular hilum .... **C. canescens**

1. Capsules winged at angles; seeds with large heart-shaped
   hilum .... **C. halicacabum**

Cardiospermum canescens Wall., Pl. Asiat. Rar. 1: 14. t. 24. 1829; Hiern in Hook.f.,
   Fl. Brit. India 1: 670. 1875; Gamble, Fl. Madras 1: 244. 1918; Pant in N.P.Singh &
   al., Fl. India 5: 355. 2000.

Climbing tendril-bearing herbs, with wiry stems and branches, stems deeply
furrowed. Leaves biternate, leaflets ovate, 1.5-6 × 1-3 cm, pubescent, acute or attenuate
at base, irregularly prominently crenate-serrate, acuminate at apex. Flowers small,
white in long-peduncled umbellate cymes, the lowest pair of pedicles transformed
into spiral tendrils; sepals 4, in two whorls, outer sepals rounded, hairy on outer
faces, inner obovate; petals 4, rounded at apex, tomentose; stamens 2-4 mm long;
ovary oblong, ca 2 mm across, tomentose, style short. Capsules at first ovate, acute,
afterwards globose, not winged; seeds black with a small, orbicular, slightly emarginate
hilum.

Common along streams in open scrubs in most districts.

Fl. & Fr.: July-December.

Mamillagudem (KMM), *RR* 105939 (BSID).

Cardiospermum halicacabum L., Sp. Pl. 366. 1753; Hiern in Hook.f., Fl. Brit. India 1:
   670. 1878; Gamble, Fl. Madras 1: 244. 1918; Pant in N.P.Singh *& al.*, Fl. India 5:
   356. 2000.

Vern. : *Buddakakara*.

Climbing herbs, stems prominently striate, faintly puberulous. Leaves biternate,
leaflets ovate-lanceolate, 1.5-3.5 × 1-2.5 cm, nearly glabrous, base acute or attenuate,
margin irregularly prominently crenate serrate, apex acuminate, stipules small, cilate.
Flowers white in axillary umbellate cymes, the lowest pair of pedicels transformed
into spiral tendrils, floral pair of tendrils circinate; sepals 4, in two whorls, outer
sepals rounded-obovate, inner rounded, larger than the outer, very thin; petals 4,
rounded. tomentose. Male flowers: stamens 2-4 mm long, filaments pilose. Female
flowers: ovary oblong, ca 2 mm across, tomentose, style 1.5 mm long, stigma trifid.
Capsules depressed; pyriform, winged at the angles; seeds black with large, heart-
shaped hilum.

Key to Varieties

1. Fruits sharply 3-lobed, not bloated .... **C. halicacabum** var. **luridum**

1. Fruits less clearly 3-lobed, bloated .... **C. halicacabum** var. **microcarpum**

Cardiospermum halicacabum L. var. **luridum** (Blume) Adelb., Blumea 6: 322. 1948. *C. luridum* Blume, Rumphia 3: 184. 1847.

Common in all districts along the streams in forests, hedges and in cultivated fields.

Fl. & Fr.: July-November.

Bendala (ADB), *GO* 4482; Anantasagar RF (RR), *MSM* & *NV* 10414; Sarlapalli (MBNR), *BR* & *TSS* 28372; Rampur (MBNR), *BSS* & *SKB* 32632.

Cardiospermum halicacabum L. var. **microcarpum** (Kunth) Blume, Rumphia 3: 185. 1847. *C. microcarpum* Kunth, Nov. Gen. Sp. 5: 104. 1821.

Occasional in forest.

Fl. & Fr.: September – December.

Gandhari (NZB), *TP* & *BR* 6345; Marrikunta (MDK), *CP* 11392; Ragammagudem RF (RR), *MSM* & *KH* 10572; Nagarjunasagar (NLG), *BVRS* 489 (NU).

DODONAEA Mill.

Dodonaea angustifolia L.f., Suppl. Pl. 218. 1782; Leenh., Blumea 28: 280. 1983; Pant in N.P.Singh & *al.*, Fl. India 5: 360. 2000. *D. burmanniana* DC., Prodr. 1: 616. 1824. *D. viscosa* auct. non (L.) Jacq., 1760; Hiern in Hook.f., Fl. Brit. India 1 : 697. 1875; Gamble, Fl. Madras 1: 253. 1918; Pullaiah & Chennaiah in Fl. Andhra Pradesh 1:223. 1997.

Vern.: *Bandara, Bandaraku Chettu.*

Gregarious evergreen shrubs or small trees, branches brown, faintly angled in young, bark fissured. Leaves simple, elliptic-oblanceolate, 5-11 × 1-3 cm, chartaceous, shining, viscid, margins outcurved, gland-dotted, base attenuate, margin entire, apex obtuse-acute, subsessile. Flowers greenish-yellow, polygamous, in lateral or terminal branched racemes; sepals oblong. Capsules compressed, 2-3-winged, yellowish-green or yellowish-brown, seeds black, globose, flattened.

Common in all districts in open scrubs, wastelands especially in hilly slopes.

Fl. & Fr.: August-January

Indaram (ADB), *GO* & *DAM* 5187; Manchippa RF (NZB), *TP* & *BR* 6011; Pochammaralu (MDK), *BR* & *CP* 11478; Thornal (MDK), *CP* 11702; Mohammadabad RF (RR), *MSM* & *KH* 10233; Anantasagar RF (RR), *MSM* & *NV* 10389; Mannanur (MBNR), *TDCK* 16810; Kodimial (KRN), *GVS* 25666 (MH); Narsapur (MDK), *KMS* 6630 (MH); Mudigonda (NLG), *BVRS* 172 (NU); Jakaram (WGL), *RKP* 10529 (BSID); Thallavakku (KMM), *RR* 112634 (BSID); Mannanur (MBNR), *TDCK* 16810.

LEPISANTHES Blume

Lepisanthes tetraphylla (Vahl) Radlk. in Sitzungsber. Math. Phys. Cl. Koenigl. Bayer. Akad. Wiss. Muenchen 8: 276. 1878; Gamble, Fl. Madras 1: 247. 1918; Pant in N.P.Singh & *al.*, Fl. India 5: 372. 2000. *Sapindus tetraphylla* Vahl, Symb. Bot. 3. 54. 1794. *Hemigyrosa canescens* (Roxb.) Blume, Rumphia 3: 166. 1849; Hiern in Hook.f.,

Fl. Brit. India 1: 671. 1875. *Molinaea canescens* Roxb. Pl. Coromandel t. 60. 1796. *Lepisanthes deficiens* (Wight & Arn.) Radlk. in Sitzungsber. Math-Phys. Cl. Koenigl. Bayer. Akad. Wiss. Muenchen 8: 276. 1878; Gamble Fl. Madras 1: 247. 1918. *Hemigyrosa deficiens* (Wight & Arn.) Bedd., Fl. Sylv. t. 231. 1872; Hiern in Hook.f., Fl. Brit. India 1: 671. 1875. *Sapindus deficines* Wight & Arn., Prodr. Fl. Ind. Orient. 111. 1834.

Vern. : *Korivi.*

Medium sized, evergreen trees, up to 10 m high, branches prominently striate when young. Leaves paripinnate, 15-25 × 10-16 cm, leaflets 4, alternate, oblong-elliptic, 15 × 9 cm, thick coriaceous, base rounded or acute, lateral nerves and reticulations prominent. Flowers white, axillary, on velvety-pubescent panicles, about 15 cm long; sepals 5, hairy outside, glabrous within, outer 2 smaller, ovate or elliptic to orbicular, oblong to transversely elliptic, ovate to obovate; petals 4, longer than sepals, with basal scales having either a hairy rim or 2 small auricles, eitire to 2-4-lobed; disc hairy or glabrous, complete or interrupted; stamens didynamous. Drupes 3-gonous, ca 2 x 1.5 cm, woody; seeds oblong, exarillate.

Common in forests in Mahabubnagar district.

Fl. & Fr.: January-April.

Mannanur (MBNR), *TDCK* 15398.

SAPINDUS L.

Sapindus trifoliatus L., Sp.Pl. 36(1): 586.1753; Hiern in Hook. f., Fl. Brit. India 1: 682. 1875; *p.p.* (non L. 1753). *S. emarginatus* Vahl, Symb. Bot. 3: 54. 1794; Gamble, Fl. Madras 1: 250. 1918; Pant in N.P.Singh *& al.*, Fl. India 5: 381. 2000.

Vern.: *Kunkudu, Kookudu.*

Trees, up to 10 m high, branches terete, striate, tomentose when young; bark rough, grey. Leaves even pinnate, leaflets 3-4 pairs, (sub) opposite, elliptic-oblong, 5-10 × 2.5-6 cm, thick, coriaceous, glabrous above, pubescent below, base cuneate or attenuate, margin entire, apex emarginate-retuse, lateral nerves about 12 pairs. Flowers brownish yellow in terminal panicles; sepals 5, ovate, hairy on outer surface; petals clawed, with tufted, white, shining hairs, glabrous on inner surfaces; stamens 8, filaments pilose; ovary 3-lobed, densely hairy when young. Drupes ovoid-tomentose, 3- lobed, wrinkled when ripe, seeds one in each locule, black.

Common in all districts, occasionally cultivated, often in forests.

Fl. & Fr.: October-February.

Khanapur (ADB), *TP* & *PVP* 4123; Jannaram (ADB), *TP* & *PVP* 4210; Nadipally (NZB), *BR* & *KH* 6446; Pochammaralu (MDK), *BR* & *CP* 11499; Central University campus (RR), *MSM* 19214; Mannanur (MBNR), *TDCK* 15398; Osmangunta (NLG), *BVRS* 423 (NU); Kinnerasani WLS (KMM), *RR* 112694, 112699 (BSID); Borapuram (MBNR), *BR* & *BSS* 32519; Mannanur RF (MBNR), *SRS* 107433 (BSID); Parnasala RF (KMM), *RC* 99044 (BSID).

Uses : Fruits used as a substitute for soap.

Figure 20: Sapindus emarginatus Vahl.
A. Twig, B. Flower, C. Sepal, D. Petal, Stamens & Disc.

SCHLEICHERA Willd. *nom. cons.*

Schleichera oleosa (Lour.) Oken., Allg. Naturgesch. 3(2): 1341. 1841; Pant in N.P.Singh & *al.*, Fl. India 5: 384. 2000. *Pistacia oleosa* Lour., Fl. Cochinch. 2: 615. 1790. *Schleichera trijuga* Willd., Sp. Pl. 4: 1096. 1805; Hiern in Hook.f., Fl. Brit. India 1: 681. 1875; Gamble, Fl. Madras 1: 248. 1918.

Vern.: *Pusku, Rakot.*

Deciduous or subdeciduous trees, up to 25 m high, branches greyish, young faintly angled, striate, brown tomentose. Leaves even-pinnate, up to 35 cm long; leaflets 3 pairs, subopposite, obovate, oblong, elliptic-ovate, 5-20 × 3-9 cm, coriaceous, nerves strongly plaited below, base cuneate or acute, margin entire, apex obtuse-rotund. Flowers pale yellow in axillary, many branched racemes; calyx small, cupular, lobed, hairy-tomentose outside; disc flat, undulate; stamens 5-8, inserted within disc, filaments ca 1.5 mm long, equal, free, pilose, anthers oblong; ovary globose, 3-loculed, ovule 1, style terete, stigma 2 or 3-lobed. Drupe ovoid, 2.5 × 1.5 cm, spiny; seeds pulpy-arillate.

Common in dry deciduous forests of most districts.

Fl. & Fr.: January-April.

Bheemaram RF (ADB), *GO* & *DAM* 5172; Neelwai (ADB), *TP* & *GO* 5442; Jalalpur (NZB), *BR* & *KH* 9654; Manchippa (NZB), *SSR* 19337; Narsapur RF (MDK), *CP* 11584; Aklaspur RF (KRN), *GVS* 22508 (MH); Mannanur (MBNR), *VBH* 84999 (BSID); Vatavarlapalli (MBNR), *VBH* 84964 (BSID).

ANACARDIACEAE

1. Leaves simple:
    2. Stamens twice the number of petals:
        3. Drupe on fleshy hypocarp .... **Anacardium**
        3. Drupe on the persistent, not enlarged calyx .... **Buchanania**
    2. Stamens as many as petals.
        4. Only one stamen fertile .... **Mangifera**
        4. All stamens fertile .... **Semecarpus**
1. Leaves compound:
    5. Leaflets 3; fertile stamens 5; branchlets spine- tipped .... **Rhus**
    5. Leaflets more than 3; fertile stamens 8-10; branchlets unarmed .... **Lannea**

ANACARDIUM L.

Anacardium occidentale L., Sp. Pl. 383. 1753; Hook.f., Fl. Brit. India 2: 20. 1876; Gamble, Fl. Madras 1: 260. 1918; Chandra & Mukherjee in N.P.Singh & *al.*, Fl. India 5: 437. 2000.

Vern.: *Jidi mamidi, Godambi.* English: Cashewnut.

Trees, up to 8 m high, branchlets glabrous. Leaves simple, obovate, 8-14 × 5-8 cm, thick coriaceous, nerves 12 pairs, plated below, glabrous above, base cuneate-rotund, margin entire, apex obtuse, sometimes retuse. Flowers yellow, streaked with pink in terminal and/or axillary panicles, unisexual (male) or bisexual, polygamous; calyx 5-lobed, sepals ovate; petals 5, liner-lanceolate; stamens 7-10, fertile, torus stipitate; carpel solitary, style eccentric, stigma simple. Drupe a reniform nut, seated on large pyrifomm fleshy body formed by the enlarged disc; seed reniform, ascending, testa membranous.

Occasionally cultivated for its edible fruits. Native of Brazil and appears to have been under cultivaion throughout tropical America before the voyage of Columbus to the New World. It seems to have been introduced into India from Brazil by the Protugese in the 16th century.

Fl. & Fr.: January-May.

Choutkoor (MDK), *TP* & *CP* 11940.

Uses: The nuts are roasted and the kernals are eaten.

BUCHANANIA Spreng.

1. Leaves oblong-elliptic, glabrous; panicles sparsely branched, glabrous .... **B. axillaris**
1. Leaves broadly ovate, pubescent below; panicles with copious branches, villous .... **B. cochinchinensis**

Buchanania axillaris (Desr.) Ramam. in Saldanha & Nicolson, Fl. Hassan Dist. 374. 1976; Chandra & Mukherjee in N.P.Singh *& al.*, Fl. India 5: 441. 2000. *Mangifera axillaris* Desr. in Lam., Encycl. 3: 697. 1792. *Buchanania angustifolia* Roxb., Pl. Coromandel t. 262. 1820; Hook.f., Fl. Brit. India 2 : 23. 1876; Gamble, Fl. Madras 1: 259. 1918.

Vern. : *Sarapappu, Peddmorli.*

Trees, up to 8 m high, branchlets stout, glabrous, deeply fissured. Leaves oblong-elliptic, 5-10 × 3-6 cm, acute at base, undulate margin, apex emarginate coriaceous, lateral nerves about 15 pairs. Flowers white in axillary and terminal panicles; calyx lobes 5, ovate, glabrous; petals 5, oblong, reflexed; stamens 10, filaments subulate, anthers oblong; disc 1.5 mm in diam., ovary sub-oval, style short, stigma truncate. Drupes compressed- globose; seeds 1, gibbose.

Occasional in hilly forest areas in Adilabad, Warangal, Karimnagar and Mahabubnagar districts.

Fl. & Fr.: June-December.

Satraguda (ADB), *GO* & *PVP* 4257; Farahabad (MBNR), *TDCK* 15345; Aklaspur (KRN), *GVS* 22509 (MH); Umamaheswaram (MBNR), *SRS* 109125 (BSID); Mallelatheertham (MBNR), *SRS* 108979 (BSID); Eturgaram (WGL), *SSR* & *KSM* 15533; Katapur (WGL), *RKP* 110847 (BSID); Nallamasidigunta (KMM), *PVS* 84099 (BSID).

Buchanania cochinchinensis (Lour.) M.R. Almeida, Fl. Maharashtra 1: 287. 1996. *Toulifera cochinchinensis* Lour., Fl. Cochinch. 1: 262. 1790. *Buchanania lanzan* Spreng., J. Bot (Schrader) 2: 234. 1800; Gamble, Fl. Madras 1: 258. 1918; Chandra & Mukherjee in N.P.Singh *& al.*, Fl. India 5: 443. 2000; Pullaiah & Chennaiah, Fl. Andhra Pradesh 1: 228. 1997. *B. latifolia* Roxb., Fl. Ind. 2: 385. 1832; Hook.f., Fl. Brit. India 2: 23. 1876.

Vern. : *Morli, Sara.*

Trees, up to 12 m high, branchlets with prominent leaf scars, sericeous. Leaves broadly ovate, 15-25 × 6-10 cm, base round, margin entire, apex obtuse-retuse; petioles 1.5-2 cm long, pubescent. Flowers creamish in axillary panicles, and shorter than leaves, pyramidal, branch axes divaricate, many, villous; calyx lobes ovate, pubescent; petals elliptic, pubescent, greenish white; stamens 10, filaments flat, anthers about as long as filaments; disc 1.5 mm in diam., deeply 10-crenated; ovary conical, style subulate, stigma simple. Drupes obliquely lentiform, black, stony.

Common in deciduous forests of all districts.

Fl. & Fr.: November- May.

Kunthala (ADB), *GO* & *DAM* 5115; Rampur (ADB), *SSR* 19306; Laxmapur RF (NZB), *TP* & *BR* 6252; Alisagar (NZB), *BR* 7286; Gangapur (MDK), *BR* & *CP* 11431; Mohammadabad RF (RR), *MSM* & *KH* 11012; Pakhal (WGL), *ANH* 15953 (MH); Mahadevpur (KRN), *SLK* 70871 (NBG); Appapur (MBNR), *SKB* & *BSS* 30598; Mannanur (MBNR), *SRS* 109122 (BSID); Thaskapally (MBNR), *SRS* 109061 (BSID); Thupakulagudem (WGL), *PVS* 76939 (BSID).

LANNEA A. Richard *nom. cons.*

Lannea coromandelica (Houtt.) Merr., J. Arnold. Arbor. 19 : 353. 1938; Chandra & Mukherjee in N.P.Singh & *al.*, Fl. India 5: 463. 2000. *Dialium coromandelicum* Houtt., Nat. Hist. Ser. 2. 2. 39. t. 2. 1774. *Odina wodier* Roxb., Fl. Ind. 2: 293. 1837; Hook.f., Fl. Brit. India 2 : 29. 1876; Gamble, Fl. Madras 1: 263. 1918.

Vern.: *Gumpina.*

Trees, up to 15 m high; devoid of leaves when in flowers, branch ash-coloured, terete, young stellately brown tomentose. Leaves odd-pinnate, crowded at the ends of branches; petiole 7-13 cm long; leaflets 4-9 × 3-4 cm, opposite, 3-8 pairs, chartaceous, ovate-caudate, base unequally rounded, margin entire, apex acute-acuminate. Flowers pale yellow in crowded, branched racemes; calyx lobes 4, ovate, ciliate; petals 4, oblong; stamens 8, ca 3 mm long; disc 8-notched; carpels in female flowers 2-2.5 mm long, ovary superior, oblong, styles 4, stigmas papillose. Drupes oblique, compressed, red when ripe.

Common in deciduous forests of most districts; often planted in avenues.

Fl. & Fr.: March-May.

Satnella river (ADB), *GO* & *PVP* 4224; Laxmapur RF (NZB), *TP* & *BR* 6265; Mothe RF (NZB), *BR* & *GO* 9067; Ghangapur RF (MDK), *BR* & *CP* 11438; Narsingi (MDK), *BR* & *CP* 11540; Lothuvagu (RR), *MSM* & *KH* 10295; Jalupenta (MBNR), *TDCK* 15311; Pakhal (WGL), *ANH* 15941 (MH); Bhongir (NLG), *BVRS* 402 (NU).

MANGIFERA L.

Mangifera indica L., Sp. Pl. 200. 1753; Hook.f., Fl. Brit. India 2 : 13. 1875; Gamble, Fl.
Madras 1: 259. 1918; Chandra & Mukherjee in N.P.Singh & *al.*, Fl. India 5: 466.
2000.

Vern.: *Mamidi.* English: Mango.

Trees, up to 20 m high, branches glabrous, terete. Leaves oblong or elliptic-
lanceolate, dark green, 10-20 × 3-6 cm, alternate, shining, coriaceous, nerves 25 pairs,
base cuneate-subacute, wavy along margins, apex acuminate. Flowers male and
bisexual, yellowish-white in terminal, tomentose panicles; pedicel 1.5-3 mm long;
calyx 5-lobed, lobes tomentose; petals elliptic, with branched ridges, tips reflexed;
stamens 5, 1 fertile, rest sterile, small, teeth like; disc distinctly 5-lobed, ovary obliquely
ovoid, style 2 mm long, lateral. Drupes globose, obliquely gibbous on one side, stone
compressed, fibrous; seeds oblong.

Common in hill forests of all districts. Universally cultivated in gardens, avenues
and 'topes' and run wild.

Fl. & Fr.: August – July.

Jaipur RF (ADB), *GO* & *PVP* 4958; Yellareddy (NZB), *TP* & *BR* 6327; Laxmapur
(MDK), *BR* & *CP* 11447; Central University campus (RR), *MSM* 19233; Appannapally
(MBNR), *SRS* 104484 (BSID).

A variety of sweet fruits from different hybrids of the species are usually cultivated
in plains.

RHUS L.

1. Armed shrubs .... **R. sinuata**
1. Unarmed shrubs .... **R. parviflora**

Rhus parviflora Roxb., Fl. Ind. 2: 100. 1832 & in DC., Prodr. 2: 70. 1825; Hook.f., Fl.
Brit. India 2: 9. 1876; Chandra & Mukherjee in N.P.Singh *& al.*, Fl. India 5: 490.
2000.

Shrubs, branchlets, undersurface of leaves and inflorescence rusty tomentose.
Leaves 5-16 cm long, 3-foliolate, leaflets elliptic, oblong-obovate, 4-16 × 2.5-5 cm,
obtuse, base cuneate, crenate in the upper half. Flowers unisexual, in axillary and
terminal panicles. Male flowers 2-3 mm in diam., pedicellate, calyx lobes ovate, petals
linear-oblong, 1 × 0.5 mm, stamens 5, 0.8 mm long, disc 0.6 mm in diam., cupular,
deeply 5-notched. Female flowers: ovary subconical. Drupes globose, 0.3-0.4 cm in
diam., epicarp glabrous, wrinkled, indehiscent.

Rare, occurs in dry stony lands of forests (Reddy & *al.*, 2008).

Fl. & Fr.: May August.

Rhus sinuata Thunb., Prodr. Pl. Cap. 52. 1794; Chandra & Mukherjee in N.P.Singh &
al., Fl. India 5: 494. 2000. *R. mysorensis* G. Don, Gen. Syst. 2: 74. 1832; Hook.f., Fl.
Brit. India 2: 9. 1876; Gamble, Fl. Madras 1: 264. 1918; Pullaiah & Chennaiah, Fl.
Andhra Pradesh 1: 230. 1997.

Shrubs or small trees, up to 3 m high, branches reddish, hirtellous, thorns about 2.5 cm long. Leaves 3-foliolate, lateral leaflets obovate, 1.5-2 × 1-1.5 cm, terminal one oblanceolate, 2.5-4 × 1.5-2 cm, coriaceous, base cuneate, margin sinuate toothed, puberulous. Flowers pale green in terminal and axillary panicles, pedicillate, polygamous; Male flowers: calyx lobes 5, ovate, pubescent outside; petals 5, oblong, subacute, glabrous; stamens 5, fertile, filaments free, equal, glabrous; disc 5-lobed; Female flowers: ovary glabrous, obconic, styles 3, stigma capitate. Drupe flattened, 5-loculed; seeds one in each locule.

In dry stony lands of Medak, Ranga Reddi, Hyderabad, Karimnagar and Mahabubnagar districts.

Fl. & Fr.: December-March.

Narsapur RF (MDK), *TP* & *CP* 11964, *RG* 106766 (BSID); Gacchibowli (RR), *MSM* 12775.

SEMECARPUS L.f.

Semecarpus anacardium L. f., Suppl. Pl. 182. 1781, var. **anacardium;** Hook.f., Fl. Brit. India 2 : 30. 1876; Gamble, Fl. Madras 1: 266. 1918; Chandra & Mukherjee in N.P.Singh & *al.*, Fl. India 5: 500. 2000.

Vern.: *Nallajeedi.* English: Marking nut.

Trees, up to 12 m high, young branches terete, tomentose, watery latex present, which on drying black. Leaves oblong-obovate, 15- 30 × 7-16 cm, rusty villous below, glabrous above, thick coriaceous, lateral nerves 18-20, reticulatios prominent, base obtuse, subacute or cordate, margin entire, apex rotund, retuse, emarginate. Flowers white with yellow tinge in terminal branched panicles; Male flowers: sessile, calyx pubescent outside; petals 5; Female flowers: ovary 1.5 mm in diam., styles densely hairy. Drupes globose-ovoid, reniform, seated on a fleshy receptcle; seed thick.

In dry deciduous forests of all districts.

Fl. & Fr.: January-December

Both RF (ADB), *GO* & *PVP* 4631; Indervelli (ADB), *GO* & *PVP* 4546; Manchippa RF (NZB), *TP* & *BR* 6049, *BR* & *SPB* 9613; Sirnapally RF (NZB), *BR* & *GO* 9096; Gangapur RF (MDK), *BR* & *CP* 11434; Dharmapur (RR), *MSM* & *KH* 10289; Kodimial (KRN), *GVS* 20128 (MH); Srinivaspur (NLG), *BVRS* 398 (NU); Aralgudem RF (KMM), *RC* 102479 (BSID); Vatavarlapally (MBNR), *VBH* 84962 (BSID); Thupakulagudem (WGL), *PVS* 76991 (BSID).

MORINGACEAE

MORINGA Adans.

1. Leaves 2-3-pinnate, the leaflets thinner and smaller, 0.5-2 x 0.3-1.3 cm, with obscure venation; petals white, cream or tinged yellowish; seeds inclusive of wings 2-3.2 cm long, wings very rarely absent ... **M. pterigosperma**

1. Leaves mostly 2-pinnate, the leaflets thicker and larger, 2-5 x 0.8-3 cm, with more distinct venation; petals yellowish with pink streaks; seeds inclusive of wings 3.5-4 cm long .... **M. concanensis**

Moringa concanensis Nimmo ex A. Gibson in Dalzell & A. Gibson, Bombay Fl. 311. 1861; Hook.f., Fl. Brit. India 2 : 45. 1876; Gamble, Fl. Madras 1: 270. 1918; Uniyal in N.P.Singh & *al.*, Fl. India 5: 515. 2000.

Vern.: *Konda munaga.*

Trees, up to 8 m high, branches terete, bark often loose. Leaves about 40 cm long, bipinnate, pinnae 5-8 pairs, opposite, leaflets odd-pinnate, 4-6 pairs, opposite, broad orbicular or elliptic, 1-3 × 1-2 cm, chartaceous, glaucous below, base round, margin entire, apex emarginate. Flowers white with purple streaks in terminal panicles; calyx thinly tomentose, segments white, oblong, reflexed; petals yellowish white, oblong, spathulate, reddish or pink-streaked; stamens 5, filaments hairy at base; staminodes 4-5. Capsules elongated, slightly 3-angled and elongated wings.

Occasional in Adilabad, Karimnagar, Warangal, Nalgonda and Mahabubnagar districts.

Fl. & Fr.: February-August.

Bheemarum RF (ADB), *TP* & *GO* 5455; Molachintapally (MBNR), *TDCK* 15315; Mallelatheertham (MBNR), BR & SKB 30635; Nagarjunakonda (NLG), *BVRS* 500 ((NU).

Moringa pterigosperma Gaertn., Fruct. 2: 314. 1791; Hook.f., Fl. Brit. India 2: 45. 1876. *Moringa oleifera* auct non Lam. 1: 398. 1785; Gamble, Fl. Madras 1: 269. 1918; Chandra & Mukherjee in N.P.Singh & *al.*, Fl. India 5: 516. 2000.

Vern.: *Munaga, Mulaga.*

Medium sized trees, up to 8 m high, trunk grey-white with longitudinal wrinkles. Leaves usually 3-pinnate, up to 40 cm long, terminal odd leaflet obovate or elliptic, 1.5-2.5 × 1-1.5 cm, glabrous, obtuse. Flowers white in terminal panicles, sweet scented; calyx tube hairy, lobes petaloid, linear-lanceolate, reflexed; petals white, the anterior erect, others reflexed; stamens 5, filaments villous at base; ovary oblong, style cylindric. Capsules long-linear, loculicidally dehiscent; woody, seeds short winged.

Cultivated in gardens and house compounds.

Fl. & Fr.: February-July.

Uses: Fruits and leaves are used as vegetables. Wood is used as floater.

Nirmal Town (ADB), *GO* & *DAM* 5031; Mohammadabad (RR), *MSM* 19234; Mahadevpur (KRN), *SLK* 70832 (NBG).

Note : Lamarck (1785) cited *Balanus myrepsica* Garacult (1764) under *Moringa oleifera.* But *B. myrepsica* being a distinct species, was later reduced to *Moringa myrepsica* (Gara.) Thell. Some authors doubting the circumscription of *M. oleifera* as given by Lamarck includes two species *M. oleifera* and *M. myrepsica* and hence the next available name *M. pterigosperma* Gaertn. (1791) be considered as valid.

FABACEAE (*nom. alter.* LEGUMINOSAE)

1. Flowers actinomorphic, calyx and corolla valvate in bud; stamens definite or indefinite .... **Mimosoideae**

1. Flowers zygomorphic, perianth segments predominantly imbricate in bud; stamens definite, 10 or fewer :

   2. Corolla ascendingly imbricate; upper petal innermost (or petals O); stamens usually free .... **Caesalpinioideae**

   2. Corolla descendingly imbricate; upper petal outermost; stamens usually united .... **Faboideae**

FABOIDEAE (*nom. alter.* PAPILIONOIDEAE)

1. Trees (except *Butea superba*) :

   2. Pods orbicular, winged .... **Pterocarpus**

   2. Pods not as above :

      3. Leaves 3-foliolate :

         4. Stems usually prickled; keel petals much shorter than the standard .... **Erythrina**

         4. Stems unarmed; keel longer than the standard .... **Butea**

      3. Leaves 5-many foliolate :

         5. Leaflets 25-30 pairs .... **Sesbania (grandiflora)**

         5. Leaflets 2-12 pairs :

            6. Leaflets opposite :

               7. Leaflets 5-9; stamens monadelphous .... **Pongamia**

               7. Leaflets 13-17; stamens diadelphous .... **Gliricidia**

            6. Leaflets alternate .... **Dalbergia** (*p.p.*)

1. Other than trees :

   8. Climbing or twining herbs or shrubs or lianas (except *Canavalia lineata*) :

      9. Leaves 5-many-foliolate (3-foliolate in *Neonotonia*) :

         10. Leaves even-pinnate .... **Abrus**

         10. Leaves odd-pinnate :

            11. Flowers solitary .... **Clitoria**

            11. Flowers other than solitary :

               12. Pods winged along the upper or both sutures :

                  13. Leaflets alternate .... **Dalbergia** (*p.p.*)

                  13. Leaflets opposite .... **Derris**

12. Pods otherwise :

    14. Pods exceeding 10 cm in length:

        15. Standard auriculate; stamens monadelphous; pods flattened, velvety, very tardily dehiscent; leaflets 7-9 .... **Milletia**

        15. Standard not auriculate, stamens diadelphous; pods glabrous, torulose, dehiscent; leaflets 11-15 .... **Endosamara**

    14. Pods not exceeding 10 cm in length .... **Neonotonia**

9. Leaves 3-foliolate:

    16. Stamens monadelphous:

        17. Inflorescence rachis not swollen at nodes .... **Teramnus**

        17. Inflorescnece rachis swollen at nodes :

            18. Calyx upper lip large, lower minute; seeds ovate, hilum linear .... **Canavalia**

            18. Calyx campanulate, teeth subequal; seeds suborbicular or transversely oblong, hilum oblong .... **Pueraria**

    16. Stamens diadelphous :

        19. Anthers dimorphic; fruit usually covered with fine needle like irritant hairs .... **Mucuna**

        19. Anthers uniform; fruit hairs not as above :

            20. Leaves gland dotted beneath :

                21. Pods fully concealed in membranous reticulately veined calyx .... **Paracalyx**

                21. Pods not as above :

                    22. Pods 2-seeded .... **Rhynchosia** (*p.p.*)

                    22. Pods many-seeded :

                          23. Pods depressed outside between seeds .... **Cajanus** (*p.p.*)

                        23. Pods not depressed outside between seeds .... **Dunbaria**

            20. Leaves not gland dotted beneath :

                24. Style bearded below the stigma :

                    25. Stigma oblique .... **Vigna** (*p.p.*)

                    25. Stigma terminal :

                        26. Styles thickened upwards .... **Lablab**

26. Styles filiform :

    27. Flowers pinkish-violet .... **Dolichos**

    27. Flowers yellow .... **Macrotyloma**

24. Styles not bearded below the stigma :

    28. Inflorescence rachis nodes not swollen

        29. Calyx teeth distinct .... **Shuteria**

        29. Calyx truncate .... **Dunbaria**

    28. Infloresence rachis nodes swollen .... **Galactia**

8. Erect or prostrate or diffuse herbs, undershrubs or shrubs :

30. Leaves 2-foliolate .... **Zornia**

30. Leaves simple or 3-many-foliolate :

    31. Anthers dimorphic:

        32. Flowers terminal or leaf-opposed racemes;
            pod not jointed .... **Crotalaria**

        32. Flowers in axillary clusters or terminal heads;
            pod jointed:

            33. Pod compressed, hooked at apex with the persistent
                base of the style; petals inserted .... **Stylosanthes**

            33. Pod turgid, apex obtuse, without persistent style;
                petals exserted .... **Arachis**

    31. Anthers uniform :

    34. Pulvinus reduced; leaves simple; pod 1-seeded .... **Cullen**

    34. Combination of characters not as above:

        35. Pod transversely jointed :

            36. Calyx glumaceous, striate .... **Alysicarpus**

            36. Calyx not as above :

                37. Pods folded within persistent calyx .... **Uraria**

                37. Pods other than folded :

                    38. Pods boat-shaped .... **Eleiotis**

                  38. Pods not as above :

                      39. Stamens 5+5 :

                          40. Leaflets 3-10 paris .... **Smithia**

                          40. Leaflets more than
                            20 pairs.............................. **Aeschynomene**

39. Stamens 9+1 or monadelphous, anthers uniform:

    41. Pods uniformly flattened; flowers distinctly exserted; calyx much shorter than the first pod-joint:

        42. Flowers hidden in persistent, orbicular bifarious bracts; stamens monadelphous ... **Phyllodium**

        42. Flowers not hidden; bracts nor orbicular:

            43. Stipules united, amplexicaul, apex biapiculate; stamens monadelphous .... **Dicerma**

            43. Stipules free, not amplexicaul, apex entire; stamens mona/ diadelphous .... **Desmodium**

    41. Pod turgid or both; flowers included or slightly exserted; calyx much longer than the first pod-joint .... **Desmodiastrum**

35. Pods not jointed :

    44. Anthers apiculate; biramous hairs present:

        45. Stamens diadelphous.................................... **Indigofera**

        45   Stamens monadelphous .... **Cyamopsis**

    44. Anthers not apiculate; biramous hairs absent :

        46. Style straight, away from standard petal .... **Rothia**

        46. Style curved towards standard petal :

            47. Stamens monadelphous .... **Mundulea**

            47. Stamens diadelphous:

                48. Upper sutures of pods straight, lower sutures indented; lateral leaflets less than 1/3 as long as the terminal .... **Codariocalyx**

                48. Upper and lower sutures of pods otherwise; leateral leaflets more than 2/3 as long as the terminal:

                    49. Leaflets and calyx with yellowish gland dots :

50. Ovules 4-many .... **Cajanus** (*p.p.*)

50. Ovules 1-2 :

    51. Pods turgid .... **Flemingia**

    51. Pods flat .... **Rhynchosia** (*p.p.*)

49. Leaflets and calyx eglandular :

    52. Leaves simple or 5-many-foliolate :

        53. Inflorescence axillary ...... **Sesbania**

        53. Inflorescence terminal or leaf-opposed .... **Tephrosia**

    52. Leaves 3-foliolate :

        54. Leaflet margin serrate or denticulate, atleast at the apex .... **Melilotus**

        54. Leaflets margins entire :

            55. Pods inflated .... **Pycnospora**

            55. Pods flattened, not inflated :

                56. Styles hairy :

                    57. Keel spiral ...... **Macroptelium**

                    57. Keel not spiral .... **Vigna** (*p.p.*)

                56. Styles glabrous .... **Pseudarthria**

ABRUS Adans.

1. Leaflets very small, to 7 × 3 mm, linear-lanceolate; -pod compressed, linear-oblong, seeds rounded .... **A. fruticulosus**

1. Leaflets large, to 4 × 1.3 cm, oblong :

    2. Pod thin, flat, smooth; 8-12-seeded, seeds compressed, dark brown with black base .... **A. pulchellus**

    2. Pod thick, wrinkled, 3-5-seeded, seeds rounded, hard, scarlet or white with a black spot .... **A. precatorius**

Abrus fruticulosus Wall. ex Wight & Arn., Prodr. Fl. Ind. Orient. 1 : 236. 1834; Baker in Hook.f., Fl. Brit. India 2 : 176. 1876; Gamble, Fl. Madras 1: 350. 1918.

Climbing shrubs, branchlets glabrous or sometimes thinly silky. Stipules linear-lanceolate. Leaves pinnate, rachis ending in a bristle, leaflets 8-12 pairs, linear-lanceolate, membranous, minutely silky pubescent, oblong, stipels absent. Flowers in racemes, purple, bracts small, deciduous, bracteoles 2 under the calyx; calyx 5-lobed, thinly silky; corolla purple; stamens 9, monadelphous; ovary subsessile. Pod glabrous, linear-oblong, compressed, 3-4.5 × 1 cm; seeds rounded, black.

Rare in Nalgonda and Mahabubnagar districts.

Fl. & Fr.: January-October

Borapur (MBNR), *BSS* & *SKB* 32502; Krishna river bank (NLG), *KMS* 9853 (MH).

Abrus precatorius L., Syst. Nat. ed. 12. 2: 472. 1767; Baker in Hook.f., Fl. Brit. India 2 : 175. 1876; Gamble, Fl. Madras 1: 349. 1918; Verdc., Kew Bull. 24 : 240. 1970. *Glycine abrus* L., Sp. Pl. 753. 1753.

Vern. : *Guruginja, Guruvenda.*

Stragglers, up to 5 m long, stems terete, sparsely pubescent when young. Leaves paripinnate, 3-8 cm long, stipules scarious, linear-subulate, leaflets 10-12 pairs, opposite, oblong, 1.5-2.5 × 0.5-0.8 cm, chartaceous, sparsely hairy on both surfaces, base rounded or narrowed, margin entire, apex apiculate. Flowers rose to pink, in axillary racemes, often clustered at ends of peduncles; calyx narrowly campanulate, tube 2 mm, pubescent, lobes truncate, obscure; corolla rose to pink, exserted, standard broadly ovate, 1.1 × 0.8 cm, wings narrow, oblong, 8 × 3 mm, keels curved, 9 × 3 mm; stamens monadelphous (9), staminal sheath 7 × 1.5 mm, filaments 3 mm; ovary subsessile, hirsute, style incurved, style and stigma glabrous. Pod thick, oblong, wrinkled, 4-5 × 0.5-1 cm, beaked at tips, hairy; seeds scarlet or white with a black spot, shining.

Common in most districts in hedges and among bushes in open lands.

Fl. & Fr. : July-December.

Jaipur RF (ADB), *GO* & *PVP* 4948; Mamidipally (NZB), *BR* & *KH* 9009; Pocharam sanctuary (MDK), *BR* & *CP* 11609; Medak (MDK), *CP* 11746; Kusumasamudram RF (RR), *MSM* & *NV* 10361; Mannanur (MBNR), *TDCK* 13621; Amrabad (MBNR), *BSS* & *SKB* 32557; Rechapalli (KRN), *GVS* 25699 (MH); Pillalamarri (NLG), *BVRS* 21 (NU); Thattilanga RF (KMM), *RC* 102411 (BSID); Kinnerasani RF (KMM), *RR* 112687 (BSID); Jakaram (WGL), *RKP* 105307 (BSID).

Abrus pulchellus Wall. ex Thwaites, Enum. Pl. Zeyl. 91. 1859; Baker in Hook.f., Fl. Brit. India 2 : 175. 1876; Gamble, Fl. Madras 1 : 350. 1918; Verdc., Kew Bull. 24 : 246. 1970.

Much-branched climbers, branchlets slender, young stems pubescent with appressed hairs, old stems glabrescent. Leaves pinnately compound, rachis 12 cm long, stipules linear-lanceolate, leaflets 9-16 pairs, oblong or oblong-obovate, 0.7-5 × 0.5-2 cm, base rounded, margin entire, apex apiculate, glabrous above, thinly silky beneath, petiolule very short. Flowers pink or pale blue, in terminal or axillary racemes, longer than the leaves; calyx ca 3.5 mm long, thinly silky; corolla pink or pale blue. Pod oblong, fulvo-puberulent, sessile, compressed, 5-7 cm long; seeds 8-12, dark brown with black base and small white aril.

Rare in Mahabubnagar district.

Fl. & Fr.: October-December.

AESCHYNOMENE L.

1. Stems glandular hispid-hirsute; leaflets with 2 to several primary veins; legume linear, slightly curved, abaxial suture undulate and indented; articles rounded .... **A. americana**

1. Stems not glandular hispid; leaflets with 1 midvein; legume linear-oblong, straight, abaxial suture straight and slightly indented; articles quadrate:

    2. Erect branched herbs; stem slender, woody; calyx glabrous; joints of pod papillose or smooth .... **A. indica**

    2. Erect undershrubs; stems stout, pith like; calyx hispid; joints of pod echinate .... **A. aspera**

Aeschynomene americana L., Sp. Pl. 713. 1753; Rudd, Reinwardtia 5: 25. 1959.

Vern.: *Seema jeelugu.*

Annual herbs, up to 50 cm high, stems erect or decumbent, pale brownish, glandular-hispid to hirsute. Leaves 1.5-7 cm long, 20-60-foliolate, petiole about 3.5 mm long, leaflets linear, subfalcate, glabrous, up to 6 × 1 mm, apex aristate, base rounded, margin ciliate-toothed. Flowers yellow in racemes, standard petal orbicular. Pod linear, short stalked, slightly curved, 2-2.5 × 0.3 cm, joints usually 6-8; seeds reniform, ca 3 × 2 mm, dark brown.

Occasional weed of ditches and marshy lands in Warangal district.

Fl. & Fr.: October – January.

Hanmakonda (WGL), *VSR* 5398 (KU).

A native of tropical America, now pantropical weed (C.S.Reddy & *al.,* 2008; C.S.Reddy and V.S.Raju, 2009).

Aeschynomene aspera L., Sp. Pl. 713. 1753; Baker in Hook.f., Fl. Brit. India 2 : 152. 1876; Gamble, Fl. Madras 1: 331. 1918.

Erect undershrubs, up to 1.5 m high, branchlets glabrescent, swollen. Leaves 12 cm long, odd-pinnate, stipules apically setaceous, lanceolate, to 1 cm, leaflets 30-35 pairs, oblong, 0.5-1 × 0.1- 0.2 cm, glabrous, base obtuse, oblique, margin entire, apex rotund-obtuse, mucronate. Flowers yellow, large, in 3 cm long racemes, peduncle hispid, bracts ovate; calyx hispid, tube to 4 mm, upper lobes round, 2 mm, lower 3.5 mm; corolla yellow, standard 1.2 × 1.5 cm, wings 8.5 × 4 mm, keels 1.8 × 0.8 cm; standard and wings hairy outside; stamens isodiadelphous (5 + 5), staminal sheath 7 mm, filaments to 7 mm; ovary tomentose, 1 cm, style 6 mm. Pod nearly straight, 5.5-7 × 0.8 cm, joints 5-7, glabrous, echinate over seeds, indented along both margins; seeds 5 or more, to 6 mm.

Common on the borders of ponds and tanks in Adialabad, Nizamabad, Medak and Mahabubnagar districts.

Fl. & Fr.: Through the year.

Servaipet (ADB), *GO* 4513; Patancheru (MDK), *BR* & *CP* 11307; Ponnala (MDK), *CP* 11371; Eklaspuram (MBNR), *DV* & *MV* 41310.

Aeschynomens indica L., Sp. Pl. 713. 1753; Baker in Hook.f., Fl. Brit. India 2: 151. 1876; Gamble, Fl. Madras 1: 331. 1918.

Annual, erect, branched herbs, up to 1 m high, stem 4 cm in diameter, woody, terete, striate, sparingly hirsute. Leaves imparipinnate, rachis glandular, stipules not setaceous; leaflets 17-25 pairs, alternate, linear-oblong, base obtuse, oblique, margin entire, apex apiculate. Flowers pale yellow with maroon streaks on the standard, in axillary racemes; calyx membranous, tube 3 mm, lobes 4-5 mm; corolla yellow to flame coloured; standard obovate, 1 × 0.7 cm, wings obtuse, 9 x 4 mm, keels 8 x 3 mm; stamens isodiadelphous (5 + 5), staminal sheath 4 × 2 mm, filaments to 6 mm, ovary linear, 4 × 1 mm, style 3 mm. Lomentum 4 cm long, flattened, 6-8-jointed, slightly falcate, upper suture of the pod straight, smooth to prominently tuberculated; seed shining.

Common marshy herb along the edges of streams, tanks and pools in most districts.

Fl. & Fr.: July-November

Potchera (ADB), *GO* & *DAM* 5076; Ibrahimpet RF (NZB), *TP* & *BR* 6117; Choutkoor (MDK), *TP* & *CP* 11938; Gangapur RF (MDK), *TP* & *CP* 12076; Kusumasamudram RF (RR), *MSM* & *NV* 10332; Pagarikutta-Narsapur (MDK), *KMS* 6075 (CAL); Pakhal (WGL), *KMS* 11620 (MH, BSID); Bhupatipur (KRN), *GVS* 22295 (BSID).

ALYSICARPUS Necker ex Desv. *nom. cons.*

1. Upper leaves 3-foliolate :
    2. Pods puberulous, 4-8 jointed .... **A. heterophyllus**
    2. Pods glabrous, 2-5-jointed .... **A. scariosus**
1. All leaves 1-foliolate or simple :
    3. Calyx shorter or slightly longer than first joint of pod :
        4. Pod moniliform, veinless :
            5. Leaves variable; pod reticulately veined,
                sparsely hairy .... **A. mahabubnagarensis**
            5. Leaves elliptic-oblong; pod nerves obscure,
                densely hairy .... **A. monilifer**
        4. Pod at most slighlty moniliform, veined :
            6. Calyx shorter than the first joint of the pod; pod
                compressed, shallowly indented on margins .... **A. hamosus**
            6. Calyx equalling or slightly longer than the first joint of
                the pod; pod turgid, not indented on margins:
                7. Peduncle to 3 cm, lax inflorescence with longer
                    distance between the pair of flowers .... **A. ovalifolius**

       7.   Peduncle to 1 cm, racemes with shorter distance between the pair of flowers .... **A. vaginalis**

3.   Calyx much longer than the first joint of pod :

      8.   Pod 4-angled .... **A. tetragonalobus**

      8.   Pod flattened or subterete (never 4-angled) :

         9.   Teeth of calyx imbricate in fruiting stage :

            10. Pods smooth .... **A. roxburghianus**

            10. Pods rugose, veined or ribbed :

               11. Pod transversely ribbed .... **A. heyenenus**

               11. Pod reticulately veined or rugose :

                  12. Racemes 15-25 cm long; pod minutely pubescent .... **A. longifolius**

                  12. Racemes 3-10 cm long; pod glabrous ..... **A. bupleurifolius**

         9.   Teeth of calyx not imbricate in fruit .... **A. pubescens**

Alysicarpus bupleurifolius (L.) DC., Prodr. 2 : 352. 1825; Gamble, Fl. Madras 1: 338. 1918. *Hedysarum bupleurifolium* L., Sp. Pl. 745. 1753.

Key to Varieties

1.   Leaves usually linear; pod with 4-8 joints .... var. **bupleurifolius**

1.   Leaves linear-oblong; pod with 1-3 joints .... var. **gracilis**

Alysicarpus bupleurifolius (L.) DC., var. **bupleurifolius**

    Prostrate or suberect herbs; 15-40 cm high, branchlets glabrous. Leaves linear, 2-7 × 0.2-0.5 cm, glabrous above, sparsely pubescent beneath, base acute, margin entire, apex apiculate, stipules lanceolate, to 9 mm. Flowers yellow with pink tinge, in axillary and terminal, lax racemes; peduncle to 2 cm, pedicel to 2 mm; calyx campanulate, tube 3 mm, lobes ciliate, longer than the first joint of the pod; corolla pink, standard 6 x 1 mm, wings 5 × 0.5 mm, keels 6 × 2 mm; stamens diadelphous (9 + 1), staminal sheath 6 mm; ovary 3 mm, style 3 mm. Pod up to 1.5 cm long, shortly exerted, apiculate, 4-8-jointed, joints smooth, slightly moniliform; seeds spherical, yellowish brown, smooth, glabrous.

    Common in forests of most districts, chiefly among long grasses.

    Fl. & Fr.: August-December

    Dichpally (NZB), *BR* & *KH* 6443; Kamareddy (NZB), *BR* & *KH* 7143; Dharmapur (RR), *MSM* 10487; Kodimial (KRN), *GVS* 25679 (MH).

Alysicarpus bupleurifolius (L.) DC. var. **gracilis** (Edgew.) Baker in Hook. f., Fl. Brit. India 2: 158. 1876; Gamble, Fl. Madras 1: 338. 1919. *p.p. Alysicarpus gracilis* Edgew., J. Linn. Soc. Bot. 9 : 312. 1869.

Prostrate or ascending herbs, up to 50 cm high. Leaves linear- oblong, 2-7 × 0.2-0.5 cm, glabrous, margin entire, apex acute. Flowers yellow with pink tinge, in racemes. Pod 1-3-jointed, smooth; seeds ovoid, smooth, shining, light brown.

Hills of Medak and Mahabubnagar districts.

Fl. & Fr.: August-December.

Way to Madikonakutta (MDK), *KMS* 6722 (CAL); Borapur (MBNR), *BSS* & *SKB* 32503 (SKU, BSID).

Alysicarpus hamosus Edgew., J. Asiat. Soc. Bengal 21: 171. 1853; Baker in Hook.f., Fl. Brit. India 2 : 157. 1876; Gamble, Fl. Madras 1: 338. 1918.

Hairy creeping herbs, up to 40 cm long, branchlets hirsute. Leaves 1-foliate, orbicular, 2-4.5 × 1.5-3.5 cm, sparsely appressed hairy, base rounded to sub-cordate, margin entire, apex mucronate; petiole to 6 mm, stipules scarious, to 7 mm, linear, stipels subulate, minute. Flowers pink or purplish- violet in pairs along the rachis in axillary racemes, pedicel to 3 mm; calyx 3 mm long, shorter than the first joint of pod, lobes linear, acute; corolla bluish-purple. Pods compressed, ca 1.4 × 0.3 cm, appressed hairy on both surfaces, 2-6-jointed.

Common in wastelands in Nizamabad, Medak, Hyderabad, Karimnagar, Nalgonda and Mahabubnagar districts.

Fl. & Fr.: October-February

Manchippa RF (NZB), *TP* & *BR* 6020; Pagarikutta, Narsapur (MDK), *KMS* 6694 (CAL); Jalapenta (MBNR), *BSS* & *SKB* 32618 (BSID); Nagarjunakonda valley (NLG), *KTH* 9825 (BSID).

Note: It is intresting species in the genus and it resembles *Desmodium* in the presence of compressed pod and on the other hand it is similar to *Alysicarpus* in having 1-foliate leaves and ebracteolate flowers. This combination of characters required experimental studies. It is an important problem for the Biosystematic studies.

Alysicarpus heterophyllus (Benth. ex Baker) Jefri & Ali, Biologia (Lahore) 12: 33. 1966. *A. vaginalis* (L.) DC. var. *heterophyllus* Benth. ex Baker in Hook. f., Fl. Brit. India 2 : 158. 1876.

Herbs, weak stemmed, decumbent, spreading on ground, up to 45 cm long, branches terete, striate, sparsely strigose. Upper leaves 3-foliolate, linear, 3-6.5 × 0.5-1 cm, sparingly hirsute below, nerves and reticulations prominent, petiole 1.2 cm long, petiolule 1 mm long; the lower leaves 1- and 3-foliolate intermixed, leaflets lanceolate to oblong-lanceolate, 1-3 × 0.8 – 1.5 cm, base and apex subacute. Flowers deep blue with red tinge, in axillary racemes; calyx hirsute, lobes acuminate; corolla bluish pink, glabrous; stamens diadelphous (9+1). Pods moniliform, ca. 1.4 cm long, puberulous, 4-8-jointed; seeds orbicular.

Occasional in open forests and waste lands in Adilabad, Warangal and Khammam districts.

Fl. & Fr.: July- December

Kotapalli RF (ADB), *TP* & *GO* 5440; 10 km towards north from Pasra (WGL), *RKP* 108199 (BSID); Krishnarajasagar (KMM), *RC* 102456 (BSID).

Alysicarpus heyneanus Wight & Arn., Prodr. Fl. Ind. Orient. 234. 1834. *A. rugosus* (Willd.) DC., Prodr. 2: 353. 1825; Baker in Hook. f., Fl. Brit. India 2 : 159. 1876; Gamble, Fl. Madras 1 : 339. 1918. *Hedysarum rugosum* Willd., Sp. Pl. 3: 1172. 1802.

Diffuse herbs, up to 50 cm high, stem villous. Leaves 1-foliolate, lower leaves elliptic-oblong, 3-4.5 × 0.5-1 cm, base obtuse, subcordate, margin entire, apex obtuse, apiculate, puberulous; petiole to 5 mm. Flowers in 10-15 cm long racemes; peduncle to 2 cm long, pedicel to 1 mm long; calyx tube 2 mm, lobes to 8 mm, ciliate; corolla dark purple-pink, standard 4.5 × 2.5 mm, wings 4 × 0.5 mm, keels 4 mm; stamens diadelphous (9+1), staminal sheath 4 mm, filaments 1.5 mm; ovary to 5 mm, style 4 mm. Pod moniliform, slightly exerted, joints 3-5, downy; seeds rugose.

Occasional in grazing grounds in Khammam, Adilabad, Medak, Karimnagar and Ranga Reddi districts.

Fl. & Fr.: September-January

Sone (ADB), *TP* & *PVP* 4006; Wankidi (ADB), *PVP* 9486; Kunavaram (KMM), *JSG* 15918 (DD); ICRISAT (MDK), *LJGV* 2599, 2823 (CAL); Central University campus (RR), *MSM* 15155; Kodimial (KRN), *GVS* 21855 (MH).

Alysicarpus longifolius (Rottl. ex Spreng.) Wight & Arn., Prodr. Fl. Ind. Orient. 1: 233. 1834; Baker in Hook.f., Fl. Brit. India 2 : 159. 1876; Gamble, Fl. Madras 1: 338. 1918. *Hedysarum longifolium* Rottl. ex Spreng., Syst. 3: 319. 1826.

Vern. : *Peddakandikaraku.*

Stout, woody glabrescent herbs to undershrubs, up to 1.25 m high, branchlets terete, slightly striate, glabrous. Leaves 1-foliate, oblong or lanceolate, 2-11 × 0.5-1 cm, glabrescent, base obtuse, margin entire, apex acute or mucronate, appressedly hairy on nerves beneath, reticulately veined; petiole to 5 mm. Flowers deep blue with red tinge in pairs on stout long-peduncled racemes in pairs, bracts conspicuous, to 1 cm long, ovate, pedicel to 2.5 mm; calyx pubescent, to 6 mm, striate, as long as or longer than the two lower joints of the pod, lobes longer than the tube. Pods moniliform, 3-6-jointed, joints of pod reticulate, puberulous, reticulately veined; seeds brown.

Common weed of cultivated fields in Medak and Mahabubnagar districts.

Fl. & Fr.: September-December.

Tekulapenta (MBNR), *BSS* & *SKB* 32626 (BSID); Irlapenta (MBNR), *BSS* & *BR* 32382 (BSID); ICRISAT (MDK), *LJGV* 2605 (CAL).

Alysicarpus mahabubnagarensis Ragh. Rao & *al.*, Asian J. Plant Sci. 1: 97. 1989.

An ascending herb, up to 50 cm high, branches few to many from base, branchlets with elongated spreading pilose hairs. Leaves 1-foliate, variable in shape, basal obovate, oblong, terminal linear, 1.5-5 × 0.75- 1.75 cm, base cordate to sub-cordate, margin entire, apex retuse to acute with a terminal bristle. Flowers yellowish in axillary and terminal racemes, bract deciduous enclosing two flower buds, to 4 mm, scarious; peduncle to 4 cm long, pedicel to 1 mm; calyx tube 1.25 mm, posterior two lobes united, 2 mm, three lobes free, 3.25 mm; corolla yellow-pinkish, standard slightly clawed, limb obovate, to 4 × 3.5 mm; stamens diadelphous (9+1), staminal sheath 3 mm, filaments uniform; ovary stipitate, densely appressed hairy, 2.5 mm, style 1.75 mm. Pod moniliform, 4-6-jointed, surface reticulately veined, sparsely hairy; seeds ovoid.

In Shadnagar in Mahabubnagar district.

Fl. & Fr.: July-September, Fr.: September-October.

Shadnagar (MBNR), *SRR* 2460 (MH).

Alysicarpus monilifer (L.) DC., Prodr. 2 : 353. 1825; Baker in Hook.f., Fl. Brit. India 2 : 157. 1876; Gamble, Fl. Madras 1 : 338. 1918. *Hedysarum moniliferum* L., Mant. Pl. 1: 102. 1767.

Vern.: *Anera.*

Prostrate herbs, up to 50 cm long, branchlets with elongate, spreading pilose hairs. Leaves elliptic to oblong, 0.7-3 × 0.5-1.8 cm, chartaceous, pubescent, base obtuse-subcordate, margin entire, apex obtuse, mucronate; petiole to 1.5 cm. Flowers pink to violet in axillary, 4-10-flowered racemes; peduncle to 2 cm long, pedicel to 1.5mm; calyx pilose, tube 2.5 mm, lobes to 4 mm; corolla violet, standard ca 6 × 2 mm, wings 5 × 2 mm, keels 6 mm; stamens diadelphous (9+1), staminal sheath 5 mm, filaments 1mm; ovary 4 mm, style 2.5 mm. Pods moniliform, ca 2 x 0.3 cm, hairs downy and hooked, nerves obscure, apex apiculate, the joints turgid, subglobose, 5- 7-seeded; seeds ovoid.

Common in all dry districts.

Fl. & Fr.: Through the year.

Rebbana (ADB), *GO* & *PVP* 4843; Pathrajampet Railway gate (NZB), *BR* & *CPR* 7155; Pocharam tank (NZB), *BR* 9502; Pocharam RF (MDK), *TP* & *CP* 12060; Mohammadabad RF (RR), *MSM* 10494; Azampur (NLG), *BVRS* 497 (NU).

Alysicarpus ovalifolius (Schumach.) J. Leonard, Bull. Jard. Bot. Etat. Brux. 24: 88. 1954. *Hedysarum ovalifolium* Schumach., Beskr. Guin. Pl. 359. 1827.

Annual, erect or prostrate herbs, up to 50 cm thigh, branchlets glabrous, striate, branches rooting at nodes with longer internodes. Leaves simple, stipules lanceolate; petiole 0.5-1 cm long, channeled on the upper side, base subcordate, margin entire, apex acute and mucronulate, puberulous on nerves beneath. Flowers in terminal or leaf-opposed lax racemes; peduncle 2-3 cm long, pedicel 1 mm long, bracts 4-5 × 1-2 mm, ovate to lanceolate; calyx tube 8-12 mm long, puberulous, teeth 3-4.5 mm long, standard pink, 4-5 × 3-4 mm, wings purplish, keels pale pink, stamens diadelphous (9+1), staminal sheath 4 mm long, filaments 3 mm long, ovary pubescent, style 3 mm, bearded with long hairs. Pod 1-25 × 0.2-0.3 cm, joints 5-7, flattened, puberulous; seeds brown, ellipsoid, compressed, with reddish patches.

Occasional in open lands in Ranga Reddi district.

Fl. & Fr.: August – December.

Kanmankalva (RR), *MSM* 10468.

Alysicarpus pubescens Law. ex Wight, Icon. Pl. Ind. Orient. t. 250. 1840; Baker in Hook.f., Fl. Brit. India 2 : 160. 1876; Gamble, Fl. Madras 1 : 339. 1918.

Erect, patently softly hairy herbs, 80-120 cm high. Leaves simple, linear-oblong or oblong-lanceolate, 2-3.5 × 1.2-1.5 cm, sparsely hairy on both surfaces, apex acute. Flowers reddish- purple, 10-16 cm long, in dense terminal white silky racemes, calyx 1.3 cm long, standard yellow, wing and keel purplish. Pod 0.4-0.6 cm long, monilifer, 3-4-jointed, joints reticulate.

Frequent near cultivated fields, occasional in waste lands in Karimnagar district (Naqvi, 2001).

Fl. & Fr.: September-December

Alysicarpus roxburghianus Thoth. & A. Pramanik, Bull. Bot. Surv. India 21: 189. 1979. *Hedysarum bupleurifolium* Roxb., Pl. Coromandel t. 194. 1804. non L. 1763. nec. Roxb., Fl. Ind. 3:346-347.

Perennial herbs or undershrubs, 30-100 cm high, much branched, diffuse or erect, internodes slender, terete, glabrous, ribbed. Leaves alternate, 1-folillate, 2.8-10 × 0.3-0.6 cm, linear, base cuneate or slightly rounded, apex acute, glabrous above, appressed glabrescent below only along midvein; petiole minute, to 1 mm. Flowers pale rose or pinkish-brown, in terminal, 7-15 cm long, lax racemes; bracts ovate; peduncle to 2 mm; calyx tube 3 mm, turbinate, lobes minute, ciliate, upper lobes emarginated; corolla pale rose. Pod subcylindric, 3-6-jointed, one side scarsely contracted between the joints, 1.5-8.0 × 2 mm, quite smooth, subglobose.

Rare in Mahabubnagar District (Raghava Rao, 1989).

Fl. & Fr.: August-November

Alysicarpus scariosus (Rottl. ex Spreng.) Grah. ex Thwaites, Enum. Pl. Zeyl. 88. 1858. var. **scariosus.** *Hedysarum scariosum* Rottl. ex Spreng., Syst. Veg. 3: 319. 1826. *Alysicarpus styracifolius* auct. non DC. 1825. *A. rugosus* (Willd.) DC. var. *styracifolius* Baker in Hook.f., Fl. Brit. India 2: 159. 1876; Gamble, Fl. Madras 1 : 338. 1918.

Diffuse herbs, up to 30 cm high, stem with a decurrent line of hairs. Leaves 1-3-foliolate, lower leaves elliptic-oblong, to 2.5 × 1.5 cm; upper ones linear-oblong, 4.5 x 0.5 cm, base cuneate, margin entire, apex obtuse, pubescent on both surfaces. Flowers purplish-blue, in 3-8 cm long racemes; calyx ca 1.2 cm long, imbricate in fruit, ciliate with tawny hairs. Pod 1-1.2 cm long, moniliform, included, joints 3-5, glabrous, broader than long, transversely ribbed.

Rare in dry deciduous forests in Adilabad and Medak districts.

Fl. & Fr.: September- November.

Bijjur RF (ADB), *GO* & *PVP* 4874; Choutkoor (MDK), *TP* & *CP* 11914.

Alysicarpus tetragonalobus Edgew., J. Asiat. Soc. Bengal, Pt. 2, Nat. Hist. 21: 169. 1853; Baker in Hook.f., Fl. Brit. India 2: 159. 1876.

Decumbent herbs, up to 30 cm long, branchlets glabrous or with a decurrent line of hairs; stipules scarious, to 5 mm, lanceolate, acute, striate, stipels caducous, minute. Leaves oblong-lanceolate, 1.8-3.7 cm long, base rounded-subcordate, margin entire, apex acute or obtuse, glabrous above, hairy beneath, petiole to 3 mm. Flowers in racemes; bracts 3 mm, ovate, lanceolate, caducous; calyx 4.5 mm, lobes lanceolate, acute, puberulous outside, corolla pink. Pod moniliform, 3-5-jointed, tetragonous, reticulate.

Rare in open forests and wastelands in Nizamabad, Karimnagar (Naqvi, 2001) Mahabubnagar (Raghava Rao, 1989) and Hyderabad (Rajagopal, 1973) districts.

Fl. & Fr.: September-December.

Kamareddy (NZB), *BR* & *KH* 7142; Gandhari RF (NZB), *BR* & *CPR* 7242.

Alysicarpus vaginalis (L.) DC., Prodr. 2 : 353. 1825. var. **vaginalis;** Baker in Hook.f., Fl. Brit. India 2 : 158. 1918; Gamble, Fl. Madras 1: 338.1918. *Hedysarum vaginale* L., Sp. Pl. 746. 1753. *Alysicarpus vaginalis* (L.) DC. var. *nummularifolius* Miq., Fl. Ind. Bot. 1: 233. 1855; Baker in Hook.f., Fl. Brit. India 2 : 158. 1876; Gamble, Fl. Madras 1: 338. 1918.

Perennial, prostrate or ascending herbs, branchlets glabrous, branches running to about 40 cm, terete, striate, pilose. Leaves unifoliolate, 0.8-2 × 0.5-1.5 cm, obovate or orbicular, cordate at base, glabrous above, sparsely puberulous below, base obtuse-cordate, margin entire, apex retuse; stipules ovate-lanceolate, parallel-nerved; petiole to 6 mm. Flowers pale pink in terminal racemes; bracts deciduous, scarious, peduncle to 1 cm, pedicel to 1 mm; calyx glabrescent, tube 3 mm, lobes to 3 mm, lobes longer than the first joint of the pod; corolla purple, standard 5.5 mm, wings 5 mm, keel 4 mm; stamens diadelphous (9+1), staminal sheath 5 mm, filaments 1 mm; ovary 3 mm, style 4 mm. Pods cylindric, falcate, about 7-jointed, joints longer than broad, tetragonous; seeds orbicular.

Common in all districts along the streams in plains and forests.

Fl. & Fr.: November-March

Bellampalli (ADB), *PVP* 9462; Kamareddy (NZB), *BR* & *KH* 7142; Sivampet (MDK), *TP* & *CP* 14013; Kanmankalva RF (RR), *MSM* 10468; Pasra (WGL), *RKP* 108105 (BSID).

ARACHIS L.

Arachis hypogaea L., Sp. Pl. 741. 1753; Baker in Hook.f., Fl. Brit. India 2: 161. 1876; Gamble, Fl. Madras 1: 230. 1918.

Vern.: *Verusanaga.* English: Groundnut.

An annual decumbent herb, stem pilose. Leaves paripinnate, leaflets 2 pairs, 2-4 × 1.5-2 cm, opposite, obovate; base cuneate, apex obtuse, mucronate; petiole 4-5 cm long; stipules 3 cm, appressed to petioles. Flowers in axillary clusters; calyx pubescent, lobes 5; corolla yellow, petals clawed, standard orbicular, wing obovate; stamens monadelphous, 8-9, filaments unequal. Pod 2-5 cm long, oblong, 1-3-seeded, indehiscent, yellowish white, geocarpic (underground).

Cultivated for its oil seed.

Fl. & Fr.: August – December.

Umamaheswaram (MBNR), *SRS* 109113 (BSID).

BUTEA Koen. ex Roxb. *nom. cons.*

1. Erect trees; lowest calyx tooth shorter than laterals; base of the pod rounded, almost semicordate .... **B. monosperma**

1. Large woody twiners; lowest calyx tooth as long as laterals; base of the pod narrowed .... **B. superba**

Butea monosperma (Lam.) Taub. in Engler & Prantl, Nat. Pflanzenf. 3(3): 366. 1894. *Erythrina monosperma* Lam., Encycl. 1: 391. 1785. *Butea frondosa* Koen. ex Roxb., Pl. Coromandel 1: 22. t. 21. 1795; Baker in Hook.f., Fl. Brit. India 2: 194. 1876; Gamble, Fl. Madras 1 : 357. 1918.

Vern. : *Modugu.* English: Flame of the forest.

Deciduous trees, up to 10 m high, branchlets densely tomentose, bark grey, trunk crooked and irregular. Leaves trifoliolate, petiole to 14 cm, petiolule to 7 mm, leaflets rhomboid or broadly ovate, 8-14 × 6-10 cm, coriaceous, nerves 7 pairs, sericeous below and glabrous above, base cuneate, margin entire, apex obtuse, retuse. Flowers yellowish red in racemes on leafless branches; calyx dark brown tomentose, lower calyx tooth slightly shorter than laterals, petals tomentose outside, standard lanceolate, silky pubescent without, 3.7 × 2.2 cm, wings falcate, adnate to keels, 4.2 × 1.2 cm, keels incurved, 4.3 × 1.3 cm, staminal sheath 3.5 cm. Pod distinctly stalked, rounded at apex, 16 × 5 cm; seed obovate.

In all dry districts both in open country and deciduous forests.

Fl. & Fr.: January-May.

Sundergiri (ADB), *GO* & *PVP* 4234; Laxmapur RF(NZB), *TP* & *BR* 6254; Gangapur RF (MDK), *BR* & *CP* 11429; Mohammadabad RF (RR), *MSM* & *KH* 10242; Kudurpalli (KRN), *SLK* 70727 (NBG); Bhongir (NLG), *BVRS* 400 (NU); Malkapur (MBNR), *SRS* 109066 (BSID); Nilagiri view point (MBNR), *SRS* 109707 (BSID).

Butea monosperma (Lam.) Taub. var. **lutea** (Witt.) Mahesh., Bull. Bot. Surv. India 3: 92. (1961) 1962; Santapau, Rec. Bot. Surv. India 16: 66. 1967. *Butea frondosa* Koen. ex Roxb. var. *lutea* Witt., Descr. List. N. & Ber. For. Circ. C.P. 75. 1916.

Vern.: *Tella moduga.*

It differs from well known *Butea monosperma* var. *monosperma* in the presence of charismatic ivory-white flower buds and bright yellow flowers.

Very rare, occurs in open forests.

Note: *Butea monosperma* var. *lutea* is endemic to Deccan Plateau of India. It was reported from areas around Karimnagar district, Peddagutta of Nizamabad district and Kummarigudem of Mallapally of Warangal district. It was an overlooked taxon and not incorporated in the Red Data Book of India. It is very rare, declared as globally endangered medicinal plant by Conservation Assessment Management Planning Workshop for Medicinal Plants of Andhra Pradesh (Jadhav & *al.,* 2001). Besides its rarity, has much demand in folk medicine. Its bark and wood scrapes are used as ethno-gynaecological medicine in tribal societies. Stem bark extract with jeer powder used for white leucorrhoea, jaundice and skin disease.

Butea superba Roxb., Asiat. Res. 3: 474. 1792; Woodr., J. Bombay Nat. Hist. Soc. 11 : 424. 1897; Baker in Hook.f., Fl. Brit. India 2 : 194. 1876; Gamble, Fl. Madras 1: 357. 1918.

Vern.: *Tiga moduga, Telga modugu.*

Large, woody twiner, bark smooth, brownish. Leaves 3-foliolate, leaflets up to 30 cm in diameter, rhomboid or broadly obovate, base cuneate, apex obtuse, sub-rugose and dull above, green and thinly hairy on nerves beneath. Flowers 9-12 cm long in racemes up to 30 cm long, orange-red; calyx lobes as long as laterals, campanulate. Pods flat, stalked, 12-15 cm long, densely rusty tomentose.

Common in the interior deciduous forest areas.

Fl. & Fr.: March- May.

Satruguda (ADB), *GO* & *PVP* 4254; Laxmapur RF (NZB), *TP* & *BR* 6255; Narsapur RF (MDK), *TP* & *CP* 14044; Mohammadabad (RR), *MSM* 19235; Kudurpalli (KRN), *SLK* 70728 (NBG); Chintalavampu (NLG), *BVRS* 396 (NU).

CAJANUS DC. *nom. cons.*

1. Twining shrubs .... **C. scarabaeoides**
1. Erect shrubs:
   2. Much branched bushy shrub, up to 2 m tall, leaflets elliptic oblong; flowers in terminal panicles; cultivated .... **C. cajan**
   2. Undershrub, 50-60 cm, leaflets oblanceolate; flowers in pairs on very short axillary peduncles; wild .... **C. lineatus**

Cajanus cajan (L.) Millsp., Publ. Field Columbian Mus. Bot. Ser. 2 : 53. 1900. *Cytisus cajan* L., Sp. Pl. 739. 1753. *Cajanus indicus* Spreng., Syst. Veg. 3 : 248. 1826; Baker in Hook.f., Fl. Brit. India 2 : 217; Gamble, Fl. Madras 1 : 369. 1918.

Vern.: *Kandi, Kandulu.*

Erect much-branched bushy shrubs, up to 2 m high. Leaves 3- foliolate, leaflets elliptic-oblong, 3-7.5 × 1.5-3 cm, soft pubescent, base obtuse, margin entire, apex acute. Flowers yellow in terminal panicles; peduncle 2-4 cm; bracts ovate- elliptic; calyx campanulate, pubescent; corolla yellow, 1.5-2 cm long, standard subobicular, with inflexed auricle, wings slightly obovate, with short auricle; keel apex obtuse, slightly inflexed; ovary hairy, ovules several, style long, linear, glabrous, stigma capitate. Pods straight, compressed between seeds, pubescent; seeds 3-5, discoid, rounded, smooth.

Important pulse crop of most districts, rarely as in escape.

Fl. & Fr.: October-March.

Sone (ADB), *TP* & *PVP* 4009.

Cajanus lineatus Maesen, Agric. Univ. Wageningen Pap. 85 (4): 143. 1985. *Atylosia lineata* Wight & Arn., Prodr. Fl. Ind. Orient. 258. 1834; Baker in Hook.f., Fl. Brit. India 2: 213. 1876; Gamble, Fl. Madras 1: 368. 1918; Keshava Murthy & Yoganarasimhan, Fl. Coorg 132. 1990.

Erect undershrubs, 30-50 cm high, branches densely clothed with soft, grey hairs. Leaves 3-foliolate, petiole 0.5-1.4 cm long, leaflets 2-3.5 × 0.8-1.3 cm, oblanceolate, base acute, apex subacute, glabrous above, densely pubescent and resinous beneath, petiolule very short. Flowers yellow, in paris on very short axillary peduncle, calyx up to 6 mm long, tube campanulate, equaling the lobes, hairy, corolla exserted, 1-1.2 cm long, stamens diadelphous (9+1); style filiform. Pod 1.6-2 × 0.5-0.6 cm, oblong, turgid, grey, silky pubescent.

Very rare, occurs as forest undergrowth.

Fl. & Fr.: August – December.

Sayeeduddin (1935) recorded from Mulug RF, Warangal district. There were no other collections from the state (C.S.Reddy, 2001).

Cajanus scarabaeoides (L.) Thours, Dict. Sci. Nat. 6: 617. 1817 (' *Cajan scarabaeoides*'); Maesen, Agric. Univ. Wageningen Pap. 85. 4: 183. 1985. *Dolichos scarabaeoides* L., Sp. Pl. 726. 1753. *Atylosia scarabaeoides* (L.) Benth. in Miq. Pl. Jungh. 242. 1852; Baker in Hook.f., Fl. Brit. India 2 : 215. 1876; Gamble, Fl. Madras 1: 369. 1918; Pullaiah & Chennaiah, Fl. Andhra Pradesh 1: 245. 1997.

Slender twining shrubs, up to 3 m long, branchlets densely pubescent, striate. Leaves 5 cm long, trifoliolate, leaflets oblong, obovate, 1-2 × 1- 1.5 cm, laterals unequally truncate at base, margins entire, excurved, apex subacute, apiculate, pubescent above, yellow glandular beneath, nerves 2, reticulation prominent beneath; petiole to 2 cm long; petiolule 1.5 mm. Flowers yellow in axillary, 2-6-flowered racemes; peduncle to 4 cm, pedicel 5 mm; calyx tube 2 mm, fulvous-pubescent, lobes linear acute, as long as the tube; corolla yellow, standard 9 × 5.5 mm, wings 8 × 2 mm, keels 9 × 2 mm; stamens diadelphous (9+1), staminal sheath 6.5 × 2 mm, filaments 2.5 mm; ovary densely woolly, 2 × 0.7 cm, style curved, 7 mm. Pod linear, tomentose, transversely depressed between seeds; seeds 4 or 5, oblong, dark with prominent bivalvular caruncle at top.

Common in forests in almost all districts.

Fl. & Fr.: July-March

Ankusapuram (ADB), *GO* 4307; Mosra (NZB), *TP* & *BR* 9074; Narsapur forest nursery (MDK), *BR* 11215; Sardhana X roads (MDK), *BR* & *CP* 11628; Lingampalle (RR), *MSM* 12759; Laxmapur (MBNR), *TDCK* 13677; Mannaram (MBNR), *STR* 40 (HH); Aklaspur (KRN), *GVS* 22487 (MH); Jakaram (WGL), *RKP* 105312 (BSID); Yenkur (KMM), *RR* 108552 (BSID); Koriguttalu (KMM), *RR* 106025 (BSID).

CANAVALIA DC. *nom. cons.*

1. Leaflets ovate, base cuneate; seeds greyish, pink or red, compressed; pods 12-24 cm long:
    2. Seed red or pink, 2-3.5 cm, hilum 1.5-2 cm long .... **C. gladiata**
    2. Seeds greyish, 8 mm long, hilum 2-4 mm long .... **C. ensiformis**
1. Leaflets broadly ovate or obovate, base rounded; seeds brown, scarcely compressed; pods to 15 cm long .... **C. africana**

Canavalia africana Dunn, Bull. Misc. Inform. Kew 1922: 135. 1922; Verdc., Kew Bull. 42: 658. 1987. *C. virosa* (Roxb.) Wight & Arn., Prodr. Fl. Ind. Orient. 253. 1834; Gamble, Fl. Madras 1: 359. 1918. *Dolichos virosus* Roxb., Fl. Ind. 3: 301. 1832. *C. ensiformis* DC. var. *virosa* (Roxb.) Baker in Hook.f., Fl. Brit. India 2: 176. 1875.

Vern.: *Thamba*

A climbing perennial, up to 8 m long, branchlets glabrous, reddish. Stipules subulate, 2 mm, stipels linear. Leaves 10-20 cm, 3-foliolate, leaflets ovate or obovate, 6-13.5 × 3-8 cm, chartaceous, base rounded, apex acute-acuminate, apiculate, margin entire, glabrous above, sparsely puberulous below, petiole to 10 cm long, petiolule to 5 mm. Flowers purple in 20 cm long racemes, peduncle to 15 cm long, pedicel to 4 mm long; calyx puberulous, tube 1.5 × 1 cm, lobes to 5 mm; corolla purple, standard broad, with two strong calli above the claw, 3 × 2 cm, wings 2.5 × 0.5 cm, keels 3 × 0.5 cm; stamens monadelphous, staminal sheath 2.5 cm, filaments to 7 mm; ovary silky, tapering into the style, 2 cm, style 6 mm. Pod 15 × 2 cm, pubescent, apically obtuse, seeds ca. 10, ellipsoid or ovoid, 1.5 × 1 cm, marbled with light and darker brown, hilum 1.2 cm long.

Common in the scrub forests in Mahabubnagar and Nalgonda districts.

Fl. & Fr.: August – January.

Canavalia ensiformis (L.) DC., Prodr. 2: 404. 1825; Hook.f., Fl. Brit. India 2: 195. 1876; Gamble, Fl. Madras 1: 359. 1918; Saver, Brittonia 16: 142. 1964. *Dolichos ensiformis* L., Sp. Pl. 725. 1753.

Vern.: *Karu thamma, Adavi thamba.*

Annual, usually bushy and erect undershrub, up to 1 m high, branchlets stout, glabrous, terete. Stipules lanceolate, stipels minute, subulate. Leaves 3-foliolate, leaflets ovate, base cuneate, margin entire, apex obtuse to acute, strigose, glabrescent. Racemes 15-20-flowered; calyx campanulate, 1.6 cm long, upper lip emarginated, lower lip 3-lobed; corolla coloured fading to white towards base, standard oblong, orbicular, notched at the apex, wings united; stamens monadelphous. Pods linear, gently curved, beaked at the tip, straw coloured when ripe, pendent, ribbed near upper suture; seeds ellipsoid, compressed, shiny white, hilum less than half of the length of the seed, 8 mm long, greyish, surrounded by orange brown border.

Occasional on field hedges in most districts

Fl. & Fr.: August – March.

Nasarullabad (NZB), *TP* & *BR* 6177; Choutkoor (MDK), *TP* & *CP* 11901.

Canavalia gladiata (Jacq.) DC., Prodr. 2 : 404. 1825. *Dolichos gladiatus* Jacq., Icon. Pl. Rar. 3(1): t. 560. 1788 & Coll. Bot. 2. 276. 1789.

Vern. : *Thammakaya.*

Stout, woody, perennial or biennial twining shrubs, up to 3 m long, branches terete, glabrous. Stipules subulate, deciduous. Leaves 3-foliolate, leaflets ovate, 6-12 × 4-6 cm, base cuneate, margin entire, apex acute, nerves and reticulations rather prominent below; petiole to 8 cm. Flowers purplish in axillary racemes; peduncle to 20 cm, pedicel to 5 mm; calyx campanulate, tube 1.5 × 1 cm, with two rounded lobes, emarginated, lower lip 3-lobed; corolla white or pink, standard 2.8 × 2.5 cm, wings oblong-spathulate, 3.2 × 0.7 cm, keels oblong, obtuse, falcate, 3 × 1.1 cm; stamens monadelphous (10), staminal sheath 2.3 × 0.8, filaments 1.1 cm; ovary pubescent, 2 × 0.3 cm, style 8 mm. Pod linear-oblong, 1.1 × 2 cm, compressed, slightly curved, beaked, thickened along both sutures, straw coloured when ripe, 8-16-seeded; seeds ellipsoid, pink or red, compressed.

Two varieties are present, one is commonly found in forests as wild and another is cultivated for its fruits in gardens.

Fl. & Fr.: September-March.

Vempalli RF (ADB), *GO* 4429; Bersaipet (ADB), *GO* & *PVP* 4579; Nasarullabad (NZB), *TP* & *BR* 6177; Saleswaram (MBNR), *BSS* & *SKB* 30733; Dharmapur (RR), *MSM* & *KH* 10545; Pakhal (WGL), *KMS* 11644 (MH); Bhadrachalam RF (KMM), *RC* 98984 (BSID).

Note: Chatterjee (1949 Indian Species of *Canavalia* p.86) has traced the history of this species and is of the opinion that the specific epithet, *ensiformis* which rightly belongs to an American plant has wrongly been used for the Indian plant. The old world plant should be known as *Canvalia gladiata* (Jacq.) DC.; true *ensiformis* has white seeds, whereas *C. gladiata* has red seeds.

CICER L.

Cicer arietinum L., Sp. Pl. 738. 1753; Baker in Hook.f., Fl. Brit. India 2: 176. 1876; Gamble, Fl. Madras 1: 349. 1918.

Vern. : *Sanagalu.*

Annual, much branched herbs; branches glandular-pubescent. Leaves imparipinnate, 2.5-6 cm long; leaflets to 9 pairs, alternate or subopposite, base cuneate, margins sharply serrate or dentate, apex mucronate. Flowers axillary, solitary, pink. Pods ovoid-oblong, turgid, 1-2-seeded.

Cultivated in black cotton soils for its seeds.

Fl. & Fr.: November – January.

CLITORIA L.

Clitoria ternatea L., Sp. Pl. 753. 1753; Baker in Hook.f., Fl. Brit. India 2 : 208. 1876; Gamble, Fl. Madras 1 : 365. 1918.

Vern.: *Sankupoolu.*

Perennial, climbing or twining herbs, up to 7 m long, branchlets appressed tomentose. Leaves odd-pinnate, leaflets 2 or 3 pairs, opposite, ovate, 2-4 × 1.5-3 cm, base cuneate, margin entire, apex obutse- retuse, stipules striate, persistent. Flowers bluish-violet with yellow streaks or white, axillary, solitary; calyx membranous, tubular, 8 mm, lobes to 7 mm; corolla deep blue or white, exserted, standard erect, emarginate, narrowed at base, 4.2 × 3.5 cm, wings falcate-oblong, adnate in the middle of the keel, 2.8 × 1.2 cm, keels short, curved, 2.1 × 0.5 cm; stamens diadelphous (9+1), staminal sheath, 1.3 × 0.5 cm, filaments to 5 mm; ovary stipitate, 9 mm, style elongate, 1.5 cm. Pod flattened or sub-cylindrical, beaked, sparsely appressed hairy, ca. 8.5 × 0.8 cm; seeds 10-15, reniform, smooth.

Common in cultivated fields on bushes and along streams in the forests, often cultivated for its showy flowers and also as fodder.

Fl. & Fr.: September-February.

Ghanapur (NZB), *BR* & *KH* 7110; Narsingi (MDK), *BR* & *CP* 11525; Nayanigutta, *MSM* & *NV* 10329; Bhupatipur (KRN), *GVS* 25703 (MH); Azampur (NLG), *BVRS* 308 (NU); Amangal to Shadnagar (MBNR), *SRS* 104576 (BSID).

CODARIOCALYX Hassk.

Codariocalyx motorius (Houtt.) H.Ohasi, J. Jap. Bot. 40(12): 367. 1965. *Hedysarum motorium* Houtt., Nat. Hist. ed. 2. 10. 246. 1779. *Desmodium motorium* (Houtt.) Merr., J. Arnold Arbor. 19: 345. 1938. *D. gyrans* (L.f.) DC., Prodr. 2: 326. 1825; Baker in Hook.f., Fl. Brit. India 2 : 174. 1576; Gamble, Fl. Madras 1: 348. 1918.

Erect herbs or undershrubs, up to 1 m high, branchlets slender, grooved, glabrous. Stipules scarious, 5 mm. Leaves 3-foliolate, leaflets linear-lanceolate or ovate-oblong, base and apex obtuse, appressed pubescent beneath, terminal leaflet 8.5-10 × 2.5-3 cm, side leaflets 1.5-2 × 0.5 cm, often moving in jerks; petiole to 1.5 cm, petiolule 2 mm. Flowers pink, solitary in the axil of broadly ovate, caducous bracts, in axillary and terminal racemes; peduncle to 5 cm, pedicel 3 mm; calyx membranous, tube 1 mm, lobes very short; corolla pink, standard broadly obovate, 0.8 × 1 cm, wings 6 × 4 mm, keels 8 × 3 mm; stamens diadelphous (9+1), staminal sheath 7 × 1.5 mm, filaments 0.5 mm; ovary sparsely pubescent, 6 × 0.5 mm, style 2 mm. Pod oblong, pubescent, indented along the lower suture, with 8-10 indistinct septa, *i.e.*, joints not separating; dehiscing continuously along lower suture.

Occasional in Khammam district.

Fl. & Fr.: August-November.

Ramavaram (KMM), *RR* 112540 (BSID).

CROTALARIA L.

1. Leaves digitately 3-5-foliolate :
    2. Pods obliquely subglobose, subquadrangular, small, sessile, 2-seeded.
        3. Pods subquadrangular, thinly silky .... **C. trifoliastrum**
        3. Pods sub-globose, pubescent or glabrescent :
            4. Erect subshrub; leaflets oblong-oblanceolate,
               sericeous on both surfaces; calyx-lobes nearly
               equal to calyx .... **C. willdenowiana**
            4. Diffuse herb; leaflets obovate-oblanceolate, glabrous
               above, appressed pubescent below; calyx-lobes 2 or
               3 times as long as calyx-tube .... **C. medicaginea**
    2. Pods oblong-cylindrical, many-seeded :
        5. Stipe of the pods more than 1 cm, pods glabrous :
            6. Leaflets 5 .... **C. quinquefolia**
            6. Leaflets 3:
                7. Herbs; corolla hardly exceeding calyx .... **C. orixensis**
                7. Shrubs; corolla twice as long as calyx .... **C. laburnifolia**
        5. Stipe of pods less than 0.5 cm; pods glabrescent .... **C. pallida**
1. Leaves simple:
    8. Plants with closely set numerous leaves and solitary flowers in
       most of the leaf axils; pods flattened .... **C. hebecarpa**
    8. Plants not as above:
        9. Pods glabrous :
            10. Pods twice as long or longer than calyx :
                11. Stipules absent:
                    12. Pod ca 2 cm long; seeds 16-20 .... **C. prostrata**
                    12. Pod 6-8 mm long; seeds 8-10 .... **C. filipes**
                11. Stipules present :
                    13. Stems with white silky hairs (villous); racemes
                        lateral, leaf-opposed .... **C. ferruginea**
                    13. Stems with white appressed pubescence;
                        racemes terminal .... **C. retusa**
            10. Pods shorter or as long as or slightly longer than calyx :
                14. Upper calyx lobes connate, except at the tip :
                    15. Flowers in 15 cm long racemes; corolla shorter
                        than calyx .... **C. montana**

15. Flowers in compact or subcapitate racemes or umbels; corolla longer or equaling calyx .... **C. nana**

14. Upper calyx lobes not connate or connate only below :

16. Corolla included .... **C. hirta**

16. Corolla equaling or slightly exceeding the calyx :

17. Pods equaling or shorter than calyx .... **C. calycina**

17. Pods longer than calyx :

18. Leaves distinctly pellucid-punctate, thick ..... **C. albida**

18. Leaves otherwise :

19. Stipules present; pod 3-4 cm long ... **C. mysorensis**

19. Stipules absent; pod ca 1 cm long ...... **C. epunctata**

9. Pods hairy :

20. Racemes all lateral, leaf-opposed :

21. Pods suborbicular .... **C. angulata**

21. Pods oblong :

22. Spreading herb up to 60 cm high; leaves not oblique at base; pods sparsely hairy .... **C. bifaria**

22. Erect herbs to 1 m tall; leaves oblique at base; pods strigosely hirsute .... **C. hirsuta**

20. Racemes terminal :

23. Flowers in single racemes :

24. Stipules semilunate; flowers purple or blue .... **C. verrucosa**

24. Combination of characters not as above :

25. Small herbs, not exceeding 15 cm; - pods silky-villous .... **C. pusilla**

25. Stout undershrubs or shrubs exceeding 60 cm :

26. Suffruticose herbs, to 50 cm high; racemes 2-3-flowered .... **C. triquetra**

26. Erect stiff, shining undershrubs to 2.5 m high; racemes 3-many-flowered .... **C. juncea**

23. Flowers in panicles :

27. Subshrubs; bracts and bracteoles margins reflexed, curved .... **C. paniculata**

27. Much branched herbs; bracts and bracteoles otherwise .... **C. ramosissima**

Crotalaria albida B. Heyne ex Roth, Nov. Pl. Sp. 333. 1821; Baker in Hook.f., Fl. Brit. India 2: 71. 1876; Gamble, Fl. Madras 1: 295. 1918; Ansari, *Crotalaria* in India 153. 2008.

Vern.: *Kondagiligichha.*

Annual herbs or woody undershrubs, 25-50 cm high, clothed with appressed hairs. Leaves subsessile, 1-2 × 0.3-0.5 cm, elliptic- oblong or oblanceolate, base cuneate, margin entire, apex obtuse-emarginate, gland dotted, glabrous above, thinly silky beneath; petiole to 1.5 mm. Racemes terminal or lateral, 2-20 cm long, laxly 6 to 25-flowered; bracts minute, linear; three calyx teeth linear, acuminate, upper two broader, subobtuse; corolla as long as calyx, yellow, standard ovate-oblong, 8 × 8.5 mm, pubescent without, wings oblong, 8 × 3 mm, keels ovate or oblong, 8 × 4 mm, beak twisted; stamens monadelphous (10); staminal sheath 3 mm, filaments 4 mm; ovary subsessile, glabrous, 3.5 mm, style pubescent, 6 mm. Pods 1.5 x 0.6 cm, oblong, glabrous, longer than calyx, prominently nerved, 6-12-seeded.

Common in forests and waste lands in Warangal, Medak, Nizamabad and Adilabad districts.

Fl. & Fr.: August-October

Servaipet (ADB), *GO* & *PVP* 4817; Bellampalli (ADB), *PVP* 9461; Kammarpally (NZB), *TP* & *BR* 6233; Kamareddy (NZB), *BR* & *KH* 7151; Majira Barrage (MDK), *BR* & *CP* 14092.

Crotalaria angulata Mill., Gard. Dict. ed. 8. no. 9. 1768; Ansari, *Crotalaria* in India 158. 2008. *C. biflora* (L.) L., Mant. Pl. 570. 1771; Baker in Hook.f., Fl. Brit. India 2 : 66.1876; Gamble, Fl. Madras 1: 292. 1918. *Astragalus biflorus* L., Mant. Pl. 273. 1771.

Prostrate, annual herbs, up to 50 cm long, branchlets hispid. Leaves simple, ovate, 1-2.5 × 0.5-1 cm, chartaceous, base rounded-obtuse, margin entire, apex obtuse, appressed hispid on both sides, petiole to 1.5 mm, stipules minute. Flowers yellow, 2-flowered in lateral racemes, rarely solitary, peduncle to 2.5 cm, pedicel to 4 mm; calyx hispid, tube 2 mm, lobes lanceolate, to 6 mm; corolla as long as or slightly longer than calyx, standard orbicular, 6 x 4 mm, wings oblong, 5 × 2 mm, keels ovate, 6 × 3 mm, beak spirally twisted; stamens monadelphous (10), staminal sheath 2 mm, filaments 3 mm; ovary shortly stipitate, 2.5 mm, style pubescent, 5 mm. Pod suborbicular, hairy, 1 cm across, turgid with stiff golden hairs; 15-20-seeded; seeds grey, smooth, shining.

In Mahabubnagar, Hyderabad, Karimnagar and Warangal districts, chiefly in open forest lands.

Fl. & Fr.: December – February.

Crotalaria bifaria L. f., Suppl. Pl. 322. 1781; Baker in Hook.f., Fl. Brit. India 2 : 69. 1876; Gamble, Fl. Madras 1: 293. 1918; Ansari, *Crotalaria* in India 161. 2008.

Much branched, spreading herb, up to 60 cm long, branchlets 4-angled when young, pubescent. Leaves simple, 2.5-5.0 × 1-2.5 cm, lower ones orbicular, upper ones ovate-lanceolate, base cuneate-obtuse, margin entire, apex subacute-obtuse, puberulous below; petiole to 2 mm, stipules lanceolate, reflexed. Racemes exceeding the leaves, peduncle to 8 cm, pedicel to 1.5 cm. Bracteoles 2, foliaceous, ovate-lanceolate, persistent; calyx teeth linear- lanceolate, acute, corolla exserted; yellow without, blue within, standard orbicular, 2.2 × 2.1 cm, wings oblong, 1.8 × 0.6 cm, apex obtuse, keels ovate, 1.9 × 0.9 cm, ciliate, beak spirally twisted; stamens monadelphous (10), staminal sheath 5 mm, filaments to 4 mm; ovary sessile, strigose, 6 mm, style 1 cm. Pod 1.2-2.5 cm long, oblong, turgid, mottled with purple, 10-12-seeded.

In Ranga Reddi, Medak, Karimnagar and Mahabubnagar districts.

Fl. & Fr.: August-October

Lingampalle (RR), *MSM* 12756; Mannanur range (MBNR), *STR* 110 (MH).

Crotalaria calycina Schrank, Pl. Rar. Hort. Monac. t. 12. 1817; Baker in Hook.f., Fl. Brit. India 2 : 72. 1876; Gamble, Fl. Madras 1 : 295. 1918; Ansari, *Crotalaria* in India 82. 2008.

Annual, erect herbs, up to 60 cm high, branchlets appressed-velvety or tomentose. Leaves simple, linear or elliptic-oblong, 3- 8 × 1.2 cm, glabrous above, pubescent beneath, base obtuse-cuneate, margin entire, apex rotund-obtuse, apiculate, stipules in pairs, linear subulate, 0.3 cm long; petiole to 2 mm. Flowers yellow, in terminal or lateral, 2-12-flowered, up to 15 cm long racemes; pedicel to 1 cm; calyx densely clothed with fulvous hairs, tube to 5 mm, lobes to 2 cm, lanceolate, free; corolla slightly exserted, yellow, standard ovate-oblong, 2.2 × 2 cm, midnerve sericeous below, wings obovate, 1.7 × 1 cm, keels obovate, 1.9 × 0.9 cm, beak spirally twisted; stamens monadelphous (10), staminal sheath 7 mm, filaments to 8 mm; ovary subsessile, 1 cm, style pubescent, 1.2 cm. Pod oblong, black when ripe, 2.5 x 0.8 cm; seeds many, pale yellow.

In open forest lands and among grasses in hill areas in Mahabubnagar, Karimnagar and Warangal districts.

Fl. & Fr.: September-December.

Jakaram (WGL), *RKP* 105320 (BSID).

Crotalaria epunctata Dalzell, Hooker's J. Bot. Kew Gard. Misc. 3 : 210. 1851; Gamble, Fl. Madras 1: 295. 1918; Ansari, *Crotalaria* in India 164. 2008. *C. albida* Heyne ex Roth var. *epunctata* (Dalzell) Baker in Hook. f., Fl. Brit. India 2: 71. 1876.

Profusely branched herb, up to 50 cm high, branchlets appressed pubescent. Leaves linear-oblanceolate, to 3.5 × 1.2 cm, base obtuse, margin entire, apex mucronate, not conspicuously pellucid-punctate, pubescent on both sides, petiole to 1 mm. Flowers yellow, in simple or branched, 2-6-flowered, lax racemes; calyx pubescent, tube 2 mm, lobes elongate, to 0.7 cm long; corolla equaling or slightly longer than calyx; standard 1.1 × 0.9 cm, wings 9 × 3.5 mm, keels 9 × 0.5 mm, beak spirally twisted; stamens monadelphous (10), staminal sheath 2.5 × 3 mm, filaments 3 mm; ovary glabrous, 3 mm, style 7 mm. Pod oblong, glabrous, about 1 cm long, seeds 6-12.

Common in deciduous forests in Adilabad and Khammam districts.

Fl. & Fr.: October-March

Bijjur (ADB), *GO* & *PVP* 4975; Vempalli RF (ADB), *TP* & *GO* 5501; Sukumamidi near Mothugudem (KMM), *EC* 7457; Edrallapalli (KMM), *RC* 98761 (BSID); Thiruppumetta RF (KMM), *RC* 99017 (BSID).

Crotalaria ferruginea Grah. ex Benth. in Hooker's Lond. J. Bot. 2 : 476, 570. 1843; Gamble, Fl. Madras 1: 292. 1918; Ansari, *Crotalaria* in India 170. 2008. *C. ferruginea* var. *pilosissima* Miq., Pl. Jungh. 205. 1852; Baker in Hook.f., Fl. Brit. India 2 : 68. 1876.

Large suberect herb, up to 50 cm high, branchlets ferrugino-tomentose. Stipules lanceolate, long acuminate. Leaves obovate-oblong, appressed-pubescent on both sides, 5 × 2 cm, base obtuse, margin entire, apex mucronate; petiole to 1 mm. Flowers larger, rusty villous, yellow, in leaf-opposed 2-8-flowered lax racemes, bracts lanceolate; calyx hispid, tube 3 mm, lobes to 1.1 cm, oblong-lanceolate; corolla not exserted, standard 1.1 cm long. Pod oblong, glabrous, 3 cm long, many-seeded.

Common in Mahabubnagar district.

Fl. & Fr.: Through the year.

Crotalaria filipes Benth., London J. Bot. 2: 475. 1843; Baker in Hook. f., Fl. Brit. India 2: 66. 1876; Gamble, Fl. Madras 1: 292. 1918; Ansari, *Crotalaria* in India 174. 2008.

A very slender trailing herb. Leaves simple, bifarious, ovate-cordate, prominently nerved, sparsely hirsute, stipules absent. Peduncles and pedicels filiform. Flowers yellow, 1-3-flowered. Pod oblong, glabrous, 8-10-seeded.

Rare, occurs in forest fringes in Karimnagar district (Naqvi and Raju, 1996).

Fl. & Fr.: August – November.

Crotalaria hebecarpa (DC.) Rudd., Phytologia 54: 28. 1983; Ansari, *Crotalaria* in India 210. 2008. *Goniogyna hebecarpa* DC., Ann. Sci. Nat. Paris 4: 92. Jan. 1825. *G. hirta* (Willd.) Ali, Taxon 16. 463. 1967; Pullaiah & Chennaiah, Fl. Andhra Pradesh 1: 286. 1997. *Hallia hirta* Willd., Sp. Pl. 3 : 1169. 1802. *Heylandia latebrosa* sensu Baker in Hook. f., Fl. Brit. India 2 : 65. 1876, non (L.) DC. 1825; Gamble, Fl. Madras 1: 280. 1918.

Prostrate annual herbs, up to 35 cm long, dichotomous branching, branchlets with white silky hairs. Leaves simple, sessile, cordate-ovate, 1-1.5 × 0.6-0.8 cm, chartaceous, base truncate, margin ciliate, apex obtuse, apiculate. Flowers yellow, with red veins, axillary, solitary; pedicel 2 mm; calyx pubescent, tube short, 2 mm, upper two lobes slightly connate; corolla yellow, exserted, petals clawed, standard obovate-orbicular, 5.5 × 6 mm, wings oblong-obovate, 4 × 2 mm, keels connate, 4 × 2.2 mm; staminal sheath 2 × 2.5 mm, filaments 2.5 mm, anthers alternately short and long; ovary sessile, 1.5 mm, style geniculate, 3mm. Pods flat, oblong, compressed, silky, tipped by persistent style; seed solitary, reniform.

Common in all districts, wastelands and open forests, rare in cultivated fields.

Fl. & Fr.: September-March.

Ankusapuram (ADB), *GO* 4371; Bhimagal (NZB), *TP* & *BR* 6204; NSF (MDK), *BR* & *CP* 11660; Anantagiri RF (RR), *TP* & *MSM* 11087; Narasampet (WGL), *KMS* 11589 (MH); Kodimial (KRN), *GVS* 21854 (CAL, MH); Kambalpally (NLG), *BVRS* 293 (NU); Amrabad (MBNR), *BSS* & *SKB* 32562 (BSID); Mannanur (MBNR), *LJGV* 3240 (CAL).

Crotalaria hirsuta Willd., Sp. Pl. 3 : 978. 1802; Baker in Hook.f., Fl. Brit. India 2 : 68. 1876; Gamble, Fl. Madras 1: 293. 1918; Ansari, *Crotalaria* in India 185. 2008.

Erect, hairy herbs, up to 1 m high. Stipules linear, minute. Leaves ovate, 4-6 × 2-4 cm, base obtuse, margin entire, apex mucronate; petiole to 1.5 mm. Flowers yellow in leaf-opposed, slender, 2-3-flowered racemes, peduncle to 1 cm, pedicel to 5 mm; calyx hirsute, tube 3 mm, lobes to 1.1. cm, lanceolate; corolla yellow, standard 1.8 × 0.9 cm, wings 1.2 × 0.5 cm, keels 1.3 × 0.6 cm, beak twisted; stamens monadelphous (10), staminal sheath 3 mm, filaments 8 mm; ovary hirsute, 5 mm, style glabrous, 1.2 cm. Pod oblong, turgid, hirsute, 2 cm long; 8-10-seeded.

Occasional in forests of most of the districts.

Fl. & Fr.: September-April

Annaram RF (NZB), *BR* & *CPR* 7188; Narsapur town (MDK), *KMS* 6674 (MH); Lingampalle (RR), *MSM* 12749; Aklaspur (KRN), *GVS* 25605 (MH).

Crotalaria hirta Willd. in Ges. Naturf. Freunde Berlin Neue Schriften 4: 217. 1802; Baker in Hook.f., Fl. Brit. India 2: 70. 1876; Gamble, Fl. Madras 1: 295. 1918; Ansari, *Crotalaria* in India 92. 2008.

Annual, spreading herb, up to 75 cm long, branchlets densely-hispid. Leaves simple, oblong-elliptic, 1.5-4 × 0.5-1 cm, chartaceous, sparsely pubescent above, appressed below, base and apex obtuse, margin entire, petiole 1 mm. Flowers yellow, in closely 1-6-flowered, terminal racemes, bracts foliaceous; calyx pubescent, tube 2 mm, lobes to 7 mm, linear, acuminate; corolla included, standard oblong-obovate, 1 × 0.6 cm, wings oblong, 6 × 2 mm, keels 7 × 3 mm, beaked; stamens monadelphous (10), staminal sheath 2 mm, filaments to 5 mm; ovary stipitate, glabrous, 4 mm, style pubescent, 5 mm. Pod oblong-ellipsoid, glabrous, faintly nerved, slightly exserted; seeds 15-20.

Common in moist deciduous and dry deciduous forests in Karimnagar, Medak and Nizamabad districts.

Fl. & Fr.: January-March

Gandhari RF (NZB), *BR* & *CPR* 7228; Aklaspur (KRN), *GVS* 22522 (CAL & MH); ICRISAT (MDK), *LJGV* 3995 (CAL, BSID).

Crotalaria juncea L., Sp. Pl. 714. 1753; Baker in Hook.f., Fl. Brit. India 2 : 79. 1876; Gamble, Fl. Madras 1: 297. 1918; Ansari, *Crotalaria* in India 95. 2008.

Erect, stiff, shining, silky brown undershrubs, up to 2.5 m high, stems and branches furrowed. Leaves exstipulate, simple, oblong-lanceolate, 3-6 × 1-1.5 cm, base acute-obtuse, margin entire, apex apiculate, petiole to 3 mm. Flowers yellow, many-flowered in terminal racemes; bracts linear, 4 mm, bracteoles subulate, to 3 mm; corolla yellow, standard ovate, 3 × 2.5 cm, wings obovate, 2 × 0.7 cm, keels 3 × 1 cm, spirally beaked; stamens monadelphous (10), staminal sheath 4 mm, filaments to 8 mm; ovary villous, 1 cm, style pubescent, to 2 cm. Pods oblong-terete, velvety, 2.5 cm long, 10-15-seeded.

Cultivated for its fibre and as a fodder plant, frequently as an escape.

Fl. & Fr.: October-May.

Navipet (NZB), *BR* & *PSPB* 9592; Rechapalli (KRN), *GVS* 22259, 25700 (CAL, MH); Tirmalagiri (NLG), *BVRS* 466 (NU); Rathamhutta hills (KMM), *RC* 99058.

Crotalaria laburnifolia L., Sp. Pl. 715. 1753; Baker in Hook.f., Fl. Brit. India 2 : 84. 1876; Gamble, Fl. Madras 1: 301. 1918; Ansari, *Crotalaria* in India 321. 2008.

Vern.: *Pedda-giligicha.*

Glabrescent, erect shrubs, 1-2.5 m high, branchlets slender, terete, glabrescent. Leaves 3-foliolate, ovate-elliptic, 3-6 × 2-4 cm, chartaceous, glabrous above, appressed pubescent beneath, base cuneate to subacute, margin entire, apex mucronate, petiole to 8 cm, petiolule to 4 mm. Flowers yellow in terminal or lateral, elongated, many-flowered racemes, bracts linear-setaceous, to 3 mm, bracteoles linear, 1 mm; calyx glabrous, tube 4 mm, lobes ovate, to 5 mm, corolla yellow, standard ovate, 3.3 x 2.9 cm, wings oblong, 2.3 × 1.7 cm, keels ovate, 3.3 × 1.4 cm; stamens monadelphous (10), staminal sheath 1.6 × 0.6 cm, filaments to 1 cm; ovary glabrous, 1 cm, with elongated stipe, style curved, to 2.3 cm; pedicel 2 cm or more. Pods glabrous, oblong, much exceeding calyx, stalk of the pod 3 cm or more; seeds 20-30.

Occasional in fields in many districts.

Fl. & Fr.: September-January

Manchippa (NZB), *BR* & *PSPB* 9607; Narsapur (MDK), *BR* & *CP* 11237; Medak (MDK), *CP* 11738; Nancherla gate (RR), *MSM* & *NV* 10393.

Crotalaria medicaginea Lam., Encycl. 2 : 201. 1790; Baker in Hook.f., Fl. Brit. India 2: 81. 1876; Gamble, Fl. Madras 2 : 299. 1918.

Key to Varieties
1. Raceme 2-3-flowered, branches divaricate; leaflets
   narrow- oblanceolate .... var. **medicagenia**
1. Racemes 6-9-flowered, branches ascending; leaflets
   obovate-oblanceolate .... var. **neglecta**

Crotalaria medicaginea Lam. var. **medicaginea;** Munk, Reinwardtia 6 : 208. 1962; Ansari, *Crotalaria* in India 303. 2008.

Diffuse herbs, up to 50 cm long, sparsely pubescent. Leaves 3-foliolate, leaflets narrow-oblanceolate, 0.5-1 x 0.2-0.5 cm, glabrous above, pubescent below, emarginate, entire; petiole to 3 mm, petiolule to 1 mm. Flowers yellow with maroon streaks in terminal, 2-3-flowered racemes. Pods sub-globose, pubescent, slightly exceeding calyx, 2-seeded.

Common in open forests.

Fl. & Fr.: July-December

Aklaspur (KRN), *GVS* 25649 (MH); Nagarjunakonda (NLG), *KTH* 9845.

Crotalaria medicaginea Lam. var. **neglecta** (Wight & Arn.) Baker in Hook. f., Fl. Brit. India 2: 81. 1876; Gamble, Fl. Madras 1: 300. 1918; Ansari, *Crotalaria* in India 307. 2008. *Crotalaria neglecta* Wight & Arn., Prodr. Fl. Ind. Orient. 192. 1834.

Diffuse herbs with ascending branches, up to 30 cm long, sparsely appressed pubescent. Leaves 3-foliolate, leaflets obovate-oblanceolate, 0.5-1 × 0.3-1 cm, base obtuse, margin entire, apex retuse or emarginate; stipules sericeous, 1 mm, deciduous; petiole to 3 mm, petiolule 1 mm. Flowers yellow in short, terminal and leaf-opposed racemes; bracts linear, 1 mm, bracteoles subulate; calyx tube 1.5 mm, lobes lanceolate, to 3 mm; corolla yellow, standard orbicular, 6 × 5 mm, wings oblong, 6 × 2 mm, keels 6 × 2mm, beak spirally twisted; stamens monadelphous (10), staminal sheath 2 mm, filaments 3.5 mm; ovary 1.5 mm, style 5 mm. Pod subglobose, pubescent, 0.3 cm across, grooved and shortly beaked, 2-seeded; seeds shining, brown.

Common weed in cultivated fields and waste lands in Ranga Reddi district.

Fl. & Fr.: July-December.

Anantasagar RF (RR), *MSM* 10506.

Crotalaria montana B. Heyne ex Roth, Nov. Pl. Sp. 335. 1821; Verdc., Man. New Guinea Leg. 581. 1979; Ansari, *Crotalaria* in India 190. 2008. *C. linifolia* auct non L. f. 1781; Baker in Hook.f., Fl. Brit. India 2 : 72. 1876; Gamble, Fl. Madras 1: 294. 1918.

Suberect herbs, up to 60 cm high, branchlets appressed-pubescent. Leaves simple, oblong-oblanceolate, linear 1.5-3 × 0.2-0.7 cm, chartaceous, punctate, glabrous above, pubescent below, base cuneate, margin entire, apex obtuse, retuse, apiculate, petiole to 1.5 mm. Flowers yellow, in terminal 15 cm long racemes; bracts subulate, 2mm, bracteoles 3 mm, calyx pubescent, tube 1 mm, lobes to 8 mm, elongate, lanceolate; corolla slightly exserted, standard broad-ovate, 1 × 0.8 cm, wings oblong, 8 × 2 mm, keels ovate-oblong, 1 × 0.4 cm, beak spirally twisted; stamens monadelphous (10), staminal sheath 2.5 mm, filaments to 5 mm; ovary sessile, glabrous, 3.5 mm, style pubescent, 7 mm, peduncle to 3 cm, pedicel to 4 mm. Pod subglobose or ovoid, glabrous, hardly or slightly exceeding calyx; seeds 8-10, black when mature.

Occasional in moist deciduous forests in Karimnagar, Mahabubnagar and Khammam districts.

Fl. & Fr.: August-November.

Katukupalli vagu- Banderevu RF (KMM), *EC* 7445; Aklaspur (KRN), *GVS* 25619 (MH); Nellippakka RF (KMM), *RC* 99088 (BSID); Poochavaram RF (KMM), *RC* 98921 (BSID).

Note: Subbarao and Kumari (1968) reported this as *Crotalaria tecta* and Naqvi (2001) reported this as first report to Andhra Pradesh and Southern India. It is identified later as *Crotalaria montana,* the type of *C. linifolia* L.f. is undoubtedly a specimen of the species usually called *C. tecta* Heyne ex Roth (Polhill, 1976; Verdcourt, 1979; Rudd, 1991) (see Ansari, 2008).

Crotalaria mysorensis Roth, Nov. Pl. Sp. 338. 1821; Baker in Hook.f., Fl. Brit. India 2 : 70. 1876; Gamble, Fl. Madras 1: 295. 1918; Munk, Reinwardtia 6 : 210. 1962; Ansari, *Crotalaria* in India 105. 2008.

Erect undershrub, up to 1.5 m high, branchlets densely silvery hairy at first, later turning brown on drying. Stipules persistent, to 6 mm, conspicuous, linear-subulate Leaves simple, sessile, alternate, linear-elliptic, oblong, ovate, obovate, 3-8 × 1-1.5 cm, base cuneate, margin entire, apex apiculate, densely browny tomentose or pilose; petiole to 1.5 mm. Flowers in terminal racemes; peduncle to 3 cm, pedicel to 9 mm; calyx sericeous, tube to 4 mm, lobes elongated, lanceolate, to 1.4 cm; corolla yellow, exserted, standard ovate-oblong, 1.8 × 1 cm, yellow with maroon streaks without, wings oblong, 1.9 × 0.4 cm, keels ovate-oblong, 1.3 × 0.6 cm, beak spirally twisted; stamens monadelphous (10), staminal sheath 4 × 4 mm, filaments to 1 cm; ovary to 6 × 3 mm, glabrous, style pubescent, to 1.3. Pod 3-4 × 1.5 cm, obovoid, stipitate, twice as long as calyx; seeds 30-35, small, maroon, black, shining, 1.5 mm.

Occasional in the hill areas of Ranga Reddi and Karimnagar districts.

Fl. & Fr.: October-January

Lingampalle (RR), *MSM* 12753; Rechapalli (KRN), *GVS* 25694 (MH).

Crotalaria nana Burm. f., Fl. Indica 156. t. 48. f. 2. 1768; Baker in Hook.f., Fl. Brit. India 2 : 71. 1876; *p.p.;* Gamble, Fl. Madras 1: 294. 1918; Munk, Reinwardtia 6: 210. 1962; var. **nana;** Ansari, *Crotalaria* in India 112. 2008.

Erect, diffuse or prostrate, villous herbs, up to 45 cm high. Stipules minute. Leaves simple, oblong or oblanceolate, 1.5-3.5 × 0.3-0.8 cm, sparsely hairy above, villous beneath, base cuneate, margin entire, apex obtuse, apiculate, exstipulate; petiole to 3 mm. Flowers yellow in compact or subcapitate racemes or umbels, terminal, axillary or leaf-opposed, bracts 1.5 mm, bracteoles subulate, to 2 mm, upper two calyx lobes connate to half of their length, lobes to 2 mm, calyx tomentose; corolla much exserted, yellowish, standard oblong, 4.5 × 2 mm, wings oblong, 3.5 × 0.5 mm, keels ovate-oblong, 4 × 2 mm, beak spirally twisted; ovary stipitate, 3 mm, style pubescent, to 3 mm. Pod 0.5 × 0.4 cm, globose-ovoid, glabrous, seeds 6-10, sub-reniform, dark brown, glossy, compressed.

Rare in moist areas and in hills, in Mahabubnagar district.

Fl. & Fr.: September-October.

Amangal to Shadnagar (MBNR), *SRS* 104584 (BSID).

Crotalaria orixensis Rottl., Willd. in Ges. Naturf. Freunde Berlin Neue Schriften 4: 217. 1803; Baker in Hook.f., Fl. Brit. India 2 : 83. 1876; Gamble, Fl. Madras 1 : 300. 1918; Ansari, *Crotalaria* in India 126. 2008.

Diffuse, herbaceous, much-branched perennial, branchlets pubescent. Stipules linear, to 3 mm, persistent. Leaves 3-foliolate, leaflets oblong-obovate, 1-3 × 0.5-1.5 cm, pubescent below, glabrous above, base cuneate, margin entire, apex mucronate; petiole to 1.7 cm, petiolule to 1 mm. Flowers yellow in lateral, lax racemes, 1-15-flowered; peduncle to 2 cm, pedicel filiform, 1.2 cm; bracts ovate, up to 4 mm, apex acuminate, reflexed, bracteoles linear, appressed to calyx; calyx tube campanulate, 2mm, lobes to 5 mm; corolla slightly exserted, yellow, standard obovate, 5 × 4 mm, wings 5 × 1.5 mm, keels 6 × 3 mm, beaked; stamens monadelphous (10), staminal sheath 2 mm, filaments 5 mm; ovary 1.5 mm, style pubescent, 3.5 mm. Pod long-stipitate, oblong, turgid, 1.2 × 0.8 cm, glabrous, seeds 8-10, ovoid.

Common in dry deciduous forests and open forests and occasional weed in most districts.

Fl. & Fr.: October-March.

Sone (ADB), *TP* & *PVP* 4003; Keslapur (ADB), *GO* & *PVP* 4553; Banjapally (NZB), *BR* 9552; Kodimial (KRN), *GVS* 21834 (MH).

Crotalaria pallida Dryand in Ait., Hort. Kew. ed. 1. 3 : 20. 1789; Ansari, *Crotalaria* in India 336. 2008. *C. striata* DC., Prodr. 2. 131. 1825; Baker in Hook.f., Fl. Brit. India 2 : 84. 1876; Gamble, Fl. Madras 1: 302. 1918. *C. striata* DC. var *acutifolia* Trimen, Handb. Fl. Ceylon 2: 19. 1894; Gamble, Fl. Madras 1: 301. 1918.

Glabrescent undershrubs, 1-2 m high, branchlets puberulous. Leaves 3-foliolate, stipules absent, leaflets broadly elliptic or obovate, 3-6 × 2-4 cm, chartaceous, punctate, glaucous below, base cuneate to subacute, margin entire, apex mucronate; petiole to 4 cm, petiolule to 2 mm. Flowers yellow in terminal or lateral, elongated racemes, many-flowered; peduncle to 6 cm, pedicel to 4 mm; calyx appressed pubescent, tube to 3 mm, lobes lanceolate, to 5 mm; corolla exserted, glabrous; standard oblong-elliptic, 1.3 × 0.8 cm, wings elliptic, 1.1 × 0.35 cm, keels 1.2 × 0.8 cm, beak not twisted; stamens monadelphous (10), staminal sheath 1cm, filaments to 5 mm; ovary pubescent, to 6 mm, style pubescent, to 8 mm. Pod glabrescent, oblong, much exceeding calyx; seeds up to 50.

Occasional in forests in Karimnagar, Medak, Ranga Reddi, Mahabubnagar and Khammam districts.

Fl. & Fr.: September-April.

Narsapur (MDK), *BR* 11237; Mannanur (MBNR), *BR* & *BSS* 30622 (SKU), 30659 (BSID); Bandi RF (KMM), *RC* 104324 (BSID).

Crotalaria paniculata Willd., Sp. Pl. 3. 980. 1802; Baker in Hook.f., Fl. Brit. India 2 : 81. 1876; Gamble, Fl. Madras 1: 299. 1918.

Key to Varieties

1. Bracts lanceolate; stipules linear, 1 cm .... var. **paniculata**
1. Bracts cordate; stipules 0 .... var. **nagarjunakondensis**

Crotalaria paniculata Willd. var. **paniculata;** Ansari, *Crotalaria* in India 241. 2008.

Erect subshrub, up to 2 m high, branchlets viscid, glandular pubescent. Leaves simple, obovate, 2-3 × 0.5-1.5 cm, chartaceous, pubescent, base attenuate, margin entire, apex obtuse, apiculate, stipules linear-subulate, 1 cm; petiole to 1 mm. Flowers yellow in lax panicles, bracts lanceolate, up to 0.7 cm long, bracteoles linear, two on separate places on the pedicel, up to 0.7 cm; calyx pubescent, tube 2 mm, lobes lanceolate, to 8 mm; corolla exserted, reddish yellow, standard obovate, 2.2 × 2.1 cm, pubescent, obovate, 1.7 × 0.7 cm, keels ovate-oblong, 1.7 × 0.9 cm; stamens monadelphous (10), staminal sheath 6 mm, filaments 8 mm; ovary sessile, 3 × 1 mm, style geniculate, 1.7 cm. Pod ellipsoid, 1 × 0.5 cm, sparsely pubescent, horned, hardly exceeding calyx; seeds 1 or 2.

Occasional in dry tracts of Nalgonda district.

Fl. & Fr.: Through the year.

Bottichalami (NLG), *KMS* 9783 (CAL, MH); Nagarjunasagar (NLG), *BVRS* 59, 302 (NU).

Crotalaria paniculata Willd. var. **nagarjunakondensis** Thoth., Bull. Bot. Surv. India 6: 67-68. 1964; Ansari, *Crotalaria* in India 243. 2008.

Undershrubs, up to 1 m high, branchlets fulvous villous. Leaves simple, alternate, exstipulate, ovate or elliptic to obovate, 3.5-6.9 × 1.5-2.9 cm, villous on both sides, base cuneate, apex acute, margin entire, lateral nerves 6-7 pairs; petiole to 3 mm. Flowers yellow, in 13-20 cm long panicled inflorescence; bracts and bracteoles distinctly cordate, cuspidate, recurved, bracts 5-6 × 3-5.5 mm, bracteoles 2, smaller than bracts; calyx tube 7-10 mm, lobes oblong, densely brown silky; corolla yellow, standard ovate-orbicular, 1.3 – 1.7 cm long with short claw, brown silky on the back, wing obovate to oblong, keels connate, much incurved; stamens monadelphous (10); ovary pilose, style incurved at base. Pod ovoid, 10-12 × 6-7 mm, inflated with a recurved persistent style, pilose; 1-2-seeded, seeds black.

Occasional in Nagarjunakonda valley.

Nagarjunakonda valley (GNT), *KTH* 9764 (CAL), *BVRS* 302 (NU).

Note: *C. paniculata* Willd. var. *nagarjunakondensis* Thoth. differs brom *C. paniculata* Willd. var. *paniculata* in having distinctly cordate bracts shortly petiolate leaves and in the absence of stipules.

Crotalaria prostrata Rottl. ex Willd., Enum Pl. Hort. Berol. 2: 747. 1809; Baker in
   Hook.f., Fl. Brit. India 2 : 67. 1876; Gamble, Fl. Madras 1: 292. 1918; Munk,
   Reinwardtia 6: 211. 1962; var. **prostrata**; Ansari, *Crotalaria* in India 197. 2008.

Prostrate herbs, about 40 cm long, branchlets terete, brown-pubescent. Leaves
simple, oblong-elliptic, 1.5-3 × 0.5-1 cm, inequilateral, chartaceous, appressed-
pubescent, nerves obscure, base oblique, margin ciliate, apex subacute, petiole 1.5
mm. Flowers yellow, in axillary or lateral, 2-4-flowered racemes, bracts subulate, 2
mm, persistent, bracteoles linear, 2 mm; calyx pubescent, tube 2 mm, lobes linear,
acuminate, to 4 mm; corolla exserted, yellow, standard oblong, 5 × 4 mm, apex deeply
emarginated, wings oblong, 4 × 1.2 mm, keels 4 × 1 mm, beak spirally twisted; stamens
monadelphous (10), staminal sheath 1.5 × 2.5 mm, filaments 2 mm; ovary sessile,
glabrous; style 3 mm, sparsely pubescent. Pods oblong, 2 × 0.5 mm exserted, shortly
stipitate, glabrous, seeds 16-20.

Occasional in Adilabad, Nizamabad, Medak, Khammam, Karimnagar and
Mahabubnagar districts.

Fl. & Fr.: August-February

Neredugunda (ADB), *GO* & *DAM* 5143; Annaram (NZB), *BR* & *CPR* 7167;
Mamidipally RF (NZB), *BR* & *PSPB* 9604; Narsapur RF (MDK), *TP* & *CP* 11977,
14082; Bandiveru RF (KMM), *EC* 7439; Aklaspur (KRN), *GVS* 25621 (MH); Kunavaram
RF (KMM), *RC* 98785 (BSID); Borapuram (MBNR), *BR* & *BSS* 32364 (BSID).

Note: All Indian workers and Index Kewensis credit Roxburgh as the author of the
   species. However it is Willdenow who had published it earlier in 1809,
   validating Rottler's epithet.

Crotalaria pusilla Heyne ex Roth, Nov. Pl. Sp. 335. 1821; Baker in Hook.f., Fl. Brit.
   India 2 : 70. 1876; Gamble, Fl. Madras 1 : 296. 1918; Ansari, *Crotalaria* in India
   203. 2008.

Diffusely branched herbs from base, up to 15 cm high, branchlets terete, brown-
pubescent. Leaves exstipulate; linear-elliptic, 1-2 × 0.2-0.4 cm, chartaceous, appressed-
sericeous on both sides, base and apex obtuse, margin entire, brown-pubescent, petiole
to 2 mm. Flowers yellow in terminal 8-10-flowered, 4 cm long racemes; peduncle to 2
cm, pedicel to 2 mm; calyx tube 0.5 mm, lobes free, lanceolate, to 2 mm; corolla exserted,
standard narrow-oblong, 3 × 0.8 mm, wings 3 × 0.5 mm, keels obovate, 3.5 × 1.8 mm,
beak spirally twisted; stamens monadelphous (10), staminal sheath 1 mm, filaments
2.5 mm; ovary subsessile, globose, 1.8 mm, hairy, style geniculate, 1.5 mm. Pod
exserted, 0.5 × 0.3 cm, ovoid-orbicular, brown-villous, seeds 4-8, subreniform, glabrous.

Common on dry waste lands and open forests of most districts.

Fl. & Fr.: September- January.

Potchera (ADB), *GO* & *DAM* 5086; Patharajampet (NZB), *BR* & *CPR* 7156;
Kamareddy (NZB), *BR* & *CPR* 7156; Pegarikutta-Narsapur (MDK), *KMS* 6626 (MH);
Mohammadabad RF (RR), *MSM* 10482; Yedupayalu RF (MDK), *RG* 113449 (BSID).

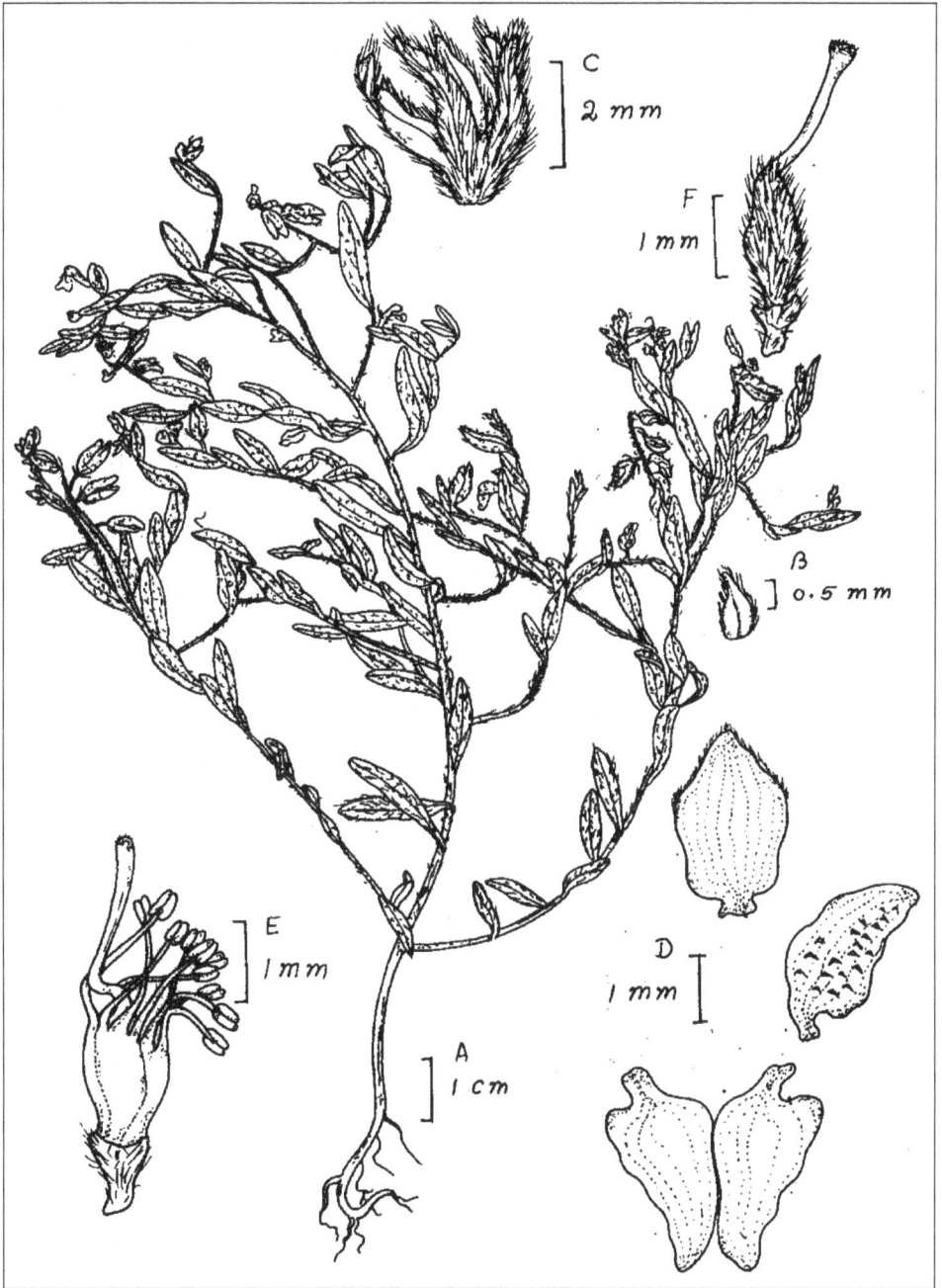

Figure 21: Crotalaria pusilla Heyne ex Roth
A. Twig, B. Bract, C. Calyx, D. Corolla, E. Staminal column with pistil, F. Fruit.

Crotalaria quinquefolia L., Sp. Pl. 716. 1753; Baker in Hook.f., Fl. Brit. India 2: 84. 1875; Gamble, Fl. Madras 1: 301. 1918; Ansari, *Crotalaria* in India 299. 2008.

Erect annual herbs, to 1 m high; branchlets with minute appressed pubescence. Leaves 5-foliolate; leaflets oblong-lanceolate, 2-9 × 0.5-2 cm, petioles to 6 cm; stipules linear lanceolate. Flowers yellow with purple striations, in terminal racemes; bracts to 5 mm long, foliaceous, lanceolate, reflexed; pedicel about 1 cm long; calyx to 1 cm long; corolla twice as long as calyx, glabrous. Pod oblong, stalked, 2-4 cm long, glabrous, many-seeded.

Rare in Karimnagar district (Naqvi, 2001).

Fl. & Fr.: August – December.

Crotalaria ramosissima Roxb., Fl. Ind. 3 : 268. 1832; Baker in Hook.f., Fl. Brit. India 2: 80. 1875; Gamble, Fl. Madras 1: 299. 1918; Ansari, *Crotalaria* in India 245. 2008.

Much branched erect herbs, up to 60 cm high, branchlets clothed with white silky hairs. Leaves exstipulate, simple, linear-oblong, lanceolate, 1-2.5 × 0.2-0.7 cm, densely silky hairy on both sides, base cuneate, margin entire, apex obtuse; petiole to 1 mm. Flowers yellow in terminal, panicled racemes, peduncle to 1.5 cm, pedicel to 5 mm, bracts ovate, acuminate, curled back, drying black; bracteoles ovate, recurved; calyx hairy, tube to 3 mm, covered with hispid secretion, lobes linear, to 7 mm; corolla yellow, standard sericeous, 1.1 × 0.5 cm, wings 7 × 2.5 mm, keels 1 × 0.4 cm, beak recurved; stamens monadelphous (10), staminal sheath 4 mm, filaments 7 mm; ovary pubescent, 1.2 × 0.3 cm, style glabrous, 8 mm. Pods sessile, ovoid, silky, as long as calyx, 1-seeded, seed brownish black, polished.

Common in wastelands and open forests Karimnagar, Warangal, Hyderabad, Khammam, Nalgonda and Mahabubnagar districts.

Fl. & Fr.: September-April.

Pakhal (WGL), *KMS* 11693 (CAL); Aklaspur (KRN), *GVS* 22488 (CAL, MH); Nagarjunakonda (NLG), *KTH* 9741 (CAL); Nagarjunasagar (NLG), *KMS* 9734 (CAL); Gowravaram (KMM), *PVS* 76948 (BSID).

Crotalaria retusa L., Sp. Pl. 715. 1753; Baker in Hook.f., Fl. Brit. India 2 : 75. 1876; Gamble, Fl. Madras 1: 293. 1918; Munk, Reinwardtia 6: 212. 1962; var. **retusa**; Ansari, *Crotalaria* in India 275. 2008.

Erect herbaceous undershrubs, up to 1.5 m high, branches furrowed, appressed white pubescent. Stipules small, subulate. Leaves obovate-oblanceolate, chartaceous, 7-10 × 1-3 cm, base narrowed, margin entire, apex obtuse or retuse; petiole 3 mm. Flowers yellow in terminal, 15-flowered racemes, peduncle to 3 cm, pedicel to 6 mm; calyx appressed pubescent, tube 5 mm, lobes lanceolate, upper lobes 8 mm, lower lobes 4 mm; corolla yellow, standard orbicular, 2.1 × 2.4 cm, wings obovate, 1.4 × 0.6 cm, keels ovate, 1.7 × 0.7 cm, beak twisted; stamens monadelphous (10), staminal sheath 1 × 0.7 cm, filaments to 7 mm; ovary 6 mm, style curved, 1.3 mm, pubescent on both sides. Pod oblong, to 4.5 × 1.6 cm, terete, exerted, shortly stipitate, glabrous, 10-15-seeded; seeds black.

Common in Adilabad, Medak, Karimnagar, Khammam and Nalgonda districts, in fields, waste places and open forest lands.

Fl. & Fr.: November-April

Vempalli (ADB), *GO* 4432; Choutkoor (MDK), *TP* & *CP* 11921; ICRISAT (MDK), *LJGV* 2594 (CAL), Aklaspur (KRN), *GVS* 22521 (CAL); Bottichalimi (NLG), *KMS* 9778 (CAL, MH); Musalivera RF (KMM), *RR* 112616 (BSID).

Crotalaria trifoliastrum Willd., Sp. Pl. 3. 983. 1802; Baker in Hook.f., Fl. Brit. India 2 : 82. 1872; Gamble, Fl. Madras 1: 300. 1918; Ansari, *Crotalaria* in India 312. 2008.

Perennial herbs, up to 1 m high with erecto-patent, slender branches, appressed pubescent. Stipules setaceous, minute. Leaves 3-foliolate, leaflets 1.2-2.5 cm long, obovate-oblong, membranous, base cuneate, margin entire, apex retuse-emarginate, glabrous above, silky below; petiole to 1.5 cm, petiolule to 1.5 mm. Racemes terminal and lateral, 20-30 cm long, 12-40-flowered, bracts setaceous; calyx pubescent, tube 2 mm, lobes linear, to 4 mm; corolla exserted, yellow, standard 1.3 × 1 cm, wings 1 × 0.4 cm, keels 1.3 × 0.4 cm, beak twisted; stamens monadelphous (10); ovary pubescent, 4 mm, style glabrous, 1 cm. Pods sessile, subquadrangular, 6 x 5 mm, beaked, thinly hairy; seeds 2.

Occasional in moist deciduous and dry places in Warangal, Nizamabad and Nalgonda districts.

Fl. & Fr.: August-September

Velutla RF (NZB), *BR* 9530; Bottichalami (NLG), *KMS* 9860 (CAL).

Crotalaria triquetra Dalzell in Hooker's J. Bot. Kew. Gard. Misc. 2: 34. 1850; Baker in Hook.f., Fl. Brit. India 2 : 71. 1876; Gamble, Fl. Madras 1: 296. 1918; Ansari, *Crotalaria* in India 145. 2008.

Suffruticose herbs, up to 50 cm high, branchlets 3-angled, obscurely pubescent. Leaves simple, elliptic-oblong or ovate-oblong, to 3 × 1 cm, obtuse, membranous, pale green, sparsely hairy beneath, stipules minute, lanceolate, reflexed. Flowers yellow in terminal and lateral laxly 2-3-flowered racemes, pedicel to 1 cm; calyx 1 cm long, silky, lobes subequal, triangular, acute; corolla exserted, pale yellow, standard broadly ovate, subacute, hairy on outside, wings oblong. Pod oblong-cylindrical, 1.5-2 cm long, thinly coated with short yellow-brown silky hairs, 15-20-seeded; seeds subreniform or nearly spherical, smooth, glabrous, dark-brown.

Rare in Mahabubnagar distircts.

Fl. & Fr.: December-February.

Crotalaria verrucosa L., Sp. Pl. 715. 1753; Baker in Hook. f., Fl. Brit. India 2: 77. 1876; Gamble, Fl. Madras 1: 297. 1918; Munk, Reinwardtia 6: 217. 1962; Ansari, *Crotalaria* in India 287. 2008.

Vern.: *Gilli gicha.*

Erect herbs, up to 1m high, branchlets 4-angled, appressed pubescent. Leaves simple, ovate-deltoid or rhomboid, 4-8 × 2.5-5 cm, base acute, margin entire, apex obtuse-acute, sparsely pubescent above, less below, stipules foliaceous, semilunate; petiole to 5 mm. Flowers pale purple with white dark purple veins or blue in terminal racemes; bracts 4 mm, subulate, bracteoles to 5 mm; peduncle to 3 cm, pedicel to 5 mm; calyx pubescent, tube 4 mm, lobes lanceolate, to 8 mm; corolla bluish white, standard broad-ovate, 1 × 1.1 cm, wings obovate, 1.1 × 0.8 cm, keels 1.3 × 0.7 cm, beak spirally twisted; stamens monadelphous (10), staminal sheath 5 x 5 mm, filaments to 7.5 mm; ovary appressed pubescent, 4 mm, style pubescent on the inner side. Pod inflated, cylindric, to 3.5 × 1 cm, villous, seeds 16-20.

Common weed of road sides, waste places, gardens and fields in all districts.

Fl. & Fr.: July-December

Kamareddy (NZB), *BR* & *KH* 7147; Narsapur forest nursery (MDK), *BR* & *CP* 11218; Siddipet (MDK), *CP* 11378; Medak (MDK), *CP* 11799; Central University campus (RR), *TP* & MSM 15148; Azampur (NLG), *BVRS* 478 (NU); Amangal to Shadnagar (MBNR), *SRS* 104567 (BSID); Kinnerasani WLS (KMM), *RR* 11298 (BSID); Katasintha (HYD), *KMS* 6013 (CAL).

Crotalaria willdenowiana DC., Prodr. 2 : 134. 1825; subsp. **willdenowiana;** Baker in
   Hook. f., Fl. Brit. India 2 : 81. 1876; Gamble, Fl. Madras 1: 300. 1918; Ansari,
   Crotalaria in India 2316. 2008.

Subshrubs, up to 75 cm high, branchlets sericeous. Leaves 3-foliolate, leaflets oblong-oblanceolate, 1-2 × 0.5 cm, sericeous, base cuneate, margin entire, apex emarginate, apiculate; stipules 4 mm, setaceous, persistent; petiole to 6 mm, petiolule to 1 mm. Flowers yellow in terminal and lateral, 3 cm long, 6-9-flowered racemes, bracts minute, linear, setaceous, persistent; calyx pubescent, tube 1.5 mm, lobes lanceolate, to 2 mm; corolla thrice the calyx, standard orbicular, 1 × 0.8 cm, wings oblong, 8 × 3 mm, keels 8 × 4 mm, beak spirally twisted; stamens monadelphous (10), staminal sheath 2 mm, filaments 6 mm; ovary sessile, pubescent, 3 mm, style pubescent, 6 mm. Pod sessile, subglobose, 0.5 cm, pubescent, twice as long as calyx, seeds 1 or 2.

Occasional in Mahabubnagar and Karimnagar (Naqvi, 2001) districts.

Fl. & Fr.: November-March.

CULLEN Medik.

Cullen corylifolium (L.) Medik., Vorles. Churfalz. Phys.-Oken. Ges. 2: 381. 1787.
   Psoralea corylifolia L., Sp. Pl. 764. 1753; Baker in Hook. f., Fl. Brit. India 2: 103.
   1876; Gamble, Fl. Madras 1: 314. 1918.

Vern: *Bapunga.*

An erect, annual undershrub, up to 1 m high, branchlets glandular-pubescent, warty. Leaves simple, ovate, 5-9 × 3.5-6 cm, chartaceous, punctate, pubescent, base obtuse, margin dentate, apex acute, petiole to 1.5 cm. Flowers purplish-violet in axillary racemes, flowers apically aggregated on elongated peduncles, subsessile, congested; bracts to 5 mm; calyx campanulate, tube to 3 mm, gland dotted, glabrous, lobes lanceolate, to 3 mm; corolla exserted, petals clawed, standard oblong-obovate, 6 × 4 m, wings oblong, 5 × 2 mm, keels connate, 4 mm; stamens diadelphous (9+1), staminal

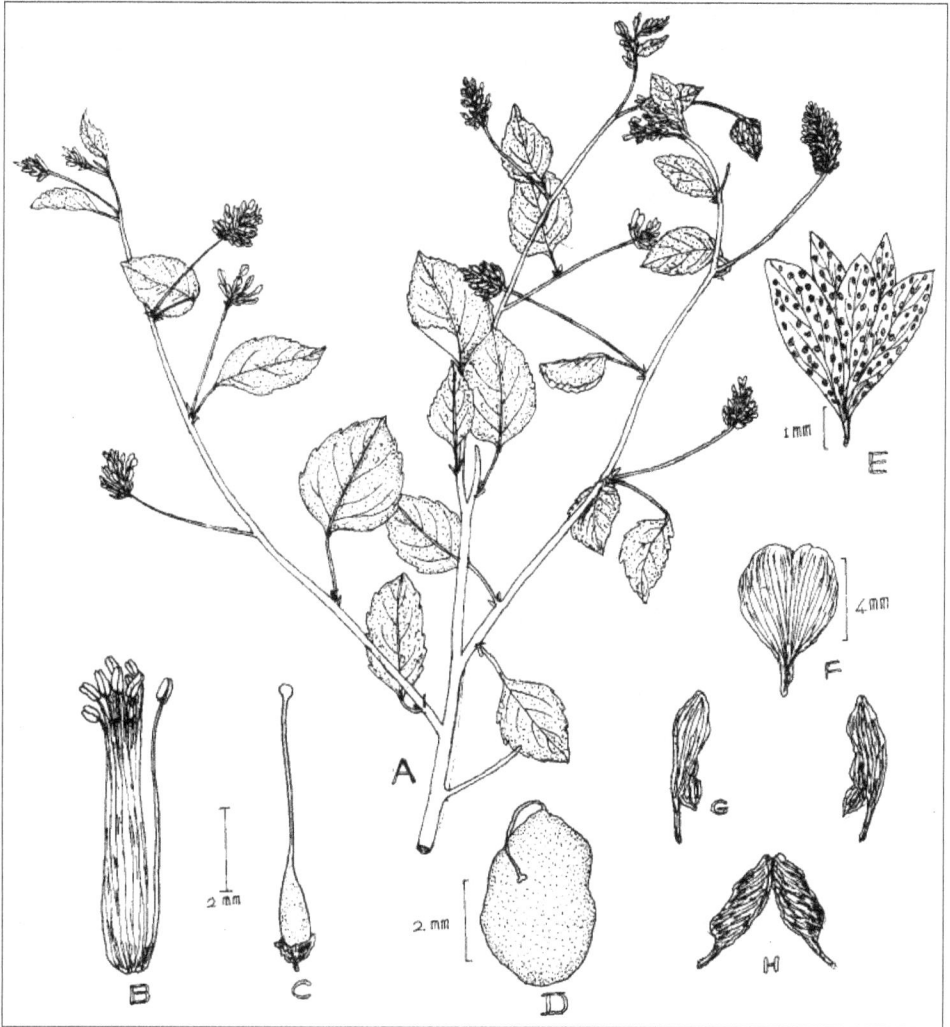

Figure 22: Cullen corylifolium (L.) Medik. (=*Psoralea corylifolia* L.)

A. Twig, B. Staminal column, C. Pistil, D. Pod, E. Calyx, F. Standard Petal, G. Wing petals, H. Keel petals.

sheath 4 mm, filaments to 1.5 mm; ovary stipitate, style 4 mm, glabrous. Pods ovoid or oblong, indehiscent; seed solitary, reniform.

Common weed of roadsides and waste places in most of the districts.

Fl. & Fr.: December-May.

Itkial (ADB), *TP* & *PVP* 4132; Kandakurthi (NZB), *BR* 7267; Gundaram (NZB), *BR* & *KH* 9687; Manjira barrage (MDK), *BR* & *CP* 14088; Naskal (RR), *MSM* & *PRSR* 15110; Nelkal (NLG), *BVRS* 451 (NU); Aklaspur (KRN), *GVS* 22525 (MH); Pakhal (WGL), *GVS* 11573 (MH); Way to Poosugundu (KMM), *PVS* 84044 (BSID).

CYAMOPSIS DC.

Cyamopsis tetragonaloba (L.) Taubert in Engl., Nat. Pflanzenfam. 3(3): 259. 1894; Gamble, Fl. Madras 1: 304. 1918. *Psoralea tetragonaloba* L., Mant. Pl. 104. 1767. *Cyamopsis psoraloides* (Lam.) DC., Mem. Legum. 230. 1826; Baker in Hook.f., Fl. Brit. India 2: 92. 1876.

Vern.: *Chowlekayalu, Matikkayalu, Goruchikkudu.*

Erect, annual herb, to 75 cm high; branchlets grooved, clothed with appressed white hairs. Leaves 3-foliolate; leaflets lanceolate or elliptic, 5-7 × 2-4 cm, margin sharply dentate, apex acute. Flowers pale pink in dense axillary racemes. Pods subtetragonous, thick, fleshy, to 15 cm long, many-seeded.

Commonly cultivated in all districts.

Fl. & Fr.: August – February.

DALBERGIA L. f. *nom. cons.*

1. Stamens 9, monadelphous; staminal tube split along the upper side only :
    2. Style short, thick; standard flat, erect; wings and keel petals
        not spurred .... **D. sissoo**
    2. Style elongate, slender; standard reflexed; wings and keel
        petals spurred :
        3. Leaflets acute; inflorescence of elongate panicles:
            4. Trees .... **D. latifolia**
            4. Stragglers .... **D. rubiginosa**
        3. Leaflets obtuse or emarginate; infloresence fascicled
            or corymbose panicles .... **D. sissoides**
1. Stamens 10; staminal tube split above and below, so that the
    stamens are in 2 bundles of 5 each. :
    5. Climbing shrubs .... **D. volubilis**
    5. Trees or erect shrubs .... **D. lanceolaria**

Dalbergia lanceolaria L. f., Suppl. Pl. 316. 1781; Baker in Hook.f., Fl. Brit. India 2 : 235. 1876; Gamble, Fl. Madras 1: 383. 1918. subsp. **lanceolaria**

Vern : *Yerra pastaru, Patsari.*

Large, deciduous, densely foliaceous trees, up to 15 m high, bark smooth, grey; branchlets terete, striate, brown tomentose when young. Leaves pinnate, leaflets 11 to 15, ovate-oblong, 4 × 1.8 cm, base cuneate-obtuse, margin entire, apex obtuse, coriaceous, lateral nerves parallel, reticulation prominent below; petiole to 3 cm long, petiolule 6 mm. Flowers in axillary panicles; pedicel 3 mm; calyx puberulous, tube 4 mm, lobes to 2 mm; corolla pale pink, standard broad ovate, 1.2 × 1 cm, wings 8 × 4.5 mm, keels 6 × 3 mm, stamens isodelphous (5+5), staminal sheath 4 mm, filaments 2.5 mm; ovary sparsely pubescent, 3 mm, style 2 mm. Pod long stipitate, oblong, tapering at both ends, to 9 × 2 cm, dark green, 1-3-seeded.

Common in dry forests of Adilabad, Medak, Ranga Reddi, Khammam, Nalgonda and Mahabubnagar districts.

Fl. & Fr.: March-December.

Bheemaram RF (ADB), *TP* & *GO* 5454; Wankidi RF (ADB), *PVP* 9487; Zaheerabad (MDK), *CP* 11588; Anantagiri (RR), *SSR* & *KSM* 18513; Rathamhutta hills (KMM), *RC* 10234 (BSID).

Key to Subspecies

1. Leaflets 9-13, oblong-obovate; standard longer
   than broad; pod short stipitate .... **D. lanceolaria** subsp. **paniculata**

1. Leaflets 11-15, ovate-oblong or orbicular, standard broader
   than long; pod long stipitate .... **D. lanceolaria** subsp. **lanceolaria**

Dalbergia lanceolaria L.f. subsp. **paniculata** (Roxb.) Thoth., Bull. Bot. Surv. India 25: 171. 1983. *Dalbergia paniculata* Roxb., Pl. Coromandel t. 114. 1799; Baker in Hook. f., Fl. Brit. India 2 : 236. 1876; Gamble, Fl. Madras 1: 383. 1918.

Vern. : *Patsaru.*

Large, densely foliaceous, deciduous trees, up to 15 m high, bark smooth, grey, branches densely pubescent. Leaves odd-pinnate, leaflets 9-13, alternate, oblong-obovate, 2-3.5 × 1-2 cm, chartaceous, base cuneate-obtuse, margin entire, apex rounded or emarginated, turn black when drying; petiole to 2 cm, petiolule to 3 mm. Flowers white in axillary, branched panicles, peduncles and pedicles rusty-pubescent; calyx rusty tomentose, tube 4 mm, lobes ciliate, to 3.5 mm; corolla white, standard ovate-oblong, 1 × 0.5 cm, wing 7.5 × 3 mm, keels 7 × 3 mm; ovary glabrous, 3 mm, style 2 mm. Pod to 7 × 2 cm, lanceolate, shortly stipitate, lanceolate, base attenuate, apex acute, faintly nerved; seeds 1-2.

Common in deciduous forests of all districts.

Fl. & Fr.: May- December

Rampur (ADB), *SSR* 19310; Bhimnagar RF (NZB), *TP* & *BR* 6209; Narsapur RF (MDK), *TP* & *CP* 11968; Eturnagaram (WGL), *SSR* 18532; Mallelatheertham (MBNR), *BR* & *SKB* 30716, 30709; Kodimial (KRN), *GVS* 20140 (MH).

Dalbergia latifolia Roxb., Pl. Coromandel t. 113. 1798; Baker in Hook. f., Fl. Brit. India 2 : 231. 1876; Gamble, Fl. Madras 1: 383. 1918.

Vern.: *Jittegi.*

Large deciduous trees, up to 20 m high, bark light grey, rough, branchlets terete, striate, glabrous. Leaves odd pinnate, leaflets 3-4 pairs, alternate, obovate-orbicular, 3.5-10 × 3.2-7.5 cm, coriaceous, glabrous above, puberulous beneath, base subacute-obtuse, margin entire, apex obtuse; petiole to 6 cm, petiolule 8 mm. Flowers in axillary, much branched panicles on old wood; peduncle 1 cm, pedicel 1 mm; calyx lobes subequal, tube 4.5 mm, lobes to 1.5 mm; corolla white with pale yellow tinge, standard 7 × 4.5 mm, obovate, wings 7 × 2.5 mm, keels 6.5 × 3.5 mm; stamens monadelphous (9), staminal sheath 5 mm, filaments 2.5 mm; ovary glabrous, 4 mm, style 2 mm. Pod oblong-lanceolate, to 4.8 × 2 cm, base obtuse, prominently nerved, glabrous; seeds 1-3.

Common in forests of all districts.

Fl. & Fr.: July-February.

Asifabad (ADB), *GO* 4478; Mudheli RF (NZB), *BR* & *CPR* 9263; Gundaram RF (NZB), *BR* & *KH* 9678; Narsapur RF (MDK), *TP* & *CP* 14068; Dharur RF (RR), *TP* & *MSM* 11096; Pakhal-near P.W.D. rest house (WGL), *ANH* 15961 (MH); Pagari Katta East-Narsapur (MDK), *KMS* 6696 (CAL); Aklaspur (KRN), *GVS* 22502 (MH).

Dalbergia sissoides Graham ex Wight & Arn., Prodr. Fl. Ind. Orient. 265. 1834; Gamble, Fl. Madras 1: 381. 1918. *D. latifolia* var. *sissoides* (Graham ex Wight & Arn.) Baker in Hook. f., Fl. Brit. India 2 : 231. 1876.

Deciduous trees, up to 10 m high. Leaves 15-20 cm long, leaflets 7-9, elliptic-oblong or obovate, 6 × 3 cm, base cuneate, margin entire, apex acute, shining above. Flowers white, in lax broad panicles; calyx glabrous; standard petal reflexed; stamens 9, monadelphous. Pod acute at apex, stipitate, reticulately veined, 9 × 1.5 cm.

Rare in Narsapur town in Medak district.

Fl. & Fr.: March-August.

Narsapur town (MDK), *KMS* 7961 (CAL).

Dalbergia sissoo Roxb. ex DC., Prodr. 2: 416. 1825; Baker in Hook.f., Fl. Brit. India 2 : 237. 1876; Gamble, Fl. Madras 1: 383. 1918.

Vern. : *Sissoo.*

Densely foliaceous, semi-evergreen trees, up to 30 m high, bark thick, grey, exfoliating in narrow longitudinal strips, branchlets pilose. Leaves imparipinnate, leaflets 3-5, alternate, obovate-suborbicular, 5-7 × 3-4.5 cm, pubescent when young, glabrous when full grown, coriaceous, base cuneate, margin entire, apex cuspidate-acuminate; petiole to 3 cm long, petiolue to 5 mm. Flowers creamish in terminal panicles, peduncle 2 cm, pedicel 1 mm; calyx tube 3.5 mm, lobes to 2 mm; corolla yellowish white, standard obovate, 7.5 × 4 mm, wings 6.5 mm, keels 7 × 1.2 mm; stamens 9, monadelphous, staminal sheath 5 mm, filaments 2 mm; ovary pubescent,

2 mm, style 4 mm. Pod oblong, 5-7.5 cm long, pale brown, glabrous, apex acute, apiculate, faintly nerved; seeds 1-4, kidney-shaped, thin, flat, light brown.

Planted as avenue tree in all districts.

Fl. & Fr.: January-September.

Asifabad (ADB), *GO* 4478; Hajipur nursery (NZB), *TP* & *BR* 6273; NSF (MDK), *BR* & *CP* 11648; Kanmankalva RF (RR), *MSM* & *KH* 10311; Aklaspur (KRN), *GVS* 20230 (CAL); Narsapur town (MDK), *KMS* 7961 (MH); Mamillgudem (KMM), *RR* 105918 (BSID); Parvathapur (MBNR), *SRS* 104468 (BSID); Jakaram (WGL), *RKP* 105323 (BSID).

Dalbergia volubilis Roxb., Pl. Coromandel t. 191. 1805; Baker in Hook.f., Fl. Brit. India 2 : 235. 1876; Gamble, Fl. Madras 1: 384. 1918.

Large, woody climbers, up to 8 m long, branches thickened at places and twisted into spiral loops, branchlets glabrous. Stipules small. Leaves imparipinnate, leaflets 7-13, oblong or obovate, 2-4.5 × 1.5-2.5 cm, base obtuse or emarginate, margin entire, apex obtuse, glabrous, coriaceous; petiole to 2.5 cm, petiolule 4 mm. Flowers in axillary and terminal panicles, peduncle to 2 cm, pedicel to 6 mm; calyx campanulate, tube 1.2 mm, lobes to 3.5 mm; corolla pale blue, or pale purple, standard suborbicular-oblong, 6.5 x 7.5 mm, wings 7 × 3 mm, keels 6 × 2 mm, stamens isodiadelphous (5+5), staminal sheath 4 × 2 mm, filaments 1 mm; ovary puberulous, to 3 mm, style 1.5 mm. Pods 4-10 × 1.5 cm, linear-oblong, narrowed at the base, obtuse at the tip, glabrous, 1 to 2-seeded; seeds ellipsoid, slightly reniform, brown.

Common in deciduous forests of Karimnagar, Warangal and Mahabubnagar districts.

Fl. & Fr.: January-May.

Mahadevpur (KRN), *SLK* 70856 (MH); Gangaram (WGL), *PVS* 76932 (BSID).

DERRIS Lour. *nom. cons.*

Derris scandens (Roxb.) Benth., J. Linn. Soc. Bot. 4 (Suppl.) 103. 1860; Baker in Hook.f., Fl. Brit. India 2 : 240. 1876; Gamble, Fl. Madras 1: 387. 1918; var. **scandens** Thoth. in Fasc. Fl. Ind. 8 : 28. 1982. *Dalbergia scandens* Roxb., Pl. Coromandel t. 192. 1805.

Vern.: *Nalla Tiga.*

Large stragglers or lianas, up to 10 m long, branches closely covered with reddish coloured, round lenticels, young parts rusty-tomentose. Stipules small. Leaves odd-pinnate, leaflets 7-13, oblong, oblong-obovate, 2-5 × 1-3, glabrous above, puberulous below, base cuneate, acute, margin entire, apex obtuse-retuse; petiole to 2 cm, petiolule to 3 mm. Flowers pale pink scented in terminal and axillary panicles; peduncle to 3 cm, pedicel slender, 5 mm, bracts small, caducous; calyx campanulate, tube 4 mm, lobes very short; corolla cream-coloured, exserted, standard obovate, broad, 8 × 6.5 mm, wings obliquely oblong, 7 × 2.5 mm, keels obtuse, incurved, 6.5 × 2.5 cm; stamens monadelphous (10), staminal sheath 8 × 2 mm, filaments 3 mm, ovary sessile, 5 mm, style 3 mm. Pod oblong, glabrous, winged along upper suture, tapering at both ends; seeds 1-4, compressed, reniform.

Common almost in all forest districts.

Fl. & Fr.: July-September.

Sirpur (ADB), *GO* & *PVP* 4769; Kundaram RF (ADB), *TP* & *GO* 5459; Servapur (NZB), *BR* 9541; Sirnapally river (NZB), *BR* & *GO* 9079; Gnanapur (MDK), *CP* 11789; Anantasagar RF (RR), *MSM* 10504; Dharur (RR), *MSM* & *TP* 11098; Pakhal (WGL), *KMS* 13195 (MH); Bhupatipur (KRN), *GVS* 22479 (MH); Rampur (MBNR), *SKB* & *BSS* 32631 (SKU, BSID); way to Gallagand (KMM), *PVS* 84070 (BSID); Parnasala RF (KMM), *RC* 98773 (BSID).

DESMODIASTRUM Prain

Desmodiastrum racemosum (Benth.) A.Pramanik & Thoth., J. Indian Bot. Soc. 65: 376. 1986. var. **racemosum.** *Alysicarpus racemosus* Benth., Linnaea 24 : 162. 1851; Gamble, Fl. Madras 1: 339. 1918. *A. belgaumensis* Wight var. *racemosus* (Benth.) Baker in Hook. f., Fl. Brit. India 1: 160. 1876.

Diffuse or erect, annual herbs, up to 40 cm high, branches covered with sparse, appressed hairs. Leaves 1-foliolate below, 3-foliolate above; stipules lanceolate, shorter than petioles, leaflets obovate or suborbicular, ca 2.5 × 1.5 cm, base round, margin entire, apex obtuse, petiole to 5 mm. Flowers rose pink, turning blue, up to 7 cm long in terminal racemes; which is covered with hooked hairs; calyx to 9 mm, lobes valvate in fruit covered with golden-brown hairs on back and along margin. Pod moniliform, slightly exserted, 0.7-1 × 0.3 cm, 2-5-jointed, glabrous.

Rare in waste lands in Ranga Reddi district.

Fl. & Fr.: August-November.

DESMODIUM Desvaux *nom. cons.*

1. Trees .... **D. oojeinense**
1. Herbs:
    2. Leaves 1-foliolate :
        3. Flowers in ca 2.5 cm long racemes .... **D. benthamii**
        3. Flowers in >3 cm long racemes :
            4. Leaves fulvous-tomentose beneath .... **D. velutinum**
            4. Leaves sparsely appressed grey-pubescent beneath ... **D. gangeticum**
    2. Leaves 3-foliolate (sometimes 1-foliolate in *D. alysicarpoides*) :
        5. Joints of pod twice longer than broad or longer .... **D. laxiflorum**
        5. Joints of pod generally as long as broad :
            6. Stipules foliaceous, subamplexicaul; pod clearly
               indented on both sutures .... **D. dichotomum**
            6. Stipules linear scarious; pod nearly straight along upper suture :
                7. Trailing herbs or shrubs .... **D. triflorum**

7. Erect herbs or undershrubs :

   8. Calyx incised halfway or less; glabrous or shortly appressed, pubescent .... **D. heterocarpon**

   8. Calyx incised more than half-way, with long hairs .... **D. alysicarpoides**

Desmodium alysicarpoides Meeuwen, Reinwardtia 6: 246. 1962. *Alysicarpus parviflorus* Dalzell in Hooker's J. Bot. Kew Gard. Misc. 3. 211. 1851; Gamble, Fl. Madras 1: 348. 1918. *Desmodium parviflorum* (Dalzell) Baker in Hook. f., Fl. Brit. Inida 2 : 172. 1876; non Mart. & Galeotti 1843.

Erect herbs, up to 30 cm high, branches sparsely pubescent. Leaves 1-3-foliolate, lateral leaflets much smaller, ovate to elliptic, to 1 × 0.5 cm, terminal leaflets ovate to elliptic, to 4 × 1.5 cm, appressed-pubescent below, glabrous above, base subcordate, margin entire, apex obtuse-round; petiole to 7 mm, petiolule 1.5 mm. Flowers violet to pink, in 40-70-flowered, lax racemes, pubescent; peduncle to 1.3 cm, pedicel 1.5 cm; bracts deciduous, 2 mm, boat-shaped; calyx tube 1.5 mm, subequal, lobes to 3.5 mm; corolla violet to pink, standard orbicular, 5 × 8 mm, wings 5 mm, keels 7 × 1 mm; stamens diadelphous (9+1), staminal sheath to 6 mm, filaments 1 mm; ovary pilose, to 4 mm, style 2 mm. Pod 1.5 cm, margins indented, articles 5 or 6, with hooked hairs, obliquely rounded, as broad as long, puberulous, much exserted from the setaceous pilose calyx lobes.

Rare in Medak district.

Fl. & Fr.: September-January

Gummadidala (MDK), *GNR* 4254

Desmodium benthamii N.P. Balakr., J. Bombay Nat. Hist. Soc. 63: 328. 1967. *D. brachystachyum* Grah. ex Benth. in Miq. Pl. Jungh. 223. 1852 non Schlecht. 1838; Baker in Hook. f., Fl. Brit. India 2 : 171. 1876; Gamble, Fl. Madras 1: 347. 1918.

Slender undershrubs, up to 60 cm high, stems caespitose, slender, glabrous except at the top, branchlets thinly appressed hairy, angular. Stipules 2 mm. Leaves simple, elliptic, 1.8-5 × 1.5-2 cm, base cordate, margin entire, apex obtuse, or emarginate, glabrous above, sparsely strigose beneath; petiole to 4 cm. Flowers deep purple, in copious, close, axillary or terminal, 2.5 cm long racemes, pedicel to 4mm; calyx white villous, teeth long, lanceolate-cuspidate, corolla twice the calyx. Pod sessile, with 1-3 joints, glabrous, not longer than calyx.

Rare in Khammam district.

Fl. & Fr.: January-July.

Vazeedu, near Venkatapuram (KMM), *VNS* 680 (MH).

Desmodium dichotomum (Willd.) DC., Prodr. 2 : 336. 1825. *Hedysarum dichotomum* Willd., Sp. Pl. 3 : 1180. 1802. *Desmodium diffusum* (Willd) DC., Prodr. 2336. 1825, *nom. illeg.,* non. DC. Prodr. 335. 1825 n. 88, quest *D. laxiflorum* Baker in Hook.f., Fl. Brit. India 2 : 169. 1876; Gamble, Fl. Madras 1: 346. 1918.

Diffuse straggling herb, up to 1 m long, with angled stems, branchlets with spreading, elongate hispid hairs. Leaves 3-foliolate, leaflets ovate-elliptic, 3-5 × 1.5-3 cm, laterals from half to nearly as long as the terminal, inequilateral, chartaceous, densely pubescent above, tomentose below, base and apex obtuse, margin entire, stipules large, auriculate, sub-amplexicaul, stipels lanceolate; petiole to 4.5 cm, petiolule to 1.5 mm, Flowers bluish in lateral and terminal lax or moderately closed up to 30 cm long, racemes. flowers 2-5 per bract, bracts persistent, ovate, 3 mm; calyx tube 1 mm, lobes as long as the tube, lobes linear, acute, ciliate; corolla bluish, standard obovate-orbicular, 3 × 2 mm, wings 2 × 0.5 mm, keels 3 × 1 mm; stamens diadelphous (9+1), staminal sheath 2.5 mm; ovary 2.5 mm, style 0.5 mm. Pod linear, moniliform, indented on both sutures, joints 6-8, reticulate with hooked hairs, articles obliquely rounded.

Common in the forests of most districts.

Fl. & Fr.: October-February.

Sone (ADB), *TP* & *GO* 4036; Keslapur (ADB), *GO* 4556; Kamareddy (NZB), *BR* 9274; Pathur (MDK), *TP* & *CP* 12109; Kanapur RF (WGL), *KMS* 11570 (CAL, BSID); ICRISAT (MDK), *LJGV* 3906 (CAL); Kodimial (KRN), *GVS* 21858 (MH).

Desmodium gangeticum (L.) DC., Prodr. 2. 327. 1825; Baker in Hook. f., Fl. Brit. India 2 : 168. 1876; incl. var *maculatum;* Gamble, Fl. Madras 1: 345. 1918; Meeuwen, Reinwardtia 6: 249. 1962. *Hedysarum gangeticum* L., Sp. Pl. 746. 1753.

Vern.: *Yetinarum.*

Erect undershrubs, up to 1 m high, branches angled, grooved, sparsely pubescent, hirtellous especially on ridges on grooves. Leaves unifoliolate, membranous, oblong, obovate or lanceolate, 4-8 × 2-4.5 cm, base rounded, margin entire, soflty pubescent beneath, less so above, stipules narrowly ovate, scarious, apex setaceous, stipels present. Flowers deep violet or white (bluish pink) in axillary and terminal panicles, up to 20 cm long, peduncle to 5 cm, pedicel to 3.5 mm. Flowers 2 per bract, bracts linear, to 3mm; calyx tube 1 mm, lobes triangular; corolla standard broadly obovate, 5 mm, wings 3 mm, keels 4 mm; stamens diadelphous (9 + 1), staminal sheath 3 mm, filaments 1 mm; ovary sessile, 3 mm, style 2 mm. Pods linear, moniliform, lower margin deeply undulate, upper straight, joints 5-6, pubescent with hooked hairs, reticulate, seeds reniform, flattened.

Occasional in the forests in most of the districts.

Fl. & Fr.: August-January.

Birsaipet (ADB), *GO* & *PVP* 4595; Manchippa RF (NZB), *TP* & *BR* 6038; Malkapur Road (NZB), *TP* & *BR* 6107; Velutla RF (NZB), *BR* 9519; Nagulabanda (MDK), *CP* 11344; Muttaypally (MDK), *CP* 11780; Anantasagar (RR), *MSM* 10507; Pegarikutta east-Aklasapur (MDK), *KMS* 6697 (MH); Ramavaram RF (KMM), *RR* 11217 (BSID);

Figure 23: Desmodium dichotomum (Willd.) DC.
A. Twig, B. Pod.

Krishnarajasagar (KMM), *RC* 102458 (BSID); Mallelatheertham (MBNR), *SRS* 108981 (BSID); Mallurgutta (WGL), *RKP* 110957 (BSID).

Desmodium heterocarpon (L.) DC., Prodr. 2: 337. 1825; var. **heterocarpon;** Meeuwen, Reinwardtia 6 : 251. 1962. *Hedysarum heterocarpon* L., Sp. Pl. 747. 1753. *Desmodium polycarpum* (Poir.) DC., Prodr. 2: 334. 1825 *p.p.,* Baker in Hook. f., Fl. Brit. India 2 : 171. 1876. incl. var. *trichocaulon;* Gamble, Fl. Madras 1: 346. 1918. *Hedysarum polycarpum* Poir. in Lam., Encycl. 6 : 413. 1805.

Scandent subshrubs, up to 75 cm high, branchlets terete, striate, pubescent when young. Stipules narrowly ovate, to 1.3 cm, scarious. Leaves trifoliolate, leaflets obovate, elliptic-oblong, 2-6 × 1-3 cm, terminal leaflets large, laterals small, lateral nerves 10, reticulations prominent, glabrous above, softly pilose below, base obtuse-truncate, margin entire, apex mucronate; petiole to 2.5 cm, petiolule 3 mm. Flowers pinkish violet in terminal, about 18-20 cm long racemes, calyx glabrous, tube 1.5 mm, lobes 2.5 mm; corolla violetish-white, standard obovate-orbicular, 5.5 × 5 mm, wings 4 × 1.2 mm, keels 6 × 2 mm; stamens diadelphous (9+1), staminal sheath 4 × 2 mm; ovary 4 mm, style 1 mm. Pods linear-oblong, faintly falcate, hooked hairs along sutures, sparsely along the surface, joints about 6, reticulate, seeds slightly flattened, reniform.

Common in Warangal and Khammam districts in forestss.

Fl. & Fr.: September – April.

Mothugudem (KMM), *EC* 7452; Pakhal RF (WGL), *KMS* 11656 (MH).

Desmodium laxiflorum DC., Ann. Sci. Nat. (Paris) 4 : 100. 1825 & Prodr. 2. 335. 1825; Baker in Hook. f., Fl. Brit. India 2 : 164. 1876; Gamble, Fl. Madras 1: 344. 1918; Meeuwen, Reinwardtia 6: 252. 1962.

Erect undershrubs, up to 1 m high, branchlets quadrangular, appressed-pubescent. Leaves 3-foliolate, leaflets ovate-elliptic or deltoid-rhomboid, 6-12 × 4-6 cm, laterals half as long, or nearly as long as the terminal, inequilateral, chartaceous, puberlous above, appressed-pubescent below, base and apex subacute-obtuse, margin wavy, stipules narrowly triangular, 5 mm, stipels 2.5 mm. Flowers bluish-white in long racemes, 2-7 in a cluster, closely placed; peduncle 5 cm, pedicel to 7 mm; bracts ovate, to 5 mm; calyx tube 1.5 mm, campanulate, 4-lobed, strigose; corolla bluish-white, standard 4 × 2.5 mm, keels blue, 5 × 2 mm; stamens diadelphous (9+1), staminal sheath 4 mm; ovary 4 mm, style 1 mm. Pod stipitate, margins indented, lower one evenly rounded, joints 6-10, more than twice as long as broad, with hooked hairs; seeds 1 mm.

Occasional in Medak and Mahabubnagar districts in forests.

Fl. & Fr.: September-February

Desmodium oojeinense (Roxb.) H. Ohasi, Ginkgoana 1: 117. 1973. *Ougeinia oojeinensis* (Roxb.) Hochr., Bull. Soc. Bot. Geneve 13 & 14: 51. 1909. *Dalbergia oojeinensis* Roxb., Fl. Ind. 3 : 220. 1820. *Ougeinina dalbergioides* Benth. in Miq. Pl. Jungh. 216. 1852; Baker in Hook. f., Fl. Brit. India 2: 161. 1876; Gamble, Fl. Madras 1: 340. 1918.

Vern. : *Tella motku.*

Moderate sized, often gregarious trees, up 15 m high, trunk short, crooked, bark black or dark-brown, deeply cracked. Leaves 3-foliolate, leaflets broadly ovate or suborbicular, 5-18 × 4-15 cm, coriaceous, glabrous above, sparsely pubescent beneath, base oblique or cordate, margin shallowly crenate, apex bluntly pointed; petiole to 6.4 cm, petiolule to 5 mm. Flowers white or rose-coloured, numerous in short fascicled racemes from the nodes of old branches; calyx campanulate, lobes distinct; corolla exserted, standard orbicular, wings slightly spured; stamens diadelphous (9+1); ovary glabrous, style incurved. Pod 5-8 cm long, 2-5-jointed, flat, reticulately veined.

Occasional in forests of Adilabad, Nizamabad, Medak and Mahabubnagar districts.

Fl. & Fr.: February-May.

Rampur (ADB), *SSR* 19304; Jalalpur (NZB), *BR* & *KH* 9653; Gangapur RF (MDK), *BR* & *CP* 11426; Farahabad (MBNR), *MVS* & *MB* 38003.

Desmodium triflorum (L.) DC., Prodr. 2 : 334. 1825; Baker in Hook.f., Fl. Brit. India 2 : 173. 1876; Gamble, Fl. Madras 1 : 347. 1918. *Hedysarum triflorum* L., Sp. Pl. 749. 1753.

Prostrate spreading herbs, up to 50 cm long, branchlets faintly angled, sparsely hirsute, rooting at nodes. Leaves trifoliolate, leaflets obovate, 4-7 × 3-8 mm, lateral equilateral, chartaceous, glabrous above, pubescent below, base cuneate, margin entire, apex obtuse-retuse, stipules obliquely ovate; petiole to 5 mm, petiolule 1 mm. Flowers purplish pink or pinkish blue, 2-5, clustered in axillary fascicles; bracts ovate, 3mm; calyx tube 1 mm, pubescent, lobes to 2 mm; corolla pinkish, standard obovate, 4.5 × 3.5 mm, wings 4 × 2 mm, keels 4 × 1.5 mm; stamens diadelphous (9+1), staminal sheath 2.5 × 0.5 mm, filaments 0.5 mm; ovary 2 mm, style incurved, 2 mm. Pods up to 1 cm, lower margin deeply indented, upper entire, joints about 4, hooked-pubescent; seeds compressed, reniform, 1.2 mm.

Common in hedges, along the rivers often in forests of all districts.

Fl. & Fr.: Throughout the year.

Jannaram (ADB), *TP* & *PVP* 4177; Mudheli RF (NZB), *BR* & *CPR* 7210; Eklaspuram (MBNR), *BR* & *MV* 41309; 28 KM from Srisailam (MBNR), *LJGV* 3284 (CAL); Pillutla RF (MDK), *BR* & *CP* 11265; Lingampalle (RR), *MSM* & *TP* 15138; Mahadevpur (KRN), *SLK* 70857 (NBG); Pakhal (WGL), *KMS* 11660 (MH).

Desmodium velutinum (Willd.) DC., Prodr. 2: 328. 1825; Meeuwen, Reinwardtia 6: 264. 1962. *Hedysarum velutinum* Willd., Sp. Pl. 3: 1174. 1802. *Desmodium latifolium* (Roxb. ex Ker.Gawl.) DC., Prodr. 2: 328. 1825; Baker in Hook.f., Fl. Brit. India 2 : 168. 1876; Gamble, Fl. Madras 1: 346. 1918. *Hedysarum latifolium* Roxb. ex Ker. Gawl. in Bot. Reg. s. t. 355. 1819.

Erect subshrubs, up to 1.5 m high, stems terete, puberulous, intermixed with recurved hairs. Leaves simple, stipulate, ovate-lanceolate, 6-12 × 4-8 cm, base truncate or slightly cordate, margin undulate, apex acuminate, sparingly hirsute on both surfaces; petiole to 1.5 cm. Flowers purple in axillary and terminal branched panicles, peduncle to 2.5 cm, pedicel to 1.5 mm; calyx tube 0.5 mm, 4-lobed, sparsely pubescent, lobes to 1mm; corolla purple, standard orbicular-obovate, 6 × 5.5 mm, wings oblong, 6 × 2.5 mm, keels slightly incurved, 6 × 2 mm; stamens diadelphous (9+1), staminal sheath 4 × 1 mm, filaments 1 mm; ovary sessile, 3 mm, style 2 mm. Lomentum joints 4-6, sparsely pubescent with hooked hairs, indented on both margins; seeds reniform, flattened.

Common in forests of Adilabad, Nizamabad and Mahabubnagar districts.

Fl. & Fr.: July-December

Sattenapalli (ADB), *TP* & *BR* 6132; Ibrahimpet RF (NZB), *TP* & *BR* 6132.

DICERMA DC.

Dicerma biarticulatum (L.) DC., Prodr. 2: 339. 1825. *Hedyasarum biarticulatum* L., Sp. Pl. 747. 1753. *Desmodium biarticulatum* (L.) F.V. Muell., Fragm. Phyt. Austr. 2. 121. 1861; Baker in Hook.f., Fl. Brit. India 2. 163. 1876; Gamble, Fl. Madras 1: 344. 1918.

Slender undershrubs; up to 60 cm high, stems slender, densely caespitose, glabrous or downy. Stipules scarious, to 5 mm, connate, 2-3-cleft. Leaves 3-foliolate, leaflets oblanceolate or oblong, 0.5-1.5 × 0.3-0.5 cm, base cuneate, margin entire, apex obtuse; coriaceous, subgabrous, nearly digitate; petiole to 4 mm, petiolule 1 mm. Flowers bright red in terminal lax racemes, the lower flowers 2-4 together; peduncle to 1.2 cm, pedicel to 2 mm; bracts subulate, to 3 mm, striate, exceeding short pedicel; calyx sericeous, tube 1.5 mm, lobes 2 mm, acute; corolla reddish, standard obovate, 6 × 4 mm, wings 5 × 1 mm, keels 4.5 × 1 mm; stamens diadelphous (9+1), staminal sheath 6 mm, filaments 1 mm; ovary 1.5 mm, style 4 mm. Lomentum usually with 2 joints, rarely round-oblong, pubescent, 1 cm long, both the sutures indented; seeds reniform, black.

Common in Warangal, Medak, Karimnagar and Ranga Reddi districts.

Fl. & Fr.: April-October.

Reddypalle (MDK), *TP* & *CP* 14076; Rudraram RF (RR), *MSM* & *TP* 12728; Central University campus (RR), *MSM* 15158.

DOLICHOS L.

Dolichos trilobus L., Sp. Pl. 726. 1753. subsp. **trilobus.** *D. falcatus* Klein ex Willd., Sp. Pl. 3. 1047. 1802; Baker in Hook.f., Fl. Brit. India 2 : 211. 1876; Gamble, Fl. Madras 1: 366. 1918.

Glabrescent, tender twiners, up to 3 m long, branchlets glabrescent. Stipules lanceolate, 4 mm, reflexed, stipels linear, 3 mm. Leaves 3-foliolate, terminal one ovate, rhomboid or deltoid, sometimes 3-lobed, 3-5.5 × 3.5-6 cm, laterals ovate, inquilateral, 2-5 × 1.5-4 cm, chartaceous, sparsely pubescent, base truncate-obtuse, margin entire, apex acute, apiculate; petiole to 5.5 cm, petiolule 1.5 mm. Flowers pinkish-violet in slender axillary racemes; peduncle to 1.5 cm, pedicel to 8 mm; bracts striate, 3 mm, bracteoles 1 mm; calyx campanulate, tube 4.5 mm, lobes ovate, upper lobes cuneate, 2 mm, lower one 3 mm; corolla pinkish-violet, exserted, standard orbicular, 1.4 × 1.5 cm, auricled at base, appendaged above claw, wings 1.9 × 0.7 cm, keels incurved, obtuse, 1.1 × 0.5 cm; stamens diadelphous (9+1), staminal sheath 1 × 0.4 cm, filaments 4 mm; ovary sessile, glabrous, 4 mm, style incurved, 6 mm. Pods linear- falcate, glabrous, apically horned; seeds 6-8, strophilate.

Common in all districts in open forests and foot hills.

Fl. & Fr.: September-March

Mosra (NZB), *TP* & *BR* 6100; Mombojipally (MDK), *CP* 11632; Gangapur (MDK), *TP* & *CP* 12090; Kanmankalva RF (RR), *MSM* 10461; Mohammadabad RF (RR), *MSM* & *KH* 10560; Kodimial (KRN), *GVS* 21880 (MH); Anangal to Shadnagar (MBNR), *SRS* 104563 (BSID).

ELEIOTIS DC.

Eleiotis monophyllos (Burm. f.) DC., Mem. Legum. 7 : 350. 1825. *Glycine monophylla* Burm. f., Fl. Ind. 161. t. 50. f. 2. 1768. *Hedysarum sororium* L., Mant. Pl. Alt. 270. 1771. *Eleiotis sororia* (L.) DC., Mem. Legum. 7 : 350. 1825; Baker in Hook.f., Fl. Brit. India 2 : 153. 1876; Gamble, Fl. Madras 1: 332. 1918.

Prostrate annual herbs, up to 50 cm long, branchlets triquetrous, glabrous. Leaves 1-foliate, obovate-orbicular, 2-3 × 2-2.8 cm, base cordate, margin entire, apex rotund-retuse, glabrous above, appressed pubescent beneath; petiole to 2.5 cm. Flowers red in 10 cm long, 18-30-flowered lax racemes, pedicels in pairs, distant, to 7 mm; bracts 2 mm, scarious, striate, deciduous; calyx tube 0.5 mm, campanulate, puberulous; corolla red, exserted, standard obovate, emarginated, 3.8 × 4 mm, wings oblong, 1.2× 2 mm, adnate to keels, keels obtuse, 3.5 × 1.5 mm; stamens diadelphous (9+1), staminal sheath 2.5mm, filaments 0.5 mm; ovary sessile, sparsely pubescent, 1.5 mm, style inflexed, 2 mm, glabrous. Pod subsessile, boat-shaped, 6 × 3 mm, pointed, reticulately veined, glabrous when ripe; seed solitary, oblong, dark-brown.

Common in Adilabad, Karimnagar, Warangal, Ranga Reddi and Mahabubnagar districts in open forests.

Fl. & Fr.: May-November

Ankurapuram (ADB), *GO* 4311; Lingampalle (RR), *MSM* & *TP* 15121; Aklaspur (KRN), *GVS* 22497 (MH).

ENDOSAMARA R. Geesink

Endosamara racemosa (Roxb.) R. Geesink, Leiden Bot. Ser. 8: 93. 1984. *Millettia racemosa* (Wight & Arn.) Benth. in Miq., Pl. Jungh. 249. 1852; Baker in Hook.f., Fl. Brit. India 2 : 105. 1876; Gamble, Fl. Madras 1: 323. 1918. *Tephrosia racemosa* Wight & Arn., Prodr. Fl. Ind. Orient. 210. 1834.

Large woody climbers, up to 20 m long, bark white, branches glabrous or thinly silky. Stipules filiform, to 1 cm, stipels setaceous. Leaves up to 30 cm long, leaflets 11-15, membranous, opposite, oblong or obovate-oblong, 5-14 × 2-4.5 cm, base obtuse, margin entire, apex acuminate, glabrous above, appressed pubescent beneath; petiole to 10 cm, petiolule 4 mm. Flowers yellowish-white, in axillary and terminal silky-brown racemes; peduncle tomentose, to 3 cm, pedicel to 3 mm; bracts very conspicuous, setaceous, subulate, to 1.5 cm; calyx silky, tube 6 mm, lobes short, to 1.5 mm; corolla standard orbicular, 1.5 × 1.5 cm, wings spurred, 1.6 × 0.5 cm, keels 1.1 × 0.5 cm; stamens diadelphous (9+1), staminal sheath 8 ×4 mm, filaments 5 mm; ovary linear, glabrous, to 7 mm, style to 7 mm. Pod linear, flat, glabrous, 10-16 cm long; seeds 4-6, ovoid, glabrous, smooth.

Occasional in dry hill forests at low level in Mahabubnagar district.

Fl. & Fr.: May-December.

Bahrapur (MBNR), *STR* 641 (MH).

ERYTHRINA L.

1. Calyx 2-lipped :
    2. Leaflets subcoriaceous, tomentose beneath, apex acute,
        margin sinuate .... **E. suberosa**
    2. Leaflets membranous, glabrescent beneath, apex
        acuminate, margin entire .... **E. subumbrans**
1. Calyx split on side half-way down or to the base :
    3. Trees unarmed; leaflets apex acuminate; calyx 5-toothed at
        tip; wings and keel petals equal, one fourth the length of
        standard; pod 6-12-seeded .... **E. variegata**
    3. Trees armed, leaflets apex subacute, calyx entire at tip;
        wings much smaller than keel petals, keels half the length
        of standard; pod 3-4-seeded .... **E. stricta**

Erythrina stricta Roxb., Fl. Ind. 3 : 251. 1832; Baker in Hook. f., Fl. Brit. India 2. 189. 1876; Gamble, Fl. Madras 1: 354. 1918.

Armed trees, up to 20 m high, bark pale, smooth, greenish after the paper exfoliations, branchlets apically stellate- pubescent, basally glabrescent, densely prickled. Leaves 3-foliolate, 10-30 cm, leaflets rhomboid-ovate, 8-12 × 9-20 cm, thin coriaceous, glabrescent, base deltoid or truncate, margin entire, apex subacute, petiole 10-13 cm, prickled or not. Flowers deep red, in 10 cm long racemes, peduncle 4-6 cm, pedicels 3 in a cluster; bracts ovate, bracteole 3 mm; calyx spathaceous 1.5 cm, entire at the tip, split half-way down, erect, glabrescent; corolla deep red, standard 5 × 2 cm, oblong-lanceolate, wings obovate, 5.5 × 3 cm, keels ovate, 2 ×0.7 cm; stamens monadelphous, staminal sheath 2.5 cm, filaments to 1.5 cm, alternately longer and shorter, vexillary stamen free; ovary pubescent, 2 cm, style suberect, 1.5 cm. Pod slightly curved, to 12 ×2 cm with slender stripes, 3-4-seeded.

Occasional in Nizamabad, Medak, Ranga Reddi districts in dry deciduous forests.

Fl.: February – March; Fr.: March onwards.

Bodhan (NZB), *BR* 7270; Manjira Barrage (MDK), *BR* & *CP* 14095; Mokarlabad (RR), *MSM* & *KH* 10261.

Erythrina suberosa Roxb., Fl. Ind. 3 : 253. 1832; Baker in Hook. f., Fl. Brit. India 2 : 189. 1876; Gamble, Fl. Madras 1: 354. 1918.

Vern. : *Mulu moduga.*

Deciduous trees, up to 10 m high, bark deeply furrowed, blaze thick yellowish, branches tomentose, prickles usually straw-coloured. Leaves 3-foliolate, 10-20 cm, leaflets broadly ovate, 5-10 × 5-11.5 cm, thin sub-coriaceous, glabrous above, tomentose below, base deltoid or truncate, margin sinuate lobed, apex acute; petiole to 7 cm, petiolule to 1 cm. Flowers red, in 8 cm long racemes; calyx campanulate, persistent, turbinate, 2-lipped; corolla red, exserted, standard 3.5 × 1.2 cm, oblong, wings 1.5 × 0.6; stamens monadelphous, the vexillary stamen free, staminal sheath 2 cm, filaments to 8 mm, alternate long and short; ovary oblong, pubescent, style to 1 cm. Pod stipitate subterete, torulose, to 10 ×0.8 cm, 2-5-seeded.

Occasional in forest in Adilabad, Karimnagar and Mahabubnagar districts, often cultivated as ornamental.

Fl.: March – May; Fr.: April – June.

Mannanur (MBNR), *STR* 490 (MH); Mahadevpur to Kataram (KRN), *SLK* 70806 (NBG).

Erythrina subumbrans (Hassk.) Merr., Philipp. J. Sci. 5 : 113. 1910. *Hyphaphorus subumbrans* Hassk., Hort. Bogor. Descr. 198. 1858.

Tall trees, 12-15 m high, branchlets often unarmed. Leaves 3-foliolate, leaflets ovate, 10-15 × 7-11 cm, membranous, base rounded, margin entire, apex acuminate, glabrescent; petiole 10-12 cm, petiolule 7 mm. Flowers red in racemes; calyx 2-lipped, 1 cm long; corolla red, standard 2.5- 3.5 × 2 cm, wings and keel petals subequal; ovary glabrous. Pod flat, broader and empty at base, turgid and 3-4-seeded towards the apex, dehiscent; seeds dark purple.

An introduced tree.

Fl. & Fr.: August-March.

ICRISAT (MDK), *LJGV* 2699 (CAL).

Erythrina variegata L., Herb. Amb. 10. 1754; var. **orientalis** Merr., Interpr. Herb. Ambion. 276. 1917. *E. indica* Lam., Encycl. 2 : 391. 1786; Baker in Hook.f., Fl. Brit. India 2 : 188. 1876; Gamble, Fl. Madras 1: 353. 1918. *E. lithosperma* Miq., Fl. Ind. Bat. 1(2) : 209. 1859; Baker in Hook. f., Fl. Brit. India 2 : 190. 1876.

Armed trees, 12-18 m high, bark smooth and greenish after the papery exfoliations, branchlets apically stellate-pubescent, basally glabrescent. Stipellae short, swollen, reflexed, finally hard. Leaves 20-30 cm, 3-foliolate, leaflets 10-15 × 8.5-12.5 cm, rhomboid-ovate, thin, coriaceous, stellate- pubescent, base deltoid or truncate-obtuse, margin entire, apex acuminate; petiole to 30 cm, petiolule to 1.5 cm. Flowers bright-red in 5-15 cm long racemes, peduncle woody, stout, 10-20 cm long, bracts triangular, bracteoles subulate, pedicels five in a cluster, calyx spathaceous, 2 × 0.6 cm, split at the base, recurved, apically 5-toothed; corolla red, standard much exserted, oblong-elliptic, 5 × 2 cm, wings and keel petals subequal; stamens monadelphous (10), staminal sheath 3.5 cm, filaments to 2.5 cm; ovary pubescent, 2 cm, style 3 cm. Pod 30 × 2 cm, beaked, somewhat curved, 6-12-seeded.

Frequently planted in gardens, grown in hedges.

Fl.: March – April; Fr.: April - July.

Mannanur to Amrabad (MBNR), *SRS* 107461 (BSID).

FLEMINGIA Roxb. ex W. Aiton & W.T. Aiton *nom. cons.*

1. Leaves 1-foliolate; flowers and fruits hidden with in foliaceous bracts .... **F. strobilifera**

1. Leaves 2-3-foliolate; flowers and fruits not hidden by bracts:

 2. Inflorescence a long-pedunculate capitulum; bracts 1.5-2 cm, persistent .... **F. involucrata**

 2. Inflorescence peduncle extremely short, racemes or panicles; bracts 0.4-1.2cm, persistent or deciduous:

  3. Peduncle conspicuous, axis slender; leaflets 7×3 cm, apex rounded or subacute .... **F. lineata**

  3. Peduncle inconspicuous, axis robust; leaflets 10-20 × 4-8 cm, apex acuminate .... **F. macrophylla**

Flemingia involucrata Benth. in Miq. Pl. Jungh. 246. 1852; Hook.f., Fl. Brit. India 2: 229. 1876. *Lespedeza involucrata* Wall., Cat. 5742. 1831-32, *nom. nud. Flemingia capitata* Zoll. ex Miq., Fl. Ind. Bat. 1, 1: 66. 1855. *Moghania involucrata* (Wall. ex Benth.) Kuntze, Revis. Gen. Pl. 1: 199. 1891.

Undershrub, 1-2 m high, branches pubescent. Leaves 3-foliolate, petiole 0.5-2 cm, stipules oblong-lanceolate, acute, caducous, leaflets narrowly elliptic or elliptic, 3.7-8.7 cm long, glabrous above, pubescent beneath, dotted with brown glands. Flowers purple, about 1.2 cm long, in dense terminal and axillary heads surrounded by oblong, 1.8-2 cm long, acuminate, dry brown bracts. Pod oblong, 5 mm long, downy, 1-seeded; seeds dark brown or black, ellipsoid, 2-3 mm long.

Rare, occurs in dry deciduous forests in shady places.

Fl. & Fr.: October- February.

Ragan & *al.* (2005) reported from forests of Warangal district.

Flemingia lineata (L.) Roxb. ex W.T.Aiton, Hort. Kew. ed. 2. 4: 350. 1812; Baker in Hook.f., Fl. Brit. India 2 : 228. 1876. *Hedysarum lineatum* L., Sp. Pl. 1054. 1753.

Small erect, branched shrubs, young branches angular, appressedly pubescent. Stipules subpersistent. Leaves 3-foliolate, laflets obovate or oblanceolate, 7 × 3 cm, glabrous above, silky- pubescent below, the lateral leaflets smaller than the terminal, base cuneate, margin entire, apex rounded or subacute; petiole to 2.5 cm, quadrangular, not winged, pubescent. Flowers pinkish, small in axillary, lax panicles; bracts linear, acute, caducous; calyx 4 mm long, lobes subequal, linear-lanceolate. Pod oblong, 1.2-5 × 0.5 cm, rounded at both ends, minutely apiculate, glandular pubescent; seeds 2, orbicular-oblong, black.

Occasional in forests of Karimnagar district.

Fl. & Fr.: October-February.

Aklaspur (KRN), *GVS* 22526 (MH).

Flemingia macrophylla (Willd.) Kuntze ex Merr., Philipp. J. Sci. 5: 130. 1910. *Crotalaria macrophylla* Willd., Sp. Pl. 3 : 982. 1802. *Flemingia congesta* Roxb. ex Ait., Hort. Kew. ed. 2. 4 : 349. 1812; Baker in Hook. f., Fl. Brit. India 2 : 228. 1876; Gamble, Fl. Madras 1: 378. 1918. *F. nana* Roxb., Fl. Ind. 3: 339. 1832; Prain, J. Asiat. Soc. Beng. 66: 441. 1876. *F. congesta* Roxb. ex Ait.f. var. *nana* (Roxb.) Baker in Hook. f., Fl. Brit. India 2: 228. 1876. *F. semialata* Roxb., Fl. Ind. 3 : 340. 1842; Gamble, Fl. Madras 1: 378. 1915. *F. congesta* Roxb. var. *semialata* (Roxb.) Baker in Hook. f., Fl. Brit. India 2 : 229. 1876. *p.p. F. prostrata* Roxb., Fl. Ind. 3: 338. 1832. *Flemingia wallichii* Wight & Arn., Prodr. Fl. Ind. Orient. 242. 1834; Baker in Hook.f., Fl. Brit. India 2 : 229. 1876; Gamble, Fl. Madras 1: 378. 1918.

Undershrubs, up to 1.5 m high, branchlets appressed pubescent. Stipules 2.5 cm, caducous. Leaves 3-foliolate, up to 18 cm, leaflets oblanceolate-elliptic, 10-20 × 4-8 cm, chartaceous, glabrous above, velvety below, terminal leaflets basally cuneate, laterals basally obtuse-round, margin entire, apex acuminate; petiole angled, not winged, glabrous, to 5 cm, petiolule to 3 mm. Flowers purplish yellow, 4 cm long, congested racemes, flowers clustered at base, lax at tip; bracts foliaceous, to 4 mm; calyx tube 1.5 mm, lobes to 6 mm; corolla purplish yellow, standard obovate, 6 ×4 mm, wings 7 × 3.5 cm; stamens diadelphous (9+1), staminal sheath 6.5 mm, filaments to 2 mm; ovary pubescent, 1.5 mm, style 6 mm. Pod oblong, 1 x 0.4 cm, pubescent and glandular, turgid, 2-seeded.

Rare in forests in Mahabubnagar and Warangal districts.

Fl. & Fr.: December-May.

Saleswaram (MBNR), *BSS* & *KP* 33337.

Flemingia strobilifera (L.) R. Br., Hort. Kew. ed. 2. 4 : 350. 1812; Baker in Hook.f., Fl. Brit. India 2 : 227. 1876; Gamble, Fl. Madras 1: 377. 1918. *Hedysarum strobiliferum* L., Sp. Pl. 476. 1753.

Erect herbs, up to 1.5 m high, branchlets striate, brown-pubescent. Stipules scarious, 8 mm, lanceolate, caducous. Leaves simple, ovate-lanceolate, 9-14 × 3-5 cm, glabrous except on nerves below, gland-dotted, silky pubescent beneath, lateral nerves 10, transverse nervules rather prominent, base round, margin entire, apex acute-acuminate, petiole to 2.2 cm. Flowers white in axillary and terminal racemes, bracts scarious; calyx tube 1.5 cm, lobes to 4 mm, setaceous; corolla exserted, white, standard obovate, 6 × 5 mm, wings oblong, 5.5 × 1 mm, keels incurved, 7 × 2 mm; stamens diadelphous (9+1), staminal sheath 6 × 1.5 mm, filaments 3 mm; ovary sessile, woolly, 2 mm, style 4 mm. Pods oblong, 1 × 0.5 cm, enclosed by bracts, 2-seeded.

Occasional in hills of Adilabad and Warangal districts.

Fl. & Fr.: February- August

Way to Kunthala (ADB), *GO* & *DAM* 5125; Pakhal (WGL), *ANH* 15934 (CAL), *RKP* 108252 (BSID); Pullelachelama (MBNR), *SRS* 109196 (BSID).

GALACTIA P. Br.

Galactia tenuiflora (Klein ex Willd.) Wight & Arn., Prodr. Fl. Ind. Orient. 206. 1834; Baker in Hook.f., Fl. Brit. India 2: 192. 1876; Gamble, Fl. Madras 1: 356. 1918. *Glycine tenuiflora* Klein ex Willd., Sp. Pl. 1059. 1802.

Key to Varieties
1. Leaflets mucronate at apex, sericeous-pubescent beneath .... var. **tenuiflora**
1. Leaflets acuminate at apex, villous beneath .... var. **villosa**

Galactia tenuiflora (Klein ex Willd.) Wight & Arn. var. **tenuiflora**

Slender twiners, branchlets appressed pubescent. Stipules setaceous, stipels minute. Leaves 3-foliolate, leaflets elliptic, 3-5 x 1.5-2 cm, thin-coriaceous, glabrous above, sericeous pubescent below, base obtuse, margin entire, apex mucronate; petiole 2-4 cm, petiolule to 3 mm. Flowers solitary or in pairs from the distant nodes of elongated racemes, peduncle to 4 cm, pedicel 1.5 mm; bracts ovate, 1.5 mm; calyx tube 3 mm, lobes to 4 mm; corolla purplish to whitish, standard 1.1 × 0.7 cm, wings 1.1 × 0.35 cm, keels 1 × 0.4 cm; stamens diadelphous (9+1), staminal sheath 5 × 1.5 mm, filaments 3 mm; ovary flat, pubescent, 3 mm, style 4 mm. Pods slightly falcate, 4.5 × 0.5 cm, thinly grey silky, seeds up to 8.

Common in dry deciduous forest in most of the districts.

Fl. & Fr.: July-December

Sirpur (ADB), *GO* & *PVP* 4928.

Galactia tenuiflora (Klein ex Willd.) Wight & Arn. var. **villosa** (Wight & Arn.) Benth., Fl. Bras. 15(1B): 143. 1862; Baker in Hook. f., Fl. Brit. India 2 : 192. 1876. *Galactia villosa* Wight & Arn., Prodr. Fl. Ind. Orient. 207. 1834; Gamble, Fl. Madras 1: 357. 1918.

Slender twiners, branchlets villous. Leaves 3-foliolate, leaflets oblong- lanceolate, 6-10 × 2.5-4 cm, thin coriaceous, glabrous above, villous below, base obtuse, margin entire, apex acuminate, mucronate. Flowers red, fewer and smaller in 10 cm long racemes. Pod 4.5 × 0.8 cm, villous; seeds 8.

Occasional in plains in Karimnagar district.

Fl.; Through the year; Fr.: November.

Aklaspur (KRN), *GVS* 25606 (MH).

GLIRICIDIA Kunth

Gliricidia sepium (Jacq.) Kunth ex Walp., Repert. Boit. Syst. 1(4): 679. 1842. *Robinia sepium* Jacq., Enum. Pl. Carib. 28. 1760.

Small deciduous trees, up to 6 m high; bark smooth, greyish, peeling in small, thin, vertical flakes; branchlets pubescent. Leaves odd-pinnate, leaflets 6-8 pairs, opposite, lanceolate-elliptic, 2-6 × 1-2.5 cm, glabrous, inequilateral, base obtuse, margin entire, apex acuminate. Flowers rosy-white or yellowish-white in axillary racemes. Pods thick, flat, narrow at both ends, 18 cm long, glabrous; seeds 9-12, compressed.

Grown in fields and as avenue tree.

Fl. & Fr.: January-May.

Rajakamp& (NZB), *BR* & *KH* 9052.

GLYCINE Willd.

Glycine max (L.) Merrill, Interp. Rumph. Herb. Amboin . 274. 1917; Verdc., Kew Bull. 24 : 256. 1970; *G. soja* auct. non Sieb & Zucc. : Baker in Hook. f., Fl. Brit. India 2 : 184. 1876.

Annual erect herb; branches patently brown-hairy. Leaves 3-foliolate; stipules ovate-lanceolate, 5-7 mm long; leaflets membranous, ovate-elliptic, 5-15 x 3-6 cm. rounded at base, acute at apex, nearly glabrous above, appressed–hairy beneath. Flowers in terminal racemes, to 3.5 cm long, hirsute; calyx 0.7-0.8 cm long, brown-hairy, upper 2 teeth almost united; corolla slightly exserted, pinkish, 0.5-0.7 long; vexillum slightly auriculate at base; wings adhering to keels. Pods slightly curved, narrowed at base, hairy, 2.5-5 x 0.8-1.2 cm, 2-4 seeded; seeds deep-brown or black, ellipsoid, 0.9-1 cm long.

Occasionally cultivated for its seeds.

Fl. & Fr.: September– December.

INDIGASTRUM Jaub. & Spach.

Indigastrum parviflorum (F. Heyne ex Wight . & Arn.) Schrire, Bothalia 22(2) 168. 1992. *Indigofera parviflora* F. Heyne ex Wight & Arn., Prodr. Fl. Ind. Orient. 201. 1834; Baker in Hook.f., Fl. Brit. India 2 : 97. 1876; Gamble, Fl. Madras 1: 311. 1918; Shashikanth & *al.,* J. Econ. Taxon. Bot. 36: 837. 2012.

Erect herbs or subshrubs, up to 60 cm high, branchlets silvery-canescent. Leaves odd-pinnate, 3-4.5 cm, leaflets 7-9, opposite, oblong-oblanceolate, 1-2.5 × 0.2-0.5 cm, membranous, appressed-canescent, base cuneate, margin entire, apex obtuse, mucronate; petiole to 1.4 cm, petiolule 1 mm. Flowers reddish-puple, in 6-8-flowered lax racemes; peduncle 2 mm, pedicel 1 mm; calyx tube 2 mm, lobes setaceous, to 1.5 cm; corolla exserted, glabrous, petals clawed, standard oblong-obovate, 5 mm, wings 4 mm, keels 5.5 mm; stamens diadelphous (9+1), staminal sheath 3.5 mm; ovary sessile, to 2 mm, style 1 mm. Pods 2-3 cm, long, appressed hairy, sub-cylindrical, deflexed, 12-16-seeded, recurved at tip.

Rare weed of moist waste lands, rice field bunds and in citrus fields in Warangal, Nalgonda and Mahabubnagar districts.

Fl. & Fr.: September- January

Nagarjuna sagar (NLG), *BVRS* 115 (NU).

INDIGOFERA L.

1. Leaves simple :
    2. Corolla as long as calyx; pod sickle-shaped, prickly .... **I. nummulariifolia**
    2. Corolla exerted; pod globose or oblong, not prickly :
        3. Branches pilose; pod 2-seeded .... **I. cordifolia**
        3. Branches silvery-canescent; pod 1-seeded .... **I. linifolia**
1. Leaves compound :
    4. Leaves 3-foliolate :
        5. Flowers solitary; leaflets linear .... **I. aspalathoides**
        5. Flowers in racemes; leaflets other than linear:
            6. Pods 2-seeded, dentately winged on either side
               of suture .... **I. glandulosa**
            6. Pods more than 3- seeded, not dentately winged :
                7. Leaflets gland-dotted or punctate :
                    8. Prostrate herbs; pod about 8-seeded;
                       racemes 3-5- flowered .... **I. prostrata**
                    8. Erect shrubs or subshrubs; pod less than 6-seeded . **I. trifoliata**
                7. Leaflets not gland-dotted beneath .... **I. trita**
    4. Leaves multifoliolate :
        9. Inflorescence of panicles; bracts foliaceous, 3-lobed .... **I. mysorensis**
        9. Inflorescence of racemes; bracts neither foliaceous nor lobed :
            10. Pods 2-4-seeded, less than 1.2 cm long :
                11. Prostrate herbs; leaflets alternate .... **I. linnaei**
                11. Erect herbs; leaflets sub-opposite .... **I. caerulea**
            10. Pods 5-many-seeded, more than 1.4 cm long :
                12. Plants viscid-pubescent .... **I. colutea**
                12. Plants not viscid pubescent :
                    13. Leaflets alternate, 11 or fewer .... **I. spicata**
                    13. Leaflets opposite, if some alternate then more than 13 :
                        14. Racemes 8-15 cm long :
                            15. Leaflets 5-11, calyx hirsute, pod tetragonous,
                                with dense spreading pubescence :
                                16. Peduncle less than 2 cm long; pod
                                    usually white hairy .... **I. astragalina**
                                16. Peduncle over 2.5 cm long; pod
                                    usually brown hairy .... **I. hirsuta**

15. Leaflets 13-23, calyx not hirsute; pod
cylindric, glabrous or nearly so .... **I. cassioides**

14. Racemes less than 6 cm long :

17. Racemes 2-10-flowered :

18. Leaflets obovate, less than 1 cm long .... **I. glabra**

18. Leaflets linear to linear-oblanceolate,
more than 1.5 cm long .... **I. karnatakana**

17. Racemes 15-many-flowered :

19. Pod linear, slightly curved, sparsely
hairy, leaflets membranous, glabrescent
on upper surface .... **I. tinctoria**

19. Pod linear, erect, grey-pubescent;
leaflets subcoriaceous, appressed
pubescent on both surfaces .... **I. wightii**

Indigofera aspalathoides Vahl ex DC., Prodr. 2 : 231. 1825; Baker in Hook. f., Fl. Brit. India 2: 94. 1876; Gamble, Fl. Madras 1 : 309. 1918.

Erect much branched undershrubs, up to 75 cm high, young branches trailing, silvery- pubescent. Stipules linear. Leaves digitately 3-foliolate, leaflets linear, to 5 × 0.5 cm, membranous, with a few obscure appressed hairs, base cuneate, margin entire, apex obtuse, apiculate. Flowers purple to brick red, solitary and axillary; pedicel to 3 mm; calyx tube 0.5 mm and lobes linear, 1 mm; corolla purple to red, exserted, standard orbicular, 4×2 mm, puberulous, without, wings 3 mm, keel 4 mm; stamens diadelphous (9+1), staminal sheath, 3.5 mm, filaments 0.5 mm, ovary sessile, 4 mm, pubescent, style glabrous, 1 mm. Legume to 1.5 × 0.2 cm, straight, slender, cylindrical, glabrescent; seeds ca 6-8, cuboid, 1mm, smooth.

Occasional near river banks in Warangal district (C.S. Reddy , 2001)

Fl. & Fr.: November-March.

Eturnagaram (WGL), *CSR* 1369 (KU)

Indigofera astragalina DC., Prodr. 2 : 228. 1825. *I. hirsuta* sensu Baker in Hook. f., Fl. Brit. India 2 : 98. 1876 *p.p.* non L. 1753; Gamble, Fl. Madras 1: 312. 1918.

Annual herbs or small shrubs, up to 1 m high, branchlets 4-angular, densely clothed with brown hairs. Stipules linear. Leaves up to 13 cm long, imparipinnate, leaflets 7-11, opposite, elliptic-oblong, to 5 × 2 cm, terminal one larger, base acute, apex apiculate, pilose on both surfaces, petiole 2.5-3 cm, shortly petiolulate. Flowers pink, axillary, in 15 cm long, many-flowered, dense racemes; peduncle less than 20 cm; calyx to 4 mm long; corolla twice as long as calyx. Pod subtetragonous, 1.5-2 cm long, shortly pointed at apex, usually white hairy; seeds 5-6, blackish-brown to pale greenish-brown, glabrous.

Occasional in Karimnagar (Naqvi, 2001), Hyderabad and Mahabubnagar districts.

Fl. & Fr.: August-December

Osmania University Campus (HYD), *B.K. Vijayakumar* 1259 (CAL).

Indigofera caerulea Roxb., Fl. Ind. 3 : 377. 1832; var. **caerulea** Ali, Bot. Not. 111 : 564.
1958. *I. argentea* L. var. *caerulea* (Roxb.) Baker in Hook. f., Fl. Brit. India 2: 99. 1876.
I. articulata sensu Gamble, Fl. Madras 1: 311. 1915.

Much branched woody herbs, up to 1 m high, branchlets appressed silvery
canescent. Stipules 1 mm, setaceous. Leaves odd-pinnate, to 9 cm, leaflets 5-9, rarely
11, sub-opposite, obovate, 1-3 × 1-2 cm, chartaceous, canescent, base cuneate, margin
entire, apex obtuse, emarginate; petiole to 2 cm, petiolule 2 mm. Flowers pink in
axillary racemes; peduncle to 5 mm, pedicel 1mm; bracts 2 mm; calyx pubescent, 1.5
mm, deltoid, lobes shorter than the tube; corolla exserted, tomentose without, standard
ovate, sessile, 4 mm, wings 3.5 mm, keels 4 mm, with short spur; stamens diadelphous
(9+1), staminal sheath 2 mm, ovary subsessile, 1.5 mm, pubescent, style 0.5 mm,
glabrous. Pods cylindrical, slightly curved upwards, silvery pubescent, slightly
torulose; seeds 3-4, orbicular, 2.5 mm, closely striate.

Common in Nalgonda district.

Fl. & Fr.: September-March.

Nagarjunasagar (NLG), *KMS* 9773 (CAL).

Indigofera cassioides Rottler ex DC., Prodr. 2. 225. 1825. *I. pulchella* auct. non Roxb.
1832; Baker in Hook. f., Fl. Brit. India 2 : 101. 1876; Gamble, Fl. Madras 1: 313.
1918.

Vern.: *Siralli vuyya.*

Deciduous grey-pubescent shrubs, up to 4 m high. Stipules minute. Leaves odd-
pinnate, to 15 cm, leaflets 13-23, sub-opposite, oblong-elliptic, 1.5-2 × 0.5-1 cm,
chartaceous, appressed pubescent, base and apex obtuse, margin entire; petiolule to
2 mm. Flowers purple-red with white patch below, in axillary racemes; peduncle to 8
mm, pedicel 2 mm; bracts acute, longer than buds; calyx silvery hairy, tube to 2mm,
lobes lanceolate, to 2.5 mm; corolla sparsely puberulous, petals clawed, standard
ovate, 1 × 0.7 cm, ciliate, wings oblong, 1× 0.3 cm, keels 1× 0.5 cm, with 1.5 mm long
spur; stamens diadelphous (9+1), staminal sheath 8 cm, filaments 3 mm; ovary
subsessile, terete, 8 mm, style incurved, 2 mm. Legumes cylindrical, 4× 0.3 cm, sharply
pointed; seeds black, about 8-10, globose, to 2.5 mm.

Common in dry deciduous forest in some districts.

Fl. & Fr.: September-May.

Sirnapalli RF (NZB), *BR* & *PSPB* 9580; Narsapur RF (MDK), *TP* & *CP* 14040;
Anantagiri RF (RR), *MSM* & *PRSR* 12776; Pakhal (WGL), *KMS* 11646 (MH, BSID).

Indigofera colutea (Burm. f.) Merr., Philipp. J. Sci. 19 : 355. 1921; Gillett, Kew Bull. 24:
484. 1970. *Galega colutea* Burm. f., Fl. Ind. 172. 1768. *Indigofera viscosa* Lam., Encycl.
3: 247. 1789; Baker in Hook.f., Fl. Brit. India 2 : 95. 1876; Gamble, Fl. Madras 1:
311. 1918.

Small erect shrubs, up to 75 cm high, branchlets viscid pubescent. Leaves odd-pinnate, to 4.5 cm, leaflets ca 7 pairs, opposite, elliptic-obovate, 0.8-1 × 0.4-0.6 cm, pubescent on both surfaces, base and apex obtuse, margin entire, stipules setaceous, 4 mm; petiolule 1 mm. Flowers red in axillary, 6-12-flowered racemes or panicles, peduncle to 2 cm, pedicel 1 mm; calyx hairy, 0.5 mm, lobes setaceous, to 1.5 mm; corolla exerted, slightly pubescent, standard oblong-elliptic, 4 × 3 mm, base narrow, wings oblong, 4 × 1.2 mm, keels obovate, 4.7 × 1.7 mm, with 1 mm spur; stamens diadelphous (9+1), staminal sheath 3 mm, filaments 1 mm; ovary 3 mm, hispid, style 1 mm. Pod cylindric, torulose, straight, 2 cm long, clothed with gland tipped and white hairs, seeds ca 15, cuboid, pitted.

Occasional in grasslands and waste places in Karimnagar district (Naqvi, 2001).

Fl. & Fr.: September – January.

Indigofera cordifolia B. Heyne ex Roth, Nov. Pl. Sp. 357. 1821; Baker in Hook.f., Fl. Brit. India 2 : 93. 1876; Gamble, Fl. Madras 1: 309. 1918.

Prostrate herbs, up to 50 cm, branchlets densely white tomentose. Stipules 2 mm. Leaves simple, ovate-cordate, 0.8-2 x 0.5-1.5 cm, sessile, chartaceous, white-tometnose, base cordate, margin entire, apex mucronate. Flowers minute, bright red, in small, sessile, 4-6-flowered racemes; pedicel 2 mm; calyx hairy, tube 2 mm, lobes setaceous, to 2.5 mm; corolla exserted, petals clawed, standard spathulate, 3.5 × 2.7 mm, densely tomentose without, base gradually tapering, wings oblong, 3 ×1 mm, keels 2.5 ×1 mm; stamens diadelphous (9+1), staminal sheath 2.5 × 1 cm, filaments 0.5 mm; ovary tomentose, 1mm, style 1.5 mm. Pods cylindrical, greyish-white pubescent, beaks straight, halves with oil glands within; seeds 2, globose, prominently pitted.

Common on stony ground and black-cotton soils in all districts.

Fl. & Fr.: June- October.

Gandhari RF (NZB), *BR* & *CPR* 7221; Choutkoor (MDK), *TP* & *CP* 11932; Rangammagudem (RR), *MSM* 10471; Gadirayal (RR), *MSM* 10516; Osmania University Campus (HYD), *B.K. Vijayakumar* 1260 (MH); Kodimal (KRN), *GVS* 25669 (CAL); Pethakutta- Narsapur (MDK), *KMS* 6774 (CAL).

Indigofera glabra L., Sp. Pl. 751. 1753; Gamble, Fl. Madras 1: 309. 1918. *I. pentaphylla* Murr., Syst. Veg. ed. 13. 564. 1774; Baker in Hook.f., Fl. Brit. India 2 : 95. 1876.

Erect herbs, up to 80 cm high, older branches brown, terete. Stipels lanceolate, hairy, persistent. Leaves 5-7-foliolate, leaflets obovate, 1 × 0.7 cm, margin entire, apex obtuse apiculate, appressed hairy on both surfaces; petiolule 1.5mm. Flowers red in 2-4-flowered, axillary, slender racemes; peduncle to 1 cm, slender, pedicel 1 mm; calyx hairy outside, tube 2.5 mm, lobes subulate; corolla red, 0.4 mm long. Pods 2 cm long, linear, pointed, glabrous, faintly torulose; seeds 10-12, black, square.

Occasional in forests of Karimnagar, Nalgonda, Ranga Reddi, Hyderabad and Mahabubnagar districts.

Fl. & Fr.: September- April.

Lingampalle (RR), *MSM* 12742 & 12760; Nagarjunakonda (NLG), *KTH* 9792 (CAL); Rechapalli (KRN), *GVS* 22241 (MH).

Indigofera glandulosa Roxb. ex Willd., Sp. Pl. 3 : 1227. 1802; Baker in Hook.f., Fl. Brit. India 2 : 94. 1876; Gamble, Fl. Madras 1: 309. 1918.

Erect branched herbs, up to 60 cm high, branchlets slender, pubescent. Leaves 3-foliolate, to 3 cm, leaflets oblanceolate, 1.5-3.5 × 0.5-0.8 cm, base cuneate, margin entire, apex obtuse, retuse, stipules setaceous, leaflets gland dotted; petiole to 1 cm, petiolule 2 mm. Flowers red in axillary, capitate racemes, calyx hairy outside, tube 1 mm, lobes setaceous, to 1.5 mm; corolla exserted, standard oblong-obovate, 2.5 mm, glandular-hirsute, wings oblong, 2.5 mm, keels 3 mm with 1.5 mm spur; stamens diadelphous (9+1), staminal sheath 3.5 mm, filaments 0.5 mm; ovary subsessile, hirsute, 2 mm, style 3 mm. Pods oblong, 5 x 2 mm, deflected pubescent, torulose, slightly winged; seeds 2, ovoid, 2 mm.

Common in open and dry deciduous forests in almost all districts.

Fl. & Fr.: October-January

Sone (ADB), *TP* & *PVP* 4017; Mamidipally (NZB), *BR* & *KH* 6425; Narsapur forest nursery (MDK), *BR* & *CP* 11221; ICRISAT (MDK), *LJGV* 2826 (CAL); Naskal (RR), *MSM* & *PRSR* 15118; Osmania University Campus (HYD), *B.K. Vijayakumar* 1266 (CAL); Aklaspur (KRN), *GVS* 22529 (MH).

Indigofera hirsuta L., Sp. Pl. 751. 1753; Baker in Hook.f., Fl. Brit. India 2 : 96. 1876; Gamble, Fl. Madras 1: 312. 1918; Gillet, Kew Bull 14 : 290. 1960.

Annual or biennial erect herbs, up to 1 m high, branches covered with brown hairs. Leaves odd-pinnate, to 10 cm, leaflets 5 pairs, opposite, elliptic-oblanceolate, 2-3.5 × 1-1.5 cm, chartaceous, densely hirsute below, sparsely villous above, base subacute, margin entire, apex obtuse, apiculate, stipules subulate-linear, to 1 cm; petiole to 3 cm, petiolule 2mm. Flowers pink in axillary racemes; peduncle over 2.5 cm, pedicel to 3 mm; calyx hirsute, tube to 0.5 mm, lobes linear, to 4 mm, setaceous; corolla as long as calyx, white pubescent without, standard broadly obovate, 5.5 mm, wings 5 mm, keels 6 mm; stamens diadelphous (9+1), staminal sheath to 4 mm; ovary sessile, 2 mm, style to 1.5 mm. Pods 1.5 × 0.4 cm, straight, tetragonous, deflexed, tomentose, mucronate; seeds ca. 5, cuboid, angular, pitted, 2 mm.

Common almost in all districts, in open forests, wastelands and roadsides.

Fl. & Fr.: September-December

Sarvaipet (ADB), *GO* 4510; Manchippa (NZB), *TP* & *BR* 6003; Mingaram RF (NZB), *TP* & *BR* 6311; Poosanpalle (MDK), *TP* & *CP* 11951; Kusumasamudram RF (RR), *MSM* & *KH* 11030; Aklaspur (KRN), *GVS* 25650 (MH).

Indigofera karnatakana Sanjappa, Taxon 32 : 120. 1983. *I. tenuifolia* Rottl. ex Wight & Arn., Prodr. Fl. Ind. Orient. 200. 1834, non Lamk. 1789; Baker in Hook.f., Fl. Brit. India 2 : 95. 1876; Gamble, Fl. Madras 1: 311. 1918.

Small erect, sparsely appressed pubescent herbs, up to 50 cm high. Leaves pinnate, up to 2 cm long, leaflets 7-11, opposite, small, linear- oblanceolate, 0.6-1.8 × 0.2-0.3 cm, appressed white-hairy, base narrow, margin entire, apex apiculate; stipules setaceous. Flowers bright red, minute, in 5-6-flowered racemes; racemes equaling or slightly exceeding the leaves; peduncle to 4 mm; calyx subulate, 3.5 mm long, lobes longer than tube; corolla bright red, standard orbicular, 3.5 mm, wings shorter than the keels, 2.5 mm, keels support filaments; ovary 2.5 mm. Pod subcylindrical, straight, 2-3 cm long, slightly torulose; 10-12-seeded.

Rare in Karimnagar district (Naqvi, 2001).

Fl. & Fr.: September – January.

Note: *Indigofera tenuifolia* Rottl. ex Wight & Arn. (1834) is antedated by *I. tenuifolia* Lamk. (1789). Hence Sanjappa (*op. cit.*) renamed it as *I. karnatakana.*

Indigofera linifolia (L.f.) Retz., Observ. Bot. 4 : 29. 1786; Baker in Hook.f, Fl. Brit. India 2: 92. 1876. incl. var. *campbelli*; Gamble, Fl. Madras 1: 309. 1918. *Hedysarum linifolium* L.f., Suppl. Pl. 331. 1781.

Vern.: *Gudlaku.*

Prostrate, annual herbs, up to 40 cm long, branchlets silvery-canescent. Leaves simple, sessile, linear, 1.5-3 × 0.1-0.4 cm, appressed-pubescent, base cuneate, margin entire, apex acute, mucronate. Flowers bright red in axillary 6-8-flowered racemes; calyx clothed with silvery hairs, tube 1 mm, lobes linear, standard ovate, 3.5 × 2.5 mm, pubescent, wings oblong, 3 × 1.2 mm, glabrous, keels 4 × 1 mm, with 1 mm spur; stamens diadelphous (9+1); ovary sessile, 1 mm, style 2 mm. Pod globose, silvery-pubescent, 2 mm; seed solitary, globose.

Common in almost all districts, in wastelands, roadsides and occasional in open forests.

Fl. & Fr.: September-December.

Sone (ADB), *TP* & *PVP* 4023; Nasrullabad RF (NZB), *TP* & *BR* 6160; Pocharam tank (NZB), *BR* 9505; Pillutla RF (MDK), *BR* & *CP* 11258; Patancheru (MDK), *BR* & *CP* 11311; Anantagiri RF (RR), *MSM* 11078; Lingampalle (RR), *MSM* 12762; Bottichalami (NLG), *KMS* 9797 (CAL); Way to Pagarikutta-Narsapur (MDK), *KMS* 6643 (MH); Kodimial (KRN), *GVS* 21873 (MH); Kinnerasani dam area (KMM), *RR* 112684 (BSID).

Indigofera linnaei Ali, Bot. Not. 111: 549. 1958. *Hedysarum prostratum* L., Mant. Pl. 102. 1767. *Indigofera enneaphylla* L., Mant. Pl. 2. 272. 1771, *nom. illeg;* Baker in Hook.f., Fl. Brit. India 2 : 94. 1876; Gamble, Fl. Madras 1: 309. 1918.

Prostrate, trailing, profusely branched, slender herbs, up to 50 cm long, branchlets appressed white hirsute. Leaves odd-pinnate, to 3.5 cm, leaflets 4 or 5 pairs, alternate, oblanceolate-obovate, 1 × 0.4 cm, membranous, densely hirsute below, base cuneate, margin entire, apex obtuse, retuse, stipules 2, falcate, scarious, caudate-acuminate. Flowers bright red in axillary 4-5-flowered, sub-capitate racemes; peduncle to 1 cm; calyx pubescent, tube 1.5 mm, lobes 2.5 mm, setaceous; corolla bright red, exserted, glabrous, petals clawed, standard obovate, 5 × 4 mm, wings 5.5 ×1.8 mm, keels 5 ×1.2

mm with 1 mm spur; stamens diadelphous (9+1), staminal sheath, 2.5 mm, filaments 0.5 mm; ovary stipitate, 2 mm, style 2.5 mm, curved. Pods cylindric, turgid, white-hirsute, seeds 2, globose, angular, 3 mm.

Common in wastelands and open forests in almost all districts.

Fl. & Fr.: May-December

Bijjur RF (ADB), *GO* & *PVP* 4883; Rajakampet (NZB), *BR* & *KH* 9053; Toopran (MDK), *BR* & *CP* 11278; Mombojipally (MDK), *BR* & *CP* 11639; Malkalcheruvu (RR), *MSM* & *KH* 10281; Moosi river bank (HYD), *KMS* 5980 (MH); Nagarjunasagar (NLG), *KMS* 19292 (MH); Khammam (KMM), *D.P. Raturi* 2517 (DD); Kodimial (KRN), *GVS* 20046 (MH); Kudurpalli (KRN), *SLK* 70751 (NBG); Billyguttalu (KMM), *RR* 105981 (BSID); North of Pasra (WGL), *RKP* 108245 (BSID).

Note: The earliest valid name for the plant is *Hedysarum prostratum* L. Mant. 1: 102. 1767. According to articles 55 and 70 of the Intern. Code of Bot. Nomencl. the correct name should be *Indigofera prostrata*. As Linnaeus did not make this combination his new name *I. enneaphylla* is illegitimate. Nor can the earliest epithet be reinstated now, as it would be a later homonym of Willdenow 1808. Hence Ali has chosen the new name "*Indigofera linnaei*".

Indigofera mysorensis Rottler ex DC., Prodr. 2: 222. 1825; Baker in Hook.f., Fl. Brit. India 2 : 102. 1876; Gamble, Fl. Madras 1: 313. 1918.

Erect undershrubs, up to 1.5 m high, branchlets viscid, villous. Leaves odd-pinnate, leaflets 5 pairs, opposite, elliptic-oblong, 0.5-1 × 0.2-0.5 cm, chartaceous, sparsely pubescent above, densely pubescent below, base and apex obtuse, margin entire; petiolule 1 mm. Flowers red in axillary panicled racemes; bracts 2 mm, foliaceous, 1-3-lobed; calyx incised to more than half-way, tube 0.5 mm, lobes lanceolate, to 2.5 mm; corolla exserted, pubescent, petals clawed, standard oblong-elliptic, 7 × 3.5 mm, wings 6 × 2.5 mm, keels 7 × 2 mm; stamens diadelphous (9+1), staminal sheath 5 × 1 mm, filaments 0.5 mm; ovary 2 mm, style 4 mm. Pod subcylindrical, oblong-angular, up to 1× 0.2 cm, sparsely pilose, generally 3-4-seeded; seeds cuboid, to 1 mm.

Common in dry deciduous forests of Mahabubnagar district.

Fl. & Fr.: July-January.

Indigofera nummulariifolia (L.) Livera ex Alston in Trimen, Handb. Fl. Ceylon 6 (Suppl.): 72. 1931. *Hedysarum nummularifolium* L., Sp. Pl. 746. 1753. *Indigofera echinata* Willd., Sp. Pl. 3 :1222. 1802; Baker in Hook. f., Fl. Brit. India 2 : 72. 1876; Gamble, Fl. Madras 1: 308. 1918.

An annual or perennial spreading hairy herb, up to 60 cm, branchlets appressed tomentose. Stipules linear, 4 mm. Leaves simple, alternate, obovate-rotund, 1-2 × 0.8-1.6 cm, base cuneate, margin entire, apex rotund, apiculate; petiole 1 mm. Flowers pink in axillary racemes; peduncle to 2.5 cm, pedicel 0.5 mm; bracts 2.5 mm; calyx tube 1 mm, lobes linear, to 2 mm; corolla red, petals with short claws, standard elliptic, 2 × 1.5 mm, pubescent without, wings 2 × 0.5 mm, keels elliptic, 2.2 × 0.5 mm;

stamens diadelphous (9+1), staminal sheath 1 mm, filaments 4 mm; ovary 1 mm, style 0.5 mm. Pods sickle-shaped, to 5 × 2 mm, beaked with hooked bristles in several rows along the broad ventral suture; seed solitary, laterally compressed, grooved in the middle.

Rare weed of rocky areas and rock crevices in most districts.

Fl. & Fr.: August-October.

Annaram RF (NZB), *BR* & *CPR* 7184; Pocharam tank (MDK), *TP* & *CP* 12051; Rampur (MDK), *TP* & *CP* 14070; Anantasagar RF (RR), *MSM* & *KH* 11018.

Indigofera prostrata Willd., Sp. Pl. 3 : 1226. 1803; Gamble, Fl. Madras 1: 310. 1918; Santapau, Rec. Bot. Surv. India 16: 55. 1967. *I. trifoliata* auct. non L.; Baker in Hook.f., Fl. Brit. India 2 : 96. 1876 *p.p.*

Flexuous, prostrate, branched herbs, up to 60 cm long, branchlets appressed-pubescent. Leaves 3-foliolate, to 3.5 cm, leaflets elliptic-obovate, 1-2.5 × 0.4-0.8 cm, chartaceous, appressed-pubescent on both surfaces, base cuneate, margin entire, apex obtuse; petiole to 1 cm, petiolule 1 mm. Flowers pinkish, in 4-8-flowered, axillary racemes; calyx pubescent, tube 0.5 mm, lobes setaceous, to 1 mm; corolla red to purple, exserted, standard elliptic-oblong, 4.5 × 3.5 mm, wings oblong, 3.5 × 1.2 mm, keels 4 × 1.5 mm; stamens diadelphous (9+1), staminal sheath 3 mm, filaments 0.5 mm; ovary sessile, 2.5 mm, style 1 mm. Pod filiform, 1.5 × 0.2 cm, sparsely hairy, terete, deflexed, slightly winged at the sides of sutures, torulose; seeds ca. 8, oblong, 0.5 mm.

Occasional in Hyderabad, Ranga Reddi and Mahabubnagar districts in waste lands and hilly regions.

Fl. & Fr.: November-February

Pargi (RR), *MSM* & *PRSR* 12798; Behind R.R. Labs (HYD), *B.K. Vijayakumar* 613 (MH).

Indigofera spicata Forssk., Fl. Aegypt.-Arab. 138. 1775. *I. endecaphylla* Jacq., Ic. Pl. Rar. 3(3): t. 570. 1789. "*hendecaphylla*" Baker in Hook.f., Fl. Brit. India 2 : 98. 1876; Gamble, Fl. Madras 1: 311. 1918.

Erect or trailing herbs, branchlets appressed-strigose or glabrescent. Leaves odd-pinnate, to 6 cm, leaflets 5 pairs, alternate, oblong-lanceolate, 1-2.5 × 0.6-1 cm, chartaceous, densely appressed pubescent below, base cuneate, margin entire, apex obtuse, apiculate, stipules lanceolate; petiole 4 mm. Flowers red, in axillary racemes; calyx appressed strigose, tube 0.5 mm, lobes setaceous, to 3 mm; corolla sparsely strigose without, standard orbicular, 4 mm, wings 4 mm, keels 5 × 2.2 mm with 1 mm long spur; stamens diadelphous (9+1), staminal sheath 4 mm; ovary sessile, 4 mm, style incurved, 1mm. Pods straight, tetragonous, 2.5 cm, deflexed, sparsely pubescent, grouped at the base of rachis, mucro sharp-pointed; seeds 6-10, oblong, to 3 mm.

Occasional in open scrub and dry deciduous forests in Ranga Reddi district.

Fl. & Fr.: August-February.

Vikarabad (RR), *B.K. Vijayakumar* 1290 (CAL).

Indigofera tinctoria L., Sp. Pl. 751. 1753; Baker in Hook.f., Fl. Brit. India 1 : 220. 1876; Gamble, Fl. Madras 1: 319. 1918. *I. sumatrana* Gaertn., Fruct. 2: 317. t. 148. t. 4. 1791; Gamble, Fl. Madras 1: 312. 1918.

Vern: *Auiri, Nili*

Woody, stiff undershrubs or shrubs, up to 2 m high, branchlets terete, faintly angular when young, appressedly pubescent. Leaves stipulate, 9-foliolate, leaflets bluish-green, ovate-oblong, 1-3 × 0.5-1.5 cm, base acute, margin entire, apex apiculate to acute, glabrescent above, apppressedly puberulous below; petiolule to 1.5 mm. Flowers pink in axillary spike like racemes; peduncle very slender, pedicel 2 mm; calyx hairy, lobes rather more than twice as long as tube, lobes lanceolate; corolla pink, standard sub-orbicular; stamens diadelphous (9+1); ovary hairy, 3.5 mm, style 0.5 mm. Pods paired, cylindric, to 2.8 × 0.2 cm, slightly falcate towards tips, faintly torulous, thick, brown; seeds 10-12, globose.

Occasional in waste lands and open forests in all districts. Generally cultivated for extraction of indigo.

Fl. & Fr.: August-November.

Sone (ADB), *TP* & *PVP* 4064; Adlur Yellareddy (NZB), *TP* & *BR* 6416; Thornal (MDK), *CP* 11708; Anantasagar RF (RR), *MSM* & *NV* 10385; Bhupatipur (KRN), *GVS* 20194 (MH); Akkannapet RF (MDK), *RG* 113404 (BSID); Jakaram (WGL), *RKP* 105306 (BSID); Rathamhutta hills (KMM), *RCS* 104236 (BSID).

Indigofera trifoliata L., Cent. Pl. 2 : 29. 1756; Baker in Hook. f., Fl. Brit. India 2 : 96. 1876 *p.p.;* Gamble, Fl. Madras 1: 310. 1918; Ali, Bot. Not. 111. 552. 1958. *I. barberi* Gamble, Bull. Misc. Inform. Kew 1918 : 222. 1918 & Fl. Madras 1: 310. 1918.

Vern: *Baragadamu.*

Erect undershrubs, up to 60 cm high, branchlets brownish pubescent. Leaves 3-foliolate, to 1.5 cm, leaflets oblanceolate, 1-1.7 × 0.2-0.4 cm, appressed pubescent, brown gland dots beneath, base cuneate, margin entire, apex obtuse, petiole shorter than leaflets, terminal leaflets sessile, stipules setaceous. Flowers red, in short 6-10-flowered racemes; calyx tube 1.3 mm, lobes setaceous, to 1 mm; corolla flame coloured, standard oblong, to 4.5 mm, pubescent. Pods linear, 1.5 × 0.2 cm, slightly 4-angled, deflexed, hoary, narrowly winged along margins; seeds 3-4, cuboid, angular.

Common in waste lands and open forests in Nizamabad, Medak, Hyderabad, Mahabubnagar, Warangal, Karimnagar and Ranga Reddi districts.

Fl. & Fr.: June-December

Bhimnagar (NZB), *TP* & *BR* 6222; Poosanpalle (MDK), *TP* & *CP* 11955; Behind RR Labs, Hyderabad city (HYD), *B.K. Vijayakumar* 613 (CAL, BSID); Kodimial (KRN), *GVS* 25668 (MH).

Note: Gamble treats *I. trifoliata* L. and *I. prostrata* Willd. as distinct species. Ali after checking the specimens of the latter identified by Gamble and kept in Kew Herbarium concluded that these are the same as *I. trifoliata* L. But here Gamble's view is being followed and treated them as separate species. *I. prostrata* is distinct from *I. trifoliolata* L. as pods are filiform, narrowly winged, more than 5-seeded.

Indigofera trita L. f., Suppl. Pl. 335. 1781; Baker in Hook.f., Fl. Brit . India 2: 96. 1876; Ali, Fl. W. Pakistan 100. 78. 1977. var **trita**

Vern.: *Nakkanaru*

Grey pubescent undershrubs, up to 1.5 m high. Leaves 3-foliolate, 4 cm long, leaflets oblong-obovate, 1.2-2.5 × 0.7-1.5 cm, appressed, silvery pubescent, base cuneate-obtuse, margin entire, apex mucronate; petiole to 1.5 cm, petiolule to 1 mm. Flowers pink, 8-12-flowered in axillary racemes, pedicel 1 mm, bracts setaceous; calyx pubescent, tube 1.5 mm, lobes 1.5 mm, subulate; corolla as long as calyx, petals clawed, standard 4 mm, orbicular, wings 4 mm, keels 5 mm; stamens diadelphous (9+1), staminal sheath 2.5 mm; ovary 2 mm, style 3 mm. Pods oblong, 4-angular, straight, 3 × 0.2 cm, deflexed or divaricate, appressed grey pubescent; seeds 6-8, rectangular, 2mm.

Occasional in forests and wastelands of many districts.

Fl. & Fr.: September- January.

Roopla thanda (NZB), *TP & BR* 6400; Gannaram (NZB), *TP & BR* 6408; Anantagiri RF (RR), *TP & MSM* 11064; Appapur (MBNR), *BSS & SKB* 30735; Kodimial (KRN), *GVS* 20116 (MH); Pakhal (WGL), *KMS* 11603; Azampur (NLG), *BVRS* 477 (NU); Amangal to Shadnagar (MBNR), *SRS* 104569 (BSID); Sathiampet to Penugadapa (KMM), *RR* 112645 (BSID); Kunavaram RF (KMM), *RCS* 98788 (BSID).

Indigofera wightii Graham ex Wight & Arn., Prodr. Fl. Ind. Orient. 202. 1834; Baker in Hook.f., Fl. Brit. India 2 : 99. 1876; Gamble, Fl. Madras 1: 313. 1918.

Erect undershrubs, up to 3 m high, branchlets silvery pubescent. Leaves odd-pinnate, to 8 × 1.5 cm, leaflets 11-23, opposite, elliptic 0.6-1 × 0.3-0.6 cm, appressed pubescent on both surfaces, base and apex obtuse, apiculate, margin entire; petiole to 1 cm, petiolule 1 mm. Flowers purple, dense, axillary, 15-25-flowered racemes; pedicel 2 mm; calyx pubescent, tube 0.5 mm, lobes triangular, 0.5 mm; corolla exserted, pubescent without, standard 6 × 3 mm, oblong, elliptic, hispid without, wings falcate, 4.5 mm, keels 6 × 2 mm with 1.5 mm long spur; stamens diadelphous (9+1), staminal sheath 1.7 mm, filaments 0.5 mm; ovary pubescent, 2.5 mm, style 1 mm. Pod linear, subcylindrical, erect, grey-pubescent, to 4 × 0.3 cm; seeds 8-12, ovoid-globose, prominently striate.

Occasional on hill slopes in Medak and Ranga Reddi districts.

Fl. & Fr.: September-April.

Vikarabad (RR), *B.K. Vijayakumar* 1293 & 1291 (CAL); Nallavally RF, Narsapur (MDK), *RG* 113477 (BSID).

LABLAB Adans.

Lablab purpureus (L.) Sweet, Hort. Brit. ed. 1. 481. 1826, var. **purpureus;** Verdc., Kew Bull. 24: 410. 1970. *Dolichos purpureus* L., Sp. Pl. ed. 2. 1021. 1763. *D. lablab* L., Sp. Pl. 725. 1753; Baker in Hook. f., Fl. Brit. India 2 : 209. 1876. *p.p.;* Gamble, Fl. Madras 1: 367. 1918.

Vern. : *Chikkudu.*

Vines, branchlets glandular-pubescent. Leaves 3-foliolate, terminal leaflets ovate-deltoid, laterals ovate, inequilateral, 5-8 × 4-7 cm, chartaceous, pubescent, base truncate-obtuse, margin ciliate, apex acuminate, apiculate. Flowers pinkish violet in axillary racemes. Pods septate, 7 × 1.5 cm; seeds 3-6.

Cultivated in fields and gardens, rarely as an escape.

Fl. & Fr.: November-May.

Nirmal (ADB), *GO* & *DAM* 5034; Nagarjunakonda valley (NLG), *KTH* 9756 (CAL).

Uses : Fruits and seeds as vegetable.

MACROPTILIUM (Benth.) Urb.

1. Creeping herb; pod torulose covered with stellate silvery hairs .... **M. atropurpureum**

1. Erect herbaceous shrubs; pod not torulose, glabrous when mature .... **M. lathyroides**

Macroptilium atropurpureum (DC.) Urb., Symb. Antill. 9. 457; 1928; Subba Rao & Gopalan, J. Bombay Nat. Hist. Soc. 77: 357. 1980. *Phaseolus atropurpureus* DC., Prodr. 2: 395. 1825.

Slender creeping herb, up to 4 m long, branchlets terete, obscurely striate, grey tomentose. Leaves 3-foliolate, stipules minute, silky, reflexed, narrowly deltoid, stipels minute, subulate; leaflets ovate or obovate, terminal one rhomboid, 1.3-3.5 × 0.7-2.9 cm, base round or truncate, margin entire, apex acute-apiculate, pubescent above, velvety below; prominently looks like butterfly wings; petiole 1.5-2 cm, petiolule 2-5 mm. Flowers dark brownish-purple in axillary racemes; peduncle to 20 cm, pedicel to 3 mm; calyx green tomentose, tube 4 mm, 3 lobes short, to 2.5 mm, cover lobes 2, narrow, to 3 mm, deltoid, acuminate; standard petal 1.2 × 1 cm, wings deeply coloured, longer than vexillum, 1.6 × 1.4 cm, keels incurved, 1.6 × 0.27 cm; stamens diadelphous (9+1), staminal sheath 1.2 cm, filaments 6 mm; ovary linear, 7 mm, style 1 cm. Pod cylindrical, to 6 × 0.25 cm, beaked, covered with stellate silvery hair, torulose; seeds ca. 10, dark brown, 3 × 1.5 mm.

Introduced as a fodder crop, long back in 1970, now it is running wild in Warangal (C.S. Reddy, 2001) Karimnagar (Naqvi, 1971) and Hyderabad districts (Ramana, 2010).

Fl. & Fr.: August-December

Macroptilium lathyroides (L.) Urb. var. **semierectum** (L.) Urb., Symb. Antill. 9. 457. 1928. *Phaseolus semierectus* L., Mant. Pl. 100. 1707; Baker in Hook.f., Fl. Brit. India 2 : 201. 1876; Gamble, Fl. Madras 1: 362. 1918.

Annual, erect, herbaceous shrubs, up to 1.5 m high, branchlets clothed with long, deflexed, deciduous hairs. Leaves 3-foliolate, leaflets elliptic-lanceolate, oblong-lanceolate, 5 × 2 cm, glabrous above, appressed-pubescent beneath, base cuneate, margin entire, apex acute, minutely apiculate. Flowers purple in spicate, 25 cm long, many-flowered racemes; calyx campanulate; stamens diadelphous (9+1). Pod linear, 8-10 cm long, cylindric, sharply beaked, silvery-pubescent when young, nearly glabrous when mature; seeds about 20, dark-brown.

Occasional in wastelands in Hyderabad district.

Fl. & Fr.: January- July.

Katasintha (HYD), *KMS* 6007 (MH).

MACROTYLOMA (Wight & Arn.) Verdc.

1. Leaflets elliptic-ovate, obtuse, margin prominently ciliate; calyx lobes lanceolate; 3-4-seeded .... **M. ciliatum**

1. Leaflets ovate-acute, margin entire; calyx lobes linear-setaceous; 5-7-seeded .... **M. uniflorum**

Macrotyloma ciliatum (Willd.) Verdc., Kew Bull. 24 : 404. 1970 & in Hook. Icon. Pl. t. 3776. 1976. 1982. *Dolichos ciliatus* Klein ex Willd. Sp. Pl. 3 : 1049. 1982; Hook.f. Fl. Brit. India 2 : 210. 1876; Gamble, Fl. Madras 1 : 367. 1918. *Macrotyloma dispermus* sensu Sanjappa

Large twiner; branches glabrous. Leaves trifoliate; leaflets ovate-rhomboid, 4-6 x 1-2 cm. inequilateral, thin coriaceous, puberlous, base and apex obtuse, apiculate, margins ciliate. Flowers axillary in racemes; calyx 5 mm long; teeth lanceolate; corolla cream; standard suborbicular; wings oblong-falcate. Pod falcate, compressed, prominantely nerved, horned; seeds 1-3, oblong-reniform.

Occasional on hedges and bushes in scrub jungles (Ramana, 2010).

Fl. & Fr.: November-March.

Macrotyloma uniflorum (Lam.) Verdc., Kew Bull. 24: 322. 1970. *Dolichos uniflorus* Lam., Encycl. 2: 299. 1786. *D. biflorus* auct. non L.; Baker in Hook.f., Fl. Brit. India 2 : 210; Gamble, Fl. Madras 1: 367. 1918.

Vern.: *Vulavalu*. English: Horsegram.

Densely villous climbers, up to 1 m long. Stipules lanceolate, to 6 mm, stipel 1.5 mm. Leaves 3-foliolate, leaflets elliptic-lanceolate, ovate-rhomboid, inequilateral, 2-3.5 × 1-2 cm, chartaceous, villous, base obtuse-cuneate, margin entire, apex acute, apiculate; petiole to 3 cm, petiolule to 4 mm. Flowers yellowish with purplish blotch in the centre, axillary, solitary or in clusters; calyx pubescent, tube 2 mm, lobes linear, setaceous, upper lobes connate, lower ones to 4 mm; corolla yellowish, standard obovate, oblong, 1.1 × 0.9 cm, wings 1 × 0.15 cm, keels 1.2 × 0.3 cm; stamens diadelphous, staminal sheath 9 × 1.5 mm, filaments to 3 mm; ovary subsessile, to 6 mm, style 5 mm. Pod oblong, to 5 × 0.6 cm, pubescent, seeds 4-6, oblong, grey.

Cultivated plant, rarely as an escape.

Fl. & Fr.: August-March.

Note: This is the "Horse Gram" cultivated for its seeds and leaves, which are used as a fodder for horses and cattle.

MEDICAGO L.

1. Pods 1-seeded .... **M. lupulina**
1. Pods 3-many-seeded:
    2. Erect perennial herbs; corolla purple, blue or white .... **M. sativa**
    2. Diffuse or trailing annual herbs; corolla yellow .... **M. polymorpha**

Medicago lupulina L., Sp. Pl. 779. 1753; Baker in Hook.f., Fl. Brit. India 2: 90. 1876.

Herbs, to 30 cm high, pubescent; branches downy, grooved. Leaves 3-foliolate; leaflets obovate-cuneate, 6 × 6 mm, apex acuminate; stipules subdentate at base; petioles 1-2 cm long. Flowers pale blue, cream when dry, in spicate, axillary racemes. Pod indehiscent, subglobose, glabrous; seed solitary, yellowish brown.

Weed of agricultural fields, sometimes cultivated as forage crop in Karimnagar district (Naqvi, 2001)

Fl. & Fr.: December – March.

Medicago polymorpha L., Sp. Pl. 779. 1753, emend. Shinners, Rhodora 58 : 310. 1956. *M. denticulata* Willd. Sp. Pl. 3 : 1414. 1802; Hook.f., Fl. Brit. India 2 : 90. 1876.

Prostrate or ascending, spreading annual herb. Leaves 3-foliolate; petiole 1.5-4 cm long; stipules laciniate, with linear, 3-4 mm long segments; leaflets obovate, 0.8-2 x 0.6-1.2 cm; base cuneate, margin slightly denticulate in upper part, apex deeply emarginate and mucronate. Flowers 1-4 in short peduncled racemes; calyx 2.2-2.5 mm long, hairy in upper part; teeth lanceolate, as long as tube; corolla yellow, 3-4 mm long. Pods with 2-4 spirals, subglobose, with spiny margins and reticulatovenose outer face, 3-4-seeded, 5-6-mm across.

Cultivated as fodder crop for rabbits and also escaped, run wild in gardens in Hyderabad district (Ramana, 2010).

Fl. & Fr.: December – May.

Medicago sativa L., Sp. Pl. 778. 1753; Baker in Hook.f., Fl. Brit. India 2: 90. 1876; Gamble, Fl. Madras 1: 304. 1918.

Pubescent herbs, to 80 cm high, densely pubescent. Leavs 3-foliolate, leaflets oblanceolate, terminal slightly larger than laterals, obtuse or sub-acute and denticulate at apex. Flowers white or blue in long peduncled, axillary racemes. Pods spirally twisted.

Cultivated as a fodder crop in Karimnagar district (Naqvi, 2001).

Fl. & Fr.: October – January.

MELILOTUS Mill.

1. Corolla white, about 0.4-0.6 cm long; pod often 2-seeded,
    0.4 cm across; biennial .... **M. albus**

1. Corolla yellow, minute, about 0.2 cm long; pod usually 1-seeded,
    0.2 cm across; annual .... **M. indicus**

Melilotus albus Desr. in Lam., Encycl. 4 : 63. 1796; Baker in Hook. f., Fl. Brit. India 2 : 89. 1876; Gamble, Fl. Madras 1: 303. 1918.

Biennial erect herbs, up to 60 cm high. Leaves pinnately 3- foliolate, leaflets obovate, oblong, 1.5-2 × 0.5-0.8 cm, base cuneate, margin serrate, apex mucronate. Flowers white, in axillary racemes. Pod 0.4 cm across, obscurely reticulate- lacunose; generally 2-seeded.

Common weed in cultivated fields in all districts.

Fl. & Fr.: January-April.

Neredugunda (ADB), *GO* & *DAM* 5102, 5103, 5104; Kandakurthi (NZB), *BR* 7274; NSF (MDK), *BR* & *CP* 11662.

Melilotus indicus (L.) All., Fl. Pedem. 1: 308. 1785; Gamble, Fl. Madras 1: 303. 1918. *Trifolium indicum* L., Sp. Pl. 765. 1753. *Melilotus parviflora* Desf., Fl. Atlant. 2 : 192. 1799; Baker in Hook. f., Fl. Brit. India 2 : 89. 1876.

Annual small herbs, up to 15 cm high. Leaves pinnately 3-foliolate, leaflets obovate-oblong, 1.5 × 0.6 cm, glabrous or with few hairs beneath, base cuneate, margin serrate, apex mucronate. Flowers pale yellow in axillary racemes, corolla minute, standard exceeding wings and keel. Pod oblong, indehiscent, 0.2 cm across, usually 1-seeded.

A cold weather weed in fields in all districts.

Fl. & Fr.: April-September.

Kadthala (ADB), *GO* & *DAM* 5007; Rudroor (NZB), *BR* & *KH* 9638.

MILLETTIA Wight & Arn.

Millettia extensa (Benth.) Baker in Hook.f., Fl. Brit. India 2: 109. 1876. *Otosema extensa* Benth. in Miq. Pl. Jungh. 249. 1852. *M. auriculata* Baker ex Brandis, For. Fl. 138. 1874; Baker in Hook.f., Fl. Brit. India 2 : 108. 1876; Gamble, Fl. Madras 1: 322. 1918.

Large woody climber, up to 7 m long, bark light brown, branchlets tomentose. Leaves odd-pinnate, 20-30 cm long, leaflets 7-9, obovate or oblong, 5-12 × 4-7 cm, terminal one larger, cuspidate, thinly silky, base obtuse, margin entire, apex acute; petiole to 10 cm, petiolule to 7 mm. Flowers yellow, in axillary racemes; peduncle to 5 cm, pedicel to 7 mm; bracts linear, minute; calyx densely silky, 3 mm, lobes short. Pod

11-14 × 2-3 cm, velvety brown, sutures thickened; seeds dark-brown, rounded and compressed.

Occasional in deciduous forests in Adilabad district.

Fl.: July-September; Fr.: October-March

Sirpur (ADB), *GO* & *PVP* 4903.

MUCUNA Adans. *nom. cons.*

1. Pods densely covered by grey bristles .... **M. pruriens**
1. Pods softly villose (velvety) .... **M. nivea**

Mucuna nivea DC., Prodr. 2: 406. 1825.; Baker in Hook. f., Fl. Brit. India 2: 188. 1876.

Quick growing, extensive twiner; branchlets glabrescent. Leaves 3-foliolate, terminal one rhomboid-ovate, acuminate, cuneate at base, laterals inequilateral, oblique at base. Flowers showy, pink-rose in 30 cm or more long pendulous panicles. Pods sigmoid, softly tomentose or velvety, 5-6-seeded.

Cultivated for its pods, used as vegetable in Karimnagar district (Naqvi, 2001).

Fl. & Fr.: September – February.

Mucuna pruriens (L.) DC., Prodr. 2 : 405. 1825; Baker in Hook. f., Fl. Brit. India 2: 189. 1876. *Dolichos pruriens* L., Herb. Amb. 23. 1754. *Mucuna prurita* Hook., Bot. Misc 2 : 348. 1831; Gamble, Fl. Madras 1: 355. 1918.

Key to Varieties
1. Leaflets white appressed pubescent beneath; calyx bristly;
    pod 5-7.5 cm long, curved like an 'S'-shape .... var. **pruriens**
1. Leaflets golden silky-tomentose beneath; calyx not bristly;
    pod 7.5-10 cm long, curved .... var. **hirsuta**

Mucuna pruriens (L.) DC. var. **hirsuta** (Wight & Arn.) Wilmot-Dear, Kew Bull. 42: 44. 1987. *M. hirsuta* Wight & Arn., Prodr. Fl. Ind. Orient. 254. 1834; Baker in Hook.f., Fl. Brit. India Fl. Brit. India 2 : 187. 1876; Gamble, Fl. Madras 1: 355. 1918; Pullaiah & Chennaiah, Fl. Andhra Pradesh 1: 300. 1997.

Annual, woody climbers, branchlets slender, densely clothed with short, fine, deflexed ferruginous or grey hairs. Leaves 3-foliolate, leaflets ovate-lanceolate, rhomboid, 6-8 × 3-5 cm, base unequally truncate, margin entire, apex acute, densely golden-silky-tomentose beneath, sparsely so above, lateral nerves about 9. Flowers pale violet, in axillary racemes; bracts small, ovate to lanceolate; calyx lobes lanceolate, corolla pale violet. Pods oblong, 7.5-10 × 1-2 cm, flattened, out-curved at tips, densely silky brown-pubescent with irritant bristles, seeds 5, rounded.

Rare along the streams in the forests in Warangal district.

Fl.: November – January; Fr.: January onwards.

Pakhal RF (WGL), *KMS* 11635 (MH); Gangaram (WGL), *PVS* 76934 (BSID).

Mucuna pruriens (L.) DC. var. **pruriens**

Vern: *Pilliadugu*

Annual, slender climbers, up to 10 m long, branchlets terete, striate, downy-pubescent. Stipules lanceolate, 6 mm, stipels subulate, to 4 mm. Leaves trifoliolate, to 35 cm, leaflets ovate-rhomboid, 6-13 x 5-10 cm, coriaceous, appressed white-pubescent, base unequally truncate, margin entire, apex mucronate, petiole to 20 cm, petiolule to 5 mm. Flowers dark purple with yellow shade on standard petal in long axillary racemes; peduncle to 6 cm, pedicel to 8 mm; calyx tube 7 mm, lobes lanceolate, to 8 mm, pubescent; standard ovate, 2 × 1.6 cm, wings 3.3 × 1.1 cm, keels 3.7 × 0.5 cm; stamens diadelphous (9+1), staminal sheath 3 × 0.6 cm, filaments 8 mm; ovary to 6 mm, style 3 cm. Pods curved at base and apex prominently S-shaped, densely silky pubescent with irritant bristles; seeds 6, orbicular, black, shining.

Occasional in forests in Adilabad, Nizamabad, Medak, Ranga Reddi, Karimnagar, Hyderabad and Mahabubnagar districts.

Fl.: December; Fr.: January – March.

Kadam river bank (ADB), *TP* & *PVP* 4124. Neela (NZB), *BR* & *GO* 9225; Maddigunta (NZB), *BR* & *PSPB* 9555; Pathur (MDK), *TP* & *CP* 12104; Kusumasamudram RF (RR), *MSM* & *KH* 11034; Bhupatipur (KRN), *GVS* 25701 (MH); Yenkur (KMM), *RR* 108537 (BSID).

MUNDULEA (DC.) Benth.

Mundulea sericea (Willd.) A. Cheval, Compt. Rend. Hebd. Seances Acad. Sci. 180: 1521. 1925. *Cytisus sericeus* Willd., Sp. Pl. 2 : 1121. 1802. *Mundulea suberosa* (DC.) Benth. in Miq. Pl. Jungh. 248. 1852; Baker in Hook.f. Fl. Brit. India 2: 110. 1876; Gamble, Fl. Madras 1: 314. 1918.

Deciduous tree, up to 6 m high, bark thick, pale yellow, deeply fissured, branchlets silky-tomentose. Leaves imparipinnate, stipulate, leaflets 8-10 pairs, opposite or sub-opposite, elliptic-lanceolate or oblong, 2-5 × 0.5-1 cm, sericeous, coriaceous, base obtuse, margin ciliate, apex acuminate, apiculate; petiole to 1.8 cm, petiolule 3 mm. Flowers pinkish-violet, in terminal, corymbose racemes; peduncle to 1 cm, pedicel to 1.5 cm; bracts 2.5 mm; calyx campanulate, tube to 6 mm, lobes triangular; petals clawed, standard orbicular, 2.2 × 1.7 cm, sericeous without, wings oblong-ovate, 1.9 × 0.6 cm, keels obovate, 2 × 1.7 cm; stamens diadelphous (9+1), staminal sheath 1.5 × 0.4 cm, filaments 4 mm; ovary 1.9 cm, style glabrous, 4 mm, incurved. Pods linear, to 10 × 1.5 cm, tomentose, compressed, velvety with short golden brown hairs, thickened at the sutures, late dehiscent; 6-12-seeded.

In dry forests on rock hills in Nalgonda and Mahabubnagar districts.

Fl. & Fr.: June-December.

Molachintapally (MBNR), *TDCK* 15308; Mannanur RF (MBNR), *SSR* & *KSM* 15533; Near Domalapenta (MBNR), *VBH* 84951 (BSID); Nagarjunasagar (NLG), *BVRS* 122 (NU).

NEONOTONIA (Wight & Arn.) J.A. Lackey

Neonotonia wightii (Wight & Arn.) J.A.Lackey, Phytologia 37: 210. 1977. *Notonia wightii* Graham ex Wight & Arn., Prodr. Fl. Ind. Orient. 208. 1834. *Glycine wightii* (Graham ex Wight & Arn.) Verdc., Taxon 15: 35. 1966; Pullaiah & Chennaiah, Fl. Andhra Pradesh 1: 286. 1997. *Johnia wightii* (Wight & Arn.) Wight & Arn., Prodr. Fl. Ind. Orient. 449. 1834. *Glycine javanica* auct. non L. : Gamble, Fl. Madras 1: 351. 1918.

Extensively climbing herbs, stems angular, slender, densely clothed with deflexed grey or fulvous hairs, young shoots silky. Stipules small, 5 mm. Leaves 3-foliolate, rachis up to 8 cm long, leaflets unequal in size, lateral leaflets 5-7 × 3.5-4.5, oblique, ovate, base oblique, margin entire, apex acute-apiculate, terminal leaflets 5-8 × 4.5-5.5 cm, base ovate-rhomboid, apex acute to acuminate, apiculate, glabrous above, appressedly hairy and pale beneath; petiole to 5 cm, petiolule 3 mm. Flowers reddish, numerous, crowded in dense axillary racemes; peduncle to 4 cm, pedicel 3 mm; calyx tube 2.5 mm, campanulate, lobes to 3 mm; corolla reddish, exserted, standard suborbicular, slightly auricled above the claw, 5 ×5 mm, wings narrow, 1.5 × 1 mm, keels short, obtuse, 3.5 ×2 mm; stamens monadelphous (10), staminal sheath 2.5 × 1 mm, filaments 1 mm; ovary subsessile, 2 mm, style incurved, 1mm. Pod linear, 2.5 × 0.5 cm flat, villous, densely appressedly hairy, septate between seeds; seeds 4-5, slightly compressed.

Common in dry deciduous and open forests in Medak district.

Fl. & Fr.: October-February.

Domadugu (MDK), *BR* & *CP* 14199.

PARACALYX Ali

Paracalyx scariosus (Roxb.) Ali, Univ. Stud. (Karachi) 5: 95. 1968. *Cylista scariosa* Roxb., Pl. Coromandel t. 92. 1975; Baker in Hook. f., Fl. Brit. India 2 : 265. 1876; Gamble, Fl. Madras 1: 371. 1918.

Woody twinners, branches densely pubescent, Leaves 3-foliolate, up to 10 cm long, terminal leaflets rhomboid-ovate, laterals ovate, inequilateral, 3-6 × 2-3 cm, pubescent, distinctly nerved beneath, base obtuse-subacute, margin entire, apex acuminate, apiculate; petiole to 5 cm, petiolule 3 mm. Flowers yellow, in 8 cm long racemes or panicles; peduncle to 9 mm, pedicel to 1 cm; bracts caducous, 2 mm; calyx pubescent, tube 5 mm, veined, accrescent, campanulate, upper lobes 3.5 × 1 cm, laterals 1 × 0.5 cm, lower lobe 2.2 × 0.9 cm; corolla enclosed within calyx, petals clawed, standard auricled at base, 1.8 × 1 cm, wings oblong, 1.4 × 0.3 cm, keels incurved, 1.5 × 0.6 cm; stamens diadelphous (9+1), staminal sheath 1.2 × 0.3 cm, filaments 6 mm; ovary subsessile, compressed, to 1 cm, style ciliate below the swollen middle, incurved, 8 mm. Pod small, ellipsoid, 5-8 × 4-5 mm, flat, pubescent; seed solitary, reniform, 3 mm.

Common in most of the districts, in open forests.

Fl. & Fr.: October-March.

Gandhari RF (NZB), *TP* & *BR* 6342; Jalalpur RF (NZB), *BR* & *KH* 9666; Gangapur RF (MDK), *TP* & *CP* 12100; Naskal (RR), *MSM* & *PRSR* 15107; Edupayalu RF (MDK), *RG* 113430 (BSID); Naskal (RR), *MSM* & *PRSR* 15107 (BSID).

PHASEOLUS L.

Phaseolus vulgaris L., Sp. Pl. 723. 1753; Baker in Hook. f., Fl. Brit. India 2: 200. 1876; Gamble, Fl. Madras 1: 363. 1918.

Vern.: *Chikkudu.*

Erect dwarf subshrubs, to 30 cm high. Leaves 3-foliolate, terminal oblong-ovate or rhomboid, laterals oblique, lanceolate, petiole up to 10 cm long; stipules basifixed, lanceolate. Flowers white, 2-3 in leaf axils. Pods about 10 cm long, subcompressed, beaked, 5-6-seeded.

Cultivated in most districts for its pod which is used as vegetable.

Fl. & Fr.: November – February.

PHYLLODIUM Desv.

Phyllodium pulchellum (L.) Desv., J. Bot. Agric. 1: 124. 1813. *Desmodium pulchellum* (L.) Benth., Fl. Hong. 83. 1861; Baker in Hook. f., Fl. Brit. India 2: 162. 1876; Gamble, Fl. Madras 1: 344. 1918. *Hedysarum pulchellum* L., Sp. Pl. 747. 1753.

Erect shrubs, up to 2 m high, stems 5-angled, striated, puberulous when young. Leaves 3-foliolate, stipules present, leaflets elliptic-ovate, terminal leaflet 6-9 × 2-3 cm, base ovate, margin wavy, apex apiculate, laterals 2-3.5 × 1.2-2 cm, coriaceous, nerves strong, parallel on the downy under surface, lateral veins 8 pairs; petiole to 5 mm, petiolule to 3mm. Flowers white, 2 or 3 enclosed in a pair of prominent orbicular bracts, in terminal branched about 10-30 cm long racemes, peduncle to 2.5 cm, pedicel 3 mm; calyx tube 2 mm, sparsely pubescent, lobes to 3 mm; corolla white, standard 7 × 4.5 mm, wings 6 × 1.5 mm, keels 7 × 2.5 mm; stamens diadelphous (9+1), staminal sheath 6 × 1.7 mm, filaments 2 mm; ovary 3 × 0.5 mm, sparsely pubescent, style curved, 6 mm. Lomentum 2-3-jointed, reticulate, 7 × 3 mm, indented on both the margins; seeds compressed, reniform.

In Nizamabad, Medak and Mahabubnagar districts as a forest undergrowth.

Fl. & Fr.: August-February.

Bhimanagar RF (NZB), *TP* & *BR* 6208; Gangapur RF (MDK), *TP* & *CP* 12095; way to Farahabad (MBNR), *GS* 18281; North from Pasra (WGL), *RKP* 108223 (BSID).

PONGAMIA Vent. *nom. cons.*

Pongamia pinnata (L.) Pierre, Fl. Forest Cochinch. t. 385. 1899; Thoth., Bull. Bot. Surv. India 3: 418. 1962. *Cytisus pinnatus* L., Sp. Pl. 741. 1753. *Pongamia glabra* Vent., Jard. Mal. t. 28. 203; Baker in Hook. f., Fl. Brit. India 2: 240. 1876; Gamble, Fl. Madras 1: 385. 1918. *Derris indica* (Lam.) Bennet, J. Bombay Nat. Hist. Soc. 68. 303. 1971. *Galedupa indica* Lam., Encycl. 2: 594. 1788.

Vern: *Kanuga.*

Evergreen trees, up to 15 m high, bark soft, greyish-green, branchlets glabrescent. Leaves odd-pinnate, leaflets 5-9, opposite, ovate-obovate, 4.5-7 × 2-4 cm, thin coriaceous, glabrous, glossy above, base obtuse-cuneate, margin entire, apex acuminate; petiole to 5 cm, petiolule to 1 cm. Flowers purplish-white in lax pseudoracemes or panicles; pedicel to 7 mm; calyx tube to 6 mm, campanulate, glabrous, lobes minute; corolla purplish-white, standard orbicular, 1.35 × 0.8 cm, wings 1.2 × 0.3 cm, keels obtuse, 1.5 × 0.3 cm; stamens monadelphous (10), staminal sheath to 7 × 3 mm, filaments 5 mm, ovary subsessile, pubescent, 5 mm, style slightly curved. Pod obliquely oblong, 4-5 × 1.5-2.5 cm, beaked, woody, compressed, indehiscent, yellowish grey when ripe, seeds 1 or 2, compressed, reddish brown.

Common along river banks, often planted near villages and as an avenue tree in all districts.

Fl. & Fr.: February-October.

Nirmal (ADB), *GO* & *DAM* 4990; Ghanapur (NZB), *BR* & *KH* 7108; Narsapur RF (MDK), *CP* 11572; Lingampalle (RR), *MSM* 19216; Mannanur (MBNR), *TDCK* 13608; Vempalli (KRN), *GVS* 20189 (MH); Dindi riverside (NLG), *BVRS* 601 (NU); Pakhal (WGL), *RKP* 108251 (BSID); Parnasala RF (KMM), *RCS* 98771 (BSID).

Note: Thothatri & Nair (Taxon 30 : 44. 1981) considered *Pongamia* and *Derris* as distinct genera.

PSEUDARTHRIA Wight & Arn.

Pseudarthria viscida (L.) Wight & Arn., Prodr. Fl. Ind. Orient. 209. 1834; Baker in Hook. f., Fl. Brit. India 2 : 154. 1876; Gamble, Fl. Madras 1: 334. 1918. *Hedysarum viscidum* L., Sp. Pl. 747. 1753.

Vern: *Nayakuponna.*

Undershrubs, up to 75 cm high, branchlets densely villous, viscid. Leaves 3-foliolate, terminal leaflets ovate-rhomboid, twice as long as lateral, laterals obliquely ovate, 4.5-9 × 2-6 cm, base cuneate-obtuse, margin ciliate, apex acuminate; petiole to 4 cm, petiolule to 3 mm. Flowers pink in pairs or clusters in axillary and terminal lax racemes; calyx campanulate, tube 1.5 mm; corolla exserted, petals clawed, standard suborbicular, 4 × 3.5 mm, wings obliquely oblong, 4.5 ×1.5 mm, keels obtuse 4.5 ×1.5 mm; stamens diadelphous (9+1), staminal sheath 3 mm, filaments 1.5 mm; ovary subsessile, flat, 4 mm, tomentose, style 2 mm, inflexed. Pod linear, oblong, to 1.5 × 0.5 cm, flat, compressed, hooked pubescent, margin undulate, non-septate, apex apiculate; seeds 4-6, finely downy.

Common in some what damp forest under-growth nearly in all districts.

Fl. & Fr.: August – January.

Bijjur RF (ADB), *GO* & *PVP* 4861; Ibrahimpet RF (NZB), *TP* & *BR* 6130; Bhimnagar RF (NZB), *TP* & *BR* 6217; Narsapur RF (MDK), *TP* & *CP* 11982, 14048; Gangapur RF (MDK), *TP* & *CP* 12070; Anantasagar RF (RR), *MSM* 10501; Kambalpally (NLG), *BVRS* 608 (NU); Pakhal RF (WGL), *KMS* 11638 (CAL); Bhadrachalam RF (KMM), *RCS* 104309 (BSID); Koriguttalu (KMM), *RR* 106026 (BSID); Pasra (WGL), *RKP* 108143 (BSID).

PTEROCARPUS L. *nom. cons.*

1. Flowers in axillary racemes; leaflets usually 3, rarely 4-5, broadly ovate or nearly orbicular; pod concavely curved between the stipe and style; restricted .... **P. santalinus**

1. Flowers in terminal panicles; leaflets 5-7, elliptic-oblong; pod convexly curved between the stipe and style; widely distributed .... **P. marsupium**

Pterocarpus marsupium Roxb., Pl. Coromandel t. 116. 1799; Baker in Hook. f., Fl. Brit. India 2: 239. 1876; Gamble, Fl. Madras 1: 385. 1918.

Vern.: *Yegisa, Yegi.*

Large deciduous trees, up to 15 m high, bark thick grey with vertical cracks, exfoliating in small, irregular scales, branchlets pubescent. Leaves odd-pinnate, leaflets 5- 7, alternate, elliptic-oblong, 8-15 × 4-7.5 cm, coriaceous, pubescent below, base obtuse-truncate, margin entire, apex emarginate; petiole to 4 cm, petiolule to 8 mm. Flowers yellow in terminal panicles, pedicel to 3 mm; calyx dark brown, pubescent, tube to 6 mm; corolla yellow, standard 1.2 × 1 cm, wings 1.1 × 0.6 cm, keels to 9 × 6 mm; staminal sheath 4 × 6 mm, stamens monadelphous becoming isodiadelphous, 5 + 5, filaments subequal, to 5 mm; ovary tomentose, 7 mm, style 6 mm. Pod stipitate, 3.5 cm across, convex, pubescent, winged; seeds 1 or 2.

Common in all forest districts, chiefly in deciduous forests.

Fl. & Fr.: May-February.

Bheemaram (ADB), *GO & DAM* 5180; Narsapur (MDK), *SSR & CPR* 18504; Naskal (RR), *MSM & PRSR* 12800; Yerrakunta (MBNR), *TDCK* 16820; Pullayapalle (MBNR), *DV & BR* 39273; Aklaspur (KRN), *GVS* 22513(MH); Kambalpally (NLG), *BVRS* 602 (NU); Rangapuram (WGL), *PVS* 76925 (BSID).

Pterocarpus santalinus L. f., Suppl. Pl. 318. 1781; Baker in Hook. f., Fl. Brit. India 2: 239. 1876; Gamble,Fl. Madras 1: 385. 1918.

Vern.: *Yerrachandanamu, Raktachandanam.*

Tall, deciduous trees, up to 20 m high, bark brownish black, divided into rectangular plates by deep vertical and horizontal cracks, branchlets glabrous. Leaves 3-foliolate, leaflets ovate-orbicular or oblong, 4-6 × 3.5-5 cm, base obtuse-subcordate, margin entire, apex emarginated; petiole to 4 cm, petiolule to 6 mm. Flowers axillary in simple or sparingly branched racemes; pedicel to 5 mm; calyx tomentose, tube 5 mm, lobes ovate; corolla yellow, standard ovate, 2.1 × 1.5 cm, wings 1.8 × 0.8 cm, keels 1.4 × 0.35 cm; stamens monadelphous (10), staminal sheath 4 ×3 mm, filaments to 5 mm; ovary stipitate, 7 mm, style 4 mm. Pods round with a broad wing, 3-4 cm in diameter, brown; seeds 1 or 2, reddish brown.

Planted occasionally in Warangal and Ranga Reddi districts .

Fl. & Fr.: April-July.

PUERARIA DC.

Pueraria tuberosa (Roxb. ex Willd.) DC., Ann. Sci. Nat., Bot. 4: 97. 1825; Baker in
Hook. f., Fl. Brit. India 2 : 197. 1876; Gamble, Fl. Madras 1: 360. 1918. *Hedysarum
tuberosum* Roxb. ex Willd., Sp. Pl. 3: 1197. 1803.

Vern: *Dari gumodi.*

A large woody climber, up to 6 m long, branchlets appressed-velvety. Leaves 3-
foliolate, leaflets ovate, 12-18 × 11-16 cm, glabrescent above, pubescent beneath, base
obtuse, margin ciliate, apex acute, mucronate, lateral leaflets unequal-sided; petiole
to 15 cm, petiolule to 7 mm. Flowers bluish-white in 15-25 cm long, dense panicles;
peduncle to 10 cm, pedicel to 4 mm; bracts small, deciduous, bracteoles caducous;
calyx campanulate, tube 5 mm, tomentose, lobes ovate, upper lobes connate, 3 mm,
lower one 4 mm; corolla exserted, petals clawed, standard ovate-orbicular, 1.5 × 1.2
cm, wings 1.2 × 0.5 cm, keels 1.1 × 0.5 cm; stamens diadelphous (9+1), staminal
sheath 1 × 0.4 cm, filaments to 3 mm; ovary subsessile, oblong, 9 × 0.1 cm, style
incurved, 3.5 mm, filiform, glabrous. Pod 5-8 cm long, membranous, flat, constricted
between the seeds, clothed with long bristly brown hairs; seeds 3-6.

Occasional in forests of Adilabad, Nizamabad and Mahabubnagar districts.

Fl. & Fr.: February-June.

Gupalapatham RF (ADB), *GO* 4399; Chedmal (NZB), *BR* & *CPR* 9259; Daravagu
(MBNR), *STR* 612 (MH).

PYCNOSPORA R. Br. ex Wight & Arn.

Pycnospora lutescens (Poir.) Schindl., J. Bot. 64 : 145. 1926. *Hedysarum lutescens* Poir.
in Lam. Encycl. 6 : 417. 1805. *Pycnospora hedysaroides* R. Br. ex Baker in Hook. f.,
Fl. Brit. India 2. 153. 1876; Gamble, Fl. Madras 1: 333. 1918.

Suberect or trailing herbs, branchlets downy-pubescent. Leaves 3-foliolate, leaflets
obovate, 1-2 × 0.5-1 cm, thin coriaceous, nerves prominent below, base cuneate, margin
ciliate, apex round, emarginate, finely pubescent on both surfaces, nerves 5-6 pairs;
petiole to 8 mm. Flowers small, purple, paired, up to 8 cm long, in terminal pubescent
and axillary lax racemes; pedicel filiform, pubescent; bracts membranous, deciduous;
calyx 3 mm long, pubescent, lobes longer than the tube, linear, acute. Pod oblong, to
1 × 0.5 cm, black when ripe, upper suture straight, lower curved; seeds 6-10.

Rare in deciduous forests in Mahabubnagar district.

Fl. & Fr.: September – December.

32 Km from Srisailam (MBNR), *LJGV* 3291 (CAL).

RHYNCHOSIA Lour. *nom. cons.*

1. Erect herbs or undershrubs :
    2. Calyx accrescent, 1-1.2 cm long in fruit, lobes oblong or
        broadly lanceolate .... **R. rufescens**
    2. Calyx not accrescent, to 0.8 cm long, lobes linear-lanceolate :

3. Leaflets acute or subacute at apex; peduncle to 1.5 cm long; pedicel articulated; pod not septate within .... **R. cana**

3. Leaflets acute to apiculate at apex; peduncle to 3 cm long; pedicel not articulated; pod septate within .... **R. suaveolens**

1. Trailing herbs or lianas :

   4. Seeds strophiolate .... **R. capitata**

   4. Seeds estrophiolate :

      5. Calyx lobes large, foliaceous .... **R. hirta**

      5. Calyx lobes acuminate, not foliaceous :

         6. Leaflets small, usually obtuse .... **R. minima**

         6. Leaflets moderately large, acute or acuminate :

            7. Pod obtuse at apex below the curved style, minutely pubescent .... **R. bracteata**

            7. Pod acute at apex below the base of the style with bulbous- based setose hairs, as well as pubescent .... **R. viscosa**

Rhynchosia bracteata Benth. ex Baker in Hook. f., Fl. Brit. India 2: 225. 1878; Gamble, Fl. Madras 1: 374. 1918.

Sarmentose or twining undershrubs, up to 4 m long, stems and branches finely striate, woody, clothed with velvety pubescence. Stipules ovate, to 4 mm, acuminate. Leaves 3-foliolate, leaflets 3.5 - 5.5 × 3.5 - 6 cm, the terminal rhomboid-orbicular, the lateral obliquely obovate, base round, margin entire, apex acuminate, thinly downy above, densely, softly grey-downy beneath; petiole to 3.5 cm, petiolule 4 mm. Flowers yellow in axillary many-flowered simple or branched peduncled racemes, much longer than leaves; bracts to 6 mm, ovate, acuminate, softly pubescent; peduncle to 4.5 cm, pedicel to 4 mm; calyx campanulate, tube to 4 mm, lobes to 1 cm, corolla yellow, standard broad, pubescent on the back without callosities at the base, auricled, 1.5 × 1.2 cm, wings 1.2 × 0.2 cm, keels 1.1 × 0.5 cm; stamens diadelphous (9+1), staminal sheath 9 × 3 mm, filaments to 5 mm; ovary sparsely pubescent, 7 × 2 mm, style curved, to 8 mm. Pod 2.5 - 4 × 1-5 cm, turgid, mucronate, narrowed at the base, slightly curved, minutely pubescent, seeds 2, dark-brown.

Very rare in Medak and Mahabubnagar districts.

Fl. & Fr.: February-April.

Rhynchosia cana (Willd.) DC., Prodr. 2: 386. 1825; Baker in Hook.f., Fl. Brit. India 2: 222. 1876; Gamble, Fl. Madras 1: 374. 1918. *Glycine cana* Willd., Sp. Pl. 3: 1063. 1802.

Undershrubs, up to 1 m high, branchlets viscid, glandular pubescent. Stipues lanceolate, ca 2 mm. Leaves trifoliolate, terminal leaflet ovate-elliptic, 2.5-4.5 × 1-2.5 cm, laterals narrow elliptic, 1-2.5 × 0.7-1.6 cm, thin coriaceous, base cuneate-obtuse, margin entire, apex obtuse-acute; petiole to 3 cm, petiolule to 2 mm. Flowers yellow with red streaks in 3-4-flowered axillary racemes; peduncle to 1.5 cm, pedicel nodding, to 6 mm, bracts ovate, 1.5 mm; calyx tube 2 mm, lobes lanceolate-acuminate; corolla yellow, exserted, standard orbicular, 7 × 6 mm, wings 6.5 × 2 mm, keels 6 × 3 mm; stamens diadelphous, staminal sheath 4 × 1.5 mm, filaments 3 mm; ovary sessile, 1.5 mm, style curved, 7 mm. Legume 1 cm long, compressed, ca 1.5 × 0.6 cm, glabrescent; seeds 2, reniform, strophiolate.

Occasional in dry forest localities in Mahabubnagar district.

Fl. & Fr.: December-April.

63 Km to Srisailam and 147 Km to Hyderabad (MBNR), *LJGV* 3944 (CAL).

Rhynchosia capitata (B.Heyne ex Roth) DC., Prodr. 2: 386. 1825; Gamble, Fl. Madras 1: 374. 1918. *R. aurea* sensu Baker in Hook. f., Fl. Brit. India 2 : 221. 1876. non DC. 1825. *Glycine capitata* B. Heyne ex Roth, Nov. Sp. 346. 1821.

Trailing annual herbs, up to 1.5 m long, branches slender with spreading pubescence. Stipules to 2.5 mm, stipels minute. Leaves 3-foliolate, to 11 cm, leaflets nearly as broad as long, rhomboid-ovate or rhomboid-obovate, laterals oblique, 2-3.5 × 1.5-3 cm, chartaceous, with a few scattered hairs on both surfaces, base subacute-cuneate, margin entire, apex rounded, strongly nerved beneath, petiole to 7.5 cm, petiolule to 2 mm. Flowers yellow in 6-10-flowered, axillary, capitate racemes, peduncles shorter than the leaves, slender, hispid, pedicel to 7 mm; bracts caducous, to 3 mm; calyx gland-dotted, tube ca. 4 mm, lobes linear, to 7 mm; corolla yellow, exserted, standard 1.3 × 0.9 cm, wings 1.1 × 0.3 cm, keels 1.2 × 0.5 cm; stamens diadelphous (9+1), staminal sheath 1 × 0.3 cm, filaments to 5 mm; ovary sericeous, 3 mm, style glabrous, to 1.1 cm. Pod orbicular, 1.5 cm in diameter, slightly compressed, sparsely pubescent, transversely striate with close parallel veins, green mottled with purple, seeds 2, strophiolate.

Common in forests of Karimnagar, Nizamabad, Medak, Nalgonda and Mahabubnagar districts.

Fl. & Fr.: August-December

Wankidi RF (ADB), *PVP* 9488; Kamareddy (NZB), *BR* 9269; ICRISAT (MDK), *LJGV* 2607 (CAL); Venkatapuram (MBNR), *BR* & *MV* 41353; Kodimial (KRN), *GVS* 21860 (MH); Kambalpally (NLG), *BVRS* 257 (NU); Murugampadu RF (KMM), *RCS* 102438 (BSID).

Rhynchosia hirta (Andrews) Meikle & Verdc., Taxon 16 : 462. 1967. *Dolichos hirtus* Andrews, Bot. Repos. 7: t. 446. 1807. *Cylista tomentosa* Roxb., Pl. Coromandel t. 221. 1811. *Rhynchosia cyanosperma* Benth. ex Baker, Fl. Trop. Africa 2 : 218. 1871; Baker in Hook. f., Fl. Brit. India 2 : 222. 1876; Gamble, Fl. Madras 1: 375. 1918.

A woody climber, up to 10 m long, branchlets with brown hairs. Stipules ovate, 1 cm, stipels setaceous, to 6 mm. Leaves 10-17 cm long, terminal leaflet ovate-elliptic, laterals ovate-rhomboid, 6-15 × 4-9.5 cm, coriaceous, velvety, rusty villous, base cuneate-obtuse, margin entire, apex acuminate, mucronate; petiole to 7 cm, petiolule 3 mm. Flowers red, yellow or white, in dense viscid racemes; bracts foliaceous, 2 × 1 cm, accrescent, tube ca 6 mm, lobes to 1.5 cm, oblong, apex obtuse; corolla scarlet, exserted, standard obovate, 1.3 × 1.2 cm, wings 1.2 × 0.4 cm, keels 1.2 × 0.4 cm; stamens diadelphous (9+1), staminal sheath 1.1 × 0.4 cm, filaments to 6 mm; ovary hispid, 1.5 × 0.6 cm, style to 1 cm. Pod ca 1.6 × 0.6 cm, constricted, as long as calyx, mucronate, grey-tomentose, oblong; seeds 2, bright blue-black, estrophiolate.

Rare in Mahabubnagar district (Raghava Rao, 1989).

Fl. & Fr.: October-February.

Rhynchosia minima (L.) DC., Prodr. 2 : 385. 1825; Baker in Hook. f., Fl. Brit. India 2: 223, 1876. incl. var. *laxiflora* (Camb.) Baker; Gamble, Fl. Madras 1: 375. 1918; Meeuwen, Reinwardtia 5: 439. 1961. *Dolichos minimus* L., Sp. Pl. 726. 1753.

Vern: *Nela alumu, Gadi chikkudu kaya*

Twining or wide-trailing annual herbs, up to 1 m long, with very slender stems, branchlets downy pubescent. Stipules lanceolate, 2.5 mm, persistent. Leaves 3-foliolate, leaflets obovate-rhomboid, 1.5-2.5 × 1.5-3 cm, thin coriaceous, pubescent, gland dotted below, base cuneate, margin entire, apex obtuse-round, apiculate; petiole to 3 mm; petiolule to 1.5 mm. Flowers yellow, 6-10- flowered in lax elongated racemes; peduncle to 2.5 cm, pedicel to 1 mm; calyx pubescent, tube ca 3 mm, lobes to 4 mm, lanceolate; corolla twice the calyx; standard oblong-ovate, 7 × 4 mm; wings 6 ×1.5 mm, keels 7 × 3 mm; stamens diadelphous (9+1), staminal sheath 5.5 × 2 mm, filaments 1.5 mm; ovary 3 mm, pubescent, style geniculate, 4 mm. Pod ca 1.5 × 0.5 cm, pubescent, slightly falcate; mostly 2-seeded.

Common weed in agricultural fields and waste lands in almost all districts.

Fl. & Fr.: September-February.

Vempalli (ADB), *GO* & *PVP* 4785; Chandoor (NZB), *BR* & *GO* 9208; Poosanpalle (MDK), *TP* & *CP* 11963; Ranjhole (MDK), *CP* 11678; Mekavaripalli (RR), *MSM* & *TP* 11043; Pakhal (WGL), *KMS* 11688 (MH); Rangammagutta (KMM), *RR* 107976 (BSID).

Rhynchosia rufescens (Willd.) DC., Prodr. 2: 387. 1825; Baker in Hook.f., Fl. Brit. India 2 : 220. 1876; Gamble, Fl. Madras 1: 373. 1918. *Glycine rufescens* Willd. in Ges., Naturf. Freunde Berlin Neue Schriften 4: 222. 1803.

Erect shrubs with trailing branches, up to 2 m long, branchlets viscid, rusty pubescent. Stipules linear, 2.5 mm. Leaves 3-foliolate, terminal leaflets ovate-elliptic, laterals narrow ovate-rhomboid, 2.5-5 × 1.5-3 cm, pubescent, thin-coriaceous, base obtuse-subacute, margin entire, apex acute-apiculate; petiole to 3 cm, petiolule ca 1.5 mm. Flowers yellow, 5-7, distantly arranged on lax racemes; peduncles slender, to 1.5 cm, pedicel to 4 mm; bracts ovate, 4.5 mm; calyx accrescent, conspicuous hiding the flowers and pod, tube ca 2 mm, foliaceous, lobes oblong or broadly lanceolate, upper one 4 mm, lower one 5 mm, pubescent; corolla slightly exceeding or equal to calyx, standard orbicular, 8 × 6 mm, wings 6.5 × 1.5 mm, keels 6 × 3mm; stamens diadelphous (9+1), staminal sheath 5 × 1.7 mm, filaments 3 mm; ovary stipitate, 2 mm, style 6 mm. Pod ovoid, apically beaked; seed solitary, subglobose, strophiolate.

Common in Medak, Ranga Reddi and Mahabubnagar districts in open forests.

Fl. & Fr.: Through the year.

Anantagiri RF (RR), *MSM* & *PRSR* 12778; Vikarabad (RR), *Ashrafunnisa* 7061 A; Appapur (MBNR), *BSS* & *MVS* 30590;

Rhynchosia suaveolens (L.f.) DC., Prodr. 2. 387. 1825; Baker in Hook. f., Fl. Brit. India 2 : 221. 1876; Gamble, Fl. Madras 1: 374. 1918. *Glycine suaveolens* L.f., Suppl. Pl. 326. 1781.

Erect undershrubs, up to 1 m high; branchlets viscid, glandular pubescent. Stipules linear, 3 mm. Leaves 3-foliolate, terminal leaflet ovate-oblong, laterals narrow, ovate-rhomboid, 6-9 × 3-4 cm, pubescent, thin coriaceous, base obtuse-subacute, margin entire, apex acuminate; petiole to 4 cm, petiolule to 2 mm. Flowers yellow in pairs, rarely solitary, axillary, peduncle to 3.5 cm, pedicel to 6 mm; bracts ovate, 3 mm; calyx glandular pubescent, tube 2.5 mm, lobes to 6 mm, lanceolate, apex setaceous; corolla exserted, standard orbicular, 1.2 ×0.9 cm, wings 1 × 0.35 cm, keels 9 × 3 mm; stamens diadelphous (9+1), staminal sheath 7 × 3 mm, filaments 2 mm; ovary 2 mm, style 7 mm. Pod oblong, pubescent, ca 1.5 × 0.7 cm, septate between seeds; exceeding calyx; seeds 2, strophiolate.

Common in most districts especially in dry forests.

Fl. & Fr.: November-March.

Darigav RF (ADB), *GO* 4465; Kammarpally (NZB), *TP* & *BR* 6245; Gummadidala hills (MDK), *BR* & *CP* 14126; Central University campus (RR), *MSM* 18160; Molachintapally (MBNR), *TDCK* 13688; Pakhal RF (WGL), *KMS* 11608 (MH); Tadwai (WGL), *RKP* 110994 (BSID).

Rhynchosia viscosa (Roth) DC., Prodr. 2: 387. 1825; Baker in Hook.f., Fl. Brit. India 2 : 225. 1876; Gamble, Fl. Madras 1: 375. 1918. *Glycine viscosa* Roth, Nov. Pl. Sp. 349. 1821.

Woody climbers, wide twining, up to 10 m long, clothed with deciduous fine short grey viscous pubescence. Leaves trifoliolate, terminal and lateral leaflets ovate-deltoid, 4-7 × 3.5-5 cm, thin coriaceous, tomentose, base cuneate-obtuse, margin entire, apex acuminate; petiole to 5 cm, petiolule to 2 mm. Flowers yellow in lax racemes; peduncle to 5 cm, pedicel ca 4 mm; bracts deciduous, 7 mm; calyx tomentose, tube 4 mm, lobes lanceolate, to 4.5 mm, apex setaceous; corolla exserted, standard obovate, 2 ×1 mm, wings 1.5 × 0.4 cm, keels 1.6 × 0.6 cm; stamens diadelphous (9+1), staminal sheath 1 cm, filaments 3.5 mm; ovary stipitate, pubescent, 3 mm, style to 1 cm. Pod oblong, 2.5 cm, sparsely pubescent, horned, exceeding calyx; seeds 2, estrophiolate.

Occasional in the hills of Adilabad, Nizamabad, and Ranga Reddi districts.

Fl. & Fr.: November-March

Indervelli (ADB), *GO* & *PVP* 4533; Dharmareddy (NZB), *BR* 9513; Central University campus (RR), *MSM* 15130; Laxmapur (MBNR), *TDCK* 13649; Mannanur (MBNR), *VBH* 86601 (BSID).

ROTHIA Pers. *nom. cons.*

Rothia indica (L.) Druce, Bot. Exch. Club Soc. Brit. Isles 3: 423. 1914. *Trigonella indica* L., Sp. Pl. 778. 1753. *Rothia trifoliata* (Roth) Pers. Syn. 2: 302. 1807; Baker in Hook.f., Fl. Brit. India 2 : 63. 1876; Gamble, Fl. Madras 1: 279. 1919. *Dillwynia trifoliata* Roth, Cat. Bot. 3: 71. 1806.

Vern.: *Vucha kura.*

Annual prostrate herbs, up to 60 cm long, branchlets hirsute. Stipules foliaceous, 5 mm, ovate. Leaves digitately 3-foliolate, leaflets oblanceolate, 1-2.3 × 0.3-0.5 cm, thin coriaceous, hirsute below, base attenuate, margin ciliate, apex obtuse; petiole to 1 cm, petiolule 1 mm. Flowers yellow, small, solitary or in short racemes from the axils of the leaves; bracts linear-lanceolate, 2 mm; calyx subequal, turbinate, tube to 4.5 mm; corolla scarcely exserted, long-clawed, standard ovate-oblong, 9 × 3 mm, wings narrow, 8 × 1.5 mm, keels falcate, 6 × 1 mm; stamens monadelphous, staminal sheath 4 × 2 mm, filaments 2 mm; ovary subsessile, 5 mm, densely hirsute, style erect, 1 mm. Pod linear, slender, nearly straight, appressed villous; seeds 20 or more, reniform.

Common in all dry districts in fields and wastelands.

Fl. & Fr.: August-April

Kamareddy (NZB), *BR* & *CPR* 7154; Central University campus (RR), *TP* & *MSM* 18160; Aklaspur (KRN), *GVS* 25618 (CAL).

SESBANIA Scop. *nom. cons.*

1. Diffuse annual herbs .... **S. procumbens**
1. Trees, shrubs or undershrubs :
   2. Flowers 10 cm long; buds falcate .... **S. grandiflora**
   2. Flowers 1.5-3.5 cm long; buds straight :
      3. Leaf rachis 12 cm long; leaflets 15-20 pairs; standard petal appendaged .... **S. sesban**
      3. Leaf rachis 25-30 cm long; leaflets 30-50 pairs; standard petal not appendaged .... **S. bispinosa**

Sesbania bispinosa (Jacq.) W. Wight in U.S.D.A. Bur. Pl. Industr. Bull. 137. 15. 1909. *Aeschynomene bispinosa* Jacq., Ic. Pl. Rar. 3: t. 564. 1792. *Sesbania aculeata* (Willd.) Poir. in Lam., Encycl. 7 : 128. 1806 *p.p., "Sesban aculeatus"*; Baker in Hook.f., Fl. Brit. India2: 114. 1876; Gamble, Fl. Madras 1: 323. 1918. *Coronilla aculeata* Willd., Sp. Pl. 3 : 1147. 1802.

An annual, erect, armed prickly shrub, up to 2 m high, watery latex present. Stipules membranous, acuminate, caducous. Leaves up to 25 cm long, leaflets 30-50 pairs, oblong, 1-2 × 0.2-0.4 cm, base rounded, margin entire, apex obtuse, glabrous; petiole to 2 cm, petiolule ca 1 mm. Flowers yellow, in lax, 3-6-flowered drooping racemes, peduncles slender, to 3 cm, pedicel filiform, 2 mm; calyx tube 5 mm, lobes fused, glands on the lobes; corolla yellow, standard orbicular, spotted purple on the back, 1.1 × 1 cm, wings 1 × 0.3 cm, keels 1 × 0.4 cm; stamens diadelphous (9+1), staminal sheath 2.5 × 0.7 cm, filaments 4 mm; ovary subsessile, style 7 mm. Pod slightly torulose, 20-25 × 4-4.5 cm, beaked, 35-40-seeded.

Common in wet places of all plain districts.

Fl. & Fr.: November- March.

Sone (ADB), *TP* & *PVP* 4026; Mosra (NZB), *TP* & *BR* 6076; Narsingi (MDK), *BR* & *CP* 11544; Manjira barrage (MDK), *BR* & *CP* 14086; ICRISAT (MDK), *LJGV* 2821 (CAL); Anantasagar RF (RR), *MSM* & *KH* 11013; Bhupatipur (KRN), *GVS* 22295 (MH); Karepally (KMM), *RR* 105968 (BSID).

Sesbania grandiflora (L.) Pers., Syn. Pl. 2: 316. 1807; Baker in Hook. f., Fl. Brit. India 2: 115. 1576; Gamble, Fl. Madras 1: 323. 1918. *Robinia grandiflora* L., Sp. Pl. 722. 1753.

Vern.: *Avise.*

Trees, up to 10 m high; bark greyish, smooth; branchlets pubescent. Stipules lanceolate, ca 1 cm. Leaves 15-30 cm, leaflets 25-30 pairs, oblong or slightly obovate, subsessile, 1- 3.5 × 0.5-1 cm, chartaceous, pubescent, base obtuse, margin entire, apex emarginate, petiole to 2 cm, petiolule 2 mm. Flowers cream-coloured in 3-5-flowered, 10 cm long racemes; peduncle to 3 cm, pedicel to 2 cm; calyx tube 2 cm, lobes triangular; corolla cream coloured, standard ca 9 × 6 cm, reflexed, wings 10 × 2.5 cm, keels 11 × 3.5 cm, stamens diadelphous (9+1), staminal sheath to 10 cm, filaments 1.5 cm; ovary to 8 cm. Pods up to 60 cm, 5 mm across, margin thick, seeds up to 30.

Cultivated in gardens or betel-vine plantations, not indigenous, probably a native of Indonesia.

Fl.: December-February, Fr.: January – March.

Sesbania procumbens (Roxb.) Wight & Arn., Prodr. Fl. Ind. Orient. 215. 1834; Baker in Hook.f., Fl. Brit. India 2: 115. 1876; Gamble, Fl. Madras 1: 323. 1918. *Aeschynomene procumbens* Roxb., Fl. Ind. 3 : 337. 1832.

Annual diffuse herbs, up to 75 cm high, more or less armed with inoffensive prickles, branches glabrous, subterete, striate. Stipules 3 mm, membranous, acute, triangular base. Leaves 4-5 cm long, leaflets 15-20 pairs, linear-oblong, ca 0.5 × 0.2 cm, base rounded, margin entire, apex obtuse, apiculate, glabrous. Flowers in short 2-4-flowered, axillary racemes; calyx 6 mm long, membranous, glabrous, lobes short, deltoid; corolla pink, 8 mm. Pod straight, erect, 5-9 cm long, not twisted, beaked, torulose; seeds 12-20.

Common weed of cultivated fields in Adilabad, Warangal and Mahabubnagar districts.

Fl. & Fr.: July-December.

Sone (ADB), *TP* & *PVP* 4026; Kajipet (WGL), *RKP* 105207 (BSID).

Sesbania sesban (L.) Merr., Philipp. J. Sci. 7 : 235. 1912. *Aeschynomene sesban* L., Sp. Pl. 714 1753. *Sesbania aegyptiaca* (Poir.) Pers., Syn. Pl. 2: 316. 1807; Baker in Hook. f., Fl. Brit. India 2 : 114. 1876; Gamble, Fl. Madras 1: 323. 1918. *Sesban aegyptiacus* Poir. in Lam. Encycl. 7 : 128. 1806.

Vern.: *Jeeluga, Sominta*

Small tree, up to 5 m high, branchlets glabrous, angled, striate, prickled or not. Stipules scarious, 7 mm, linear, acute, caducous. Leaves to 12 cm, leaflets 15-20 pairs, opposite, oblong, 0.9-2 × 0.2-0.5 cm, glabrous, base and apex obtuse, apiculate, margin entire; petiole 1 cm, rachis prickled or not; petiolule 1 mm. Flowers brown to purple without, golden yellow within, in 6-10 cm long, 7-9-flowered, lax, slender, axillary pendulous racemes; peduncle to 2.5 cm, pedicel filiform, to 9 mm long; calyx tube 4 mm, lobes deltoid, much shorter than the tube, glabrous; corolla standard obovate-orbicular, 1.5 × 1 cm, wings 1.2 × 0.7 cm, keels 1.5 ×0.5 cm; stamens diadelphous (9+1), staminal sheath 1 cm, filaments 4 mm; ovary subsessile, ca 1 cm, style 5 mm. Pods linear-cylindric, pendulous, torulose, beaked, up to 30 cm long, the sutures not much thickened; seeds 20-30, oblong, greenish-brown, smooth, glabrous, polished.

Cultivated or found run wild in Nizamabad and Medak districts.

Fl. & Fr.: November-April.

Manchippa (NZB), *BR* & *KH* 9037; Devenpally (NZB), *BR* & *KH* 9630; Gummadidala (MDK), *BR* & *CP* 11488.

SHUTERIA Wight & Arn. *nom. cons.*

Shuteria involucrata (Wall.) Wight & Arn., Prodr. Fl. Ind. Orient. 207. 1834; Thuan, Adansonia (ser.) 12 : 298. 1972. var. **glabrata** (Wight & Arn.) H. Ohashi, J. Jap. Bot. 50: 305. 1975. *Glycine involucrata* Wall., Pl. As. Rar. 3: 22. t. 241. 1832. *Shuteria glabrata* Wight & Arn. *loc. cit. S. vestita* Wight & Arn. *loc. cit.;* Baker in Hook.f., Fl. Brit. India 2 : 181. 1876; Gamble, Fl. Madras 1: 350. 1918.

Twining herbs, branches strigose-villous. Stipules scarious, striate, to 5 mm, stipels subulate, to 6 mm. Leaves pinnately 3-foliolate, terminal one ovate-rhomboid, to 6 × 6 cm, lateral pairs ovate, 2.5-4 × 1.5-3 cm, base cuneate, margin ciliate, apex obtuse, mucronate; petiole 6 cm, petiolule 2 mm. Flowers purplish in 3 cm long, dense, axillary racemes, flowers solitary or in pairs on rachis; bracts lanceolate, 4 mm, bracteoles 2 mm; pedicel 2 mm; calyx gibbous, tube 3 mm, lobes ovate, to 1 mm; corolla exserted, petals clawed, standard orbicular, 8 × 5 mm, wings obliquely oblong, 7 × 2 mm, keels 1 × 0.5 cm; stamens diadelphous (9+1), staminal sheath 7 mm, filaments to 2 mm; ovary stipitate, 6 mm, style incurved, glabrous, 2 mm. Pods linear, 2.5-3 × 0.4-0.5 cm, hairy, flat, slightly curved, the valves twisted when open; seeds 5-7, ellipsoid, compressed, brownish-black, estrophiolate.

Occasional in rocky situations on bushes in Hyderabad and Mahabubnagar districts.

Fl. & Fr.: December-April.

SMITHIA W. Aiton *nom. cons.*

1. Flowers 5-8, long peduncled, racemose .... **S. sensitiva**
1. Flowers 1-2, short peduncled, axillary .... **S. conferta**

Smithia conferta Sm. in Rees. Cyclop. 33. n. 2. 1816; Gamble, Fl. Madras 1: 329. 1918. *S. geminiflora* Roth, Nov. Pl. Sp. 352. 1821; Baker in Hook. f., Fl. Brit. India 2 : 149. 1876 incl. var. *conferta*.

Annual, diffuse herbs, up to 120 cm high, not bristly. Stipules large, persistent, scarious. Leaves even pinnate, small, nearly sessile, leaflets 4-7 pairs, oblong, to 1.1 × 0.3 cm, upper floral leaves narrower, margin entire, apex rounded or acute, midrib ending in a seta, obtuse or subacute, densely bristly on the margins and on the midrib beneath. Flowers yellow, solitary or 2 in the axils of the upper leaves which are crowded at the ends of branches; bracteoles ovate, 5 mm, bristly; calyx lobes 7.5 mm, acute, with numerous parallel veins, bristles along the midrib and top; corolla exserted, standard broadly oblong, 5 mm, wings and keels oblong; stamens isodiadelphous (5+5); ovary deeply jointed. Pod 3-6-jointed, moniliform, 7.5 mm, turgid, joints small, turgid, round, smooth.

Occasional in Adilabad district.

Fl. & Fr.: August-December

Bersaipet (ADB), *GO* & *PVP* 4606.

Smithia sensitiva Ait, Hort. Kew. 3: 496. 1789; Baker in Hook.f., Fl. Brit. India 2 : 148. 1876; Gamble, Fl. Madras 1: 329. 1918.

Much branched diffuse herb, up to 60 cm long, not bristly, branchlets woody. Leaves 2 cm long, even pinnate, leaflets 3-10 pairs, oblong, to 0.6 × 0.25 cm, rachis bristly-ciliate on the midrib below. Flowers yellow in short racemes; peduncle slender, to 2.5 mm, pedicel short; bracteoles 3 mm; calyx with acute lobes, with a few short deciduous bristles, nerved, fruiting calyx with included pods. Lomentum 4-6-jointed, joints papillose.

Rare in Narsapur forest of Medak district (Narasimha Rao, 1986).

Fl. & Fr.: August-March.

SOPHORA L.

1. Leaflets densely grey-silky beneath; corolla purple; pod moniliform, not winged .... **S. velutina**
1. Leaflets with a few obscure hairs beneath; corolla bright yellow; pod furnished with four distinct wings .... **S. interrupta**

Sophora interrupta Bedd., Ic. t. 165. 1868-1874; Baker in Hook. f., Fl. Brit. India 2 : 251. 1878; Gamble Fl. Madras 1: 389. 1918.

Vern: *Adivibillu*

Tall erect shrubs, up to 1.5 m high, branchlets minutely appressed grey pubescent. Leaves odd-pinnate, leaflets 19-29, elliptic, 3-4 × 1.2-1.6 cm; base obtuse, margin entire, apex emarginate, glabrous above, with a few obscure hairs beneath; petiole to 1 cm, petiolule to 1.5 mm. Flowers yellow, distinctly peduncled, shorter than the leaves in racemes, pedicels twice the length of the calyx; calyx to 7 mm long, subtruncate, lobes minute, strigose; corolla yellow, standard orbicular with a slender claw. Pod glabrous, 7.5-10 cm long, 5-6-seeded, the joints with four distinct wings separated by distinct constrictions.

Rare in Nallamalais in Mahabubnagar district.

Fl. & Fr.: December – February.

Farahabad way (MBNR), *GS & L.Md. Bakshu* 18288; Pullaechelama (MBNR), *SRS* 109192 (BSID).

Sophora velutina Lindl., Bot. Mag. 14: t. 1185. 1828; Grierson & Long, Fl. Bhutan 1: 652. 1987. *S. glauca* Lesch. ex DC., Ann. Sci. Nat. (Paris) 4 : 98. 1825, non Salisb.1796; Baker in Hook. f., Fl. Brit. India 2 : 249. 1878; Gamble, Fl. Madras 1: 389. 1918.

Pretty shrubs, up to 3 m high, branches grey or velvety-brown. Stipules linear, to 1 cm, stipels 0. Leaves imparipinnate, rachis 15 cm long, leaflets 21-31, subopposite or alternate, oblong-lanceolate, or elliptic-lanceoalte, 3-5 × 0.5-1.5 cm, pubescent below, base obtuse-truncate, margin entire, apex acute; petiole to 1.5 cm, petiolule 2 mm. Flowers purple in terminal racemes; pedicel to 5 mm; bracts linear, 1 cm, bracteoles 0; calyx campanulate, tube 6 mm, lobes deltoid-ovate, to 2 mm, corolla exserted, petals clawed, standard obovate, 1 × 0.7 cm, wings oblong, 1.4 × 0.5 cm, keels 1 × 0.2 cm, cohering sometimes with a mucro; stamens monadelphous (10), filaments unequal, to 1 cm, ovary stipitate, woolly, 8 mm, style incurved, glabrous, 4 mm. Pod moniliform, 8 cm long, velvety, not winged; seeds 5-6, ellipsoid, ca 8 × 4 mm, white, estrophiolate.

Rare in forests in Warangal district (C.S. Reddy, 2001).

Fl. & Fr.: May-January.

Pakhal (WGL), *CSR* 870 (KU).

STYLOSANTHES O. Swartz.

1. Ascending or prostrate herb; flowers 5-15 in spikes .... **S. humilis**
1. Diffuse, rigid, undershrubs, flowers 3-5 in terminal heads .... **S. fruticosa**

Stylosanthes fruticosa (Retz.) Alston in Trimen, Handb. Fl. Ceylon 6 (Suppl.): 77. 1931. *Arachis fruticosa* Retz., Observ. Bot. 5: 26. 1789. *Stylosanthes mucronata* Willd., Sp. Pl. 3. 1166. 1802, *nom. illeg.;* Baker in Hook. f., Fl. Brit. India 2 : 148. 1876; Gamble, Fl. Madras 1: 326. 1918.

Diffuse, rigid, pubescent undershrubs, up to 55 cm long, root stock woody, branchlets silky-villous. Leaves pinnately trifoliolate, leaflets elliptic or elliptic-lanceolate, 1.5-2.5 × 0.2-0.5 cm, chartaceous, prominently nerved, base acute, margin entire, apex sharply mucronate, stipules adnate downwards to the petiole and sheathing; petiole to 7 mm, petiolule to 1 mm. Flowers yellow, sessile, 3-5 in terminal heads; bracts scarious, linear, lanceolate, 6 mm; calyx tube being relatively slender, lobes subconnate; petals shortly clawed, inserted at the throat of calyx tube, exserted, standard orbicular, 5 × 5 mm, wings oblong-obovate, 4 × 2 mm, keels incurved, 4 × 2 mm; stamens monadelphous (10), staminal sheath 2.5 × 1.5 mm, filaments 1.5 mm; ovary subsessile, to 3 mm, style 4 mm. Pod oblong, ca 4 mm, beaked, compressed, 1-2-jointed, tipped with the falcate indurated lower part of the style; seeds 2, compressed.

Common almost in all districts in dry localities.

Fl. & Fr.: November-February.

Mannanur (MBNR), *KSM* 16001; Sangareddy tank (MDK), *BR* & *CP* 11299, 11624; Burugupally (MDK), *CP* 11765; Narsapur town (MDK), *KMS* 6715 (MH); Gacchibowli (RR), *MSM* 12768; Pakhal (WGL), *RKP* 111593 (BSID); Akkannapet RF (MDK), *RG* 113407 (BSID).

Stylosanthes humulis Kunth, Nov. Gen. Sp. 6: 506. 1823. *S. sundaica* Taub., Verh. Bot. Brand 32: 21. 1890.

Annual or short lived perennial, ascending or prostrate herb, to 70 cm high, with short hairs along one side of the stem and often scattered short bristles on stem and nodes. Leaves trifoliolate, leaflets lanceolate or elliptic, apex acute, terminal leaflet to 15 × 3.5 mm, glabrous. Flowers bright yellow in crowded, 5-15-flowered spikes; standard petal 3-4 mm diam. Fruit a biarticulate pod, terminated by a persistent style; seeds yellowish to brown and purple black.

Introduced as fodder crop, often as an escape.

Fl. & Fr.: September – January.

Musalivera RF (KMM), *RR* 112613 (BSID).

TAVERNIERA DC.

Taverniera cuneifolia (Roth) Arn., Nov. Acta. Caes, Leop.-Carol. Nat. Cur. 18(1): 332. 1836. *Hedysarum cuneifolium* Roth, Nov. Pl. Sp. 357. 1821. *Taverniera nummularia* sensu Baker in Hook.f., Fl. Brit. India 2: 140. 1876 (non DC.) 1825.

A much branched, erect undershrub, branches grooved. Leaves 1-foliate, leaflet obovate or suborbicular, mucronate, thick, glabrous. Flowers in short, axillary racemes, becoming more prominent towards the end of branches, calyx 3.5-4 mm long, teeth equaling the tube, corolla reddish or flesh-coloured, 2-3 times longer than calyx, standard with dark purple veins. Pods 1-2-jointed, joints ovoid, transversely and subreticulately rugose and echinate.

Occasional among dry grasslands in Mahabubnagar district.

Fl. & Fr.: August – January.

TEPHROSIA Pers. *nom. cons.*

1. Leaves 1-foliate .... **T. strigosa**
1. Leaves compound :
   2. Stipules spinescent .... **T. spinosa**
   2. Stipules not spinescent :
      3. Pod villous .... **T. villosa**
      3. Pod pilose or glabrescent but not villous :
         4. Inflorescence axillary or leaf-opposed; style bearded :
            5. Leaflets obovate, base cuneate, terminal leaflet not more than 2.5 cm long .... **T. maxima**
            5. Leaflets oblong-elliptic, base obtuse, terminal leaflet 3-10 cm long .... **T. tinctoria**
         4. Inflorescence terminal or leaf-opposed; style not bearded :
            6. Stems procumbent; racemes normally not more than 3- flowered .... **T. pumila**
            6. Stems not procumbent; racemes normally more than 3-flowered .... **T. purpurea**

Tephrosia maxima Pers., Syn. Pl. 2 : 329. 1807; Gamble, Fl. Madras 1: 329. 1918. *Tephrosia purpurea* (L.) Pers. var. *maxima* (Pers.) Baker in Hook. f., Fl. Brit. India. 2 : 113. 1876.

Prostrate herbs, to 70 cm long, branchlets appressed pubescent to sparsely pilose or glabrescent. Stipules lanceolate, 4 mm. Leaves odd-pinnate, leaflets 6-9 pairs, oblanceolate, 1-1.5 × 0.3-0.6 cm, thin coriaceous, puberulous or glabrous, base cuneate, margin entire, apex truncate-obtuse, retuse, mucronate; petiole to 5 mm, petiolule to 1 mm. Flowers purple to pink in leaf-opposed axillary racemes (interrupted with the basal cluster of flowers in the axils of leaves); peduncle to 2.5 cm, pedicel to 2.5 mm; bracts ovate, 2.5 mm; calyx pubescent, tube 2 mm, lobes lanceolate, appressed pubescent, corolla bright purple, standard obovate, 1.5 × 1 cm; stamens diadelphous (9+1), staminal sheath 1.2 cm, filaments to 3 mm, ovary sparsely appressed pubescent, 8.5 mm, style pubescent, 7 mm. Pod straight, sparsely pubescent, 5.5 × 0.5 cm, continuous within; seeds 10-12, orbicular, strophiolate.

Common in Hyderabad, Warangal, Karimnagar (Naqvi, 2001), Mahabubnagar, and Ranga Reddi districts.

Fl. & Fr. : December.

Central University campus (RR), *MSM* 15159.

Tephrosia pumila (Lam.) Pers., Syn. Pl. 2 : 330. 1807. *Galega pumila* Lam., Encycl. 2 : 599. 1788. *Tephrosia purpurea* (L.) Pers. var. *pumila* (Lam.) Baker in Hook. f., Fl. Brit. India 2 : 113. 1876. *p.p. T. procumbens* (Buch.-Ham) Benth., Gen. Index to Trans. Linn. Soc.101. 1866; Gamble, Fl. Madras 1: 318. 1918. *Galega procumbens* Buch.-Ham., Trans. Linn. Soc. 13 : 546. 1822.

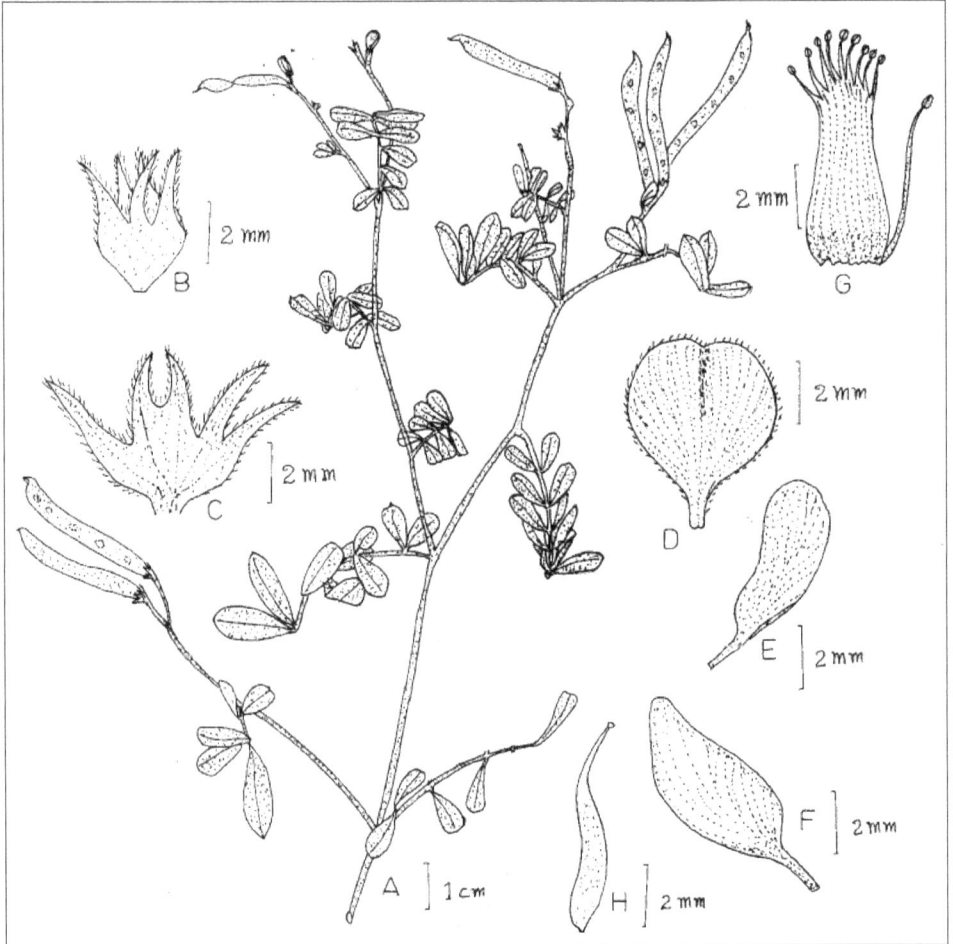

Figure 24: Tephrosia pumila (Lam.) Pers.

A. Twig, B. Calyx, C. Calyx split open, D. Standard petal, E. Wing petals, F. Keel petals, G. Staminal column, H. Pistil.

Prostrate herbs, up to 75 cm long, branchlets sericeous-velutinous. Stipules setaceous, 4 mm. Leaves odd-pinnate, leaflets 4-6 pairs, obovate-oblanceolate, 1-1.5 × 0.3-0.7 cm, chartaceous, appressed-tomentose, base cuneate, margin entire, apex truncate, emarginate, mucronate; petiole to 4 mm, petiolule to 1 mm. Flowers white or purplish, 1-3 in leaf-opposed lax pseudoracemes; peduncle 3 cm long, pedicel 1.5 mm; bracts 4 mm; calyx scattered-pubescent, tube 4 mm, lobes lanceolate; standard petal tomentose, suborbicular, 6 × 5.5 mm, wings oblong, 6.5 × 2 mm, keels 5.5 × 3 mm; stamens diadelphous (9+1), staminal sheath 8.5 mm, filaments 2 mm; ovary densely tomentose, 4 mm, style glabrous, 2 mm. Pod straight, 3.5 × 0.4 cm, pubescent, stiff-hispid, continuous within, seeds 9-11, trapezoid, strophiole subapical on the seed.

Common in all districts on wastelands, in open forests and among grass pastures.

Fl. & Fr. : July-November

Ankusapuram (ADB), *GO* 4308; Pathrajampet Railway gate (NZB), *BR* & *CPR* 7157; Choutkoor (MDK), *TP* & *CP* 11935; Central University campus (RR), *MSM* 15157; Nagarjunasagar (NLG), *BVRS* 76 (NU).

Tephrosia purpurea (L.) Pers., Syn. Pl. 2 : 329. 1807; Baker in Hook.f., Fl. Brit. India 2 : 112. 1876. *p.p.;* Gamble, Fl. Madras 1: 320. 1918; subsp. **purpurea.** *Cracca purpurea* L., Sp. Pl. 752. 1753. *Tephrosia hamiltonii* Gamble, Fl. Madras 1: 320. 1918; Pullaiah & Chennaiah, Fl. Andhra Pradesh 1: 318. 1997.

Vern.: *Vempali, Tella vempali, Bonta vempali.*

Glabrescent, erect undershrubs, up to 1 m high. Stipules lanceolate, 5 mm. Leaves odd-pinnate, leaflets 4-9 pairs, obovate, 0.8-2 × 0.3-0.5 cm, pubescent, base cuneate, margin entire, apex mucronate; petiole to 1 cm, petiolule to 2 mm. Flowers bluish-pink to purple in leaf-opposed pseudoracemes; peduncle to 1.5 cm, pedicel to 4 mm; bracts ternate, to 2 mm; calyx pubescent, tube 25 mm, lobes lanceolate, to 3 mm; corolla bluish-pink to purple, standard broad, orbicular, sericeous, 1.1 × 0.8 cm, wings 1.1 × 0.4 cm, keels 7 × 3 mm; stamens diadelphous (9+1), staminal sheath 6 ×3 mm, filaments 1 mm; ovary appressed pubescent, to 5 × 0.7 mm, style glabrous, 3 mm. Pod to 4 × 0.4 cm, downy-puberulous, slightly falcate, seeds 5-7, strophiole in the middle of seed.

Very common in almost all plains districts in wastelands and roadsides.

Fl. & Fr. : Through the year.

Devapur (ADB), *GO* & *PVP* 4250; Mosra forest nursery (NZB), *TP* & *BR* 6073; Pathrajampet Railway gate (NZB), *BR* & *CPR* 7158; Nathyanipally (MDK), *BR* & *CP* 11253; Madikonakutta-Narsapur (MDK), *KMS* 6740 (MH); Mohammadabad RF (RR), *MSM* & *KH* 10244; Lingampalle (RR), *MSM* 12736; Central University campus (RR), *MSM* 15137; Molachintapally (MBNR), *TDCK* 16823; Kodimial (KRN), *GVS* 20113 (MH); Pakhal (WGL), *KMS* 13185 (MH); Yekur (KMM), *RR* 1088549 (BSID); Rangapur (WGL), *PVS* 76957 (BSID).

Tephrosia spinosa (L. f.) Pers., Syn. Pl. 2 : 330. 1807; Baker in Hook. f., Fl. Brit. India 2: 112. 1876; Gamble, Fl. Madras 1: 320. 1918. *Galega spinosa* L. f., Suppl. 385. 1781.

Shrubs, up to 50 cm high, branchlets grey-canescent. Stipules spinose. Leaves odd-pinnate, up to 3 cm long, leaflets 4 or 5 pairs, obovate-oblanceolate 0.5-1.5 × 0.3-0.5 cm, appressed-pubescent, base cuneate, margin entire, apex truncate, emarginated, petiole o 5 mm, petiolule to 2 mm. Flowers red, in axillary fascicles; calyx 4.5 mm long, lobes as long as tube. Pod flat, ca 3 × 0.35 cm, curved, compressed, glabrescent, spinescent, dehiscing by both sutures; 5-8-seeded, seeds ellipsoid.

Occasional in Warangal district.

Fl. & Fr. : August-December

Tephrosia strigosa (Dalzell) Sant. & Mahesh., J. Bombay Nat. Hist. Soc. 54 : 805. 1957. *Macronyx strigosus* Dalzell, Hooker's J. Bot. Kew Gard. Misc. 2: 35. 1850. *Tephrosia tenuis* Wall. ex Dalzell & Gibson, Bombay Fl. 61. 1861; Baker in Hook.f., Fl. Brit. India 2: 111. 1876; Gamble, Fl. Madras 1: 318. 1918.

Caespitose, spreading herb, up to 40 cm long, branchlets appressed pubescent. Leaves simple, linear-lanceolate, 2-7 × 0.3-0.6 cm, pubescent on both surfaces, base and apex obtuse, apiculate, margin entire, stipules subulate, 2 mm, deltoid; petiole to 3 mm. Flowers pink, solitary or paired in axils; pedicel filiform, to 1 cm; calyx tube 1.5 mm, lobes to 1 mm; corolla pink, standard ovate, 4 × 1.75 mm, keels 3.5 ×1.75 mm, wings 3 × 1.75 mm; stamens diadelphous (9+1), staminal sheath 2.5 mm, filaments 5 mm; ovary glabrous, 2 × 0.5 mm, style 1 mm. Pod linear, ca 2.1 ×0.3 cm, smooth, compressed, appressed-pubescent, seeds ca 7, oblong- ellipsoid, sometimes mottled.

Occasional in Hyderabad, Mahabubnagar, Warangal, Karimnagar and Nalgonda districts.

Fl. & Fr.: July-February

Rechapalli (KRN), *GVS* 25690 (MH); Nagarjunasagar (NLG), *BVRS* 92 (NU).

Tephrosia tinctoria Pers., Syn. Pl. 2 : 329. 1807; Baker in Hook.f., Fl. Brit. India 2 : 111. 1876; Gamble, Fl. Madras 1: 319. 1918; Saldanha, Fl. Karnataka 1 : 498. 1984.

Herbs or undershrubs, up to 1 m high; branchlets faintly angled, striate, brown pubescent. Stipules ovate-deltoid, 8 × 2 mm. Leaves odd-pinnate, leaflets 7-9(11), terminal one twice as long as laterals, oblong-elliptic, 1-5.5 × 1-2.5 cm, thin-coriaceous, tomentose beneath, base obtuse, margin entire, apex emarginate, apiculate, lateral nerves parallel, reticulations prominent; petiole to 5 mm, petiolule to 2 mm. Flowers orange in axillary racemes, ca 5-flowered; peduncle to 10 cm, pedicel to 2 mm; calyx campanulate, silky hirsute, 7 ×2 mm, lobes lanceolate; standard petal sericeous, 1 × 0.9 cm, wings obovate, 8 × 3 mm, keels 8 ×3.5 mm; stamens diadelphous, staminal sheath 7 mm, filaments 2 mm; ovary velvety, 6 mm, style 3 mm. Pod linear, ca 5.5 × 0.5 cm, flattened, pubescent; seeds 8-10, dark brown, flattened, strophiolate.

Rare in Ranga Reddi district.

Fl. & Fr. : October-December.

Note: There are at least 4 binomials associated with this taxon-*Tephrosia canarensis* Drummond, *T. roxburghiana* Drummond, *T. senticosa* Persoon and *T. tinctoria* Persoon. All the above mentioned species have been segregated on the basis of stem pubescence, stipule shape, leaf indumentum, number of leaflets and seed markings. The field observations indicate a considerable variation in these characters within populations and even in the same plant. So on this basis Saldanha (1984) grouped the plants denoted by the 4 binomials under *T. tinctoria.* For the present Saldanha's (*loc. cit.*) interpretation is followed, and all the three taxa that are available in Telangana *viz., T. senticosa* Persoon, *T. roxburghiana* Drummond and *T. tinctoria* are included in one taxon *viz. T. tinctoria.*

Tephrosia villosa (L.) Pers., Syn. Pl. 2 : 329. 1807; Baker in Hook.f., Fl. Brit. India 1 : 113. 1876; Besman & Haas, Blumea 28 : 476. 1983. *Galega villosa* L., Syst. Nat. ed. 10. 2 : 1172. 1758. *Tephrosia hirta* (Buch.-Ham.) Benth., Gen. Index to Trans. Linn. Soc. Lond. 101. 1866; Gamble, Fl. Madras 1 : 319. 1918. *Galega hirta* Buch.-Ham., Trans. Linn. Soc. London 13 : 546. 1822.

Vern: *Nugu vempali*

An erect silky-villous undershrub, up to 1 m high. Leaves odd-pinnate, leaflets 5-9 pairs, obovate-oblanceolate, 1-2 × 0.4-0.8 cm, chartaceous, glabrous above, sericeous below, base cuneate, margin entire, apex obtuse, retuse, stipules subulate; petiole to 1 cm, petiolule 2 mm. Flowers pink in axillary racemes, flowers paired on rachis; peduncle to 3.5 cm, pedicel to 3 mm; bracts linear-subulate; calyx hairy, tube 3 mm, lobes lanceolate, setaceous, ciliate, lobes to 5mm; standard petal orbicular, sericeous, 1.2 × 1.1 cm, wings oblong-obovate, 1.1 × 0.4 cm, keels beaked, 8 × 4.5 mm; stamens diadelphous (9+1), staminal sheath 7 mm, filaments 3 mm; ovary 4 mm, style flattened, glabrous, 4 mm. Pods falcately curved upwards, 3 × 0.5 cm, tomentose, seeds 6-8, strophiolate.

Common in all plains districts, in wastelands and by road sides.

Fl. & Fr. : June-October.

Vempalli (ADB), *GO & PVP* 4733; Malkapur – Nizamabad road (NZB), *TP & BR* 6104; Yancha (NZB), *BR & PSPB* 9589; Choutkoor (MDK), *TP & CP* 11936; Poosanpalle (MDK), *TP & CP* 11961; Kusumasamudram RF (RR), *MSM & KH* 11024; Kodimial (KRN), *GVS* 20038 (MH); Narsapur Town (MDK), *KMS* 6669 (MH); Pakhal (WGL), *KMS* 13182 (MH); Mahadevpur (KRN), *SLK* 70841 (MH); Thunnikicheruvu RF (KMM), *RCS* 104281 (BSID).

TERAMNUS P. Br.

Teramnus labialis (L. f.) Sprengel, Syst. Veg. 3 : 235. 1826; Baker in Hook.f., Fl. Brit. India 2 : 184. 1876; Gamble, Fl. Madras 1: 352. 1918. *Glycine labialis* L. f., Suppl. Pl. 325. 1781.

Slender twining herbs, branchlets appressed tomentose. Stipules 3 mm. Leaves 3- foliolate, terminal leaflets obovate-lanceolate, laterals inequilateral, 3-4 × 2-3 cm, thin coriaceous, pubescent below, base cuneate, margin entire, apex obtuse-acute, mucronate; petiole to 1.5 cm, petiolule to 3 mm. Flowers reddish, minute, in axillary lax racemes, solitary or 2-3 in a fascicle on rachis; pedicel 4 mm; bracts linear, 3 mm; calyx campanulate, tube 2.5 mm; corolla exserted, petals clawed, standard obovate, 5 × 3 mm, wings oblong, 4.5 × 1.5 mm, keels 3.5 × 1.5 mm; stamens monadelphous (10), anthers alternately perfect and sterile, staminal sheath 2 × 1.5 mm, filaments 1 mm; ovary sessile, 2.5 mm, style short, 0.5 mm. Pod 5 x 0.8 cm, linear, straight, or curved slightly, glabrous when mature, shortly beaked; seeds 2-9, smooth, polished, reddish-brown or dark-brown, oblong, rounded at ends.

Common in all districts.

Fl. & Fr.: August-December.

Jannaram (ADB), *TP* & *PVP* 4187; Bijjur RF (ADB), *GO* & *PVP* 4872; Bhimnagar RF (NZB), *TP* & *BR* 6213; Gannaram (NZB), *TP* & *BR* 6407; Pathur (MDK), *TP* & *CP* 12112; Narsapur town (MDK), *KMS* 6748 (CAL); Mohammadabad RF (RR), *MSM* & *KH* 10598; Rechapalli (KRN), *GVS* 22278 (MH); Kunavaram RF (KMM), *RCS* 98780 (BSID); Appanapally RF (MBNR), *SRS* 104475 (BSID).

TRIGONELLA L.

Trigonella foenum-graecum L., Sp. Pl. 777. 1753; Baker in Hook.f., Fl. Brit. India 2: 87. 1876; Gamble, Fl. Madras 1: 303. 1918.

Vern.: *Menthulu.*

Aromatic, erect herbs, to 30 cm high; branchlets glabrous. Leaves 3-foliolate; stipules ovate; leaflets oblanceolate-obovate, apex obtuse, base cuneate, margin distantly dentate or entire. Flowers pale yellow-white, solitary or in pairs in leaf axils. Fods 5-8 cm long, oblong, beaked, prominently nerved; seeds many, yellowish-brown.

Commonly cultivated throughout the state.

Fl. & Fr.: February – April.

Adlur Yellareddy (NZB), *TP* & *BR* 6417; Kadthala (ADB), *GO* & *DAM* 5028.

Note: Leaves used as vegetable, seeds used as condiment.

URARIA Desv.

1. Leaves 5-9-foliolate .... **U. picta**

1. Leaves 1- and 3-foliolate intermixed :

    2. Racemes much laxer; lomentum joints 4-6, brown or black,
        minutely hairy .... **U. rufescens**

    2. Racemes dense hairy; lomentum joints usually 2,
        smooth .... **U. lagopodioides**

Uraria lagopodioides (L.) DC.. Prodr. 2 : 324. 1825, " *lagopoides* "; Hook.f., Fl. Brit. India 2: 156. 1876; " *lagopoides* "; Gamble, Fl. Madras 1: 336. 1918. *Hedysarum lagopodioides* L., Sp. Pl. 1198. 1753. *Uraria alopecuroides* Sweet, Hort. Brit. ed. 2. 148. 1830; Gamble, Fl. Madras 1: 336. 1918. *U. repanda* Wall. ex Benth. in Miq., Pl. Jungh. 213. 1832; Baker in Hook.f., Fl. Brit. India 2: 156. 1876.

Prostrate or spreading herbs, up to 35 cm long, branchlets pubescent. Leaves 3- and 1-foliolate occur on the same plant, leaflets cordate, ovate, 4-6.5 × 3-4.5 cm, base cordate, margin undulate, apex mucronate, glabrescent above, finely pubescent beneath. Flowers pink, pale purple in dense cylindrical hairy racemes; calyx lobe much longer than the tube, the lower teeth of calyx very long, the upper very small, persistent; corolla nearly as long as calyx. Pod 2-jointed, smooth and faintly reticulate.

Occasional in Nizamabad district.

Fl. & Fr.: July-October.

Bhimnagar RF (NZB), *TP* & *BR* 6216; Gandhari RF (NZB), *BR* & *CPR* 7219, 7245.

Uraria picta (Jacq.) Desv., J. Bot. Agric. 1 : 123. t. 5. f. 19. 1813; Baker in Hook.f., Fl. Brit. India 2 : 155; Gamble, Fl. Madras 1: 336. 1918. *Hedysarum pictum* Jacq., Coll. 2: 262. 1789.

Vern.: *Pingerragadda.*

Erect undershrubs, up to 90 cm high, branchlets pubescent. Stipules 1.2 cm, triangular, striate, stipels linear, to 5 mm. Upper leaves 5-9-foliolate, lower leaves simple, leaflets linear-oblong, 4-14 × 1.5-2.5 cm, glabrous above, minutely pubescent beneath, base subcordate, margin entire, apex acute, mucronate; petiole to 7.5 cm, petiolule to 2 mm. Flowers purple, in 10-20 cm long, terminal cylindrical racemes; bracts to 5 mm, nerved, acuminate, ciliate; calyx tube 3 mm, lobes subulate, much longer than tube; corolla purple, standard 6 mm; stamens diadelphous (9 + 1). Pod 1.2 × 0.3 cm, glabrous, 3-6-jointed, with persistent calyx; seeds strongly reticulate.

Common in interior forests in most districts.

Fl. & Fr.: August- November

Sirpur (ADB), *GO* & *PVP* 4976; Velutla RF (NZB), *BR* 9521; Saleswaram (MBNR), *BR* & *SKB* 30734 (BSID); Aklaspur (KRN), *GVS* 25636 (MH); Pakhal RF (WGL), *KMS* 13190 (MH, BSID); Narsapur RF (MDK), *RG* 113475 (BSID); North of Pasra (WGL), *RKP* 108222 (BSID).

Uraria rufescens (DC.) Schindl., Repert. Spec. Nov. Regni Veg. 21 : 14. 1925; Van Meeuwen, Reinwardtia 5 : 453. 1961. *Desmodium rufescens* DC., Ann. Sci. Nat. (Paris) 4 : 101. 1825. *Uraria hamosa* Wall. ex Wight & Arn., Prodr. Fl. Ind. Orient. 222. 1834; Baker in Hook.f., Fl. Brit. India 2 : 156. 1876; Gamble, Fl. Madras 1: 336. 1918.

Straggling undershrubs, up to 1 m high, branchlets downy villous. Stipules lanceolate, to 3 mm, acuminate, stipels 4 mm, foliaceous. Leaves 3-foliolate, leaflets oblong, ovate or elliptic, 4-10 × 2-6 cm, tomentose below, chartaceous, base truncate-obtuse, margin ciliate, apex obtuse, mucronate; petiole to 3 cm, petiolule to 2 mm. Flowers pink or purplish in long panicled racemes; peduncle to 3.5 cm, pedicel to 5 mm; bracts 2.5 mm, bracteoles 0; calyx campanulate, teeth subequal, deltoid cuspidate; corolla pink or purplish, standard orbicular, 4 mm, wings oblong, falcate, keels incurved, adherent to wings; stamens diadelphous (9+1), staminal sheath 3.5 mm;

ovary stipitate, style filiform, 5 mm. Pod 4-6-jointed, brown or black, minutely hispid, segments inflated.

Common in Mahabubnagar district in dry forest undergrowth.

Fl. & Fr. : July-November.

Jalapenta (MBNR), *BSS* & *SKB* 32611.

VIGNA Savi *nom.cons.*

1. Keel not prolonged into a beak .... **V. unguiculata**
1. Keel prolonged into a spiral :
    2. Leaflets distinctly lobed :
        3. Stipules lanceolate, peltate, leaflets deeply 3-7- lobed .... **V. aconitifolia**
        3. Stipules oblong or ovate-oblong, not distinctly peltate, leaflets shallowly 3- lobed .... **V. trilobata**
    2. Leaflets entire or obscurely lobed :
        4. Pods glabrous .... **V. dalzelliana**
        4. Pods hirsute or villous :
            5. Racemes capitate .... **V. mungo**
            5. Racemes sub-umbellate clusters .... **V. radiata**

Vigna aconitifolia (Jacq.) Marechal, Bull. Jard. Bot. Natl. Belg. 39 : 160. 1969. *Phaseolus aconitifolius* Jacq., Obs. 3 : 2. t. 52. 1768; Baker in Hook.f., Fl. Brit. India 2: 202. 1876; Gamble, Fl. Madras 1: 363. 1918.

Suberect or diffuse, nearly glabrous herbs. Stipules lanceolate. Leaves 3-foliolate, leaflets palmately 3 or more-lobed, lobes lanceolate-elliptic, 3-5.5 × 0.4-0.6 cm, chartaceous, appressed-pilose, base obtuse, margin entire, apex acute; petiole to 5 cm, petiolule to 1 mm. Flowers pale to bright yellow in short condensed, axillary, 5-7-flowered, capitate racemes; peduncle 2-9 cm, pedicel 5-8 mm; bracts to 4 mm; calyx campanulate, tube to 2 mm, pubescent, lobes 7.5 mm; corolla yellow, standard orbicular, 4 × 7 mm, base appendaged, wings obliquely obovate, 5 × 3 mm, keels deeply falcate, 5.5 × 1.5 mm; stamens diadelphous, staminal sheath 4.5 ×1.5 mm, filaments subequal, to 1.5 mm; ovary terete, to 4 mm, style 3 mm, incurved, upper part beaked. Pods 2-6 cm long, subcylindric, glabrous or nearly so, turgid; seeds 3-8, brown, smooth, glabrous.

Cultivated as a fodder crop and also as an escape.

Fl. & Fr. : August-January.

Gandhari (NZB), *BR* & *CPR* 7229; Rampur (MBNR), *BSS* & *KP* 32639.

Vigna dalzelliana (Kuntze) Verdc., Kew Bull. 24: 558. 1970. *Phaseolus dalzellianus* Kuntze, Revis. Gen. 1: 202. 1891. *P. dalzelli* T. Cooke, Fl. Bombay 1: 376. 1902; Gamble, Fl. Madras 1 : 363. 1918.

Slender, twining or creeping herbs, branchlets grey hairy when young, glabrescent when mature, rooting at nodes. Stipules minute, lanceolate, peltate. Leaves 3-foliolate, leaflets ovate-rhomboid, 3-4.5 × 1-1.2 cm, minutely lobed or unlobed, base cuneate-rounded, margin entire, apex acuminate; petiole 1-4 cm, petiolule to 2 mm. Flowers yellow in 5-6-flowered axillary racemes; peduncle to 5 cm, pedicel short; bracts linear, to 6 mm, longer than calyx; calyx glabrous, 2-3 mm long, lobes shorter than the tube, lobes lanceolate, obtuse. Pod glabrous, more or less compressed, to 3 cm, beaked, 8-10-seeded.

Rare in hills of Khammam district.

Fl. & Fr. : August-December.

Khammam hills (KMM), *RR* 113801 (BSID).

Vigna mungo (L.) Hepper, Kew Bull. 11 : 128. 1956. *Phaseolus mungo* L., Mant. Pl. 1 : 101. 1767; Baker in Hook.f., Fl. Brit. India 2 : 203. 1878; Gamble, Fl. Madras 1 : 363. 1918.

Vern.: *Minumulu.* English: Blackgram.

Erect herbs, up to 40 cm high, branchlets strigose. Stipules ovate-oblong, 7-12 × 4-5 mm, peltate. Leaves 3-foliolate, leaflets not lobed, oblong-lanceolate or ovate, terminal leaflet ca 7 × 4 cm, laterals ca 5 x 3 cm, scattered-strigose, base obtuse, margin entire, apex acute; petiole to 10 cm, petiolule to 4 mm. Flowers yellow in axillary or leaf-opposed capitate racemes; peduncle 1.5-4 cm, pedicel short; bracts lanceolate, 4 mm, acuminate; calyx 3 mm long, lobes linear. Pod 4-6 cm long, subcompressed, hairy, about 10- seeded.

Native of India, cultivated as a fodder crop and for its seeds, often found as an escape.

Fl. & Fr.: August-December.

Vigna radiata (L.) Wilczek, Fl. Congo Belge 6 : 386. 1954; Verdc., Kew Bull. 24 : 558. 1970.var. *radiata. Phaseolus radiatus* L., Sp. Pl. 725. 1753; Gamble, Fl. Madras 1 : 363. 1918. *P. mungo* L. var. *radiatus* (L.) Baker in Hook.f., Fl. Brit. India 2 : 203. 1876.

Vern.: *Pacha pesalu, Pesalu.* English: Greengram.

Erect herbs, up to 40 cm high, branchlets spreading, hispid. Stipules peltate, 8 mm, ciliate, stipels to 3 mm. Leaves 3-foliolate, up to 15 cm long, leaflets not lobed, oblong-lanceolate or rhomboid, 6.5-10 × 2-5.5 cm, glabrous above, sparsely hispidulous below along nerves, base oblique, obtuse- rounded or cuneate, margin ciliate, apex acuminate; petiole to 10 cm, petiolule 4 mm. Flowers yellow in subumbellate cluster; peduncle to 4 cm, pedicel to 3 mm; bracts peltate, 8 mm, bracteoles linear; calyx glabrous, tube to 2.5 mm, lobes 1.8 mm; corolla exserted, petals clawed, standard orbicular, 1.6 × 1.2 cm, auricled, wings 1.2 × 0.21 cm, base narrow, keels curved, narrow, appendagd, only on the left side of keel, to 5 mm; stamens diadelphous (9+1), staminal sheath to 1 cm, filaments to 8 mm; ovary to 5 mm, pubescent, style incurved, to 1 cm, apically bearded. Pod 4-6 × 0.5 cm, straight, puberulous; seeds 10-15, green.

Cultivated, often found as an escape.

Fl. & Fr.: May-December.

Sirnapally RF (NZB), *BR* & *PSPB* 9583; Aklaspur (KRN), *GVS* 25645 (MH).

Vigna radiata (L.) Wilczek var. **sublobata** (Roxb.) Verdc., Kew Bull. 24: 559. 1970. *Phaseolus sublobatus* Roxb., Fl. Ind. 3: 288. 1832.

Stem twining or prostrate. Leaflets frequently lobed. Pods and seeds usually smaller, towards the lower limits of the range given.

Cultivated, often found as an escape.

Fl. & Fr.: September – December.

Central University campus (RR), *MSM* 15189; Krishnarajasagar (KMM), *RCS* 102459 (BSID).

Vigna trilobata (L.) Verdc., Taxon 17 : 172. 1968 & Kew Bull. 24 : 560. 1970. *Dolichos trilobatus* L., Mant. Pl. 1 : 101. 1767; Baker in Hook.f., Fl. Brit. India 2 : 262. 1876. *Phaseolus trilobatus* (L.) Schreb., Nov. Actorum Acad. Caes. Leop Carol Nat. Cur. 4 : 132. 1770; Gamble, Fl. Madras 1: 363. 1918.

Vern.: *Pillipesara.*

Prostrate or twining herbs, branchlets angular, finely ribbed, pubescent or glabrous. Stipules oblong or ovate-oblong, to 1 cm, stipels small, ovate. Leaves 3-foliolate, leaflets palmately 3-lobed, mid lobe large, obovate-spathulate, lateral lobes oblique, small, 1.5-3.5 ×1-1.5 cm, glabrous or sparsely hairy, base and apex obtuse, margin ciliate; petiole to 6 cm, petiolule 2 mm. Flowers yellow in capitate racemes; peduncle to 30 cm; bracts 7 mm, bracteoles 3mm; calyx tube deeply-lobed, lobes 4 mm; corolla yellow, standard orbicular, 4 mm, wings falcate, 5 × 2 mm, keels obovate, 6 ×3 mm, auricled; stamens diadelphous (9+1), staminal sheath 4 mm, filaments subequal, to 2 mm; ovary subsessile, to 5 mm, pubescent, style to 5 mm, apex bearded. Pods 2.5-5 cm long, subcylindric, sparsely pubescent or glabrous; seeds 6-10, dark brown, glabrous, smooth.

Common in all plains districts, in fallow lands, waste places, river banks.

Fl. & Fr.: June-February.

Sone (ADB), *TP* & *PVP* 4022; Kammarapally (NZB), *TP* & *BR* 6230; Pathur (MDK), *TP* & *CP* 12166; Sivampet (MDK), *TP* & *CP* 14012; Rampur (MBNR), *BSS* 32640; Kodimial (KRN), *GVS* 20102 (MH); Pakhal (WGL), *KMS* 11689 (MH); Jangao (NLG), *BVRS* 386 (NU); Kammam – Dhaniyayigudem (KMM), *RR* 108087 (BSID).

Vigna unguiculata (L.) Walp., Repert. Bot. Syst. 1 : 779. 1843. subsp. **cylindrica** (L.) Verdc., Kew Bull. 24: 544. 1970. *Phaseolus cylindricus* L., Herb. Amb. 23. 1754. *Dolichos unguiculatus* L., Sp. Pl. 725. 1753. *Vigna catjang* (Burm. f.) Walp., Linnaea 13: 533. 1839; Gamble, Fl. Madras 1: 365 . 1918. *Dolichos catjang* Burm. f., Fl., Ind. 161. 1768.

Vern.: *Alasandalu.* English: Cowpea.

Suberect annual herb, up to 1 m high, branches glabrous. Stipules peltate, to 1.5 cm, acuminate, stipels ovate, 1.5 mm. Leaves 3-foliolate, up to 15 cm long, leaflets ovate-deltoid, angled or shallowly-lobed, 4-8 × 3-6 cm, glabrous, base oblique, truncate, margin entire, apex acute, acuminate, apiculate; petiole to 10 cm, petiolule to 4 mm. Flowers axillary in umbellate clusters; peduncle to 17 cm, pedicel to 4 mm; calyx campanulate, tube 3.5 mm, lobes 2-4 mm. Pod oblong, ca 10 × 0.3 cm, straight, sparsely puberulous, seeds 10-15, brown or white, 6-7 mm.

Cultivated for its pods and seeds and often found as an escape.

Fl. & Fr. : December - May.

Rangapuram (WGL), *PVS* & *NRR* 76923 (BSID).

ZORNIA Gmel.

1. Leaflets lanceolate, auricle of bract punctate; articles of pod 3 x 2 mm with retrosely scabrid prickles; standard petal cordate above claw; annuals .... **Z. gibbosa**

1. Leaflets broadly ovate; auricle of bract epunctate; articles of pod 5 x 3.5 mm, with glochidiate prickles; standard petal narrow above claw; perennials .... **Z. diphylla**

Zornia diphylla (L.) Pers., Syn. Pl. 2 : 318. 1807; Baker in Hook.f., Fl. Brit. India 2 : 147. 1876. *p.p.;* Gamble, Fl. Madras 1 : 325 . 1918. *Hedysarum diphyllum* L., Sp. Pl. 747. 1753. *Zornia diphylla* (L.) Pers. var. *zeylonensis* (Pers.) Baker in Hook. f., Fl. Brit. India 2 : 148. 1876. *Z. zeylonensis* Pers., Syn. 2 : 318. 1807; Gamble, Fl. Madras 1: 325. 1918.

Perennial, ascending or prostrate herbs, up to 30 cm long, branchlets glabrescent. Stipules epunctate to sparsely punctate, 1 m, lanceolate, acute, strongly nerved. Leaves 2-foliolate, leaflets broadly ovate, ca 1.5 × 1 cm, glabrous, base and apex obtuse, apiculate, margin entire, petiole to 3 m. Flowers pale yellow in 1-13-flowered axillary racemes, peduncle to 5 cm, pedicel filiform; calyx membranous, tube 2 mm, the upper lobes 2 mm, lower ones 3 mm; corolla twice as long as the calyx, standard petal cordate above claw, 8 ×5 mm, wings 6 × 2 mm, keels 7 ×2 mm; stamens monadelphous (10), staminal sheath 5 mm; ovary pubescent, 2 mm, style erect, 6 mm. Pod ca 5 × 3.5 mm, joints 3, articles 3-5, apically with glochidiate prickles.

Occasional in open meadows on gravely soil and in waste places in Medak district.

Fl. & Fr. : July-December.

Pilutla RF (MDK), *RG* 106798 (BSID).

Zornia gibbosa Span., Linnaea 15 : 192. 1841. *Z. diphylla* auct non (L.) Pers. 1807; Baker in Hook.f., Fl. Brit. India 2 : 147. 1876 *p.p.;* Gamble, Fl. Madras 1 : 325. 1918.

Prostrate herbs, up to 40 cm long, branchlets pubescent. Stipules foliar, peltate, glandular, 1 × 0.4 cm. Leaves 2-foliolate, leaflets lanceolate, 1.5-3 × 0.3-0.9 cm, sparsely pubescent, base obtuse, margin entire, apex acute, apiculate; petiole to 1.5 cm. Flowers yellow in 3-12-flowered axillary racemes; peduncle to 3 cm; bracts to 1 cm with punctate auricles; calyx tube 2 mm, upper lobes ovate, 4 mm, lower ones lanceolate, 6 mm; corolla yellow, standard cordate above claw, 6 × 6 mm, wings 6 × 3 mm, keels 5 × 3 mm; stamens monadelphous, staminal sheath 4 mm, filaments 2 mm; ovary densely pubescent, 3.5 mm, style incurved, 4 mm. Pod to 1.2 × 0.2 cm, articles 6, 3 × 2 mm, with retrorsely scabrid prickles.

Common in waste lands, open forests, and scrub jungles in most of the districts.

Fl. & Fr.: September-November.

Kotapalli RF (ADB), *TP* & *GO* 5426; Manchippa RF (NZB), *TP* & *BR* 6019; Borgaun (NZB), *BR* 9283; Sangareddy tank (MDK), *BR* & *CP* 11298; Dharur RF (RR), *MSM* & *TP* 11092; Lingampalle (RR), *MSM* 12741; Way to Pegarikutta-Narsapur (MDK), *KMS* 6624 (MH); Pakhal RF (WGL), *KMS* 11700 (MH); Tirmalagiri (NLG), *BVRS* 468; Mahadevpur (KRN), *SLK* 70904 (NBG); Koriguttalu (KMM), *RR* 106030 (BSID).

Note: The species *Zornia gibbosa* Span. due to its extensive pantropic distribution in Africa, India and East Asia, growing in a variety of habitat and climate, presents extremely variable forms and has confused several earlier workers. Noteboom (1961) accepted a composite species of the genus *Zornia i.e. Z. diphylla* (L.) Pers. which includes *Z. gibbosa* Span. also. But Mohlenbrock (1961) in his monograph on the genus *Zornia* considered *Z. diphylla* (L.) Pers. as a perennial and separate from *Z. gibbosa* Span. which is an annual form, former growing in Ceylon and the latter in India, Burma, Malysia, China, Australia and New Guinea. Subsequent to Mohlenbrock's contention, what occurs in India was *Z. gibbosa.* S.K. Wagh (1964) has definitely pointed out that in India both *Z. diphylla* and *Z. gibbosa* occur. Recently, Chakravarthi and Jain have summarised stating that 4 species *viz., Z. walkeri, Z. diphylla, Z. gibbosa* and *Z. quilonensis* occur in India. Ravi (1979) stated that there are three species of *Zornia* in South India *viz. Z. diphylla, Z. gibbosa* and *Z. quilonensis;* among them *Z. quilonensis* endemic to Kerala only.

Vicia faba L. is cultivated as vegetable.

CAESALPINIOIDEAE

1. Leaves 1-pinnate :
   2. Leaflets 2, palmately nerved :
      3. Petals absent; pods compressed, not woody, 1- seeded .... **Hardwickia**
      3. Petals present; pods flattened, woody, many-seeded .... **Bauhinia**
   2. Leaflets 4-numerous, reticulately nerved :
      4. Petals absent .... **Saraca**
      4. Petals present :
         5. Petals 3; pods with pulpy mesocarp .... **Tamarindus**
         5. Petals 5; pods without pulpy mesocarp:

6. Filaments of three abaxial antisepalous stamens sigmoidly curved and many times longer than their anthers which are dorsifixed, subversatile and introrsely dehiscent by slits; the filaments of 2 abaxial antepetalous and of the remaining 5 adaxial stamens straight and shorter with anthers dehiscing mostly by basal pores; pods indehiscent; extrafloral glands absent .... **Cassia**

6. Filaments of all stamens straight or simply incurved, either shorter than or not over twice as long as their anthers which dehisce terminally by slits or pores; pods various, dehiscent or indehiscent; extrafloral glands present or absent:

   7. Bracteoles 2, inserted in the middle on pedicel or above; androecium actinomorphic; pods elastically dehiscent, valves coiling; extrafloral glands, when present, disc or cup-shaped, rarely flat .... **Chamaecrista**

   7. Bracteoles absent; androecium zygomorphic; pods indehiscent or inertly dehiscent through one or both sutures; if dehiscent through one suture then pods follicular; when dehiscent through both sutures then valves tardily separating but valves not coiling or valves breaking up into 1-seeded joints .... **Senna**

1. Leaves 2-pinnate :

   8. Plants armed :

      9. Stragglers or lianas; pods samaroid .... **Pterolobium**

      9. Shrubs or small trees; pods not samaroid :

         10. Leaflets reduced to scales; petals not glabrous .... **Parkinsonia**

         10. Leaflets not reduced to scales; petals glabrous .... **Caesalpinia**

   8. Plants unarmed :

      11. Bark rough; flowers golden yellow .... **Peltophorum**

      11. Bark smooth; flowers other than golden yellow .... **Delonix**

BAUHINIA L.

1. Fertile stamens 10:

   2. Flowers white in colour; leaflets connate for two-third of their length .... **B. racemosa**

   2. Flowers other than white in colour:

      3. Flowers yellow in colour; leaflets connate for about half their length .... **B. tomentosa**

3. Flowers pale yellow in colour; leaflets connate for
    three-fourths of their length or even more .... **B. malabarica**

1. Fertile stamens 1- 5:

    4. Fertile stamen 1, long, nearly free from other nine sterile
        stamens .... **B acuminata**

    4. Fertile stamens 3-5:

        5. Flowers other than yellow:

            6. Flowers rose in colour; leaflets connate about halfway
                up and sometimes overlapping .... **B. purpurea**

            6. Flowers pink in colour; leaflets connate for about
                two-thirds up .... **B. variegata**

        5. Flowers yellow in colour:

            7. Flowers yellow to pale yellow in colour; leaflets
                connate for twothirds up or higher .... **B. vahlii**

            7. Flowers yellow with purple patches; leaflets connate
                almost to the apex .... **B. semla**

Bauhinia acuminata L., Sp. Pl. 376. 1753; Baker in Hook.f., Fl. Brit. India 2 : 276. 1878; Gamble, Fl. Madras 1: 408. 1919.

Vern.: *Kanchana.*

Shrub, 2-3 m high. Leaves 3.7-6.2 cm long, lobed to one-third their length, lobes acute or acuminate, 7-11-nerved. Petiole ca 3.7 cm long. Flowers white, in short racemes, 2.7-7.5 cm long; calyx 2.5-3.7 cm long, spathaceous, long acuminate or beaked; corolla 6.2-7.5 cm across. Pods 10-12.5 × 1.8 cm, beaked, widest above and tapering downwards, ca 7-seeded, ridged on each side of the upper suture.

Often planted in gardens, also runs wild.

Fl. & Fr.: July- February.

Bauhinia malabarica Roxb., Fl. Ind. 2 : 321. 1832; Baker in Hook. f., Fl. Brit. India 2: 217. 1878; Gamble, Fl. Madras 1: 407. 1919. *Piliostigma malabaricum* (Roxb.) Benth. in Miq. Pl. Jungh. 261. 1852.

Vern.: *Pulishinta, Pedda ari.*

Deciduous trees, about 20 m high; bark brown, rough; stems rather rough, glaucous when young. Leaves about 10-16 cm; leaflets 11 × 13 cm, adnate to three-fourths the length, oblong, cordate at base, glabrous and glaucous beneath, nerves 9 from base, reticulatious prominent. Flowers pale yellow, in axillary racemes; pedicel long, slender; calyx spathaceous, with 5 equal triangular teeth, 8-9 × 2-3 mm; petals spathulate, equal and similar, prominently nerved, obovate, 1.6 × 3 mm, stamens 10, alternately 5 long, fertile and 5 short, sterile, anthers oblong, long anther to 1.5 mm, small anther 1.25 mm; ovary compressed, to 3.25 mm, slightly pubescent. Pods nearly straight, 25 × 2 cm, flattened, rugulose, puberulous; seeds 20, ovoid, flattened.

Occasional in Warangal and Mahabubnagar districts.

Fl. & Fr.: August- December.

Appapur (MBNR), *BR* & *SKB* 30656; Eturnagaram (WGL), *SSR* 18540; North of Pasra (WGL), *RKP* 108207 (BSID).

Bauhinia purpurea L., Sp. Pl. 375. 1753; Baker in Hook. f., Fl. Brit. India 2 : 284. 1878; Gamble, Fl. Madras 1: 407. 1919. *Phanera purpurea* (L.) Benth. in Miq. Pl. Jungh. 262. 1852.

Vern.: *Kanchanam*

Trees, up to 15 m high, bark grey to brown, branchlets warty. Leaves suborbicular, 5.5-11.5 × 6-12 cm, subcoriaceous, deeply lobed, leaflets oblong, apex obtuse or subacute, connate about half way up and sometimes overlapping, 9-11-nerved, pointed below, base sub-cordate, margin entire, petiole to 3 cm, stipules triangular. Flowers rose to pink in terminal or axillary racemes or panicles, peduncle stout, to 5 cm, pedicel to 2.5 cm, bract ovate, to 4 cm, calyx spathaceous, 2.5 × 1 cm, tube turbinate, petals 5, equal and similar, obovate, 1.9 × 0.5 cm, narrow at base, obtuse, stamens 5, fertile only 3, filaments to 6 mm, anther oblong, to 6 mm, ovary compressed, grooved, 2.5 cm. Pods 30-40 × 1-2 cm, linear, flat, apiculate, reddish-brown, woody, pendent, apex horned, puberulous; seeds 10, globose, smooth, glabrous, bright brown.

Occasional in Nizamabad, Ranga Reddi, Khammam and Mahabubnagar districts in dry deciduous forests, chiefly along water courses, often cultivated in gardens.

Fl.: September – February; Fr.: February – March.

Jalalpur RF (NZB), *BR* & *KH* 9661; Lingampalle (RR), *MSM* 12758; Farahabad (MBNR), *TDCK* 16829; Rollapadu (KMM), *RR* 108072 (BSID); Tiruppumetta RF (KMM), *RCS* 99018 (BSID).

Bauhinia racemosa Lam., Encycl. 1: 390. 1785; Baker in Hook. f., Fl. Brit. India 2 : 276. 1878; Gamble, Fl. Madras 1: 407. 1919. *Piliostigma racemosa* (Lam.) Benth. in Miq. Pl. Jungh. 262. 1852.

Vern. : *Arechettu.*

Trees, about 10 m high; bark white, rough; branchlets densely tomentose. Leaves ovate-orbicular, 1.5-3.5 × 3-3.5 cm, thin coriaceous, leaflets connate to two thirds of their length, 7-8-nerved, shortly pubescent below, glabrous above, base cordate, margin entire, apex deeply emarginated; petiole to 1 cm. Flowers creamy yellow in terminal and leaf-opposed racemes; pedicel to 2 mm; bracteoles subulate, to 1 mm; calyx spathaceous, 5 × 2 mm, glabrescent, apex cleft, petals 5, equal and similar, 5 ×1 mm, narrow at base, stamens 10, filaments to 1.5 mm, anther oblong, 3 mm, ovary compressed, grooved, to 4 mm. Pods linear-oblong, 12-16 × 1.5-2 cm, woody, slightly curved, dark-brown or brownish black, indehiscent; seeds about 15, flattened, ovoid, brown, smooth, polished.

Common in all districts in dry deciduous forests.

Fl. & Fr.: March-November.

Sattenapalli RF (ADB), *TP* & *PVP* 4091; Nasrullabad (NZB), *TP* & *BR* 6164; Sirnapally RF (NZB), *BR* & *GO* 9090; Narsapur RF (MDK), *TP* & *CP* 14066; Pakhal (WGL), *KMS* 13163 (CAL); Anantasagar RF (RR), *MSM* & *NV* 10415; Molachintapally (MBNR), *TDCK* 15322; Kambalpally (NLG), *BVRS* 603 (NU); Rechapalli (KRN), *GVS* 20156 (MH); Rollapadu (KMM), *RR* 108011 (BSID); Lakshmipuram RF (KMM), *RCS* 99006 (BSID); Tupakulagudem (WGL), *PVS* 76987 (BSID).

Bauhinia semla (Buch.-Ham. ex Roxb.) Wunderlin, Taxon 25: 362. 1976. *B. retusa* Buch.-Ham. ex Roxb., Fl. Ind. 2 : 322. 1832; Baker in Hook. f., Fl. Brit. India 2: 279. 1878; Gamble, Fl. Madras 1: 407. 1919.

Vern.: *Nirpa.*

Moderate sized deciduous tree; bark dark brown, branchlets long, slender. Leaves broader than long, 9-11 ×10-15 cm long, rigidly coriaceous, glabrous, base subcordate, margin entire, apex deeply emarginate, 7-11-nerved. Flowers yellow with purple streaks in terminal panicles; calyx limb splitting into 2 or 3 segments, prominently nerved, minutely appressed pubescent, petal clawed, 6 × 2 mm, hairy outside, blade orbicular, pale yellow, marked with dark-purple veins, stamens 5, fertile 3, filament 1.5 mm, anther 0.75 mm, oblong; ovary 2 mm, style terminal, 1 mm. Pod 10-15 cm, broad, flattened, gradually widening to an obtuse tip, late dehiscing; seeds 6-8.

Occasional in Warangal, Karimnagar and Nizamabad distircts.

Fl. & Fr.: September-February.

Gundaram RF (NZB), *BR* & *KH* 9676.

Bauhinia tomentosa L., Sp. Pl. 375. 1753; Baker in Hook. f., Fl. Brit. India 2: 275. 1878; Gamble, Fl. Madras 1: 406. 1919.

Vern.: *Kanchini.*

Large shrubs or small trees, 5-8 m high, tomentose in younger parts. Leaves orbicular, 3-5.5 × 3-7 cm, leaflets connate for about half their length, chartaceous, 7-nerved, glabrous above, densely pubescent below, base truncate or subcordate, margin entire, apex obtuse, subacute, petiole to 2.5 cm. Flowers yellow in axillary and terminal 2-3-flowered racemes, pedicel to 2.5 cm; calyx spathaceous, lobes 5, oblong, 8 × 3 mm; petals 5, obovate, all are not equal in size, in average 7 × 2.75 mm; stamens 10, pubescent at base, 5 stout are fertile, short, thin, 5 sterile, long, filaments to 2 mm; ovary pubescent, substipitate, stipe to 1 mm, style to 6 mm. Pods oblong, to 11 × 2 cm, flattened, sutures raised, dehiscent, glabrous; seeds 8-12, oblong, smooth, glabrous, to 8 × 8 mm.

Common in Mahabubnagar district, often planted for ornament.

Fl. & Fr.: July-December.

Saleswaram (MBNR), *BR* & *SKB* 34465, *KP* & *BR* 39206.

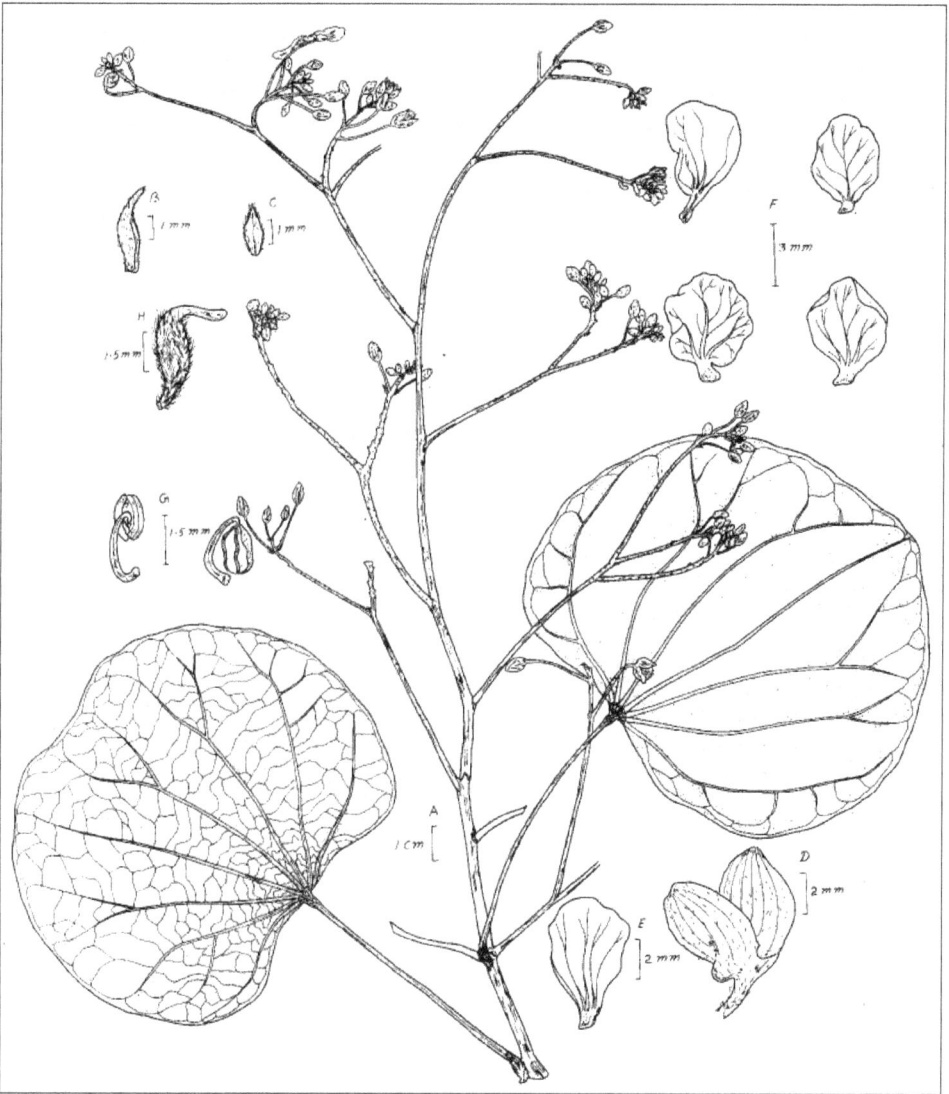

Figure 25: Bauhinia semla (Buch.-Ham. ex Roxb.) Wunderlin
A. Twig, B. Bract, C. Bracteole, D. Calyx lobe, E. Corolla lobes, F. Stamens, G. Pistil.

Bauhinia vahlii Wight & Arn., Prodr. Fl. Ind. Orient. 297. 1834; Baker in Hook. f., Fl. Brit. India 2 : 279. 1878; Gamble, Fl. Madras 1: 408. 1919. *Phanera vahlii* (Wight & Arn.) Benth. in Miq. Pl. Jungh. 263. 1853.

Vern. : *Addaku.*

Gigantic lianas, stems woody, girth up to 60 cm in diameter, tendrils opposite, circinate. Leaves broader than long, about 15-20 × 20-25 cm, connate one third of the way, base deeply cordate, margin entire, glabrous or sparsely puberulous above, tomentose beneath, nerves 11-15 from base. Flowers white in terminal corymbs, under side of the inflorescence clothed with dense ferrugineous, rarely grey tomentum; bracteoles 2, 2 × 1 mm, calyx densely tomentose, limb into 2 broadly ovate lobes, 8 × 3 mm; petals densely villous on the back, to 2.2 × 1.3 cm, equal and similar, stamens 3, filaments glabrous, style and stigma villous. Pods ca 27 × 6 cm, flattened, oblong, rusty tomentose, woody; dehiscent; seeds about 6, flattened, 2.5-3 × 2-2.5 cm, brackish in colour.

In deciduous forests of Adilabad, Ranga Reddi, Hyderabad and Mahabubnagar districts.

Fl. & Fr.: March-July.

Madharam RF (ADB), *TP* & *GO* 5478; Anantagiri RF (RR), *MSM* & *PRSR* 12786; Farahabad (MBNR), *TDCK* 13673; Nilgiri viewpoint (MBNR), *SRS* 108990 (BSID).

Bauhinia variegata L., Sp. Pl. 375. 1753; Baker in Hook. f., Fl. Brit. India 2: 284. 1878; Gamble, Fl. Madras 1: 406. 1919.

Vern. : *Mandari.*

Trees, about 15 m high, bark grey, with vertical cracks, branches terete, brown tomentose when young. Leaves about 15 cm long, leaflets 10.5-11.5 × 8.5-9.5 cm, adnate to about two-thirds up, ovate-oblong or rounded, glabrescent, nerves 11 from base, transverse nervules prominent, base deeply cordate, margin entire, apex obtuse, mucronate at the cleft; petiole 3-4 cm; stipules triangular, 1-2 mm. Flowers white or pink in terminal panicles; pedicel to 2.5 cm, bract ovate, to 4 mm; calyx spathaceous, 5-toothed, 4 × 2.5 cm; petals 5, 5ᵗʰ petal to 5.3 × 1.8 cm, 4 petals to 4.7 × 1.4 cm, 4 petals light purple variegated, obovate, stamens 10, 5 long fertile, alternating with 5 short sterile; ovary to 1.5 cm, pubescent, style to 8 mm. Pods 18-20 × 1.5-2 cm, flattened, seeds about 15, slightly flattened.

Occasional in dry deciduous forests in Karimnagar, Hyderabad and Mahabubnagar districts and often cultivated for its beautiful appearance in flower.

Fl. & Fr.: August-February.

Farahabad (MBNR), *SSR* 15535, *SRS* 109191 (BSID); Aklaspur (KRN), *GVS* 22496 (MH).

CAESALPINIA L.

1. Armed climbers:
    2. Pods prickly; pinnae 4-6 pairs .... **C. bonduc**
    2. Pods not prickly; pinnae 10-12 pairs .... **C. decapetala**
1. Unarmed trees or shrubs or weakly armed:
    3. Flowers cream-coloured, in short dense panicles; pod coiled;
        leaflets 25-30 pairs .... **C. coriaria**
    3. Flowers yellow crimson coloured, in long racemes; pod not
        twisted; leaflets 8-12 pairs .... **C. pulcherrima**

Caesalpinia bonduc (L.) Roxb., Fl. Ind. 2 : 362. 1832 emend Dandy & Exell., J. Bot. 76
: 175. 1938; Saldanha & Nicolson, Fl. Hassan (reprint) 871. 1978. *Guilandia
bonducella* (L.) Flem., Asiat. Res. 11: 159. 1810. *nom. illeg;* Baker in Hook. f., Fl. Brit.
India 2 : 254. 1878. *Caesalpinia crista* auct; Gamble, Fl. Madras 1: 393. 1919.

Vern. : *Gachakaya.*

Scandant, straggling, perennial shrubs, up to 8 m high, branchlets downy-
pubescent, prickles recurved. Leaves bipinnate, pinnae 4-6 pairs, prickled, leaflets 7-
9 pairs, ovate-elliptic, 2-4 × 1-2 cm, glabrous above, puberulous below, base truncate,
rotund, margin entire, apex obtuse, mucronate, stipules foliaceous; petiole ca 1 mm.
Flowers pale yellow in terminal and supra axillary spicate racemes, peduncle to 4
cm, prickled; pedicel 0.5-1.5 cm; bracts 1.5-2 cm long, linear-lanceolate, deciduous;
calyx tube campanulate, to 5 mm, gibbous at base, lobes 5, unequal, oblong-obovate,
4.5 – 5 × 1 mm, tomentose; petals 5, oblanceolate, upper ones smaller, 5 × 2 mm,
laterals 5-6 × 1.5 – 2.5 mm, apex obtuse, marked with yellow colour; stamens 10,
declinate, alternate, 5 long, to 5 mm, 5 short, to 4 mm, filaments attenuate, base
glandular-villous, to 3 mm, ovary stipitate, globose, to 3 mm, style short, to 2 mm,
pubescent. Pods 5-8 × 2.5-5, ovoid-oblong, somewhat flattened, densely clothed with
sharp prickles, dehiscent; seeds 1-2, oblong, smooth, polished, lead-coloured.

In most districts in hedges and on wastelands near villages.

Fl. & Fr.: August-November.

Chandoor (NZB), *BR* & *GO* 9213; Choutkoor (MDK), *TP* & *CP* 11926;
Mekavaripalli (RR), *MSM* & *TP* 11050; Laxmapur (MBNR), *TDCK* 15340; Tirmalgiri
(NLG), *BVRS* 467 (NU).

Caesalpinia coriaria (Jacq.) Willd., Sp. Pl. 2 : 532. 1799; Gamble, Fl. Madras 1: 394.
1919. *Poinciana coriaria* Jacq. Select. Amer. 123. t 175. f. 36. 1763.

Vern.: *Divi-divi.*

Unarmed trees, up to 15 m high, branchlets warty. Leaves bipinnate, 7-10 cm
long, pinnae 3-9 pairs, sub-opposite, leaflets 25-30 pairs, oblong, 0.4-0.7 × 0.1-0.3 cm,
punctate, glabrous, base oblique, cordate, margin entire, apex emarginate; petiole to 2
cm; petiolule 0. Flowers greenish-yellow in terminal panicles; peduncle to 1 cm;

pedicel to 1.5 mm; calyx-tube campanulate, short, lobes 5, oblong, lower one to 3 mm, subequal to other; petals 5, cream, ovate-orbicular, to 4 mm, clawed, upper one smaller, to 3.5 mm; stamens 10, declinate, filaments subequal, 3.5-4 mm, basally villous; ovary stipitate, 3 mm, glabrous, style suberect, to 4 mm. Pods twisted, 3-4 × 1-2 cm, glabrous; seed flattened, oblong, to 6 × 4.5 mm.

Planted along roads and in gardens.

Fl.: May – June; Fr.: August –November.

Vikarabad (RR), *MSM* & *PRSR* 12781.

Caesalpinia decapetala (Roth) Alston in Trimen, Handb. Fl. Ceylon 6 (Suppl.): 89. 1931. *Reichardia decapetala* Roth, Nov. Pl. Sp. 212. 1821. *Caesalpinia sepiaria* Roxb., Fl. Ind. 2 : 360. 1832; Baker in Hook. f., Fl. Brit. India 2: 256. 1876; Gamble, Fl. Madras 1: 394. 1919.

Vern.: *Korinda.*

Armed stragglers, up to 5 m high, branchlets matty-tomentose, prickles recurved on the stem and leaf rachis. Leaves bipinnate, pinnae 10-12 pairs, leaflets 10 pairs, elliptic or oblong-obovate, 1-1.5 × 0.3-0.5 cm, chartaceous, puberulous above, pubescent below, base subacute, cuneate, margin entire, apex obtuse, rotund; petiole to 3.5 cm; rachis and petiole prickled. Flowers bright yellow in axillary and terminal racemes; pedicel to 3 cm, jointed, prickled; bracts lanceolate, to 4 mm; calyx-tube campanulate, to 3 mm, lobes 5, oblong, lower one 7 × 3.5 mm, laterals to 8 × 4.5 mm; petals 5, golden yellow, upper one spathulate, reflexed, to 7 × 7 mm, lateral equal to lower, orbicular, 6.5 ×2.5 mm, clawed; stamens 10, filaments subequal, 7 mm, base pubescent, ovary sessile, to 5 mm, style attenuate, to 2 mm. Pods 7 × 3 cm, straight or slightly curved, linear-oblong, slightly turgid, reticulately veined, slightly winged, constricted between seeds; seeds 6-8, oblong-globose, 8 ×6.5 mm, smooth.

Occasional in Nizamabad and Ranga Reddi districts in hedges and open bushy places.

Fl. & Fr.: October-April.

Patharajampet (NZB), *BR* 9700; Pargi (RR), *MSM* & *PRSR* 12799.

Caesalpinia pulcherrima (L.) Sw., Observ. Bot. 166. 1791; Baker in Hook. f., Fl. Brit. India 2: 255. 1878; Gamble, Fl. Madras 1: 394. 1918. *Poinciana pulcherrima* L., Sp. Pl. 380. 1753.

Shrubs or small trees; bark greyish brown, prickles few and scattered on the branches. Leaves bipinnate, alternate, 11-14 cm long, pinnae 4-9 pairs, leaflets 6-12 pairs, 1-1.5 × 0.5-0.7 cm, sessile, close, oblong, rarely obovate; base attenuate, margin entire, apex obtuse, nerves prominent, glabrous above, glabrescent beneath. Flowers yellow or reddish-yellow in terminal racemes; pedicel more than twice the length of flower; sepals 5, unequal, oblong-obovate, 0.6-0.8 × 3.5 – 4 mm, lower one 1 cm long, glabrous; petals 5, oblanceolate, upper one smaller, 9 × 5 mm, laterals 1.1 – 1.2 ×0.8-0.9 cm, apex obtuse; stamens 10, declinate, subequal, filaments bright red, 3-4 times the length of the corolla, base glandular-villous, 3.5-4 mm; ovary stipitate, globose, style short, glabrescent, to 4 mm. Pods nearly straight, 6-11 × 1-2 cm, broadly linear, flat, glabrescent; seeds 8-10, obovate-oblong, smooth, glabrous, brown, 1-2 × 0.75-0.85 mm.

Ornamental, very common in gardens, also as an escape.

Fl. & Fr.: Throughout the year.

Manchippa (NZB), *TP* & *BR* 6057; Medak (MDK), *CP* 11720; Anantasagar RF (RR), *MSM* & *NV* 10384.

CASSIA L.

1. Flowers yellow, in 20-40 cm long, pendent branched racemes; seeds ca 50 .... **C. fistula**

1. Flowers pinkish or rose, in axillary corymbose racemes; seeds 16-20:

    2. Petals upto 1.5 cm long; long filaments without globular swellings; anthers without distinct longitudinal sutures or rim .... **C. roxburghii**

    2. Petals exceeding 1.5 cm in length; long filaments with globular swellings; anthers with distinct longitudinal sutures or rim .... **C. roxburghii**

Cassia fistula L., Sp. Pl. 377. 1753; Baker in Hook. f., Fl. Brit. India 2 : 261. 1878; Gamble, Fl. Madras 1: 400. 1919.

Vern.: *Rela.*

Deciduous trees, about 10 m high; bark pale, smooth when young, darker and rough when old; branchlets glabrous. Leaves about 25-40 cm long, leaflets 4-8 pairs, sub-opposite, oblong-broadly ovate, inequilateral, 8-20 × 4-8 cm, thick coriaceous, nerves plaited, arching, glabrous above, pubescent below, base subacute-cuneate, margin entire, apex gradually tapering, emarginated; petiole 3-5 cm, rachis and petiole eglandular, petiolule 0.5-1 cm, stipules caducous. Flowers bright to golden-yellow in 20-40 cm long lax, pendent, branched racemes; pedicel 3-7 cm; calyx-lobes 5, reflexed, ovate, ca 1 × 0.5 cm, apex obtuse; petals obovate, 6-9.5 ×3.5-5.5 mm; stamens 10, antheriferous, upper 3 short, with erect filaments to 5 mm and basifixed anther to 1 mm, lower 3 large with curved filaments to 7 mm and versatile anthers with curved beak to 2.5 mm; ovary appressed pubescent, 8 mm long, style to 2 mm. Pods 40-60 × 2-4 cm, cylindric, indehiscent, dark-blackish-brown; seeds about 50, orbicular, flattened.

Common in deciduous forests of all districts, frequently planted in gardens and avenues.

Fl. & Fr.: March-December

Kadam RF (ADB), *TP* & *PVP* 4073; Devapur RF (ADB), *GO* & *PVP* 4245; Kubgal RF (NZB), *TP* & *BR* 6201; Akkannapet (MDK), *BR* & *CP* 11553; Mohammadabad RF (RR), *MSM* & *KH* 10230; Anantagiri (RR), *SSR* 18511; Mannanur (MBNR), *TDCK* 13611; Vikarabad (RR), *LJGV* 3797 (CAL); Kambalpally (NLG), *BVRS* 605 (NU); Kodimial (KRN), *GVS* 20074 (MH); Rollapadu (KMM), *RR* 108025 (BSID); Mahabubnagar (MBNR), *SRS* 109072 (BSID).

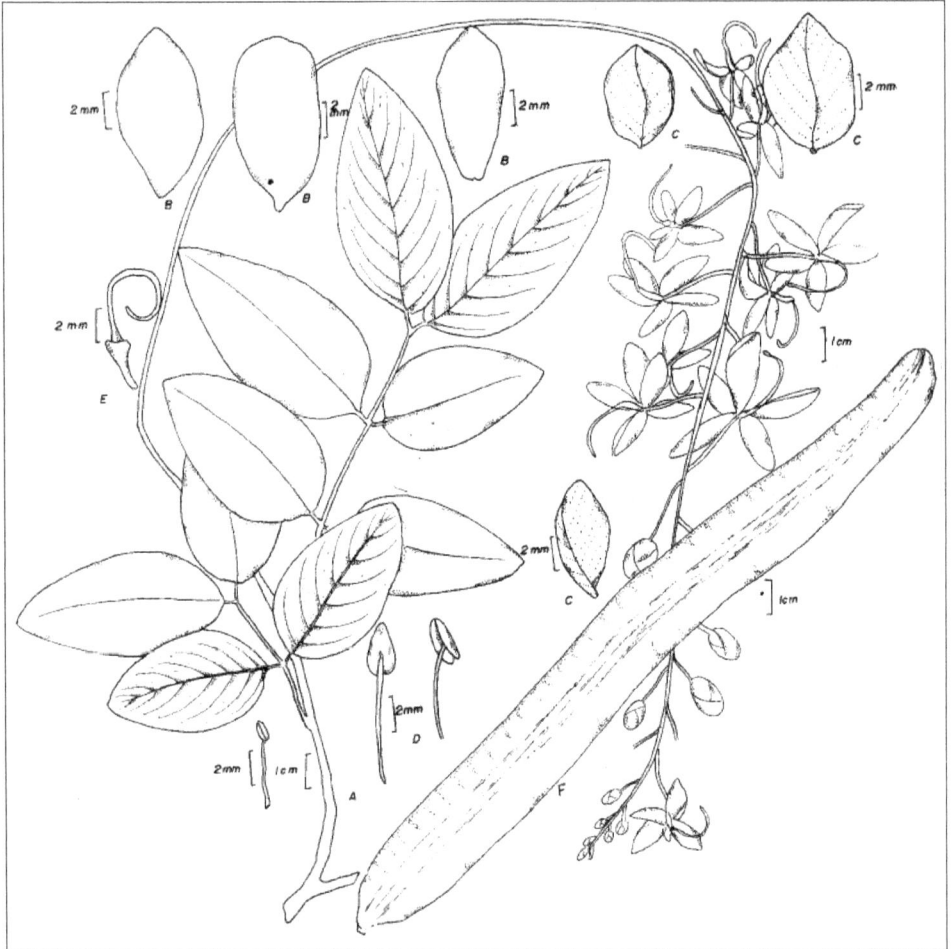

Figure 26: Cassia fistula L.

A. Twig, B. Sepals, C. Petals, D. Stamens, E. Pistil, F. Pod.

Cassia javanica L., Sp. Pl. 379. 1753; Baker in Hook. f., Fl. Brit. India 2 : 249. 1878.

Deciduous trees, 6-20 m high; branchlets glabrous or pubescent. Leaf rachis 4-30 cm long, eglandular, pubescent; leaflets 3-14 pairs, opposite or subopposite, 1.8-4 × 1-2.7 cm, oblong, base mucronate, apex truncate-emarginate, pubescent on both sides; petioles 1.5-4.5 cm long; petiolules 1-3 mm long; stipules 10-25 mm long, foliaceous, crescent-shapted. Flowers pink or red, in corymbs; bracts ovate, bracteoles 2, axillary; sepals 5-7 mm long, ovate-oblong, pubescent; petals 2-3 cm long, spathulate to obovate, sparsely pubescent; stamens 10, all fertile, 3 longer filaments recurved, with globular swelling about or above the middle; ovary shortly stipitate, pubescent, stigma subapical, punctiform. Pod 20-60 × 1.5-2 cm, cylindric, terete, indehiscent, thin-valved, glabrous; seeds oavate-suborbicular, flat.

Cultivated in gardens and along the roads for ornamental flowers and as avenue tree; also naturalized at many places (Singh, 2001).

Fl. & Fr.: April – November.

Cassia roxburghii DC., Prodr. 2: 489. 1825. *C. marginata* Roxb., Fl. Ind. 2: 338. 1832; Hook.f., Fl. Brit. India 2: 262. 1878; Gamble, Fl. Madras 1: 401. 1919.

Small or middle sized trees, to 10 m high; bark brownish, rough, branchlets tomentose. Leaves 15-25 cm long, rachis glabrescent, leaflets 8-15 pairs, sub-opposite, oblong-elliptic or ovate, inequilateral, 2-35 × 1-1.5 cm, coriaceous, woolly below, glabrous above, young leaves densely yellow pubescent, base subacute, margin entire, apex obtuse; petiole to 1 cm, rachis and petiole eglandular, petiolule to 2 mm; stipules ovate, muriculate, caducous. Flowers pinkish or rose, in axillary corymbose racemes; peduncle to 3 cm; pedicel to 1 cm; bracts ovate, 0.5-1 cm, bracteoles to 2.5 mm; calyx lobes 5, ovate, 0.4-1.1 cm, apex obtuse; petals 5, ovate-oblong, 0.8-1.1 × 0.8-0.9 cm, concave, pubescent without, claw to 2 mm; stamens 10, all fertile, upper 2 short, with erect filaments to 3 mm and erect anthers 5-8 mm, 2 medium with erect filaments to 1 cm and anthers with curved beak to 4 mm; ovary stipitate, 2 cm, grooved, pubescent, style to 6 mm. Pod oblong, terete, 20-30 × 2-3 cm, torulose, indehiscent; seeds 16-20, transverse.

Grown as ornamental.

Fl. & Fr.: August – December.

Siddpet – Medak Road (MDK), *CP* 11710.

CHAMAECRISTA Moench

1. Fertile stamens 5-7:
    2. Leaves bifoliolate .... **C. rotundifolia**
    2. Leaves more than 2-foliolate:
        3. Leaflets 2 pairs; pod covered with viscid glandular hairs .... **C. absus**
        3. Leaflets 10-24 pairs; pod pubescent without glandular hairs:
            4. Foliar glands stipitate .... **C. pumila**
            4. Foliar glands sessile .... **C. leschenaultiana**
1. Fertile stamens 10, rarely with 1-3 staminodes .... **C. mimosoides**

Chamaecrista absus (L.) H.S. Irwin & Barneby, Mem. New York Bot. Gard. 35(2): 664. 1982; V. Singh, Monogr. Cassiinae 55. 2001. *Cassia absus* L., Sp. Pl. 376. 1753; Baker in Hook. f., Fl. Brit. India 2 : 265. 1878; Gamble, Fl. Madras 1: 403. 1919.

Erect annual herbs, up to 60 cm high, stem and leaves clothed with grey bristly viscous hairs. Leaves pinnately compound, petiole to 4 cm, leaflets 2 pairs, digitate, ovate-oblong, or ovate-elliptic, inequilateral, 2.5-4 × 1-2.5 cm, glabrous above, pubescent below, base cuneate, margin entire, apex mucronate, petiolule to 2 mm; rachis with subsessile glands between leaflets, stipules linear, subulate, to 6 mm. Flowers pure yellow or tinged red, in terminal or leaf-opposed racemes; peduncle to 2 cm, pedicel 0.6-1 cm; bract ovate-subcordate, 2.5 mm, bracteole linear, to 1.5 mm, petals 5, broadly obovate, subequal, to 5.75 × 2.5 mm, claw 5 × 1.75 mm; stamens 5, 2-5 mm, antheriferous, unequal, filaments erect, 1.25-1.5 mm, anthers of 3 kinds, large 2.75 mm, 1 medium 2.5 mm, 2 short 0.75-1.5 mm; ovary subsessile, terete, to 4 mm, densely long-strigose, style slightly curved, glabrous, to 1.25 mm. Pods 3-4.5 × 0.5-0.6 cm, flat, oblique, covered with stiff glandular hairs; seeds 5-7, black, ovoid, shining.

Common in all districts, on wastelands and in open forests.

Fl. & Fr.: August-February.

Sirpur (ADB), *GO* & *PVP* 4895; Arsapally (NZB), *BR* & *KH* 6480; Kamareddy (NZB), *BR* & *CPR* 7153; Pathur (MDK), *TP* & *CP* 12128; Narsapur RF (MDK), *TP* & *CP* 14043; Anantagiri RF (RR), *TP* & *MSM* 11066; Tirmalgiri (NLG), *BVRS* 471 (NU); Aklaspur (KRN), *GVS* 25624 (MH); Koriguttalu (KMM), *RR* 106041 (BSID).

Chamaecrista leschenaultiana (DC.) Degener, Fl. Hawaiiensis Fam. 169 b. 1934. *Cassia leschenaultiana* DC., Mem. Soc. Phys. Geneve 2: 132. 1824.

Herbs, up to 1.5 m high. Leaves 7-9 cm long, leaflets 8-24 pairs, oblong, 1-2 × 0.2-0.3 cm, chartaceous, sessile, glabrous or nearly so; gland solitary, borne in the middle of the petiole, discoid, flat; stipules broadly linear, 10-15 mm long. Flowers in supra-axillary racemes. Pods flat, pubescent, strap-shaped, 10-12-seeded.

Rare as forest undergrowth in Medak district.

Fl. & Fr.: August – October.

Medak cheru (MDK), *RG* 113517 (BSID).

Chamaecrista mimosoides (L.) Greene, Pittonia 4: 27. 1899; V. Singh, Monogr. Cassiinae 55. 2001. *Cassia mimosoides* L., Sp. Pl. 379. 1753; Baker in Hook. f., Fl. Brit. India 2: 266. 1878; Gamble, Fl. Madras 1: 403. 1919.

Prostrate herbs with long and slender branches, branchlets densely hispid. Leaves up to 10 cm long, leaflets 25-40 pairs, oblong-elliptic, inequilateral, 2.5-6 × 0.5-1 mm, chartaceous, glabrescent, base and apex obtuse, margin entire, apiculate; petiole to 1.5 mm, with a sessile gland at the top, petiolule reduced, stipules auriculate, lanceolate, to 1.5 mm. Flowers yellow, axillary, solitary or 2-3-fascicled; peduncle to 6 mm, bracts stipular, bracteoles very small, scarious, 1.5 mm, pedicel very slender, hairy; calyx-lobes 5, linear-lanceolate, 5-6 × 1-2 mm, pilose, apex acute; petals 5, ovate, 5-5.5 × 2-4 mm, clawed; stamens 10, all antheriferous, alternately long and short, long ones to 5 mm, short ones 3.5-4 mm; ovary to 6 mm, sericeous, style to 1.75 mm, style slightly pubescent. Pods short-stipitate, flat, strap-shaped, 3-5 × 0.5 cm, sericeous, compressed downy- tomentose, obtuse, dehiscent; seeds 15-20, ellipsoid, 2-3 mm, longitudinal.

Common in waste lands, open forests and scrub jungles in all districts.

Fl. & Fr.: October-May.

Gupalpatnam (ADB), *GO* 4394; Narsapur RF (MDK), *TP* & *CP* 14050; Farahabad (MBNR), *MVS* & *MB* 38655; Pagarikutta-Narsapur (MDK), *KMS* 6758 (MH); Lingampalle (RR), *MSM* 12746; Rechapalli (KRN), *GVS* 25691 (MH).

Chamaecrista pumila (Lam.) V. Singh, J. Econ. Taxon. Bot. 16(3): 600. 1992 & Monogr. Cassinae 84. 2001. *Cassia pumila* Lam., Encycl. 1: 651. 1785; Baker in Hook. f., Fl. Brit. India 2: 266. 1878; Gamble, Fl. Madras 1: 403. 1919.

Vern.: *Nalla jeeluga.*

Prostrate or decumbent herbs, 15-40 cm long, branchlets downy villous. Leaves 2- 4 cm, leaflets 10-25 pairs, linear, oblong, 0.8-1 × 0.2-0.4 cm, chartaceous, glabrous, sparsely villous below, nerves prominent, middle nerve closer to upper margin, base cuneate, margin ciliate, apex obtuse, mucronate; petiole to 3.5 mm with a basal stipitate gland, petiole and petiolule reduced, stipules lanceolate, to 3 mm. Flowers pale to bright yellow, solitary or geminate, axillary; calyx lobes 5, lanceolate, 3.5-4 × 0.05-0.1 cm, densely pubescent, apex acute; petals 5, ovate, 3.25-3.5 × 1 mm, clawed; stamens 5, all antheriferous, erect, filaments to 1.5 mm, empty anthers 1-5 mm; ovary subsessile, to 3 mm, densely strigose, style 0.5-1 mm. Pods 2-4 × 0.3-0.4 cm, linear-oblong or linear-lanceolate, pale-dark brown, pubescent; seeds 10- 15, brown, rounded or slightly acute at ends.

Common in most of the dry districts in waste lands and open forests.

Fl. & Fr.: August-November.

Ankusapuram (ADB), *GO* 4314; Sirnapalli (NZB), *BR* & *PSPB* 9586; Narsapur RF (MDK), *BR* & *CP* 14108; Mohammadabad RF (RR), *MSM* 10454; Madimalakala (MBNR), *TDCK* 16838; Amrabad (MBNR), *BR* & *BSS* 32545 (BSID); Venkatapuram (MBNR), *MV* & *DV* 41354; Krishnarajasagar (KMM), *RCS* 102455 (BSID); Jakaram (WGL), *RKP* 105276 (BSID).

Chamaecrista rotundifolia (Pers.) Greene, Pittonia 4: 31. 1899. *Cassia rotundifolia* Pers., Syn. Pl. 1: 456. 1805; C.S.Reddy & K.N.Reddy, J. Econ. Taxon. Bot. 28(1): 73-74. 2004.

A herbaceous, subwoody, short-lived perennial or self-regenerating legume, stems prostrate to semi-prostrate, 30-80 cm long. Leaves bifoliate, small, stipules lanceolate-cordate, 4-11 mm long, ciliate or glabrous, up to 1 cm long; petiole short, 3-8 mm long, not exceeding the stipule, not eglandular, pubescent like the stems; leaflets asymmetrically subrotund to broadly ovate, rounded apically, 0.5-3 cm long. Flowers one or two, axillary, small, yellow. Pedicels more or less filiform, 1.5 – 3.5 cm long, longer than the leaves, sepals lanceolate, ciliate, up to 5 mm long, petals obovate, about 6 mm long, glabrous, sessile, fertile stamens 5, somewhat unequal, anther linear-oblong, glabrous. Pods linear, flat, 1.5-4 cm × 3-5 mm, elastically dehiscent, blackish-brown when ripe; seeds obliquely transverse in pod, rectangular, flattened.

Rare, weed of sandy soils in Medak district (C.S.Reddy and K.N.Reddy, 2004). Native of Mexico.

Fl. & Fr.: March – August.

Nagunur (MDK), *CSR* 2003 (KU, EPTRI).

DELONIX Rafin.

1.   Bark black; petals yellowish-white; pods 8-15 cm long .... **D. elata**
1.   Bark grey to pale brown; petals red; pods 25-50 cm long .... **D. regia**

Delonix elata (L.) Gamble, Fl. Madras 1: 396. 1919. *Poinciana elata* L., Cent. Pl. 2: 16. 1756; Baker in Hook. f., Fl. Brit. India 2: 260. 1878.

Vern.: *Sunkesula, Chitti-kesar.*

Trees, up to 12 m high, bark black; branchlets warty puberulous. Leaves 15-20 long, pinnae 6-9 pairs, 4-6 cm long, leaflets 10-22 pairs, oblong-elliptic, 0.5-1 × 0.1-0.3, chartaceous, puberulous, base cuneate, margin entire, apex obtuse; petiole to 4 cm, petiolule 0; stipules caducous. Flowers yellowish-white in terminal racemes; peduncle to 1.5 cm, pedicel to 2.5 cm, puberulous; bracts small, caducous; calyx 5-lobed, to 5 mm, thick, spathaceous in bud, lanceolate-elliptic, 1.8-2 × 0.3-0.5 cm; petals 5, obovate, suborbicular, 8-9 × 6.5-8 mm, shortly clawed, margins curled and fimbriate; stamens 10, much exserted, subequal, filaments subulate, to 4 cm and empty anthers to 5 mm, villous at base; ovary to 8 mm, subsessile, densely pubescent, flat, style to 5 mm. Pods 8-15 × 2.5-3 cm, pale to dark brown, linear or oblanceolate, glabrous, reticulate, beaked, sutures grooved; seeds ca 10.

Cultivated as avenue tree, occasionally wild.

Fl. & Fr.: December- June.

Molachintapally (MBNR), *TDCK* 15302; Nagarjunakonda (NLG), *KTH* 9842 (CAL); Karimnagar (KRN), *SLK* 70916 (NBG); Amangal (MBNR), *SRS* 104573 (BSID).

Delonix regia (Bojer ex Hook.) Raf., Fl. Tellur. 2: 92. 1836; Gamble, Fl. Madras 1: 396. 1919. *Poinciana regia* Boj. ex Hook. in Bot. Mag. t. 2884. 1829.

Vern.: *Thurai.*

Deciduous trees, up to 12 m high, bark grey to pale-brown, branchlets warty puberulous. Leaves 10-30 cm long, pinnae 8-20 pairs, leaflets 12-30 pairs, oblong, 0.8-1 × 0.4-0.5, glabrous or nearly so, base oblique, margin entire, apex obtuse; petiole to 4 cm, petiolule 0; stipules caducous. Flowers red, terminal, simple or branched racemes; calyx lobes 5, obovate, suborbicular, 1.5-1.6 × 0.9-1.2 cm with elongated claw, margins not fimbriate; stamens 10, much exserted, stamens nearly equal to or shorter than petals, filaments subulate, to 5.3 cm, and empty anthers 5 mm, villous at base, anther linear; ovary densely pubescent, subsessile, flat, to 9 mm, style to 6 mm, filiform. Pods 25-50 × 3-4 cm, broadly linear, woody, dark-brown or reddish-brown, flat, beaked; seeds ca 8, oblong, glabrous, smooth white or creamy-white.

Planted in gardens and along road sides.

Fl. & Fr.: April – August.

Digwal (MDK), *CP* 11682.

HARDWICKIA Roxb.

Hardwickia binata Roxb., Pl. Coromandel t. 209. 1819; Baker in Hook. f., Fl. Brit. India 2 : 270. 1878; Gamble, Fl. Madras 1: 412. 1919.

Vern.: *Yepi.*

Large deciduous trees, up to 30 m high, bark grey, rough, branchlets glaucous. Leaves 2-foliolate, leaflets ovate-trapezoid, 2.5-5 x 1.5-3 cm, thick coriaceous, glabrous, 3-6-nerved from base, base oblique, cuneate, margin entire, apex obtuse, petiole to 2 cm. Flowers pale yellow, in axillary and terminal panicled racemes, pedicel to 3 mm; calyx lobes 5, petaloid, greenish-cream, oblong-ovate, oblong-ovate, 6-8 × 1.25-1.75 mm, spreading; petals 0; stamens 10, subequal with filaments 1.75 mm and empty anthers 2.25 mm, anthers ovoid; ovary sessile, ovoid, to 5 mm, slightly glabrous, style to 3 mm, filiform. Pod 6.5-7.5 × 1-1.6 cm, flattened, winged, glabrous; seed 1, generally placed at top, orbicular, flattened.

Common in Adilabad, Karimnagar, Nalgonda and Mahabubnagar districts in deciduous forests.

Fl. & Fr.: August-January.

Kotpalli RF (ADB), *GO* 4518; Molachintapally (MBNR), *TDCK* 15301; Mahadevpur to Kataram (KRN), *SLK* 70809 (NBG); Kambalpally (NLG), *BVRS* 604 (NU); Umamaheswaram (MBNR), *SRS* 10915 (BSID).

PARKINSONIA L.

Parkinsonia aculeata L., Sp. Pl. 375. 1753; Baker in Hook. f., Fl. Brit. India 2: 260. 1878; Gamble, Fl. Madras 1: 397. 1919.

Vern.: *Simatumma.*

Armed shrubs, up to 3.5 m high, branchlets tomentose, spines erect, woody. Leaves 2-pinnate, up to 35 cm, main rachis short, spiny, pinnae 2-3 pairs, long, with flattened rachis bearing many very small leaflets about 50 pairs, oblanceolate, 0.3-0.6 × 0.1- 0.3 cm, base cuneate, margin entire, apex obtuse, petiole short; stipules spinescent, stipels 0. Flowers yellow in lax terminal racemes; peduncle to 2.5 cm; bracts caducous, bracteoles 0; pedicel to 1.5 cm; calyx tube short, lobes 5, oblong, 5.5-6 × 2 mm, subequal, subclavate; petals 5, obovate-suborbicular, unequal, 6 × 5.6 mm, upper one to 6 × 6 mm, claw short, to 6 × 2 mm, stout and villous; stamens 10, free, filaments subequal, 3.5-4 mm, and empty anthers 1.5 mm, subulate, densely villous; ovary sessile or shortly stalked, to 5 mm, style attenuate, to 2 mm. Pods 8-9 × 0.7-1 cm, torulose, turgid above seeds, finely nerved, glabrous, attenuate at apices, dehiscent; seeds 2-6, oblong, smooth, dark brown.

In all districts throughout the plains, more common in arid and semiarid regions.

Fl. & Fr.: July-December.

Khanapur (ADB), *GO* & *BR* 9408; Dharmaram (NZB), *BR* & *KH* 6459; Hugelli (MDK), *CP* 11680; Gachibowli (RR), *MSM* 12767; Hyderabad city (HYD), *SSR* 19341; Molachintapalli (MBNR), *PRSR* & *TDCK* 13641; Dhararam (MBNR), *TDCK* 13641 (BSID); ICRISAT (MDK), *LJGV* 2691 (CAL).

PELTOPHORUM (Vogel) Benth. *nom. cons.*

Peltophorum pterocarpum (DC.) Backer ex K. Heyne, Nutt. Pl. Ned. Ind. 2: 755. 1927. *Inga pterocarpum* DC., Prodr. 2: 441. 1825. *Peltophorum ferrugineum* Benth., Fl. Austral. 2: 270. 1864.

Vern.: *Pachasunkesula.*

Trees, up to 20 m high, bark brown; younger parts rusty-browny or greyish-tomentose. Leaves 15 cm long, alternate, pinnae 6-12 pairs, leaflets 6-17 pairs, oblong, 1-2 × 0.5-0.8 cm, glabrous, broadest on the upper side of the midrib, base semi-cordate, margin entire, apex obtuse, nerves distinct. Flowers bright-yellow, in 10-35 cm long terminal and axillary reddish- brown panicles; calyx lobes 5, oblong, 6-7.25 × 4-5 mm, subequal, subclavate; petals 5, red in colour, obovate-suborbicular, subequal, 0.6-1 ×0.5-0.7 cm, ferruginous hairs on back; stamens 10, subequal, free, filaments 3-4 mm, empty anthers to 4 mm, subulate, filaments with tufts of silky hair at base, to 2 mm, ovary subsessile, 3 mm, glabrescent, style attenuate, to 7 mm. Pods flat, indehiscent, 4-10 × 1.5-2.5 cm, lanceolate, dark-brown, woody; seeds obovate-oblong, compressed, smooth, glabrous.

Planted as an avenue tree.

Fl. & Fr.: March-August.

Jaipur RF (ADB), *GO* & *PVP* 4945; Machareddy (NZB), *BR* & *KH* 9055; Siddipet –Hyderabad road (MDK), *CP* 11380; Bhadrachalam RF (KMM), *RCS* 104270 (BSID).

PTEROLOBIUM R. Br. ex Wight & Arn. *nom. cons.*

Pterolobium hexapetalum (Roth) Santapau & Wagh, Bull. Bot. Surv. India 5: 108. 1963. *Reichardia hexapetala* Roth, Nov. Pl. Sp. 210 1821. *Pterolobium lacerans* Wall. ex Wight & Arn., Prodr. Fl. Ind. Orient. 283. 1834. *P. indicum* A. Rich., Tent. Fl. Abyss. 1: 247. 1847; Baker in Hook. f., Fl. Brit. India 2: 259. 1878; Gamble, Fl. Madras 1: 395. 1919.

Vern.: *Korintha, Pariki.*

Straggling shrubs, to 6 m high; branches dark pink, prickly, glabrous. Leaves 2-pinnate, 12-18 cm long, prickly, pinnae 6-9 pairs, leaflets 8- 10 pairs, ovate-oblong, 1-1.5 × 0.5-1 cm, chartaceous, glabrous, base cuneate-truncate, margin entire, apex obtuse or retuse; petiole to 3 cm, petiolule 0; stipules and stipels caducous. Flowers white with a red tinge in axillary or terminal racemes; peduncle to 5 cm, pedicel to 2 cm; bract subulate, bracteoles 0; calyx-tube to 5 mm, persistent, lobes 5, unequal, 3-4 × 1.25 mm, obovate-oblanceolate, hooded; petals 5, subequal, 4 × 1-1.25 mm, spreading, suberect, basally villous, anthers uniform, to 1 mm; ovary sessile, to 2 mm, style filiform, to 3 mm. Pods samaroid, dark green below (enclosing the single seed) with reddish terminal wing above, oblong, indehiscent, 3-4 × 1 cm; seed solitary, at base flattened, 7 × 5 mm.

Occasional in Ranga Reddi, Nalgonda and Mahabubnagar districts, in open places over bushes and small trees.

Fl. & Fr.: July-December

Saikunta (RR), *MSM* 10492; Kakalakonda (NLG), *KCJ* 396 (MH); Umamaheswaram (MBNR), *SRS* 109149 (BSID).

SARACA L.

Saraca asoca (Roxb.) de Wilde, Blumea 15: 393. 1968. *Jonesia asoca* Roxb., Asiat. Res. 4: 365. 1799. *Saraca indica* auct. non L. 1769; Baker in Hook. f., Fl. Brit. India 2 : 271. 1878; Gamble, Fl. Madras 1: 409. 1919.

Vern. : *Asoka.*

Small trees or large shrubs, 8-10 m high, bark greyish, rough. Leaves even pinnate, 25 cm long, leaflets 4-6 pairs, oblong, 15 × 7 cm, coriaceous, base rounded or cuneate, slightly oblique, apex acute to acuminate; petiolules short, stout, wrinkled, stipels intrapetiolar, deciduous. Flowers brilliant orange-yellow, pink with age in short dense often lateral corymbose panicles; peduncle stout, pedicels very short, red, glabrous; bracts ovate, subulate, bracteoles 2, appearing like calyx, spathulate-oblong, subacute, ciliolate, amplexicaul, coloured, calyx passing from yellow to orange and finally red, 08-1.1 × 0.3-0.4 cm, obovate-oblong; petals 0; stamens usually 7, rarely 3-4, much exserted, filaments filiform, 1.2-1.8 cm, anthers purple; ovary long stipitate, glabrous, 7 mm, style curved, to 4 mm. Pod flat, oblong, coriaceous or almost woody; seeds obovate-orbicular, compressed; seeds 4-8, obovate-orbicular, compressed, ca 1.5 cm.

Rarely cultivated (Rajagopal, 1973).

Fl. & Fr.: All seasons.

Saleswaram (MBNR), *BR, SKB* & *SS* 33333, 34466.

SENNA Mill.

1. Foliar glands absent on the petiole or rachis:
   2. Fertile stamens 10; staminodes absent .... **S. siamea**
   2. Fertile stamens 7; staminodes 3:
      3. Pods with a ridge of longitudinal crest-like plaite or crenate-margined longitudinal wing along the middle of each valve:
         4. Herbs or undershrubs; pods up to 5 cm long, reniform .... **S. italica**
         4. Shrubs or small trees; pods 12-18 cm long, straight .... **S. alata**
      3. Pods without plaite or wings on the valves:
         5. Ovary glabrous .... **S. montana**
         5. Ovary pubescent or velvety .... **S. alexandrina**

1. Foliar glands present on the petiole and / or rachis:

   6. Petiole with a single gland at base, rachis eglandular :

      7. All parts softly tomentose .... **S. hirsuta**

      7. All parts glabrous :

         8. Leaflets 3-5-pairs, elliptic-ovate .... **S. occidentalis**

         8. Leaflets 6-10 pairs, oblong-lanceolate .... **S. sophera**

   6. Petiole eglandular; rachis with glands between the leaflets :

      9. Antheriferous stamens 10 .... **S. surattensis**

      9. Antheriferous stamens 6-7 :

         10. Stipules large, cordate, persistent, leaflets 8-12 pairs ...... **S. auriculata**

         10. Stipules linear, deciduous, leaflets 3-5 pairs:

            11. Plants glabrous or pubescent; peduncles 1 to 2-flowered; pedicels 1-3 cm long; pods 5-20 cm long, terete, glabrous or pubescent :

               12. Pods subterete or tetragonous; seeds transverse . **S. obtusifolia**

               12. Pods flat, compressed; seeds longitudinal .... **S. tora**

            11. Plants rufous hairy; peduncles 3 to 4-flowered; pedicels 2-4 mm long; pods 3-3.5 cm long, subcompressed, rufous hairy .... **S. uniflora**

Senna alata (L.) Roxb., Fl. Ind. 2 : 349. 1832. *Cassia alata* L. Sp. Pl. 378; Hook.f. Fl. Brit. India 2 : 264. 1878; Gamble, Fl. Madras 1 : 404. 1919.

Evergreen shrub, 2-3 m high; leaf scars persistent. Leaves 20-25 cm long; leaflets 5-6 pairs; petioles 4-5 cm long; rachis and petiole eglandular, deeply grooved; stipule lunate, 5 mm long; lower leaflets oblong-elliptic; upper ones broadly obovate,3-6 x 2-3 cm. inequilateral, base oblique, truncate, margins entire, apex obtuse, apiculate. Flowers yellow turning orange, in dense terminal racemes; bract spatheceous, enclosing bud. Calyx lobes oblong, obtuse. Petals ovate, to 2 cm long. Pod shortly stipitate, oblong, thin, subterete, base and apex obtuse, margins with 2 longitudinal crenulate wings.

Planted in gardens and along roadsides for its showy racemes in Karimnagar (Naqvi, 2001) and Hyderabad districts (Ramana, 2010) and naturalized.

Fl. & Fr.: November-February.

Senna alexandrina Mill., Gard. Dict. ed. 8. No. 1. 1768; V. Singh, Mongr. Cassinae 104. 2001. *Cassia senna* L., Sp. Pl. 377. 1753. *C. angustifolia* Vahl, Symb. Bot. 1: 29. 1790; Baker in Hook. f., Fl. Brit. India 2 : 264. 1878; Gamble, Fl. Madras 1: 404. 1919.

Shrubs, 1-1.5 m high, glabrous, bark pale-yellowish-brown. Leaves alternate, 6-8 cm long, leaflets 4-7 pairs, opposite, linear-lanceolate, 1-3 × 0.5-0.8 cm, glabrous above, appressed pubescent below, base cuneate, margin entire, apex acute, apiculate; petiole to 1.5 cm, petiole and rachis eglandular, petiolule to 1 mm, stipules 1.5 mm, spreading or reflexed. Flowers greenish-yellow in 6-8 cm long, terminal or axillary, branched or simple racemes; peduncle to 2.5 cm, pedicel to 5 mm; bract ovate, to 2 mm; bracteoles to 1.5 mm; calyx lobes 5, spathulate, concave, 8-9 × 3-3.75 mm; petals 5, orbicular-ovate, 8-9 × 2-4 mm, claw to 2 mm; stamens 10, upper 3 staminodes 7-9 mm, antheriferous ones 7, lower large with filaments 2 mm and curved anthers to 1 mm, one medium with filaments to 1 mm and anther to 1 mm, 4 short, with filaments 0.5 mm and anthers to 4 mm; ovary stipitate, to 9 mm, densely yellowish pubescent, style 3 mm. Pod 3-6 × 1.5-2 cm, oblong, rarely obovate, greenish-black, shortly beaked at apex; seeds ca 8, ovoid, membranous, 7 × 4 mm, longitudinal.

It is cultivated for its cathartic leaves and pods, sometimes as an escape (Ramana, 2010).

Fl. & Fr.: July-September.

Senna auriculata (L.) Roxb., Fl. Ind. 2: 349. 1832; V. Singh, Mongr. Cassinae 119. 2001. *Cassia auriculata* L., Sp. Pl. 379. 1753; Baker in Hook. f., Fl. Brit. India 2: 263. 1878; Gamble, Fl. Madras 1: 402. 1919.

Vern.: *Thangedu.*

Erect bushy shrubs, 1.5-2.5 m high, bark greyish, branchlets finely pubescent, stipules large, foliaceous, reniform, persistent, cordate at base with one end produced into a tail. Leaves pinnately compound, 5-9 cm long, leaflets 6-12 pairs, oblong-obovate, 1-2 × 0.5-1 cm, thin coriaceous, glabrous above, sparingly puberulous beneath, base cuneate-truncate, margin entire, apex obtuse-retuse, mucronate; petiole to 1.5 cm, eglandular, rachis with linear, erect, stipitate glands opposite to all leaflets, petiolule to 4 mm. Flowers yellow with orange veins in axillary and terminal corymbs, often panicled; calyx lobes 5, ovate, outer ones smaller than the inner, outer sepals 0.7-0.8 × 0.4-0.6 cm, inner 1.1-1.8 × 0.9-1.1 cm, obtuse; petals 5, 2 larger 1.5 × 1 cm, 3 smaller 1.3 × 0.6 cm, ovate-orbicular; stamens 10, upper 3 staminodes with filaments 3 mm and empty anther 1.2 cm, antheriferous ones 7, lower 2 small, with filaments 1-2 mm and curved anthers 3 mm, one medium with filament to 2 mm and anther 4 mm, 4 long with filaments 1-2 mm and anthers 1.2 mm; ovary stipitate, falcate, to 1.7 cm, pubescent, style to 1 cm. Pods 10-12 × 2 cm, linear, flattened, torulose, mucronate, stipitate, puberulous; seeds about 10-15, ovoid, laterally compressed.

Common in most of the districts on dry stony hills and abundant in plains.

Fl. & Fr.: Almost throughout the year.

Sadarmat (ADB), *TP* & *PVP* 4102; Kagaznagar Road (ADB), *GO* 4369; Malkapur – Nizamabad road (NZB), *TP* & *BR* 6114; Nathyanipalle (MDK), *BR* & *CP* 11248; Kusumasamudram RF (RR), *MSM* & *KH* 11021; Mannanur (MBNR), *TDCK* 13620; Way to Pagarikutta-Narsapur (MDK), *KMS* 6657 (CAL); Azampur (NLG), *BVRS* 484 (NU); Kodimial (KRN), *GVS* 20091 (MH); Pakhal (WGL), *KMS* 11564 (MH); Ramavaram RF (KMM), *RR* 112659 (BSID); Nawapet road, Shadnagar (MBNR), *SRS* 104480 (BSID).

Senna hirsuta (L.) H.S.Irwin & Barneby, Phytologia 44 (7): 499. 1979 & Mem. New York Bot Gard. 35(1): 425, 434. 1982; Singh, Monogr. Cassinae 137. 2001. *Cassia hirsuta* L., Sp. Pl. 378. 1753; Baker in Hook. f., Fl. Brit. India 2 : 263. 1878; Gamble, Fl. Madras 1: 401. 1919.

Diffuse undershrubs, branches softly tomentose, grooved. Leaves 15-20 cm long, with a gland at base of petiole, leaflets 4-6 pairs, ovate-elliptic, 3-8.5 × 1.5-3 cm, chartaceous, appressed hirsute, base cuneate, margin entire, apex acuminate; petiole to 3 cm, with a gland at base; rachis eglandular, petiolule to 4 mm; stipules linear, subulate, about 1.5 cm. Flowers yellow in pairs, in leaf-axils or flowers in corymbose panicles; peduncle to 1 cm; pedicel to 2 cm; bract linear, to 1.5 cm, calyx lobes 5, 0.8-1 × 0.3-0.6 cm, obtuse, ovate; petals 5, obovate, 7.5-9-4-6 mm; stamens 10, upper 3 staminodes, to 3 mm, antheriferous ones 7, lower 2 large, with suberect filaments to 6 mm and apically beaked, anthers to 7 mm, 5 medium erect filaments 1-2.5 mm and anthers 3-4.5 mm, ovary subsessile, to 3.5 mm, tomentose, style short, 1.5-2 mm, glabrous. Pod to 15 × 0.3 cm, ribbed, hirsute, slender, flattened, dehiscent, many-seeded.

Native of Tropical Amercia and is naturalised in a few areas.

Fl. & Fr.: June-January

Mothugudem (KMM), *PRSR* 15428.

Senna italica Mill., Gard. Dict. ed. 8. no. 2. 1768; subsp. **italica**; V. Singh, Monogr. Cassiinae 147. 2001. *Cassia italica* (Mill.) Lam. ex Andr., Fl. Pl. Anglo Egypt. Sudan 2: 117. 1952. *C. obtusa* Roxb., Fl. Ind. 2 : 344. 1824; Gamble, Fl. Madras 1: 403. 1919. *C. obovata* Collad., Hist. Cass. 92. t. 15A. 1816; Baker in Hook. f., Fl. Brit. India 2 : 264. 1878.

Diffuse perennial herbs, up to 60 cm high, branchlets glabrescent. Leaves 4-6 cm long, leaflets 4-8 pairs, obovate-oblong, 1.5-2.5 × 0.8-1 cm, membranous, very glaucous, base obtuse, margin entire, mucronate at apex, nerves rugose; petiole to 1 cm, eglandular, rachis with linear, erect, stipules linear, to 5 mm, foliaceous, persistent. Flowers pale yellow in axillary lax racemes, peduncle to 2 cm, pedicel to 2 mm; bract stipular, bracteole linear, 1 mm; calyx lobes 5, oblong-obovate, 6.5-9 × 1-4 mm, obtuse; petals 5, ovate, subequal, 5-5.5 ×1.25-3 mm, claw to 1.5 mm, distinctly nerved; stamens 10, very unequal, upper 2 staminodes 0.5 mm, antheriferous ones 8, lower 2 large, with filaments 1 mm and empty anthers 1 mm, 3 medium with filaments 1.25 mm and anthers 1.5 mm, 3 short, with filaments 0.25 mm and empty anther 1 mm; ovary subsessile, appressed-pubescent, to 5 mm, style to 1.5 mm. Pods 2.5-4 × 1.5 cm, flat, falcate, papery, crested longitudinally above the seeds, narrowed suddenly at both ends; seeds 6-12, ovoid, flat.

Occasional in Hyderabad and Mahabubnagar districts in open lands.

Fl. & Fr.: April-December.

Haipur hills (MBNR), *TDCK* 13619; Mukkidugundam (MBNR), *BR* & *TSS* 28327.

Senna montana (B. Heyne ex Roth) V. Singh, J. Econ. Taxon. Bot. 16(3): 600. 1992 & Monogr. Cassiinae 154. 2001. *Cassia montana* B. Heyne ex Roth, Nov. Pl. Sp. 214. 1821; Baker in Hook. f., Fl. Brit. India 2: 264. 1878; Gamble, Fl. Madras 1: 401. 1919.

Vern.: *Pagadi tangedu, Konda tangedu.*

Small trees or large shrubs, 10 - 15 m high, bark dark in colour. Leaves 10-20 cm long, leaflets 10-12 pairs, oblong-elliptic, 2-4 × 1-2 cm, chartaceous, nearly glabrous above, hairy beneath, base and apex obtuse, margin entire; petiole to 1.5 cm, rachis and petiole eglandular, petiolule 2 mm, stipules minute, caducous. Flowers yellow in axillary and terminal corymbose racemes; peduncle to 2.5 cm, bracts stipular, bracteoles linear, to 4 mm, pedicel to 2 mm; calyx lobes 5, ovate, 8-9 × 2.5-7 mm, claw 6 × 5 mm, apex obtuse; petals 5, ovate, 7-9 × 2.5-6 mm; stamens 10, upper 3 staminodes, to 7 mm, lower 2 large with erect filaments, to 1.5 mm and anthers 1 mm, 5 medium with filaments 2 mm and anthers 3-4 mm; ovary stipitate, to 1 cm, style to 3 mm. Pod straight, short stipitate, flat, to 9.5 × 1 cm, compressed, membranous, strongly nerved, sutures thin, glabrous; seeds ca 20, ovoid, 4.5 mm.

Common on dry stony hills of Medak district.

Fl. & Fr.: Through out the year.

Akkannapet to Lakshmipuram (MDK), *RG* 106722 (BSID).

Senna obtusifolia (L.) H.S. Irwin & Barneby, Mem. New York Bot. Gard. 35(1): 252. 1982; V. Singh, Monogr. Cassiinae 164. 2001. *Cassia obtusifolia* L., Sp. Pl. 377. 1753. *C. tora* sensu Baker in Hook. f., Fl. Brit. India 2: 263. 1878 *p.p.;* Gamble, Fl. Madras 1: 401. 1919. *C. tora* (L.) var. *obtusifolia* (L.) Haines Bot. Bihar & Orissa 304. 1922.

Vern.: *Tantipu.*

Annual herbs, up to 60 cm high, crisped hairy in young branches. Leaves 6-8 cm long with subulate glands between the two lowest pair of the leaflets, leaflets 3-5 pairs, sessile, obovate- oblong, 2-4.5 × 1.5-2.5 cm, unequally rounded at base, thin coriaceous, glabrous above, pubescent below, base cuneate, rotund, margin entire, apex obtuse, apiculate; petiole to 2 cm, petiolule to 2 mm; stipules falcate, 0.6-1 cm. Flowers golden yellow or with reddish tinge, solitary or geminate, axillary; calyx lobes 5, ovate, 6.5-8 × 2-5.5 mm, pubescent, apex obtuse; petals 5, obovate, subequal, 1-1.4 ×0.5-0.7 cm, claw to 1 cm; stamens 10, upper 3 staminodes to 6.5 mm, antheriferous ones 7; 2 lower one small with erect filaments to 2.7 mm and anthers 1.75 mm, abruptly tapering at the ends, one small with erect filament to 2.75 mm and anther 1 mm, 4 short ones with filaments to 2.75 mm and oblong anthers 2.75 mm; ovary subsessile, to 1.5 cm, curved, pubescent, style to 4 mm. Pods 5-9 × 0.4-0.6 cm, linear, quadrangular, puberulous, sutures prominent; seeds about 20, faintly tetragonal, longitudinal.

Common in all plains districts, on way sides, and waste places, fallow lands and in forest under growth and also on hills.

Fl. & Fr.: September-December

Vempalli (ADB), *GO* & *PVP* 4729; Mosra - Malkapur Road (NZB), *TP* & *BR* 6090; Ramayampet (MDK), *BR* & *CP* 11507; Dharur (RR), *MSM* 11093; Gacchibowli (RR), *MSM* 12770; Mothugudem forest (KMM), *PRSR* 15429; Parnapalli (HYD), *KMS* 5986 (MH); Kambalpally (NLG), *BVRS* 239 (NU); Rechapalli (KRN), *GVS* 22243 (MH).

Senna occidentalis (L.) Link, Handbuch 2: 140. 1831; V. Singh, Monogr. Cassiinae 170. 2001. *Cassia occidentalis* L., Sp. Pl. 377. 1753; Baker in Hook. f., Fl. Brit. India 2 : 262. 1878. Gamble, Fl. Madras 1: 401. 1919.

Vern.: *Kashindha.*

Erect, glabrous undershrubs, up to 3 m high, branchlets glabrous. Leaves 10-15 cm long, glandular above petioles, leaflets 3-5 pairs, elliptic-ovate, 3-6 × 1-3 cm, membranous, glabrous above, thinly hairy beneath, base rounded or cuneate, margin entire, apex obtuse or acute, apiculate; petiole 2.5-4 cm with a sessile globose gland at its base, rachis eglandular, petiolule 1-1.5 mm, stipules 3-4.5 cm, caducous. Flowers bright reddish yellow in axillary and terminal corymbose racemes; peduncle to 2 cm; pedicel to 1.5 cm; bracts linear-lanceolate, 7 mm; calyx-lobes 5, ovate, 7-8.5 × 2-3 mm, apex obtuse, mucronate; petals 5, oblong-obovate, distinctly nerved, 8-8.5 × 3-5 mm, claw 3 mm; stamens 10, upper 3 staminodes, with filaments 5 mm and empty anthers to 5 mm, antheriferous ones 7, 3 medium with flat filaments to 1.5 mm and erect anthers to 3.5 mm, 4 short with thin filaments to 3 mm and anthers 1 mm; ovary subsessile, to 1.2 cm, densely appressed-pubescent, style to 2.5 mm. Pods short stipitate, slightly falcate, 10-12 x 1 cm, laterally compressed, linear-oblong, brown with yellow margins; seeds 20-25, subcylindric, 6.5 × 5 mm, greenish brown, smooth, glabrous.

Common in all plains districts, by road sides and on the wastelands.

Fl. & Fr.: Almost throughout the year.

Sattenapalle RF (ADB), *TP* & *PVP* 4087; Nasarullabad RF (NZB), *TP* & *BR* 6162; Pathur (MDK), *TP* & *CP* 12115; Rangammagudem (RR), *MSM* & *KH* 10323; Gourammasari (MBNR), *TDCK* 13693; Mothugudem (KMM), *PRSR* & *SSR* 15427; Moosi river bank (HYD), *KMS* 5979 (MH); Rayagir (NLG), *BVRS* 16 (NU); Bhupatipur (KRN), *GVS* 22285 (MH); Subedari (WGL), *RKP* 105258 (BSID); Mamillagudem (KMM), *RR* 105929 (BSID).

Senna siamea (Lam.) H.S. Irwin & Barneby, Mem. New York Bot. Gard. 35(1): 98. 1982; V. Singh, Mongr. Cassinae 193. 2001. *Cassia siamea* Lam., Encycl. 1: 648. 1785; Baker in Hook. f., Fl. Brit. India 2: 264. 1878; Gamble, Fl. Madras 1: 402. 1919.

Vern.: *Seematangedu.*

Moderate-sized trees, up to 10 m high with light-blackish-brown, rough, longitudinally fissured bark; branchlets glabrescent. Leaves 10-15 cm long, leaflets 7-14 pairs, elliptic-oblong, 2-5 × 1.5-2 cm, coriaceous, minutely petiolulate, glabrous above, pilose below, nerves prominent, base and apex obtuse, margin entire; petiole to 2.5 cm, rachis and petiole eglandular, petiolule to 3 mm, stipules subulate, to 1 mm, caducous. Flowers pale-yellow in axillary and terminal corymbose racemes; bracteole to 6 mm, obovate; calyx lobes 5, ovate, outer 2 smaller, 6-6.5 × 3-4 mm, inner 3 large, 7-

8 × 4.25-6 mm, concave, apex obtuse; petals 5, ovate-elliptic, 6.5-9 × 4-6 mm, claw to 1.2 × 6 mm; stamens 10, upper 3 staminodes with erect filaments, 7.5-8 mm and empty anthers to 3 mm, antheriferous ones 7, lower 2 large, with filaments to 2.5 mm and empty anthers to 5 mm, one medium with filaments 1.5 mm and anther to 3 mm, other 4 short, with filaments 2 mm and anthers 5 mm; ovary sessile, to 7.5 mm, deeply grooved, pubescent, style 2-5 mm. Pod long- stipitate, strap-shaped, flat, 20-25 × 1-1.5 cm, compressed, woody with thick sutures; seeds 20-30.

Planted as avenue tree and naturalised.

Fl. & Fr.: July-December.

Adilabad town (ADB), *GO* & *PVP* 4251; Biknoor (NZB), *BR* & *CPR* 7249; Siddipet – Medak Road (MDK), *CP* 11711; Mohammadabad RF (RR), *MSM* & *KH* 10209; Aklaspur (KRN), *GVS* 22537 (MH); Umamaheswaram (MBNR), *SRS* 109155 (BSID).

Senna sophera (L.) Roxb., Fl. Ind. 2: 347. 1832; var. **sophera**; V. Singh, Monogr,. Cassiinae 199. 2001. *Cassia sophera* L., Sp. Pl. 379. 1753; Baker in Hook. f., Fl. Brit. India 2: 262. 1878 [incl. var. *purpurea* (Roxb.) Baker]; Gamble, Fl. Madras 1: 402. 1919.

Vern. : *Pyditangedu.*

Diffuse undershrubs, up to 3 m high, branchlets nearly glabrous. Leaves 10-15 cm, with a solitary, conical gland near the base, rachis grooved, leaflets 6-10 pairs, oblong-lanceolate, 2-4 × 0.8-1.2 cm, glaucous, base cuneate, margin entire, apex acuminate, mucronate, petiolules 1-1.5 mm long. Flowers yellow in terminal and axillary, corymbose racemes; pedicel to 0.5 mm long, pubescent, bracts small, green, ovate, caducous; calyx lobes 5, obtuse, 6-8 × 5.5-6 mm, green; petals 5, subequal, ovate, 9-9.5 × 3-4.5 mm; stamens 10, the upper 3 are reduced to staminodes, with filaments 2 mm, empty curved anthers 7 mm, the remaining 7 usually perfect, lower with filaments 1 mm and empty anthers 4.5 mm, 4 lateral ones of which one is sometimes reduced to a staminode, with filaments 3 mm and anthers 1 mm; ovary stipitate, 5 × 1 mm, glabrous, incurved, style 1 mm, glabrous. Pods 7-12 × 0.6-0.8 cm, linear-oblong, apiculate, septate; seeds 30-40, broadly ovoid, flat, dark brown.

Common in all districts, by road sides and on wastelands.

Fl. & Fr.: July-January.

Gowraram RF (NZB), *TP* & *BR* 6378; Maddigunta (NZB), *BR* & *PSPB* 9557; Manchanpalle (RR), *TP* & *MSM* 11100; ICRISAT (MDK), *LJGV* 2634 (CAL); Annaram (NZB), *BR* & *CPR* 7191; Mohammadabad (RR), *MSM* & *KH* 10596; Amangal to Shadnagar (MBNR), *SRS* 10456 (BSID)

Key to Varieties

1. Glands finger-like, spindle-shaped or ovoid-cylindric .. **S. sophera** var. **sophera**
1. Glands hemispheric, dome-shaped or subglobose .... **S. sophera** var. **purpurea**

Senna sophera (L.) Roxb. var. **purpurea** (Roxb. ex Lindl.) Singh, J. Econ. Taxon. Bot. 16(3): 600. 1992. *Cassia purpurea* Roxb. ex Lindl., Bot. Reg. t. 856. 1824. *Cassia sophera* L. var. *purpurea* (Roxb. ex Lindl.) Benth., Trans. Linn. Soc. London 27: 533. 1871; Baker in Hook.f., Fl. Brit. India 2: 263. 1878.

Gland solitary, 0.5-1 mm in diam., globose to somewhat oblong or dome-shaped, borne usually below the middle on the petiole, rarely above the middle or between the lowest pair of leaflets in some leaves.

Occasional in Nizamabad and Ranga Reddi districts.

Fl. & Fr.: August – February.

Annaram (NZB), *BR* & *CPR* 7191; Mohammadabad (RR), *MSM* & *KH* 10596.

Senna surattensis (Burm.f.) H.S. Irwin & Barneby, Mem. New York Bot. Gard. 35(1): 81. 1982; V. Singh, Monogr. Cassiinae 215. 2001. *Cassia surattensis* Burm.f., Fl. Ind. 97. 1768. *C.suffruticosa* Koen. ex Roth, Nov. Pl. Sp. 213. 1821. *C. glauca* Lamk. var. *suffruticosa* (Koen. ex Roth) Baker in Hook. f., Fl. Brit. India 2: 265. 1878 *nom. illegit.*; Gamble, Fl. Madras 1: 402. 1919. *C. glauca* Lam., Encycl. 1: 647. 1785; Baker in Hook. f., Fl. Brit. India 2: 265. 1878; Gamble, Fl. Madras 1: 403. 1919.

Vern.: *Kondatangedu, Mettangedu.*

Shrubs or small trees, 5-10 m high, branchlets sparsely pubescent. Leaves 12-15 cm long, rachis pale, puberulous with a clavate gland between each of the 2-3 basal pairs of leaflets, stipules short, linear, acute, falcately curved, caducous, leaflets 4-6 pairs, elliptic, base cuneate, margin entire, apex obtuse, petiole to 2.5 cm, eglandular, petiolule to 1.5 mm. Flowers yellow in axillary corymbose racemes; peduncle to 4.5 cm; pedicel to 2 cm, bracts linear, lanceolate, to 6 mm; calyx lobes 5, ovate, unequal, outer 3 smaller, 5-7 × 1-3 mm, inner larger, 0.9-1 × 0.5-0.6 cm, glabrous, membranous, reticulately veined, oblong or suborbicular; petals 5, ovate, 1-1.2 × 0.6-1 cm, clawed, broadly oblong, obtuse; stamens 10, all antheriferous, subequal, 2+1+4+3, filaments long 1.5 mm, medium 1 mm, smaller 0.75 mm, anthers large 7 mm, medium 3.5 mm, smaller 2 mm; ovary stipitate, to 1 cm, tomentose, style to 4 mm, curved. Pod flat, thin, transversely barred, 15-20 × 1-1.5 cm, with a few scattered hairs seeds 25-30, oblong-ellipsoid, 7 × 4 mm, beaked, compressed, smooth, dark brown, shining, with a shallow oblong pit on each of the flat faces.

Rare in forests of Karimnagar and Ranga Reddi districts.

Fl. & Fr.: August-December.

Kodimial (KRN), *GVS* 21891, 25676 (MH); Anantagiri RF (RR), *MSM* & *PRSR* 12788.

Senna tora (L.) Roxb., Fl. Ind. 2: 340. 1832; V. Singh, Monogr. Cassiinae 222. 2001. *Cassia tora* L., Sp. Pl. 376. 1753; Baker in Hook. f., Fl. Brit. India 2: 263. 1878 *p.p.;* Gamble, Fl. Madras 1: 401. 1919.

Vern.: *Tantipu.*

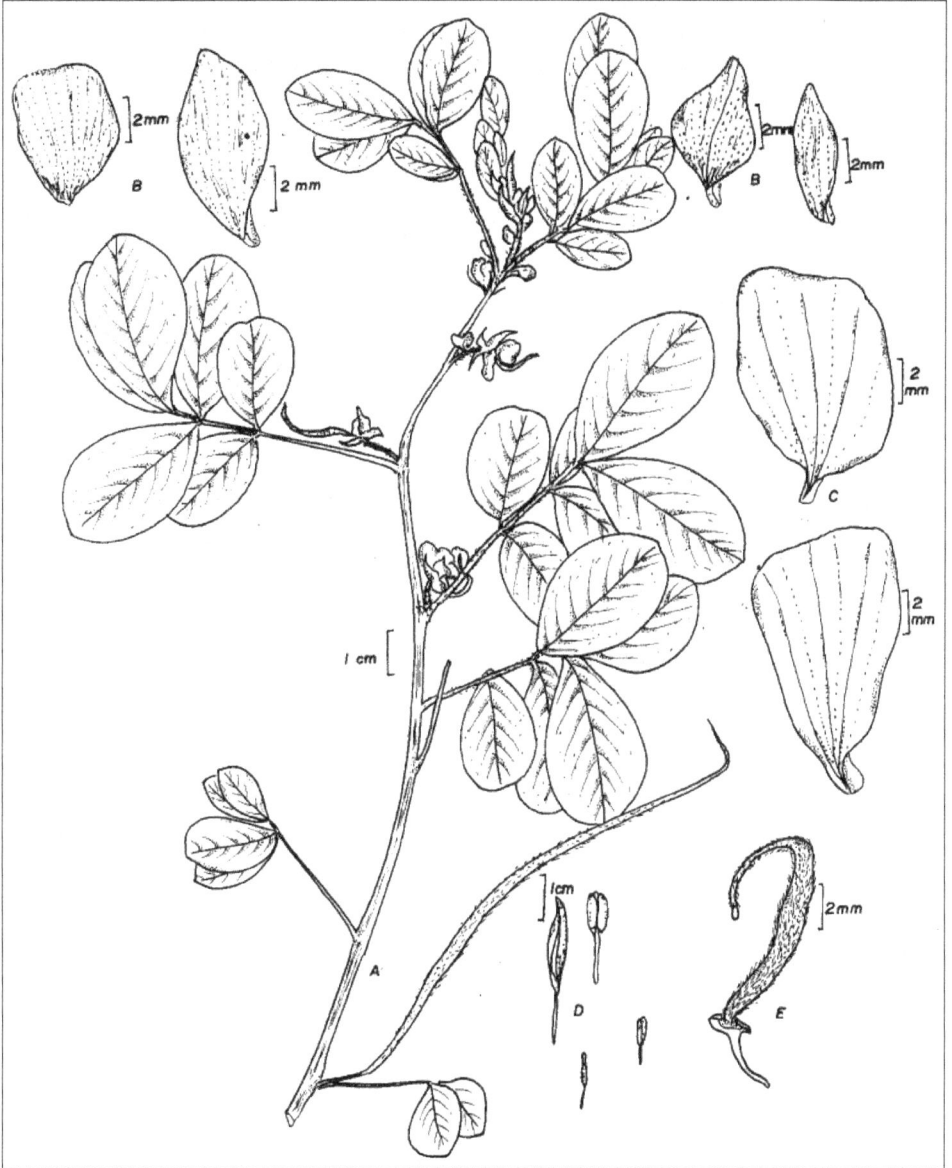

Figure 27: Senna tora (L.) Roxb. (= *Cassia tora* L.)
A. Twig, B. Sepals, C. Petals, D. Stamens, E. Pistil.

Annual herbs, up to 60 cm high, crisped hairy in young branches. Leaves 6-8 cm long with subulate glands between the two lowest pair of the leaflets, leaflets 3-5 pairs, sessile, obovate- oblong, 2-4.5 × 1.5-2.5 cm, unequally rounded at base, thin coriaceous, glabrous above, pubescent below, base cuneate, rotund, margin entire, apex obtuse, apiculate; petiole to 2 cm, petiolule to 2 mm; stipules falcate, 0.6-1 cm. Flowers golden yellow or with reddish tinge, solitary or geminate, axillary; calyx lobes 5, ovate, 6.5-8 × 2-5.5 mm, pubescent, apex obtuse; petals 5, obovate, subequal, 1-1.4 × 0.5-0.7 cm, claw to 1 cm; stamens 10, upper 3 staminodes to 6.5 mm, antheriferous ones 7; 2 lower one small with erect filaments to 2.7 mm and anthers 1.75 mm, abruptly tapering at the ends, one small with erect filament to 2.75 mm and anther 1 mm, 4 short ones with filaments to 2.75 mm and oblong anthers 2.75 mm; ovary subsessile, to 1.5 cm, curved, pubescent, style to 4 mm. Pods 5-9 × 0.4-0.6 cm, linear, quadrangular, puberulous, sutures prominent; seeds about 20, faintly tetragonal, longitudinal.

Common in all plains districts, on way sides, and waste places, fallow lands and in forest under growth and also on hills.

Fl. & Fr.: September-December

Vempalli (ADB), *GO* & *PVP* 4729; Sirpur (ADB), *GO* & *PVP* 4926; Mosra Road (NZB), *TP* & *BR* 6090; Ramayampet (MDK), *BR* & *CP* 11507; Dharur (RR), *MSM* & *TP* 11093; Mothugudem forest (KMM), *PRSR* 15429; Chukkalagundam (MBNR), *TDCK* 13672; Molachintapalli (MBNR), *BR* & *TSS* 28221; Parnapalli (HYD), *KMS* 5986 (MH); Jakaram (WGL), *RKP* 105330 (BSID).

Note: Raju and Rao (1986) found that *Cassia tora* is an admixture of *C. tora* and *C. obtusifolia.* However, the former always has a subulate gland between each of the 2 lower pairs of leaflets, while the later, between the leaflets of the lowest pair only (Naqvi, 2001).

Senna uniflora (Mill.) H.S. Irwin & Barneby, Mem. New York Bot. Gard. 35 (1): 258. 1982. *Cassia uniflora* Mill., Gard. Dict. ed. 8. no. 2. 1768; Raghavan, Bull. Bot. Surv. India 22: 225. 1982. *C. sericea* Swartz., Prodr. 66. 1788 & Fl. Ind. Occ. 2(1): 724. 1788; Saldanha, Fl. Karnataka 1: 386. 1984.

Annual, erect herb, up to 50 cm high, younger parts rufous hairy. Stipules linear-lanceolate, leaf rachis with clavate glands between the leaflets, leaflets (2)-3-(4) pairs, obovate, obtuse, apiculate, hairy on both faces, 2.5-3.5 × 1.2-2 cm, the lowest pair smallest, lateral nerves 4-5 pairs. Flowers 3-5 in axillary racemes, peduncle ca. 1 cm long, sepals 5, free to the base, 3-4 × 1.5-3 mm, subequal, ovate-oblong, petals 5, yellow, clawed, 4.25-5 × 2-3.5 mm, subequal, ovate-oblong, stamens 10, upper 3 staminodes, 1-1.5 mm long, other 7 perfect, subequal, pistil hairy, 6 mm long, ovary 4 mm long, sessile, style and stigma 2 mm long. Pod linear, densely hairy when young, greenish-yellow, subcompressed, 3-3.5 cm long, 7-9-seeded; seeds subquadrangular, rhomboidal, 3 mm long, smooth, dull-brown.

Invasive weed, found in agricultural fields, waste lands and open forests.

Fl. & Fr.: August – January.

Note: C.S.Reddy & *al.* (2000) reported it from Mahabubnagar district, Gopalan & *al.* (2006) reported rom Medak district. Now it has become a problematic weed in Andhra Pradesh (C.S.Reddy & *al.,* 2008).

TAMARINDUS L.

Tamarindus indica L., Sp. Pl. 34. 1753; Baker in Hook. f., Fl. Brit. India 2 : 273. 1878; Gamble, Fl. Madras 1: 409. 1919.

Vern.: *Chinta.* English: Tamarind.

Trees, about 15-30 m high, bark fissured, grey, branchlets glabrous. Leaves even pinnate, 8 cm long, leaflets 12-16 pairs, oblong, 2 × 1 cm, chartaceous, glabrous or puberulous, base and apex obtuse, margin entire; petiole to 7 mm, petiolule reduced; stipules caducous. Flowers yellow with pink striations in terminal, few-flowered racemes; calyx tube narrowly turbinate, to 7 mm, lined by disc, lobes 4, subequal, oblong, 6-8 × 2.75-3.5 mm; petals 3, oblong-oblanceolate; stamens 3, monadelphous, filaments to 6 mm, base pubescent, empty anthers 2 mm; ovary stipitate, to 5 mm, tomentose, style attenuate, 3 mm. Pod linear-oblong, falcate, thick, somewhat compressed, with brittle epicarp, pulpy mesocarp and leathary septate endocarp; seeds 1-many, quadrangular, brown, shining.

Common in all plains districts cultivated and self-sown. It is also planted as an avenue tree.

Fl. & Fr.: March- September.

Nirmal-Nizamabad Road (ADB), *GO* & *DAM* 4989; Tekrial (NZB), *TP* & *BR* 6423; Ramayampet (MDK), *BR* & *CP* 11511; Central University campus (RR), *MSM* 19213; Rachakonda (NLG), *BVRS* 553 (NU).

MIMOSOIDEAE

1. Flowers tetramerous; stamens 4-8 .... **Mimosa**
1. Flowers pentamerous; stamens 10 or more :
   2. Stamens definite :
      3. Flowers in globose heads :
         4. Trees :
            5. Pinnae never more than one pair; fruit woody .... **Xylia**
            5. Pinnae more than one pair; fruit membranous or coriaceous:
               6. Branchlets glabrescent; bark smooth; pinnae
                  3-8 pairs, leaflets 10-15 pairs; pods dehiscent .... **Leucaena**
               6. Branchlets pubescent, bark rough, pinnae
                  20-40 pairs, leaflets 60-80 pairs, pods indehiscent .... **Parkia**
         4. Herbs or undershrubs:
            7. Prostrate herbs .... **Neptunia**
            7. Erect undershrubs .... **Desmanthus**

    3.   Flowers in elongate spikes :

        8.   Plants unarmed .... **Adenanthera**

        8.   Plants armed:

            9.   Spikes dimorphic, bicoloured, upper yellow flowers bisexual, lower sterile with large staminodes; pods thin, twisted .... **Dichrostachys**

            9.   Spikes uniform, concolourous; usually all flowers bisexual; pods turgid, straight .... **Prosopis**

  2. Stamens indefinite :

    10. Unarmed trees; flowers always in globose heads .... **Albizia**

    10. Mostly armed trees, shrubs or woody climbers; flowers in spikes or heads forming panicles :

        11. Pinnae 1-paired; stamens mondelphous; pods falcate, often spirally twisted .... **Pithecellobium**

        11. Pinnae more than 1-paired, stamens free, pods flat, seldom turgid or torulose .... **Acacia**

ACACIA Mill.

Note: Most of the floras dealing with this plant give Willdenow as the author of the genus. However, it was Ph. Miller who in 1754 published it in his Gard. Dict. abridged edition No. 4, Vol. 1, 1754. Adanson used it in 1763; Willdenow in 1806.

1.   Climbing shrubs or stragglers:

  2. Leaflets 35 60 pairs .... **A. pennata**

  2. Leaflets 8-30 pairs :

    3. Flower heads pink; leaflets 8-12 pairs .... **A. intsia**

    3. Flowers white or yellow; leaflets more than 15 pairs:

        4.   Leaflets 17-20 pairs; flowers pale yellow; pods thin, flat, coriaceous .... **A. caesia**

        4.   Leaflets 20-25 pairs; flowers white; pods thick, fleshy, much wrinkled, when dry .... **A. sinuata**

1.   Erect shrubs or trees:

  5. Phyllodes falcate .... **A. auriculiformis**

  5. Phyllodes absent:

    6. Flowers in cylindrical spikes :

        7. Branches pubescent:

           8. Pinnae 15-25 pairs; leaflets 50-60 pairs .... **A. polyacantha**

           8. Pinnae 3-5 pairs; leaflets 8-15 pairs .... **A. senegal**

7. Branches glabrous :

    9. Stipular spines more than 2 cm long; corolla 5 times as long as calyx .... **A. horrida**

    9. Stipular spines less than 1 cm long; corolla 2-3 times as long as calyx :

       10. Calyx and petals villous .... **A. catechu**

       10. Calyx and petals glabrous:

          11. Leaflets 14-20 pairs, more than 2 mm broad .... **A. ferruginea**

          11. Leaflets 30-40 pairs, less than 1 mm broad ...... **A. chundra**

6. Flowers in globose heads :

    12. Heads in terminal panicles .... **A. leucophloea**

    12. Heads axillary, solitary or fascicled :

       13. Pods grey-pubescent, moniliform .... **A. nilotica**

       13. Pods glabrous, not moniliform :

          14. Pods cylindric, turgid :

             15. Pods straight .... **A. farnesiana**

             15. Pods twisted; petiole eglandular .... **A. planifrons**

          14. Pods flattened, dry

             16. Leaves glabrous; peduncles filiform .... **A. eburnea**

             16. Leaves pubescent (at least beneath); peduncles stout .... **A. tomentosa**

Acacia auriculiformis A. Cunn. ex Benth., London J. Bot. 1: 377. 1842; Sen & Naskar, Bull. Bot. Surv. India 7: 31. 1868; Chakrabarty & Gangopadhyay, J. Econ. Taxon. Bot. 20: 603. 1996.

Unarmed medium sized trees, about 15 m high, bark whitish, branches glabrous and glaucous, slightly angular. Phyllodes alternate, elliptic, falcate, 10-14 × 2.5-3.5 cm, thick-coriaceous, with 4-6 principal sub-parallel nerves tapering at the ends, glabrous, base attenuate, margin entire, apex obtuse; petiole to 2 cm, stipules to 1.5 mm. Flowers yellow in solitary or paired, axillary spike(s), to 8 cm, peduncle to 1 cm; calyx 5-lobed, campanulate, 2 mm, acute, halfway down; corolla 5-toothed, lanceolate, 2 mm; stamens numerous, filaments to 3 mm, basally connate; ovary subsessile, globose, to 1 mm, pubescent, style short. Pod coiled or spirally twisted, 3-6 mm, margin wrinkled at outer side, entire at inner side; seeds 5-7, brown.

Introduced, commonly planted in hills and under social forestry.

Fl.: December – February; Fr.: February-April.

Vempalli (ADB), *GO* 4456; Bijjur Guest House (ADB), *GO* & *PVP* 4973; Medak (MDK), *CP*11732; Appapur (MBNR), *TDCK*16834; Aralgudem RF (KMM), *RCS*102476 (BSID).

Acacia caesia (L.) Willd., Sp. Pl. 4 : 1090. 1806; Gamble, Fl. Madras 1: 428. 1919; Chakrabarty & Gangopadhyay, J. Econ. Taxon. Bot. 20: 604. 1996. *Mimosa caesia* L., Sp. Pl. 522. 1753 *p.p. Acacia intsia* Willd. var. *caesia* (L.) Baker in Hook. f., Fl. Brit. India 2: 297. 1878. *A. intisa* auct. non Willd.; Baker in Hook. f., Fl. Brit. India 2: 297. 1878. *A. torta* (Roxb.) Craib, Bull. Misc. Inf. Kew 1915: 410. 1915; Gamble, Fl. Madras 1: 428. 1919. Pullaiah & Chennaiah, Fl. Andhra Pradesh 1: 356. 1997. *Mimosa torta* Roxb., Fl. Ind. 2: 566. 1832. *Albizia sikharamensis* Sahni & Bennet, Indian Forest. 101: 337. 1975; Pullaiah & Chennaiah, Fl. Andhra Pradesh 1: 356. 1997.

Shrubby climbers, 10-15 m high, branches prickly, sparingly grooved, puberulous when young. Leaves alternate, 10-15 cm long, bipinnate, pinnae 6-8 pairs, leaflets not overlapping, 17-20 pairs, oblong, 1-1.5 × 0.3-0.5 cm, glabrous above, glabrescent beneath, base unequally truncate, margin entire, apex obtuse, apiculate, prickled on the underside with a convex elongate gland at base, rachis prickled, with glands opposite to lower most and 2 upper pairs of pinnae; petiole to 6 cm, prickled. Flowers pale yellow in head inflorescence, in terminal panicles; peduncle to 4 cm, bracts lanceolate, apex cuspidate, 0.5-1 cm, bracteoles to 1 mm; calyx tube campanulate, to 2 mm, 5-lobed, pubescent; petals ovate, 2.25 mm; stamens numerous, 2.5 mm, basally connate; ovary stipitate, oblong, to 4 mm, pubescent, style to 4 mm. Pod stipitate, 7-15 × 2-2.5 cm, thin, flat, glabrous, coppery, horned, base and apex obtuse; seeds 9-12, brown, flat.

Common in most of the dry districts in deciduous forests.

Fl. & Fr.: August-June.

Devapur (ADB), *GO*4236; Rajganjala RF (ADB), *GO*4386; Neelwai RF (ADB), *TP* & *GO* 5461; Sarvapur (NZB), *BR*9540; Gandhari RF (NZB), *TP* & *BR* 6337; Mothe RF (NZB), *BR* & *GO* 9069; Burugupally (MDK), *BR* & *CP*11615; Pochammaralu (MDK), *CP*11758; Mohammadabad RF (RR) *MSM* & *KH*10258; Gadirayal RF (RR), *MSM* & *KH*10258; Khammam to Dhanyayigudem (KMM), *RR* 108511 (BSID); Karkalpadu RF (MBNR), *SRS*104520 (BSID).

Note: Wight & Arnott pointed out a confusion arising from Linnaeus (Sp. Pl. 522. 1753) who established his *Mimosa intsia* from Rheede's figure and of course characterized it in nearly the same terms as his *M. caesia*. C. Ohasi (Enum. Fl. Pl. Nepal 2: 103. 1979) has included both the above names under *Acacia intsia* (L.) Willd., Sp. Pl. 4: 1091. 1806.

Acacia catechu (L.f.) Willd., Sp. Pl. 4 : 1079. 1806; Baker in Hook. f., Fl. Brit. India 2: 295. 1878; Gamble, Fl. Madras 1: 427. 1919; Chakrabarty & Gangopadhyay, J. Econ. Taxon. Bot. 20: 606. 1996. *Miomosa catechu* L. f., Suppl. Pl. 439. 1781.

Moderate-sized deciduous trees, 5-7 m high; bark greyish brown, rough; branches dark-brown, smooth, glabrous. Leaves bipinnate, stipular spines 0.5 cm long, petiole gland near the middle; rachis 10 cm long, with gland between the upper most pair of pinnae, pinnae 20-25 pairs, leaflets 30-35 pairs, linear-oblong, 2-2.5 × 1-1.5 mm, base cordate, margin entire, apex subacute. Flowers white in axillary spikes; peduncle 2-3 cm long; calyx 4-teethed, to 1 mm, campanulate, hairy without, teeth deltoid, ciliate; corolla lobes 4, to 2 mm, lobes ovate-oblong, subacute, pubescent; stamens numerous, free, filaments 3 mm, anthers minute; ovary 0.75 mm, sessile, glabrous, style 2.25 mm, filiform. Pod flat, smooth, acute at both ends, grey, dehiscent; seeds many.

Occasional in dry deciduous forests in all districts.

Fl. & Fr.: July-December.

Burugupally (MDK), *BR* & *CP* 11616; Nagarjunakonda (NLG), *BVRS* 97 (NU); Kudurpalli (KRN), *SLK* 70778 (NBG); Thattilanga RF (KMM), *RCS* 102415 (BSID); Rangapur (WGL), *PVS* & *NRR* 76928 (BSID).

Acacia chundra (Roxb. ex Rottl.) Willd., Sp. Pl. 4 : 1078. 1806; Baker in Hook. f., Fl. Brit. India 2: 295. 1878; Gamble, Fl. Madras 1: 428. 1919; " *sundra*"; Pullaiah & Chennaiah, Fl. Andhra Pradesh 1: 348. 1997. *Mimosa chundra* Roxb. ex Rottl. in Ges. Naturf. Freunde Berlin Neue Schriften 4: 207. 1803.

Vern. : *Sandra*.

Moderate sized trees, branches smooth, glabrous. Leaves alternate, 10-15 cm long, stipular spines short, hooked; pinnae 10-20 pairs, leaflets 30-40 pairs, elliptic, to 4 × 1 mm, glabrous, base oblique, margin entire, apex subacute, glabrous with glands at the basal and the uppermost pinnae. Flowers white in axilary spikes; calyx tube 5-lobed; corolla slightly longer than calyx, glabrous; stamens numerous, basally connate. Pod stalked, flat, glabrous, subacute, suture wavy, depressed between seeds; seeds 3-6.

Common in most of the districts in scrub forest.

Fl.: July-September; Fr.: December onwards.

Sirpur RF (ADB), *GO* & *PVP* 4909; Ibrahimpet RF (NZB), *TP* & *BR* 6124; Anantagiri RF (RR), *TP* & *MSM* 11074; Nelikal (NLG), *BVRS* 453 (NU).

Acacia eburnea (L. f.) Willd., Sp. Pl. 4 : 1081. 1806; Baker in Hook. f., Fl. Brit. India 2: 293. 1878. *p.p.;* Gamble, Fl. Madras 1: 426. 1919; Chakrabarty & Gangopadhyay, J. Econ. Taxon. Bot. 20: 609. 1996. *Mimosa eburnea* L. f., Suppl. Pl. 437. 1781. *Acasia campbellii* Arn., Nova Acta Phys. Med. Acad. Caes. Leop. Carol. Nat. Cur. 18: 333. 1836; Gamble, Fl. Madras 1: 426. 1919.

Trees, up to 8 m high, bark of young branches purplish, branchlets glabrous, warty, dilated at nodes. Leaves 3-5 in a cluster, to 2 cm, pinnae 6-8 pairs, leaflets 8-10 pairs, ovate, to 2.3 × 1 mm, pubescent, base obtuse, margin entire, apex subacute; rachis with glands at the basal and the upper most pinnae, stipular thorns unequal, straight, to 2 cm long. Flowers orange-yellow in heads, clustered in the axils of leaves; calyx-tube campanulate, 5-toothed, to 0.5 mm; petals 5, 1.5 mm; stamens numerous, filaments to 3 mm, basally connate; ovary stipitate, to 0.5 mm, style to 4 mm. Pods very thin, flat, curved, glabrous, 5-10-seeded; seeds ovoid.

Common in open habitats, waste places, along road sides and in protected areas in Mahabubnagar, Hyderabad and Nizamabad districts.

Fl. & Fr.: November-March.

Laxmapur (NZB), *TP* & *BR* 6297; Hyderabad, *PVP* & *MVR* 1117 (BSID).

Acacia farnesiana (L.) Willd., Sp. Pl. 4 : 1083. 1806; Baker in Hook. f., Fl. Brit. India 2: 292. 1878; Gamble, Fl. Madras 1: 425. 1919; Chakrabarty & Gangopadhyay, J. Econ. Taxon. Bot. 20: 610. 1996. *Mimosa farnesiana* L., Sp. Pl. 521. 1753.

Vern.: *Kusturi.*

Thorny shrubs or small trees, up to 8 m high, branchlets warty, slender, often zig-zag; stipular spines unequal, 2-2.5 cm long, straight. Leaves 5 cm long, 5 in a cluster, pinnae 5 pairs, leaflets 10-15 pairs, elliptic, 5 × 1 mm, overlapping, glabrous, base truncate, margin entire, apex obtuse, petiole with a gland near the middle, rachis stiff-pubescent, eglandular, stipular thorns unequal, straight. Flower heads creamy white, fragrant; peduncle 0.75-1 mm long, pubescent; calyx 5-toothed, to 1.3 mm, teeth short, acute, triangular; petals 5, to 2 mm long, lobes very short, pubescent, stamens numerous, free, filaments to 4 mm, anthers small; ovary stipitate, 1 mm long, style filiform, to 3 mm long. Pods 5-8 cm, sub-cylindric, dehiscent, obtuse or hooked at apex, slightly curved, faintly striate, pulpy, turgid, brown; seeds 4-6, dark brown.

Common in most of the districts.

Fl. & Fr.: August-March.

Gandhari (NZB), *TP* & *BR* 6352; Devanpally (NZB), *BR* & *KH* 9634; Pochammaralu (MDK), *BR* & *CP* 11500; Mokarlabad (RR), *MSM* & *KH* 10262; Domalapenta (MBNR), *TDCK* 16826; Narayanpur (NLG), *BVRS* 443 (NU).

Acacia ferruginea DC., Prodr. 2: 458. 1825; Baker in Hook. f., Fl. Brit. India 2: 295. 1878; Gamble, Fl. Madras 1: 425. 1919; Chakrabarty & Gangopadhyay, J. Econ. Taxon. Bot. 20: 611. 1996.

Vern.: *Ansandra.*

Deciduous trees, up to 10 m high; branchlets glabrous, armed with prickles. Leaves alternate, 7-10 cm, pinnae 6-8, 6 cm long, leaflets 15-25 paris, oblong, 0.3-1 × 0.1-0.3 cm, glaucous, base oblique, margin sparsely ciliate, apex subacute, rachis terete, glabrous, with glands opposite to the two uppermost pinnae; petiole gland above middle. Flowers creamy-white in spikes, axillary, solitary or in pairs; peduncle 2-2.5 cm; bract small, caducous, bracteoles caducous; calyx 5-lobed, to 1 mm, puberulous without; petals 5, to 1 mm; stamens numerous, 4 mm, basally connate; ovary stipitate, terete, to 1 mm, style to 3 mm. Pod stipitate, lanceolate, 6-15 × 2 cm, grey, glabrous, apiculate; seeds 3-6, grey.

Common in scrub forests in Adilabad, Karimnagar and Mahabubnagar districts.

Fl. & Fr.: February-June.

Neelwai RF (ADB), *TP* & *GO* 5447.

Acacia horrida (L. f.) Willd., Sp. Pl. 4 : 1082. 1806; Chakrabarty & Gangopadhyay, J. Econ. Taxon. Bot. 20: 613. 1996. *Mimosa horrida* L. f., Suppl. Pl. 438. 1781. *Acacia latronum* (L. f.) Willd., Sp. Pl. 4 : 1077. 1806; Baker in Hook. f., Fl. Brit. India 2 : 296. 1878; Gamble, Fl. Madras 1: 427. 1919. *Mimosa latronum* L. f., Suppl. Pl. 438. 1781.

Low trees, forming an umbrella-like top when old, 6-8 m high, young branches dark brown, striate, glabrous. Leaves bipinnate, 2-4 cm long, rachis sparsely puberulous, pinnae 2-5 pairs, leaflets 10-12 pairs, elliptic or linear-ovate, 2-4 × 1 mm, glabrous, base oblique, truncate, margin sparsely ciliate or entire, apex obtuse; petiole 2-3 cm with a gland near the middle, rachis pubescent, eglandular, stipular thorns unequal, white, hollow, 1- 4 cm. Flowers pale yellow, in elongated spikes; peduncle to 1 cm, bract and bracteoles caducous; calyx 4-lobed, campanulate, 1 mm, glabrous; petals 4, 2 mm, lobes triangular, divided one fourth of the way down; stamens numerous, free, filaments 4 mm, basally connate; ovary subsessile or slightly stalked, 0.75 mm, style 2.25 mm, filiform. Pods 3 × 1-5 cm, flattened, recurved, glabrous, rich brown colour, apex horned; seeds 2-4, ovoid, to 5 mm.

Common in forests in Nalgonda district.

Fl. & Fr.: August-February.

Nagarjunasagar (NLG), *KMS* 9772 (CAL).

Acacia intsia (L.) Willd., Sp. Pl. 4: 1091. 1806. *Mimosa intsia* L., Sp. Pl. 522. 1753.

Vern.: *Korintha.*

Armed stragglers, 6-8 m high, bark greyish, rough, branchlets pubescent, prickles short, hooked. Leaves bipinnate, 10-12 cm long, pinnae 4-7 pairs, 4-9 cm, leaflets 8-12 pairs, oblong-elliptic, 0.5-1.5 × 0.3-0.4 cm, over-lapping, mid-nerve closer to upper margin, base truncate, margin entire, apex obtuse, acute, petiole to 3 cm, prickled, rachis prickled. Flower heads pink, 1.5 cm across, 3-5 in a cluster on racemes; peduncle to 3 cm; bracteole linear-spathulate; calyx 4-lobed, 1mm, teeth very short, ciliolate; corolla 4-toothed, to 2 mm, connate at base, glabrous without; stamens 10-12, to 7 mm, pistillode O; stamens 8, filaments 6.5 mm, exserted, ovary stipitate, terete, 1 mm, style to 6 mm, filiform. Pods stipitate, falcate, 8 × 2 cm, jointed, thin, spinous along margin, apices obtuse; seeds 5-8, flat, ovoid, beaked, dark brown.

Common in Adilabad, Medak, Hyderabad and Ranga Reddi districts; usually in grassy savannahs.

Fl.: July-October; Fr.: October –January.

Medak (MDK), *CP* 11727; Panapalli RF (KMM), *RR* 112516 (BSID); Pasra (WGL), *RKP* 108141 (BSID).

Acacia leucophloea (Roxb.) Willd., Sp. Pl. 4 : 1083. 1806; Baker in Hook. f., Fl. Brit. India 2: 294. 1878; Gamble, Fl. Madras 1: 427. 1919; Chakrabarty & Gangopadhyay, J. Econ. Taxon. Bot. 20: 611. 1996. *Mimosa leucophloea* Roxb., Pl. Coromandel t. 150. 1800.

Vern. : *Tellatumma.*

Trees, up to 15 m high, bark grey and smooth when young, dark brown and rough when old, branches pale, striate, puberulous when young. Leaves alternate or clustered, 2.5-4 cm, pinnae 7-10 pairs, leaflets 20-30 pairs, elliptic, 2 × 0.5 cm, slightly falcate, glabrous above, pubescent below, base oblique, margin ciliate, apex acute, petiole 1 cm with a gland near the middle or at base; rachis densely pubescent, with glands opposite to 4 upper most pinnae, stipular thorns short or long, to 1 cm, grey, straight. Flowers pale yellow in heads, in terminal panicles; peduncle 0.6-1 cm; bracts ovate, minute, bracteoles in the middle of peduncle; calyx 5-lobed, to 0.5 mm, teeth short, pubescent, teeth apex ciliate, ovary stipitate, to 0.5 mm, style 2.5 mm, filiform. Pods sessile, 10-15× 0.5-1 cm, thick, straight, turgid, rusty tomentose; seeds 10-15, angular, ovoid, to 6 × 4 mm, pale brown.

Common in dry deciduous forests in all districts.

Fl. & Fr.: August-February

Pulikunta RF (ADB), *PVP* 9470; Manchippa (NZB), *TP* & *BR* 6039; Ghanapur (NZB), *BR* & *KH* 7117; Ponnala (MDK), *CP* 11393; Lingampalle (RR), *MSM* 12747; Sri Rangapur (MBNR), *TDCK* 13675; Narayanpur (NLG), *BVRS* 442 (NU); Aklaspur (KRN), *GVS* 22503, 25616 (MH); Bhadrachalam RF (KMM), *RCS* 104285 (BSID); Pasra (WGL), *RKP* 108147 (BSID).

Acacia nilotica (L.) Willd. ex Del. subsp. **indica** (Benth.) Brenan, Kew Bull. 12: 84. 1957; Chakrabarty & Gangopadhyay, J. Econ. Taxon. Bot. 20: 611. 1996. *A. arabica* (Lam.) Willd. var. *indica* Benth., London J. Bot. 1 : 500. 1842. *A. arabica* auct. non (Lam.) Willd. 1806; Hook. f., Fl. Brit. India 2: 293. 1878; Gamble, Fl. Madras 1: 425. 1919.

Vern. : *Nallatumma*.

Trees, 12-15 m high, young branches sparsely puberulous, bark dark brown, fissured. Leaves alternate, 4-7 cm long, stipular spines straight, to 5 cm long, rachis grey-pubescent, generally with a gland between the first pair of pinnae and one between the terminal pair, pinnae 5-10 pairs, 3.5 cm long, leaflets 15-25 pairs, subsessile, elliptic, glabrous, ca 4 × 1 mm, base oblique, obtuse, margin entire, apex rotund. Flower heads globose, yellow, in axillary fascicles; calyx 5-toothed, 1 × 2 mm, glabrous without; corolla 5-toothed, 2 ×3 mm, glabrous without; stamens numerous, filaments 4 mm, basally connate; ovary stipitate, terete, to 1 mm, style to 4 mm. Pods 10-18 ×1-1.5 cm, stipitate, glaucous, moniliform, constricted at apex; seeds about 15.

Common in almost all districts in dry localities.

Fl. & Fr.: August-April.

Penganga river (ADB), *GO* & *PVP* 4287; Kamareddy (NZB), *TP* & *BR* 6414; Mombojipally (MDK), *BR* & *CP* 11633; Gadirayal RF (RR), *MSM* & *KH* 10251; Katasintha (HYD), *KMS* 6006 (MH); Mukkigundam (MBNR), *BR* & *TSS* 28876; Aklaspur (KRN), *GVS* 22519 (MH); Nellippakka RF (KMM), *RCS* 99029 (BSID).

Acacia pennata (L.) Willd., Sp. Pl. 4 : 1090. 1806; Baker in Hook. f., Fl. Brit. India 2: 297. 1878 *p.p.;* Gamble, Fl. Madras 1: 429. 1919; Chakrabarty & Gangopadhyay, J. Econ. Taxon. Bot. 20: 620. 1996. *Mimosa pennata* L., Sp. Pl. 522. 1753.

Vern.: *Karusikaya, Mullugorintha.*

Prickly, glabrescent, climbing shrubs, rarely small trees, 4-6 m high; branchlets yellowish-tomentose. Internodal thorns short, erect or recurved. Leaves alternate, 7-14 cm long, petiole gland oblong, below middle of the petiole, rachis with gland between upper 1-3 pairs of pinnae, pinnae 12-14 pairs, leaflets 45-60 pairs, overlapping, elliptic, 5 × 1 mm, glabrous, base truncate, margin ciliate, apex acute, midnerve very close to dorsal margin, rachis pubescent, grooved, obscurely prickled, stipular thorns 0. Flowers white in globose heads, 1-5 cm across, 1-3 in a cluster, in terminal or axillary racemes; peduncle to 2 cm; bracts lanceolate, to 8 mm, bracteoles 1 mm, glabrous, divided one third of the way down, lobes linear-lanceolate, acute; corolla 5-toothed, 2 mm; stamens numerous, very shortly connate at base, 4 mm long, anthers small; ovary stalked, to 1 mm, villous, style filiform, to 3 mm. Pods 14-18 × 2.6 cm, flattened; glabrous, sutures prominent; seeds 8-14, ovoid-oblong, flat, dark brown.

Common in Adilabad, Medak, Warangal and Mahabubnagar districts in moderately dry forests, in ravines and along streams.

Fl.: May – August; Fr.: August-December.

Darigav (ADB), *GO* & *PVP* 4573; Gangapur RF (MDK), *BR* & *CP* 11444; Pakhal (WGL), *KMS* 13149 (MH); Pulichilmal (MBNR), *SRS* 109163 (BSID).

Acacia planifrons Wight & Arn., Prodr. Fl. Ind. Orient. 276. 1836; Baker in Hook. f., Fl. Brit. India 2: 293. 1878 *p.p.;* Gamble, Fl. Madras 1: 426. 1919; Chakrabarty & Gangopadhyay, J. Econ. Taxon. Bot. 20: 620. 1996.

Vern.: *Goduguthumma.*

Small trees with umbrella like crown, 4-6 m high, bark thick, dark grey in older parts, purplish in young. Leaves 3-5 in a cluster, 2-3 cm long; stipular spines to 3 cm long; petiole 1 cm long, with a gland near middle, rachis glabrous, with glands opposite to the basal and two uppermost pinnae, pinnae 3-6 pairs, up to 1 cm long, leaflets 10-14 pairs, linear-oblong, 0.2-0.4 cm long, glabrous, base oblique, margin entire, apex round. Heads yellow, 1 cm across, in axillary fascicles; peduncle to 1 cm, bracteoles erect, below the middle of the peduncle; calyx 5-lobed, 1 mm, campanulate, teeth very short, acute, corolla 5-toothed, 2 mm long, lobes short, acute, villous; stamens numerous, filaments 4 mm long, exserted, free; ovary sessile, 1 mm, style filiform, to 3 mm. Pods subcylindric, 2 cm long, acute, turgid, circinate, glabrous; seeds 2-3, obovate.

Common in Nizamabad district in scrub forest, planted elsewhere.

Fl. & Fr.: January-March.

Borgaun (NZB), *BR* & *KH* 9041; Sathiyapet RF (KMM), *RR* 112653 (BSID).

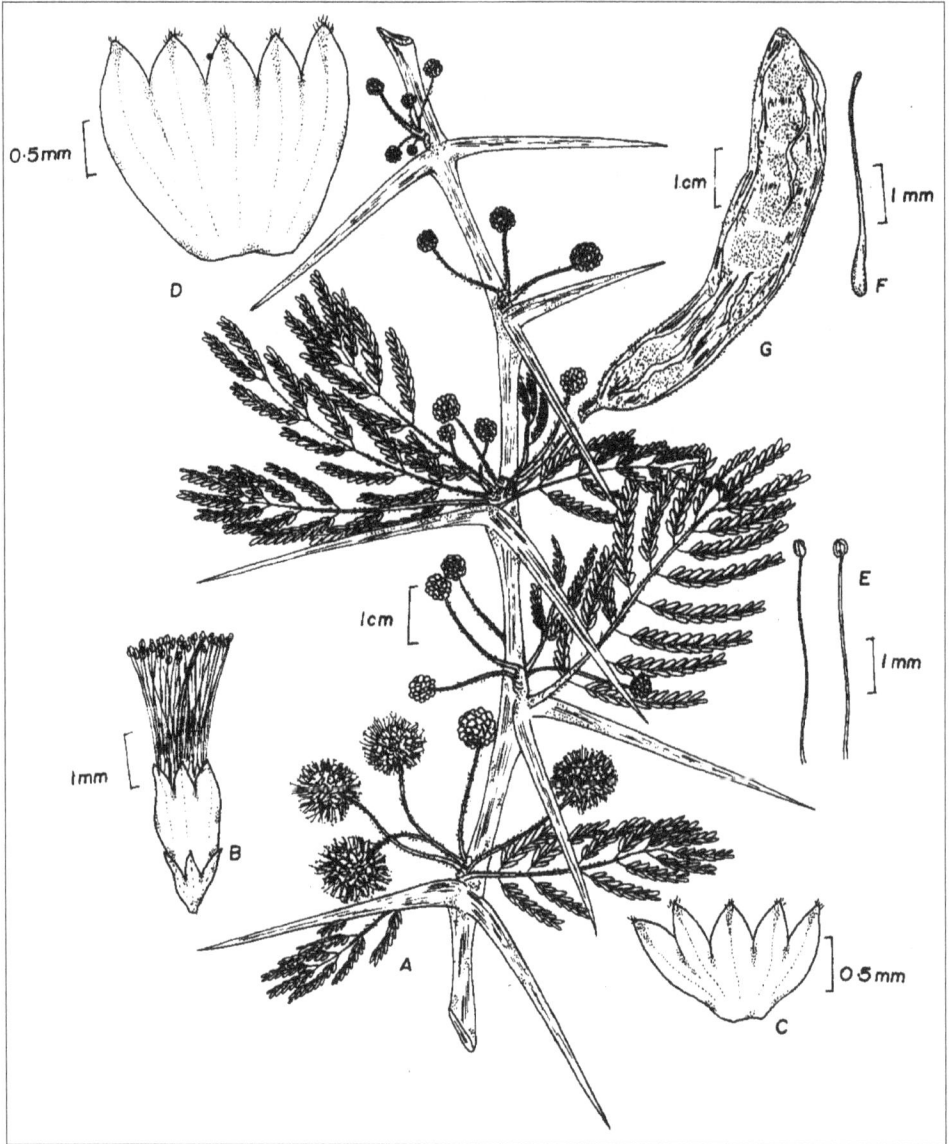

Figure 28: Acacia planifrons Wight & Arn.

A. Twig, B. Flower, C. Calyx split open, D. Corolla split open, E. Stamens, F. Pistil, G. Pod.

Acacia polyacantha Willd., Sp. Pl. 4. 1079. 1806; Brenan, Kew Bull. 11: 195. 1956. *A. suma* (Roxb.) Buch-Ham. ex Voigt, Hort. Sub. Calc. 260. 1845; Baker in Hook.f., Fl. Brit. India 2 : 294. 1878; Gamble, Fl. Madras 1: 427. 1919. *Mimosa suma* Roxb., Fl. Ind. 2 : 563. 1832.

Trees, up to 15 m high, branchlets yellowish-pubescent. Leaves alternate, 12-20 cm long, stipular spines straight, 0.7 cm long, petiole gland near middle, rachis with gland between upper 3-4 pairs of pinnae, rachis pubescent, softly prickled, pinnae 15-25 pairs, 5 cm long, leaflets 50-60 pairs, overlapping, elliptic, 5 × 0.5 mm, densely pubescent, base truncate, margin ciliate, apex subacute, obtuse, midnerve near distal margin. Flowers creamy-white, in 8-10 cm long axillary spikes, solitary or paired. Pods 10-15 × 1.5 cm, grey, woody, oblong, prominently nerved, base and apex obtuse, indehiscent; seeds 5-8, black, ovoid-spherical, compressed, flat.

Common in Nizamababad district.

Fl. & Fr.: March-October.

Jalalpur (NZB), *BR* & *KH* 9668.

Acacia senegal (L.) Willd., Sp. Pl. 4(2): 1077. 1806. *Mimosa senegal* L. Sp. Pl. 521. 1753.

English: Gum Arabic tree.

Medium sized prickly tree; bark greyish, peeling in papery flakes; prickles just below the nodes, with the middle one hooked downwards and the lateral ones curved upwards. Leaves bi-pinnate, leaflets linear-elliptic-oblong. Flowers pale yellow, in axillary spikes. Pods oblong; seeds orbicular, brownish.

Occurs wild in Mahavir Harinavanasthali National Park in Hyderabad district. Probably it might have been introduced during Nizams period, now it is naturalized and self propagating. This species is not reported in local flora from South India (Prasanna & *al.,* 2012).

Fl.: June; Fr.: October – March.

Hyderabad, *PVP* & *MVR* 918 (BSID).

Acacia sinuata (Lour.) Merr., Trans. Amer. Philos. Soc. n.s. 24: 186. 1935; Chakrabarty & Gangopadhyay, J. Econ. Taxon. Bot. 20: 625. 1996. *Mimosa sinuata* Lour., Fl. Cochinch. 653. 1790. *Acacia concinna* (Willd.) DC., Prodr. 2: 464. 1825; Baker in Hook. f., Fl. Brit. India 2: 296. 1878; Gamble, Fl. Madras 1: 429. 1919. *Mimosa concinna* Willd., Sp. Pl. 4 : 1039. 1806. *Acacia rugata* Buch.-Ham. ex Benth., London J. Bot. 1 : 514. 1842; Gamble, Fl. Madras 1: 429. 1919. *A. concinna* var. *rugata* (Benth.) Baker in Hook. f., Fl. Brit. India 2: 297. 1878.

Vern. : *Shikakai.*

Prickly climbing shrubs, 15-20 m high; branchlets apically yellowish-tomentose, basally glabrescent, warty, internodal thorns hooked. Leaves alternate, 10 cm long, petiole gland below the middle, rachis prickled with a gland between upper most pair of pinnae, stipular thorns 0, pinnae 8-15 pairs, not overlapping, 4 cm long, leaflets 20-25 pairs, linear, 3-5 × 1.7-2 mm, pubescent, base oblique, margin entire, apex obtuse, apiculate, midnerve near distal margin. Flower heads creamy-white, 1 cm across, in terminal and/or axillary racemes or panicles; peduncle 2-3 cm, bracts ovate, to 5 mm, bracteoles to 1 mm; calyx 5-lobed, to 3 mm; corolla 5-toothed, 3-5 mm; stamens numerous, to 4 mm, basally connate; ovary stipitate, terete, to 1 mm, glabrous, style 4-5 mm. Pods 12-15 × 2-2.5 cm, fleshy, wrinkled when dry, often indented on sutures, depressed between seeds; seeds 10-14, brown.

Occasional in forests in Mahabubnagar district.

Fl. & Fr.: September- March.

The pods are used as a detergent.

Acacia tomentosa Willd., Sp. Pl. 4 : 1087 1806; Baker in Hook. f., Fl. Brit. India 2: 294. 1878; Gamble, Fl. Madras 1: 426. 1919; Chakrabarthy & Gangopadhyay, J. Econ. Taxon. Bot. 20: 626. 1996.

Small trees, 9-12 m high, bark yellowish, with large lenticels, branchlets and leaf rachises densely clothed with fine grey pubescence. Leaves up to 5 cm long, stipular spines 3 cm long; petiole gland below the lowest pair of pinnae, rachis ending in a short spine; pinnae 6-12 pairs, 2.5 cm long, leaflets 20-30 pairs, oblong, 2-2.5 × 1 mm, grey green, membranous, sessile, pubescent on both sides. Flower heads greenish-white, 1-2 cm across, axillary, bracts near the middle of the peduncle; calyx 5-lobed, campanulate, rounded, pubescent, ciliolate; corolla 5-toothed, greenish, free or very shortly connate at base, ovary very shortly stalked, glabrous, style short, filiform. Pods flat, thin, slightly falcate, 10-15 × 1-1.5 cm, thick stipitate, shortly stalked, dry dehiscent, slightly depressed between the seeds; seeds 6-10, ellipsoid, compressed.

Rare in the Adilabad district in waste lands and hedges.

Fl. & Fr.: June-October.

Sone (ADB), *TP* & *PVP* 4034; Indervelli (ADB), *GO* & *PVP* 4542.

ADENANTHERA L.

Adenanthera pavonina L., Sp. Pl. 384. 1753; Hook.f. Fl. Brit. India 2 : 287. 1878; Gamble, Fl. Madras 1 : 418. 1919.

Vern. : *Bandi gurivenda.*

Deciduous unarmed tree, 6-15 m high. Leaves 25-30 cm long; rachis grooved without glands; petioles 8 cm long; pinnae 4-5 pairs, 10-15 cm long; leaflets 10 pairs, oblong-elliptic, subopposite, 2.5-3 x 1.5 cm, base truncate, margins entire, apex obtuse, apiculate. Flowers pale yellow, scented, in short peduncled racemes, racemes spiciform, 10-20 cm long, axillary or at the ends of branchlets; calyx campanulate, 5-toothed; corolla 3 mm long, segments lanceolate; stamens 10, free; filaments filiform. Pod narrow, 20 x 1.5 cm. dark brown when mature, falcately curved, 8-13-seeded; seeds globose, shiny, bright scarlet red, 5 x 5 mm.

Planted in gardens.

Fl. & Fr .: January-March.

ALBIZIA Durazz.

1. Petiole eglandular; pods woody .... **A. saman**

1. Petiole glandular; pods not woody:

   2. Leaflets large, usually more than 1 cm long:

      3. Flowers sessile:

         4. Midrib towards upper margin; calyx less than
1.5 cm, pubescent .... **A. odoratissima**

         4. Midrib towards lower margin; calyx more than
1.5 cm long, glabrous .... **A. procera**

      3. Flowers pedicelled:

         5. Leaflets broadly oblong, curved upwards, slightly
pubescent beneath .... **A. lebbeck**

         5. Leaflets oblong, slightly falcate, appressed pubescent
when young, ashy-grey when dry .... **A. thompsoni**

   2. Leaflets small, usually less than 1 cm long:

      6. Stipules foliaceous; leaflets acute at apex .... **A. chinensis**

      6. Leaflets appressed pubescent, midnerve nearly central;
pod pubescent, broader .... **A. amara**

Albizia amara (Roxb.) Boivin in Encycl. 19: 34. 1838; Baker in Hook. f., Fl. Brit. India 2: 301. 1878; Gamble, Fl. Madras 1: 432. 1919; Chakrabarty & Gangopadhyay, J. Econ. Taxon. Bot. 20: 583. 1996. *Mimosa amara* Roxb., Pl. Coromandel t. 122. 1799.

Vern.: *Nallrange.*

Unarmed deciduous trees, up to 10 m high; bark greenish, smooth, branchlets densely yellowish or grey-pubescent. Leaves 12 cm long, petiole gland near middle or above it, rachis with gland between 4 upper pinnae, pinnae 6-10 pairs, leaflets 15-25 pairs, narrow-elliptic, overlapping, 8 × 2.5 mm, appressed pubescent or glabrescent, base subacute, margin sparsely ciliate, apex obtuse, middle nerve nearly central; stipules minute, linear, subulate, 1 cm. Flower heads creamy-white to pale-yellow, 1-1.5 cm across, solitary or 2-3 fascicled in upper axils; pedicel to 1 mm; bracts to 5 mm, caducous; calyx 5-lobed, to 2 mm, acute, ciliate at apex; staminal tube to 9 cm, stamens many, filaments to 9 mm, basally connate; ovary stipitate, to 2 mm, style to 7 mm. Pods flat, compressed, greyish-brown, 8-20 cm, faintly veined, straight or wavy along margins, base and apex rotund, indehiscent; seeds 6-8, ovoid, 1 × 0.5 cm, brick red.

Common in most of the districts in dry forests.

Fl.: March- May; Fr.: June onwards.

Kammarpally (NZB), *TP* & *BR* 6206; Machareddy RF (NZB), *BR* & *KH* 9050; Narsingi (MDK), *BR* & *CP* 11533; Near Inspection Bungalow (MDK), *KMS* 7947 (MH); Mohammadabad RF (RR), *MSM* & *KH* 10210, 10223; Aklaspur (KRN), *GVS* 25607 (MH); Yadagir (NLG), *BVRS* 547 (NU); Badaburthikota (MBNR), *SRS* 109034 (BSID).

Albizia chinensis (Osbeck) Merr., Amer. J. Bot. 3: 575. 1916; Chakrabarty & Gangopadhyay, J. Econ. Taxon. Bot. 20: 584. 1996. *Mimosa chinensis* Osbeck, Dagbok. Ostind. Resa 233. 1757. *Albizia stipulata* (Roxb.) Boivin in Encycl. 19: 33. 1838; Baker in Hook. f., Fl. Brit. India 2: 300. 1878. *Mimosa stipulata* Roxb., Fl. Ind. 2 : 549. 1832; "*stipulacea*". *Albizia marginata* (Lam.) Merr., Philipp. J. Sci 5: 23. 1910; Gamble, Fl. Madras 1: 433. 1919.

Vern.: *Bandichinduruge, Bandichoragu.*

Large deciduous trees, up to 20 m high, bark grey, horizontally furrowed; branchlets densely yellowish pubescent. Leaves 18-22 cm long, petiole gland below middle of petiole; rachis with gland between base of most pinnae; pinnae 10-13 pairs, 4.5-10 cm, stipules very large, 2 × 2 cm, ovate, deciduous, leaflets 15-25 pairs, sessile, narrow-elliptic, 5-7 × 1.5-2 mm, pubescent, base very oblique, truncate, margin ciliate, apex acute, midnerve very close to the upper margin. Flower heads 1-few, 2 cm across, in axillary or terminal racemes; peduncle to 1 cm; bracts stipular, persistent, pedicel to 1 mm; calyx 5-lobed, to 3 mm, campanulate, pubescent half way down, lanceolate; stamens many, filaments to 8 mm, basally connate; ovary stipitate, to 4 mm, style 1 cm. Pods flat, compressed, 15-25 × 3-4 cm, thin, brown, straight or wavy margin, often wrinkled over seeds; seeds 8-12, ovate, much compressed, dark brown, smooth.

Occasional in hill regions of the deciduous forests, in Medak, Ranga Reddi and Mahabubnagar districts.

Fl. & Fr.: April-December.

Anantagiri RF (RR), *MSM* & *SSR* 18510; Molachintapally (MBNR), *TDCK* 15303.

Albizia lebbeck (L.) Benth., London J. Bot. 3: 87. 1844, *p.p.* (as '*lebbek*') & Trans. Linn. Soc. London 30: 562. 1875, *p.p.*; Baker in Hook. f., Fl. Brit. India 2: 298. 1878; Gamble, Fl. Madras 1: 432. 1919; Chakrabarty & Gangopadhyay, J. Econ. Taxon. Bot. 20: 589. 1996. *Mimosa lebbeck* L., Sp. Pl. 516. 1753. *Acacia lebbeck* (L.) Willd., Sp. Pl. 4: 1066. 1806.

Vern.: *Dirisena.* English: East Indian Walnut.

Large deciduous trees, 14-18 m high, bark brownish grey; branchlets puberulous to glabrous. Leaves 15-25 cm, rachis 10 cm long, with a gland near base of the petiole and one gland below the uppermost pair of pinnae, pinnae usually 2 pairs, 10-15 cm long, leaflets 4-8 pairs, basal ones oblong, terminal obovate, 2-4 × 1-2.5 cm, thin coriaceous, glabrous above, pubescent beneath, base oblique, obtuse, margin entire, apex obtuse-retuse. Flower heads solitary or 2-3- fascicled, white-cream, in upper axils; peduncle 4-6 cm; pedicel to 3 mm; calyx 5-lobed, 4 mm, turbinate, pubescent without; corolla 5-toothed, greenish, lanceolate, to 9 mm, pubescent without; staminal tube to 1.8 cm, stamens many, filaments pink or greenish, to 1 cm; ovary subsessile, to 3 mm, style to 7 mm. Pods oblong or elliptic, flat, compressed, 10-20 × 4-5 cm, glabrous, yellow, straight or wavy along margin, slightly tumid and reticulate above seeds; seeds 4-12, spherical or reniform, smooth, glabrous, yellowish brown.

Common in dry deciduous forests in most of the districts, cultivated and run wild. It is also commonly grown as avenue tree along road sides.

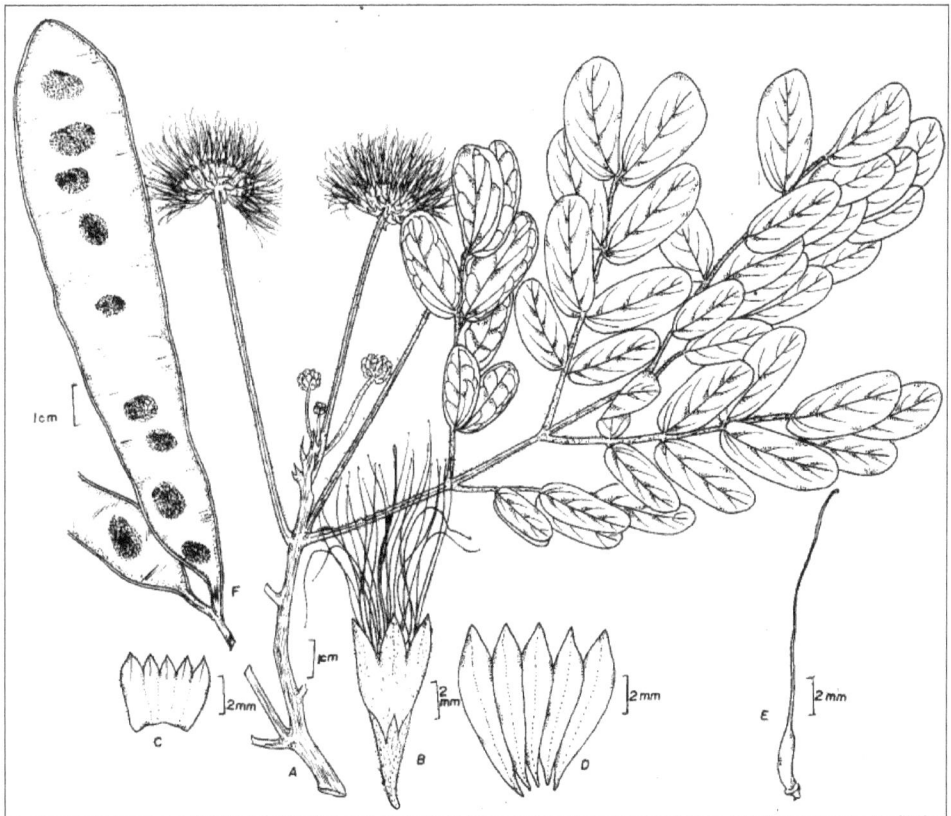

Figure 29: Albizia lebbeck (L.) Benth.

A. Twig, B. Flower, C. Calyx split open, D. Corolla split open, E. Pistil, F. Pods.

Fl. & Fr.: July-March.

Kundaram RF (ADB), *GO* 4493; Neela (NZB), *BR* & *GO* 9222; Lingampalle (RR), *MSM* 19224; Nelikal (NLG), *BVRS* 454; Kodimial (KRN), *GVS* 20044 (MH); Narsapur RF (MDK), *RG* 113467 (BSID); Appanapally RF (MBNR), *SRS* 104469 (BSID).

Albizia odoratissima (L. f.) Benth., London J. Bot. 3 : 88. 1844, var. **odoratissima;** Baker in Hook. f., Fl. Brit. India 2: 299. 1878; Gamble, Fl. Madras 1: 431. 1919; Chakrabarty & Gangopadhyay, J. Econ. Taxon. Bot. 20: 593. 1996. *Mimosa odoratissima* L. f., Suppl. Pl. 437. 1781.

Vern.: *Chinduga.*

Large unarmed evergreen trees, up to 20 m high. Leaves 15-20 cm, petiole gland near base, rachis with gland between upper most pairs of pinnae, pinnae 2-5 pairs, 10-15 cm, leaflets 8-15 pairs, elliptic-oblong, 1-2.5 × 0.5-1 cm, puberulous or glabrescent, base oblique, truncate, margin entire, apex obtuse, midnerve closer to the upper margin. Flower heads globose, 1-2 cm across, flowers scented, white, sessile in umbellate to corymbose panicles; calyx 5-lobed, to 1 mm, pubescent without, ciliate; corolla toothed, to 4.5 mm, pubescent without, acute, ciliate; staminal tube to 9 mm, stamens many, filaments to 9 mm; ovary stipitate, 1.5 mm, style to 4.5 mm. Pods brown, flat, straight, smooth, reticulate above seeds; seeds 8-12, ovoid, smooth, glabrous, yellow or yellowish- brown.

Common in forests in all districts.

Fl.: April-June; Fr.: June - November.

Bheemaram (ADB), *GO* & *DAM* 5186; Nasarullabad RF (NZB), *TP* & *BR* 6144; Narsapur RF (MDK), *TP* & *CP* 14063; Kanmankalva RF (RR), *MSM* & *KH* 10304, *MSM* 11069; Farahabad (MBNR), *TDCK* 15364; Mannanur (MBNR), *SKB, BSS* & *VSL* 30602, *GS* 18277; Mahadevpur (KRN), *SLK* 70863 (NBG); Kunavaram RF (KMM), *RCS* 98783 (BSID).

Albizia procera (Roxb.) Benth., London J. Bot. 3 : 89. 1844; Baker in Hook. f., Fl. Brit. India 2 : 299. 1878; Gamble, Fl. Madras 1: 432. 1919; Chakrabarty & Gangopadhyay, J. Econ. Taxon. Bot. 20: 583. 1996. *Mimosa procera* Roxb., Pl. Coromandel t. 121. 1798.

Vern.: *Tella chinduga.*

Very large deciduous trees, up to 30 m high, bark yellowish white; young branches pale, lenticellate. Leaves 20 cm long, rachis glabrous with gland near base of petiole; pinnae usually three pairs, 15-18 cm long, leaflets 6-12 pairs, oblique, ovate, 2-6 × 1.5-4 cm, broad and rounded at base on the upper side, narrower and attenuate on the lower side of the midrib, margin entire, apex obtuse, pubescent beneath. Flowers sessile in globose heads, heads 1.5-2 cm across, in panicles; calyx 5-lobed, campanulate or tubular, glabrous without, teeth distinct, triangular; corolla 5-toothed, connate, funnel-shaped, pubescent without, teeth lanceolate, twice as long as calyx; stamens many, monadelphous at the base; ovary sessile or shortly stalked, style filiform. Pods orange-brown, reticulate above the seeds, dehiscent, 5-9 × 2-2.5 cm; seeds 10-12, smooth, glabrous, yellow or yellowish-brown.

Occasional in deciduous forests along river banks in Adilabad, Nizamabad, Karimnagar and Mahabubnagar districts.

Fl.: April-June; Fr.: July-December.

Boath (ADB), *GO* & *PVP* 4628; Rampur (ADB), *SSR* 19305; Gopalpet (NZB), *BR* 9509; Mahadevpur (KRN), *SLK* 70823 (NBG); Rampur (ADB), *SSR* 19305 (BSID).

Albizia saman (Jacq.) F.Muell., Select. Pl. ed. 2. 12. 1876. *Samanea saman* (Jacq.) Merr., J. Wash. Acad. Sci. 6: 47: 1916. *Mimosa saman* Jacq. Fragm. Bot., 15. t. 19. 1800. *Enterlobium saman* (Jacq.) Prain ex King, J. Asiat. Soc. Bengal 66: 252. 1898; Gamble, Fl. Madras 1: 435. 1919.

Vern.: *Nidra ganneru.*

Large unarmed tree, 15-20 m high, bark whitish, rough, branchlets yellowish-tomentose. Leaves alternate, bipinnate, 10-16 cm long, pinnae 2-4 pairs, 6-8 cm long; leaflets 6-7 pairs, basal pairs oblong-elliptic, 1-2.5 × 0.7-1.5 cm, upper ones obovate, inequilateral, 1-3.5 × 0.7-2.5 cm, puberulous or glabrous above, pubescent beneath, base truncate, cuneate, margin entire, apex obtuse, apiculate, petiole 3.5-4.5 cm, eglandular, rachis tomentose, with glands opposite to almost all pinnae, petiolule reduced or 0; stipules linear, 3-5 mm, caducous, stipels 0. Flower heads pinkish brown in terminal or axillary corymbose panicles, 2 cm across; peduncle 4-6 cm, bracts lanceolate, 0.6-1 cm, bracteoles linear, to 5 mm; pedicel to 1.5 mm; calyx 5-lobed, to 6 mm, pubescent, triangular, connate from above middle; corolla 5-toothed, to 6 mm, connate from above middle, densely pubescent; staminal tube 2.9 mm, stamens many, monadelphous, filaments to 2.6 cm, apex pinkish, base white, anthers small, eglandular; ovary stipitate, to 6 mm, style to 2.7 cm, glabrous. Pods linear-oblong, 10-13 × 1-1.5 cm, slightly twisted, woody, septate between seeds, dehiscent, brownish black; seeds 8-10, ovoid.

Frequently planted, in wastelands, parks and as avenue tree.

Fl.: January – June; Fr.: May onwards.

Mothugudem forest (KMM), *PRSR* 15426; Central University campus (RR), *MSM* 19202; Zaheerabad – Hyderabad road (MDK), *CP* 11683; Eturnagaram (WGL), *RKP* 110813 (BSID).

Albizia thompsoni Brandis, Indian Forest. 25: 284. 1899; Gamble, Fl. Madras 1: 432. 1919; Chakrabarty & Gangopadhyay, J. Econ. Taxon. Bot. 20: 596. 1996.

Large deciduous trees, 12-13 m high; bark black, rough; branchlets pubescent. Leaves 15-20 cm long, petiole gland above middle of the petiole, rachis tomentose with a gland between the uppermost pairs of pinnae, pinnae 3-6 pairs, 15-20 cm long, leaflets 15-20 pairs, overlapping, oblong, slightly falcate, 1-1.5 × 0.3-0.5 cm, broadest on the lower side of the midrib, rounded or semicordate at base, margin entire, apex obtuse, appressed pubescent when young, ashy grey when dry. Flowers pale yellow with pink anthers in globose heads, peduncle 2-4 cm, pedicel 3 mm; calyx campanulate, 5-lobed, puberulous; corolla 5-toothed, to 3.5 mm, pubescent without, lanceolate; staminal tube 6 mm, stamens many, filaments 5 mm, monadelphous at the base;

ovary 1.5 mm, style to 3.5 mm, filiform. Pods moderately thick, 10-12 × 2.5-4 cm, rich brown, dehiscent; seeds 6-8, brown.

Rare in Mahabubnagar district in dry deciduous forests.

Fl. & Fr.: April-October.

Appapur RF (MBNR), *SSR* 15545, *SKB* & *BSS* 30599; Farahabad (MBNR), *TDCK* 15364; Saleswaram (MBNR), *BR* & *SKB* 30724.

DESMANTHUS Willd.

Desmanthus virgatus (L.) Willd. Sp. Pl. 4 : 1047.1805; Hook.f. Fl. Brit. India 2 : 290. 1875. *Mimosa virgata* L., Sp. Pl. 1502.1753.

Much branched, glabrous undershrub, 60-80 cm high. Leaves 5-7 cm long; petioles short; stipules subulate – filiform, 2-4 mm long; pinnae 2-4 pairs, 2.5-4 cm long; leaflets 10-20 pairs, oblong to linear – oblong, 5-8 x 1-1.5 mm, more or less rounded at both ends. Flowers in 1-2-nate, axillary, peduncled heads; peduncles 1-2 cm long, elongating in fruit; calyx minute campanulate, 5-toothed; petals nearly free, oblong, twice as long as the calyx, pale yellow; stamens 10; filaments distinct, filiform. Pods linear, straight, 6-8 x 0.3 cm, thin, compressed, beaked, dark-brown, glabrous, 15-30-seeded; seeds quadrangular, 3-3.5 mm across, red-brown in the center.

Grown as fodder crop and also naturalized in Hyderabad district (Ramana, 2010).

Fl. & Fr.: June-August.

DICHROSTACHYS (DC.) Wight & Arn. *nom. cons.*

Dichrostachys cinerea (L.) Wight & Arn., Prodr. Fl. Ind. Orient. 271. 1834; Baker in Hook. f., Fl. Brit. India 2: 288. 1878; Gamble, Fl. Madras 1: 419. 1919; subsp. **cinerea** var. **cinerea;** Brenan & Brummit, Bot. Soc. Brot. (Ser. 2) 39. 110. 1965; Kosterm. in Dassan. & Fosb., Rev. Handb. Fl. Ceylon. T. 466. 1980. *Mimosa cinerea* L., Sp. Pl. 520. 1753, non L. 517. 1753.

Vern.: *Velthuru.*

Armed shrubs or trees, up to 6 m high; bark grey or light brown; branchlets densely pubescent, spine tipped. Leaves alternate or 3 in a cluster, 5-10 cm long, bipinnate, rachis pubescent, with glands between all pairs of pinnae; pinnae 5-10 pairs, leaflets 10-20 pairs, sessile, linear-oblong, 4 × 1 mm, close, pubescent beneath, base obtuse, margin entire, apex acute; stipules 2, minute, stipels 0. Flowers in cylindric, peduncled spikes on short axillary branches; calyx 5-lobed, 0.05 mm, campanulate, pubescent without; corolla 5-toothed, 1.5 × 0.5 mm, gland crested, exserted; ovary stipitate, pilose, to 0.1 mm, style 0.3 mm; sterile flowers with elongate staminodes to 1mm, with eglandular anthers. Pods linear, flat, 4-5 cm long, coiled when ripe, continuous within, indehiscent or opening from apex; seeds 4-6, ovoid, compressed.

Common in scrub jungles and open forests in most of the districts .

Fl. & Fr.: May-December.

Ankusapuram RF (ADB), *GO* 4375; Narsapur RF (MDK), *TP* & *CP* 14057; Central University campus (RR), *MSM* & *TP* 15141; Banjara Hills (HYD), *LJGV* 3155 (CAL); Nagarjunakonda (NLG), *BVRS* 78 (NU).

LEUCAENA Benth.

Leucaena leucocephala (Lam.) De Wit, Taxon 10: 53. 1961. *Mimosa leucocephala* Lam., Encycl. 1: 12. 1783. *Leucaena glauca* auct. non Benth. 1842; Baker in Hook. f., Fl. Brit. India 2: 290. 1878; Gamble, Fl. Madras 1: 419. 1919.

Vern.: *Subabul*

Deciduous trees, 6-20 m high or large shrubs, bark smooth, greyish; branchlets glabrescent. Leaves 10-20 cm long, bipinnate, pinnae 3-8 pairs, leaflets 10-20 pairs, linear-oblong, 0.5-1.5 × 0.2-0.3 cm, membranous, glabrous above, sparsely hairy beneath, base rounded, margin entire, apex acute. Flowers white in globose heads, heads solitary in the axils of leaves; peduncles 1-1.5 cm long, slender, pubescent, slightly thickened in pod; calyx 5-lobed, 3 mm long, whitish, spathulate; stamens 10, 0.7 mm long, much exserted anthers not gland crested; ovary shortly stalked, 3 mm, slightly hairy at one side, style 3 mm. Pods 10-15 × 1-1.5 cm, flat, glabrous, linear-oblong, strap-shaped, pale to dark-brown; seeds 20-30, ovoid.

Cultivated for green manure, fodder, fuel in afforestation; it is frequently found in hedges and near villages.

Fl.: November – March; Fr.: Throughout the year.

Mandamarri (ADB), *PVP* 9441; Dharmaram (NZB), *BR* & KH 6460; Siddipet (MDK), *CP* 11375; Gacchibowli (RR), *MSM* 15192; Parnasala RF (KMM), *RCS* 99042 (BSID); Eturnagaram (WGL), *RKP* 110815 (BSID).

MIMOSA L.

1. Herbs; pinnae of the leaves 1-2 pairs; stamens 4 .... **M. pudica**
1. Erect shrubs or small trees or stragglers; leaves more than 2- pinnate; stamens 8 :
   2. Pods stipitate:
      3. Prickles along the sutures .... **M. prainiana**
      3. Prickles absent along the sutures .... **M. rubicaulis**
   2. Pods subsessile:
      4. Pods pubescent; seeds 4-8, ellipsoid .... **M. hamata**
      4. Pods glabrous, seeds 3-6, flattened .... **M. polyancistra**

Mimosa hamata Willd., Sp. Pl. 4 : 1033. 1806; Baker in Hook. f., Fl. Brit. India 2: 291. 1878; Gamble, Fl. Madras 1: 421. 1919.

Prickly shrubs, up to 2 m high, branchlets pubescent. Leaves bipinnate, 4-5 cm long; pinnae 4-5 pairs, 1.5-2 cm long; leaflets 6-10 pairs, obovate-oblong, 4-6 × 2.5-3 mm, some what hairy beneath; base oblique, apex acute, mucronate. Heads globose, pink, axillary, solitary and in short terminal racemes, slender, downy, often with a few prickles; bracteole solitary, linear-spathulate, ciliate at the apex; calyx 5-toothed, to 0.09 mm, campanulate; corolla 4-toothed, pink, to 3 mm, divided nearly half-way down, connate at base, glabrous without; stamens 8, much exserted, filaments to 5 mm, filiform, connate at base, anther minute, 0.5 mm; ovary subsessile, to 1 mm, pubescent or slightly hairy, style to 4 mm. Pods flat, 3-6 × 1-1.5 cm, strongly falcate, reddish-brown, nearly glabrous, prickly along with both sutures; seeds 4-8, light-reddish-brown, ellipsoidal, glabrous, smooth.

Common in most of the districts in scrub forests in Medak, Ranga Reddi and Nalgonda districts.

Fl. & Fr.: June-November.

Sardhana X roads (MDK), *BR* & *CP* 11622; Sirpur RF (ADB), *GO* & *PVP* 4929; Pathur RF (MDK), *TP* & *CP* 12125; Laxmapur (MBNR), *TDCK* 13670; Nagarjunakonda (NLG), *BVRS* 98 (NU).

Mimosa polyancistra Benth., Trans. Linn. Soc. London 30: 422. 1875; Gamble, Fl. Madras 1: 421. 1919.

Vern.: *Errasandra, Pariki.*

Straggling shrubs, up to 2 m high, branches prickly, terete, glabrous, prickles whitish, to 1 cm, straight or slightly curved. Leaves 15 cm long, bipinnate, pinnae 5-8 pairs, 4 cm long, leaflets 4-5 pairs, linear-ovate, oblong, obovate, 1.2 × 1.1 cm, faintly puberulous, base rounded, apex slightly mucronate. Flowers pink in globose heads, 1-2 cm across; bracts small, deciduous, bracteoles minute; pedicel short; calyx 4-lobed, to 0.7 mm, shortly toothed, campanulate; corolla 4-toothed, to 2 mm long, divided at apex, connate at base; stamens 8, much exserted, filaments to 6 mm, filiform, anthers minute; ovary subsessile, glabrous, globose, to 1 mm, pubescent at base, style 5 mm. Pods to 8 × 1.5 cm, pointed at tips, flattened, prickly, 3-6-jointed, each 1-seeded; seeds orbicular, flattened.

Occasional on dry clayey soil in Karimnagar district.

Fl. & Fr.: August-December.

Aklaspur (KRN), *GVS* 22507 (MH).

Mimosa prainiana Gamble, Fl. Madras 1: 421. 1919 & Kew Bull. 1920 : 5. 1920.

Straggling thorny shrubs, to 6 m high branchlets pubescent, with numerous, closely curved to straight, short prickles. Leaves bipinnate, 5-8 cm long, main rachis slender, grooved, with hooked prickles, stipules minute, linear, to 2 mm; pinnae 5-7 pairs, 2 cm long; leaflets 7-8 pairs, oblong, 5- 6 × 2-2.5 mm, touching each other, 1-3 mm apart, base rounded, margin entire, apex slightly mucronate. Flowers pink in axillary pedunculate heads, 1-few, 1 cm across; bracteole minute; calyx 4-lobed, to 0.5 mm, teeth very short, campanulate; corolla 4-toothed, to 2 mm, glabrous without, connate at base, divided nearly half way down; stamens 8, filaments ca 8 mm, exserted, filiform, basally connate, anthers to 0.2 mm; ovary stipitate, slightly pubescent at base or almost glabrous, terete, 1 mm, style to 8 mm. Pods 6-8 × 1.5 cm, 6-8-jointed, prickly along the sutures, minutely pubescent, slightly curved; seeds ovoid.

Occasional in most of the districts in scrub forests.

Fl. & Fr.: August-December.

Varni (NZB), *TP* & *BR* 6174; Sirnapally RF (NZB), *BR* & *PSPB* 9581; Kusumasamudram RF (RR), *MSM* & *NV* 10358; Sardanha (MDK), *BR* & *CP* 11622; Pakhal (WGL), *KMS* 13145 (MH); Edupayalu RF (MDK), *RG* 113428 (BSID).

Mimosa pudica L., Sp. Pl. 518. 1753; Baker in Hook. f., Fl. Brit. India 2. 291. 1878; Gamble, Fl. Madras 1: 421. 1919.

Vern. : *Lajjavanthi.*

Spreading herbs; up to 50 cm long, branchletes glabrescent or hispid; prickles short, erect or curved, to 4 mm. Leaves 4-6 cm, sensitive, pinnae 1-2 pairs, 4-6 cm, leaflets 15-20 pairs, elliptic-oblong, 4-6 × 1-1.5 mm, overlapping, hispid, base truncate-obtuse, margin ciliate, apex acute, rachis and petiole hispid, stipules to 5 mm, linear-lanceolate, acute, bristly. Flower heads pink, axillary, 2 or 3 in a cluster, oblong-globose; peduncle 2-3 cm; bracteoles 2-3, linear, to 4 mm, bristly; calyx 4-lobed, to less than 1 mm, campanulate; corolla 4-toothed, 1.5 mm, triangular, connate at base; stamens 4, much exserted, filaments to 5 mm; ovary sessile, glabrous, to 1.5 mm, style to 5 mm. Pods clustered, flat, 1-1.5 × 0.5-0.7 cm, slightly undulate, 2-5-jointed, bristly along margins; seeds 2-5, compressed, ovoid.

All hot, moist localities like river banks, bunds of arable lands, follow fields, water courses etc.

Fl. & Fr.: July-November.

Kamareddy (NZB), *BR* & *KH* 9626; Central University campus (RR), *MSM* 19206; Tekulapenta (MBNR), *BSS* & *BR* 33326; Azampur (NLG), *BVRS* 494 (NU); Panapalli RF (KMM), *RR* 112502 (BSID).

Mimosa rubicaulis Lam, Encycl. 1(1): 20. 1783; Baker in Hook. f., Fl. Brit. India 2. 291. 1878; Gamble, Fl. Madras 1: 421. 1919. *M. intsia* sensu auct non L.; Pullaiah & Chennaiah, Fl. Andhra Pradesh, 1: 359.1997.

Vern.: *Korntha.*

An erect straggling shrub 2–3 m tall, with thorny branches, bark greyish, rough, branchlets pubescent, prickles short, hooked, to 5 mm. Leaves bipinnate, 9-13 cm long, main rachis slender, grooved, closely set with numerous hooked prickles, stipules linear, to 4 mm, pinnae 4-8 pairs, 4-9 cm long, shortly stalked, the rachis without prickles; leaflets 8-12 pairs, oblong-elliptic, 0.5-1.5 × 0.3-0.4 cm, overlapping, base truncate, margin entire, apex obtuse, stipels minute; petiole to 3 cm, prickled. Flower heads pink, to 2 mm, connate at base, glabrous without, stamens 10-12, to 7 mm, pistillode 0; stamens 8, filaments 6.5 mm, exserted; ovary stipitate, terete, 1 mm, style to 6 mm, filiform. Pod stipitate, falcate, 8 × 2cm, jointed, thin, spinous along margins, strongly nerved, apices obtuse, horned, seeds 5-8, flat, ovoid, beaked, dark brown.

Occasional in scrub forest in Khammam and Medak districts.

Fl.: July – October; Fr.: November – January.

Khammam (KMM), *RR* 113867 (BSID); Nagsanpally RF (MDK), *RG* 104191 (BSID).

NEPTUNIA Lour.

1. Annual; aquatic; rachis eglandular; flower heads oblong .... **N. oleracea**

1. Perennial; terrestrial; rachis with gland between lowest pair of
pinnae; flower heads globose .... **N. triquetra**

Neptunia oleracea Lour., Fl. Cochinch. 2: 654. 1790; Baker in Hook. f., Fl. Brit. India 1: 285. 1878; Gamble, Fl. Madras 1: 416. 1919.

Vern.: *Niruthalavapu, Nidrayam.*

Annual free floating herbs; stem terete, glabrous, hardly branching, nodes with numerous; feathery rootlets, and white spongy floats on the internodes. Leaves alternate, bipinnate, 6 cm long, pinnae 2-3 pairs, 3-5 cm, leaflets 8-15 pairs, narrowly oblong, 3-8 × 1.5-2 mm, inequilateral, glabrous, base truncate, margin entire, apex obtuse; petioles 1-2 cm. Flowers yellow, minute, in oblong heads, axillary, lower flowers replaced by yellow staminodes, peduncle 6-10 cm long; calyx 5-lobed, 2 mm, membranous, campanulate, acute, calyx in male 1mm; corolla 5-lobed, 2 mm, membranous, campanulate, acute, glabrous without (corolla in male 2.5 mm long); stamens 10, to 2 mm long, much exserted, free, filaments slender, subequal, anthers gland-crested, staminodes in neuter flowers flattened, to 7 mm; ovary stipitate, 2 mm, style to 5 mm long, filiform. Pods stipitate, flat, 2.5 × 1 cm, oblong, reddish-brown, septate between seeds, dehiscing along valves; seeds 4-6, ovoid, compressed, brown.

Common in ponds and tanks in Warangal and Khammam districts.

Fl. & Fr.: December-April

Bandirevu RF (KMM), *RCS* 104323 (BSID).

Neptunia triquetra (Willd.) Benth., J. Bot. (Hooker) 4 : 355. 1842; Baker in Hook. f., Fl. Brit. India 2: 286. 1878; Gamble, Fl. Madras 1: 416. 1919. *Desmanthus triquetrus* Willd., Sp. Pl. 4 : 1045. 1806.

Perennial, prostrate glabrous herbs, somewhat woody at base, stem slender, compressed, more or less angled. Leaves 2-6 cm long, abruptly 2-pinnate; pinnae 1-3 pairs, 5 cm long, the lower pair with a small gland in between, leaflets 10-20 pairs, sessile, elliptic-oblong, 5 × 2 mm, glabrous, base oblique, margin ciliate, apex rounded. Flower heads yellow, globose, 1-1.5 cm across, solitary, axillary; sterile flowers few or absent, bracteole to 1 mm; pedicel minute; calyx 5-lobed, campanulate, to 1mm; corolla 5-toothed, free at apex, connate at base, to 3 mm; stamens 10, much exserted, filaments slender, to 4 mm, subequal, anther small, gland crested; ovary stalked, 1.5 mm, glabrous, style to 3.5 mm, filiform. Pods oblong, 1-2 × 0.5 cm, flat, beaked at apex, dark-brown; seeds 4-6, oblong or obovate-oblong, compressed, dark-brown, shining, glabrous, smooth.

Common in Medak, Mahabubnagar and Nalgonda districts among grasses.

Fl. & Fr.: May- November.

Sarvaipet (ADB), *GO* & *PVP* 4812; ICRISAT (MDK), *LJGV* 3867 (CAL); Uttanur (MBNR), *MV* & *DV* 40723; Azampur (NLG), *BVRS* 493 (NU).

PARKIA R.Br.

Parkia biglandulosa Wight & Arn., Prodr. Fl. Ind. Orient. 279. 1834; Hook. f., Fl. Brit. India 2: 289. 1878.

Unarmed tree, 15-20 m high, bark black, rough, branchlets pubescent. Leaves alternate, bipinnate, 10-14 cm long, pinnae 20-40 pairs, 5-9 cm, leaflets 60-80 pairs, oblong, 2-4 × 0.2-0.5 mm, inequilateral, glabrous above, puberulous below, base truncate, margin ciliate, apex acute-obtuse; petiole to 3 cm, with oblong basal gland on the upper side, rachis pubescent, gland opposite to 6 upper pinnae, stipules minute, stipels 0. Flower heads large, globose, axillary, solitary, 4-5.5 × 3.5-5 cm, pendulous, peduncle 6-10 cm, bract clavate, to 1.2 cm, apex dilated; calyx 5-lobed, infundibuliform; corolla 5-toothed, cream, subequal to calyx, basally connate; staminal tube long, stamens 10, monadelphous, filaments long, anthers eglandular; ovary stipitate, glabrous, style short. Pod with an elongated stipe, 1.4-6 cm, oblong-linear, compressed, brown, dehiscent, slightly curved when dry, seeds 6-13; ellipsoid, compressed.

Planted along road sides, in gardens in towns and Hyderbad.

Fl.: January – March; Fr.: March onwards.

PITHECELLOBIUM Mart. *nom. cons.*

Pithecellobium dulce (Roxb.) Benth., London J. Bot. 3: 199. 1844; Baker in Hook. f., Fl. Brit. India 2: 302. 1878; Gamble, Fl. Madras 1: 434. 1919. *Mimosa dulcis* Roxb., Pl. Coromandel t. 99. 1798 & Fl. Ind. 2: 556. 1832.

Vern.: *Simachinta.*

Armed trees, up to 15 m high, bark smooth, greyish; branchlets densely tomentose, stipular thorns straight, erect, to 1.5 cm, stipels linear, 1 mm. Leaves bipinnate, petiole 2.5 cm, with a solitary apical concave gland, pinnae 2 pairs, leaflets paired, obliquely oblong, 2-3 × 0.5-1 cm, glabrous, base and apex obtuse, margin entire, apiculate, petiolule to 1 mm. Flowers white in heads, heads in terminal and axillary panicles; peduncle 1-1.5 cm, bract small; calyx 5-lobed, campanulate, 1 mm, pubescent, corolla 5-toothed, cream, to 2.5 mm, connate above the middle, valvate, densely tomentose without; staminal tube to 6 mm, stamens many, monadelphous, filaments to 4.5 mm, anthers eglandular; ovary sessile, to 1 cm, glabrous, style 1.5 mm. Pods strap-shaped, 13-15 × 0.1-1 cm, torulose, spiraly twisted, submoniliform, turgid, dehiscent; seeds black, 5-7, broadly ovate, covered by reddish-white aril.

Common in all dry plains districts, cultivated as a hedge plant, occasionally run wild in wastelands and along road sides.

Fl. & Fr.: November-June.

Luxettipet (ADB), *GO & DAM* 5415; Mamidipally (NZB), *BR & KH* 9011; Zaheerabad (MDK), *CP* 11598; Central University campus (RR), *MSM* 19205; Aklaspur (KRN), *GVS* 22536 (MH); Bhadrachalam RF (KMM), *RCS* 98980 (BSID); Eturnagaram (WGL), *RKP* 110814 (BSID).

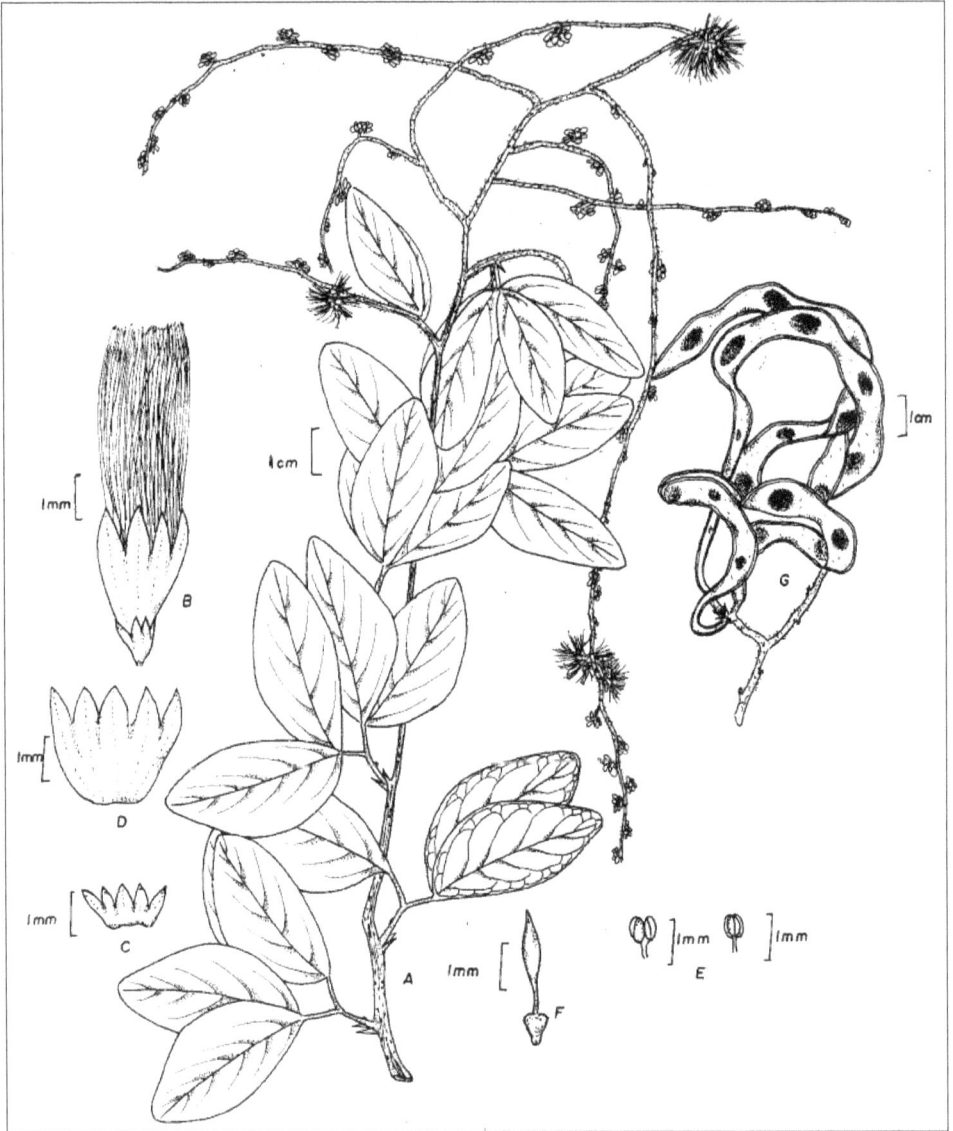

Figure 30: Pithecellobium dulce (Roxb.) Benth.
A. Twig B. Flower, C. Calyx split, D. Corolla split open, E. Anthers, F. Pistil, G. Pods.

PROSOPIS L.

1. Pinnae only 1 pair; corolla pilose up to above middle .... **P. glandulosa**

1. Pinnae 1-3 pairs; corolla other than pilose:

    2. Leaflets 15-20 pairs; pods compressed .... **P. juliflora**

    2. Leaflets 7-12 pairs; pods cylindric .... **P. cineraria**

Prosopis cineraria (L.) Druce in Bot. Exch. Club. Soc. Brit. Isles. 3: 422. 1914. *Mimosa cineraria* L., Sp. Pl. 517. 1753, "*cinerea*". *Prosopis spicigera* L., Mant. Pl. 68. 1767; Baker in Hook. f. Fl., Brit. India 2: 288. 1878; Gamble, Fl. Madras 1: 421. 1919.

    Vern. : *Jammi.*

    Armed trees, up to 10 m high; bark thick, grey, rough with deep fissures and cracks; branches with scattered spines. Leaves bipinnate, 3-5 cm long, pinnae 2 pairs, 3.5 cm long; leaflets 8-12 pairs, sessile, oblong, 0.6-1.2 × 0.3-0.5 cm, appressed hairy, base oblique, margin entire, apex rounded. Flowers yellow in 6-10 cm long slender, simple or branched spikes; calyx 5-lobed, to 2 × 3 mm, cup-shaped, membranous; petals 5, 2-3 mm; stamens 10, free, exserted, filaments 4 mm, anthers gland-tipped; ovary sessile, to 2.5 mm, glabrous, style filiform, to 1.5 mm. Pods straight or curved, 5-10 cm long, glabrous, moniliform, pendent; seeds 10-15, dull brown, oblong.

    Common on dry stony lands, waste places, in open forest in all districts.

    Fl. & Fr.: October-June.

    Central University campus (RR), *MSM* 19212; Laxmapur (MBNR), *TDCK* 15335; Nelikal (NLG), *BVRS* 456 (NU).

Prosopis glandulosa Torrey, Ann. Lyceum Nat. Hist. New York 2: 192. t.2. 1827. var. **torreyana** (Benson) Johnston, Brittonia 14: 82. 1962; Burkart, J. Arnold Arbor. 57: 452. 1976.

    Small tree, up to 4 m tall, bark grey, branchlets glabrous, spines axillary, to 1.5 cm, straight. Leaves bipinnate, petiole 5-6 cm, eglandular, pinnae 1 pair, 14-16 cm long, leaflets 12-14 pairs, linear, oblong 2-3.5 × 0.3-0.5 cm, glabrous, apex acute, margin entire, base connate, petiole short, 1 mm, rachis with a solitary gland, stipules spinescent, stipels 0. Flower heads creamish-yellow, to 7.5 cm long, axillary spikes, peduncle 1-1.5 cm, pedicel 0.2 mm, flowers to 1 mm across, calyx 5-lobed, campanulate, to 1.25 mm, glabrous without; corolla 5-toothed, cream, 3 x 3 mm, pilose on apex up to above middle, united above middle; stamens 10, free, filaments to 6 mm, anthers 0.2 mm with a deciduous gland; ovary stipitate, to 2 mm, pubescent, style 3.5 mm. Pods straight or slightly curved, compressed, beaked, 14-18 × 0.5 – 1 cm, indehiscent; seeds more than 10.

    Introduced and naturalized.

    Fl. & Fr.: January – April.

    Bhadrachalam RF (KMM), *RCS* 98983 (BSID).

Prosopis juliflora (Sw.) DC., Prodr. 2: 447. 1825. *Mimosa juliflora* Sw., Prodr. 85. 1788.

Vern.: *Sarkar thumma.*

Armed shrubs or small trees, 8-10 m high; branchlets glabrous; spines axillary, 1-1.5 cm long, straight. Leaves alternate or clustered, bipinnate, pinnae 1-2 pairs, 4-5 cm long, leaflets 15-20 pairs, oblong, 0.8-1.4 × 0.2- 0.3 cm, sessile, glabrous, base and apex obtuse, margin entire. Flowers yellow in 2-8 cm long, axillary, pendent spikes; stipules spinescent; peduncle 2-2.5 cm; bracteole linear, to 1.5 mm; pedicel to 0.5 mm; calyx 5-lobed, campanulate, 1 × 0.7 mm, triangular; corolla 5-toothed, 1.5 × 1 mm, pilose within, united below middle; stamens 10, free, filament 4 mm, anther with a deciduous gland; ovary stipitate, to 2 mm, pubescent, style to 1.5 mm. Pods drupaceous, 8-15 × 0.8-0.9 cm, straight or slightly curved, pendent, indehiscent; seeds 15-19, ovoid.

Very common in waste places and road sides in all districts.

Fl.: October – March; Fr.: February – May.

Jannaram (ADB), *TP* & *PVP* 4209; Indalwai (NZB), *TP* & *BR* 6410; Mambojipally (MDK), *BR* & *CP* 11451; Patancheru (MDK), *CP* 11692; Mohammadabad (RR), *MSM* 19236; Kodimial (KRN), *GVS* 21894 (MH).

XYLIA Benth.

Xylia xylocarpa (Roxb.) Taub., Bot. Centralbl. 47. 395. 1891; Gamble, Fl. Madras 1: 417. 1919. *Mimosa xylocarpa* Roxb., Pl. Coromandel t. 100. 1798. *Xylia dolabriformis* Benth., J. Bot. (Hooker) 4: 417. 1842.

Vern. : *Chennangi.*

Unarmed deciduous trees, up to 15 m high; bark reddish grey, rough; branchlets brown tomentose. Leaves bipinnate, 15-20 cm long, petiole 4.5 cm long, with a gland at tip; pinnae 2, terminal on the petiole, 10-15 cm long, leaflets 3-5 pairs, elliptic-lanceolate, 4-7 × 1.5-2.5 cm, sub-coriaceous, glabrous, base rounded, margin entire, apex acuminate, glands present in between each pair of leaflets. Flowers yellowish white, sessile, in dense globose heads; calyx 5-toothed, tubular, 2 mm; corolla 5-toothed, to 4 mm; stamens 10, free, filaments 7 mm long, slender, exserted, anthers crested when young with a stalked deciduous gland; ovary sessile, to 2 mm, style filiform, 8 mm. Pods flat, 10-16 × 4-4.5 cm, thick, woody, oblong, falcate, rusty-tomentose, septate between the seeds, dehiscent; seeds 6-10, oblong-ellipsoid, brown, smooth polished.

Common in forests of Mahabubnagar, Warangal, Karimnagar and Adilabad districts.

Fl. & Fr.: November-January.

Inchapalli RF (ADB), *GO* & *DAM* 5150; Jaipur RF (ADB), *GO* & *PVP* 4947, *SSR* & *KSM* 18523; Pakhal RF (WGL), *ANH* 15936 (MH), *RKP* 108284 (BSID).

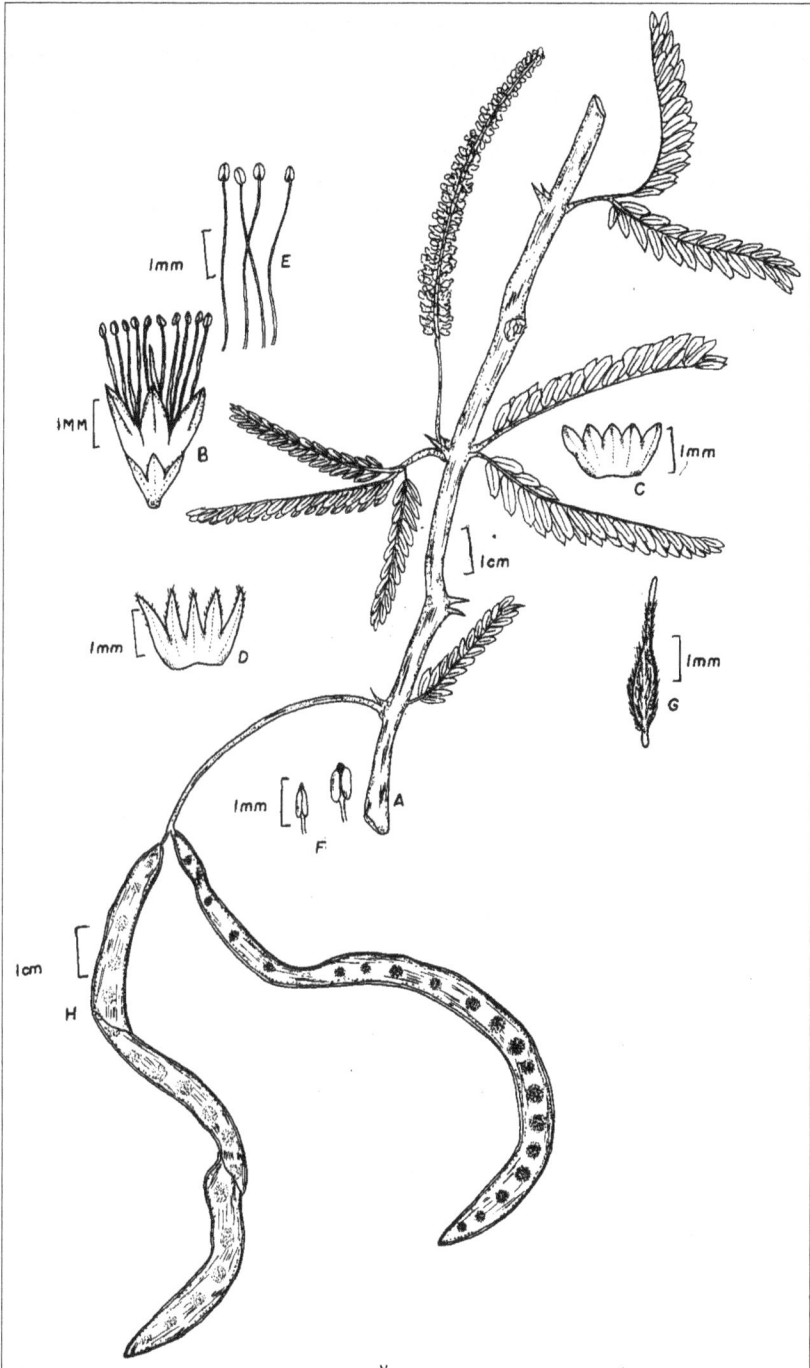

Figure 31: Prosopis juliflora (Sw.) DC. (*Prosopis chilensis* auct.)

A. Twig, B. Flower, C. Calyx split open, D. Corolla split open, E. Stamens, F. Anthers, G. Pistial

ROSACEAE

ROSA L.

1. Deciduous shrubs .... **R. multiflora**
1. Evergreen shrubs:
    2. Branches recurving :
        3. Leaflets 3-5; flowers deep rose .... **R. setigera**
        3. Leaflets 5-7; flowers white .... **R. canina**
    2. Branches erect:
        4. Petals in two whorls .... **R. damascena**
        4. Petals in many whorls... **R. chinensis**

Rosa canina L., Sp. Pl. 1 : 491. 1753.

Shrub with recurving branches; prickles stout, hooked. Leaflets 5-7, oval or elliptic, doubly serrate, glabrous or slightly pubescent or somewhat glandular beneath. Flowers pink on usually glabrous pedicels. Sepals reflexed, caducous.

Grown in house yards, and gardens as an ornamental (Ramana, 2010).

Fl. & Fr .: Almost round the year.

Rosa chinensis Jacq., Observ. Bot. 3 : 7. t. 55. 1768. *Rosa indica* L., Sp. Pl. 492. 1753; Hook.f., Fl. Brit. India 2 :364. 1878.

Armed shrub; branches slender, usually armed with scattered stout compressed more or less hooked prickles. Leaflets 3-5, broadly ovate, ovate-oblong, acuminate, serrate, shining and dark green above, pale beneath; stipules narrow, glandular ciliate. Flowers usually several, less often solitary, on long glandular stalks, crimson or pink or rarely whitish, not or slightly fragrant. Fruits not seen.

Grown in house yards and gardens as an ornamental (Ramana, 2010).

Fl. & Fr .: Almost round the year.

Rosa damascena Mill., Gard. Dict. ed. 8 : *Rosa* no. 15. 1768. Hook.f., Fl. Brit. India 2 : 364. 1878.

Scandent shrub usually with numerous, stout hooked prickles. Leaflets 5-7, ovate-oblong, serrate, more or less pubescent beneath; stipules scarsely dilated, sometimes pectinate; petioles prickly. Flowers in usually corymbose inflorescence, sepals deciduous, corolla double, red or pink or white, sometimes striped.

Grown in house yards, and gardens as an ornamental (Ramana, 2010).

Fl. & Fr .: March-October.

Rosa multiflora Thunb., Syst. Veg. ed. 14 : 474. 1784; Hook.f. Fl. India 2 : 364. 1878.

Decidious shrub with vigorous, long recurving or climbing branches. Leaflets 9, obovate to oblong, acute or obtuse, serrate, pubescent. Flowers in corymbs, white. Sepals ovate, abruptly acuminate.

Grown in house yards, and gardens as an ornamental (Ramana, 2010).

Fl. & Fr .: More or less throughout the year.

Rosa setigera Michx., Fl. Bor.-Amer. 1 : 295. 1803.

Shrub with prickly branches; branches slender, recurving or climbing. Leaflets 3-5, ovate to oblong-ovate, shortly acuminate, serrate, tomentose beneath. Flowers solitary or 2-3 together on short peduncles, deep rose, fading to whitish, not fragrant.

Grown in house yards, and gardens as an ornamental (Ramana, 2010).

Fl. & Fr .: September-October.

VAHLIACEAE

VAHLIA Thunb. *nom. cons.*

Vahlia digyna (Retz.) Kuntze, Revis. Gen. Pl. 227. 1891; Bridson, Kew Bull. 30: 177. 1975. *Oldenlandia digyna* Retz., Observ. Bot. 4 : 23. 1786. *Bistella digyna* (Retz.) Bullock, Acta Bot. Nerrl. 15: 84 & 85. 1966. *Vahlia viscosa* Roxb., Fl. Ind. 2 : 89. 1832; C.B.Clarke in Hook. f., Fl. Brit. India 2 : 399. 1878; Gamble, Fl. Madras 1: 447. 1919.

Annual prostrate herb, 15-30 cm long, stem glandular pubescent. Leaves oblong-lanceolate 0.5-2 × 0.3-0.5 cm, base obtuse- truncate, margin entire, apex acute, subsessile. Flowers yellow, 1-2 in most of the upper axils of the leaves, subsessile; calyx-tube cupular, 2 mm, sparsely pubescent, lobes 5; petals 5, triangular, to 1 mm, midrib distinct; stamens 5, included, filaments pubescent, 1 mm long, basal scale present; ovary 2 mm, styles 2, up to 1 mm. Capsule ca 3 mm in diameter, sparsely puberulous, dehiscing at apex; seeds many, very minute, ovoid.

Occasional on riverbeds and moist places in Nizamabad and Karimnagar districts.

Fl. & Fr.: February-April.

Kandakurthi (NZB), *BR* 7259.

CRASSULACEAE

1. Calyx gamosepalous; flowers in pendent panicles .... **Bryophyllum**
1. Calyx lobes free to base; flowers in erect panicles .... **Kalanchoe**

BRYOPHYLLUM Salisb.

Bryophyllum pinnatum (Lam.) Oken., Allg. Naturgesch. 3: 1966. 1841; Gamble, Fl. Madras 1: 451. 1919. *Cotyledon pinnata* Lam., Encycl. 2: 141. 1876. *Kalanchoe pinnata* (Lam.) Pers., Syn. 1: 446. 1805. *Bryophyllum calycinum* Salisb., Parad. Lond. t. 3. 1805; C.B. Clarke in Hook. f., Fl. Brit. India 2: 413. 1878.

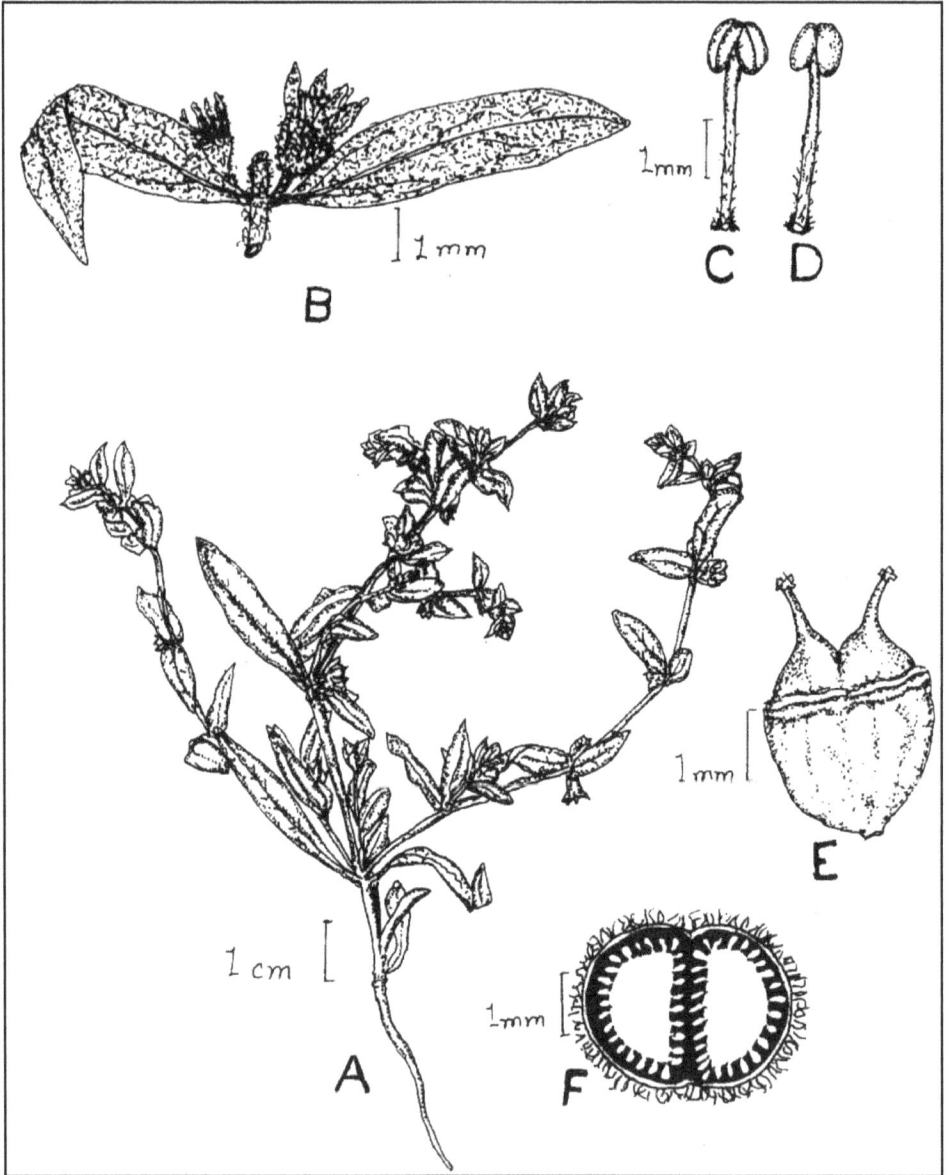

Figure 32: Vahlia digyna (Retz.) Kuntze
A. Habit, B. Flower-enlarged, C&D. Stamens, E. Pistil, F. Ovary c.s

Semisucculent herbs, up to 1 m high, branchlets glabrous, swollen at nodes. Leaves decussate, crowded in young shoots, distant, in adult ones, often 3-5-foliolate, oblong or ovate-elliptic, 6-17 × 4-8 cm, thick coriaceous, glabrous, greenish, base oblique, obtuse-round, margin crenate, apex obtuse-subacute. Flowers yellow, pendulous in paniculate cymes; peduncle to 10 cm, bracts foliar; bracteoles linear, to 6 mm; pedicel nodose, to 1.5 cm; calyx-tube lobed, 2.5-3 × 1-1.5 cm, lobes 4, triangular; corolla to 5 cm, base greenish, 8-folded, glandular with apex reddish, lobes 4, ovate, 1 × 0.6 cm; stamens 8, filaments to 3.5 cm, hypogynous scales 4, ovary 1 cm, styles 3 cm. Follicles to 1.5 cm, seed numerous.

It is an introduced plant found in gardens and often run wild in many places, in plains of most districts.

Fl. & Fr.: January-March.

KALANCHOE Adans.

Kalanchoe lanceolata (Forssk.) Pers., Syn. Pl. 1: 446. 1805. var. **lanceolata.** *Cotyledon lanceolata* Forssk., Fl. Aegypt.-Arab. 89. 1775. *Kalanchoe floribunda* Wight & Arn., Prodr. 359. 1834; C.B. Clarke in Hook. f., Fl. Brit. India 2: 414. 1878; Gamble, Fl. Madras 1: 451. 1919.

Succulent herbs, up to 1.5 m high, upper parts usually glandular-hairy. Leaves decussate, elliptic-obovate or oblong-elliptic, thick-coriaceous, 5-8 × 2-4 cm, brownish, base cuneate, margin crenate, apex obtuse-rounded; petioles to 20 cm. Flowers yellowish-white in dense panicles, peduncle to 3 cm; bracts foliar, bracteoles linear, 1.5 mm; pedicel erect, to 8 mm; calyx tube lobes free almost to the base, lobes 4, 7 × 2.5 mm, glabrous; corolla 1.8 cm long, lobes 4, lanceolate, 8.5 × 4.5 mm, acuminate; stamens 8, filaments to 1 cm, hypogynous scales 4, linear, 2 mm, styles 4, to 4 mm. Follicle to 1 cm, many-seeded.

Occasional in forest in Mahabubnagar district.

Fl. & Fr.: July-October.

Mallalatheertham (MBNR), *TDCK* 15358.

DROSERACEAE

DROSERA L.

1. Acaulescent; all leaves in a basal rosette, stipulate, spathulate-cuneate; styles 5 .... **D. burmannii**
1. Caulescent; leaves cauline, estipulate, linear; styles 3 .... **D. indica**

Drosera burmannii Vahl, Symb. Bot. 3: 50. 1794; C.B. Clarke in Hook. f., Fl. Brit. India 2: 424. 1878; Gamble, Fl. Madras 1: 452. 1919.

Acaulescent herbs with a basal rosette of obovate-orbicular, stipulate leaves. Leaves 5-8 × 3.5-6 mm, reddish-green, glandular tentacles elongate, base attenuate or subtruncate-obtuse. Scape terminal, circinate, few-many-flowered. Flowers white, peduncle 4-12 cm; bracts linear, to 1.5 mm; pedicel 2-5 mm; calyx 5-lobed, lobes elliptic-acute; petals 5, obovate; stamens 5, filaments 3.5 mm; ovary ovoid, 1.5 mm, styles 5, apically shortly lobed, to 2 mm. Capsules ca 1.5 mm long, 5-valved; seeds numerous, minute, black, reticulate.

Occasional in Medak, Ranga Reddi, Hyderabad and Khammam districts, in suitable damp places .

Fl. & Fr.: July-April.

Pocharam tank (MDK), *TP* & *CP* 12026; Mohammmadabad RF (RR), *MSM* & *KH* 11019; Murugampadu RF (KMM), *RCS* 102453 (BSID); Medak to Bodhan (MDK), *RG* 113534 (BSID).

Drosera indica L., Sp. Pl. 282. 1753; C.B.Clarke in Hook. f., Fl. Brit. India 2: 424. 1878; Gamble, Fl. Madras 1: 452. 1919.

Slender caulescent herbs, up to 8 cm high, with glandular hairy stem. Leaves estipulate, lower leaves deflexed, to 2.5 cm, upper ones alternate, narrow linear, 1-3 cm long, shortly petioled, glandular tentacled above, glabrous below. Flowers white in leaf-opposed racemes; peduncle to 2 cm, pedicel to 1mm, bracts linear, to 1.5 mm; petals 5, narrowly spathulate, slightly longer than the calyx; stamens 5, filaments 3 mm, ovary ovoid, 2 mm, styles 3, bifid to the base, to 2 mm. Capsules 3-valved; seeds minute, obovoid, shortly ridged and with raised reticulations.

Occasional in Medak, Ranga Reddi and Hyderabad districts and in wet places in the hills.

Fl. & Fr.: November-March.

Nagulabanda (MDK), *CP* 11355; Pocharam tank (MDK), *TP* & *CP* 12025; Mohammadabad RF (RR), *MSM* & *KH* 11020.

COMBRETACEAE

1. Trees or shrubs :
    2. Flowers in globose heads; fruit to 0.8 cm across, 2-winged .... **Anogeissus**
    2. Flowers in spikes or racemes; fruit above 1 cm across, entire or 3-5-winged .... **Terminalia**
1. Stragglers, lianas or climbers :
    3. Petals 0; calyx limb accrescent in fruit .... **Calycopteris**
    3. Petals 4-5; calyx limb not accrescent in fruit .... **Combretum**

ANOGEISSUS (DC.) Guill. & Perr.

1. Leaves elliptic or suborbicular, obtuse-rotund at apex, rounded or cordate at base; flower heads usually in cymes; fruit broadly winged, wings entire; bark grey, smooth .... **A. latifolia**
1. Leaves elliptic-lanceolate, acute at apex, narrowed at base; flower heads usually solitary on bracteate peduncles; fruit fairly broadly winged, wings dentate; bark dark grey, rough .... **A. acuminata**

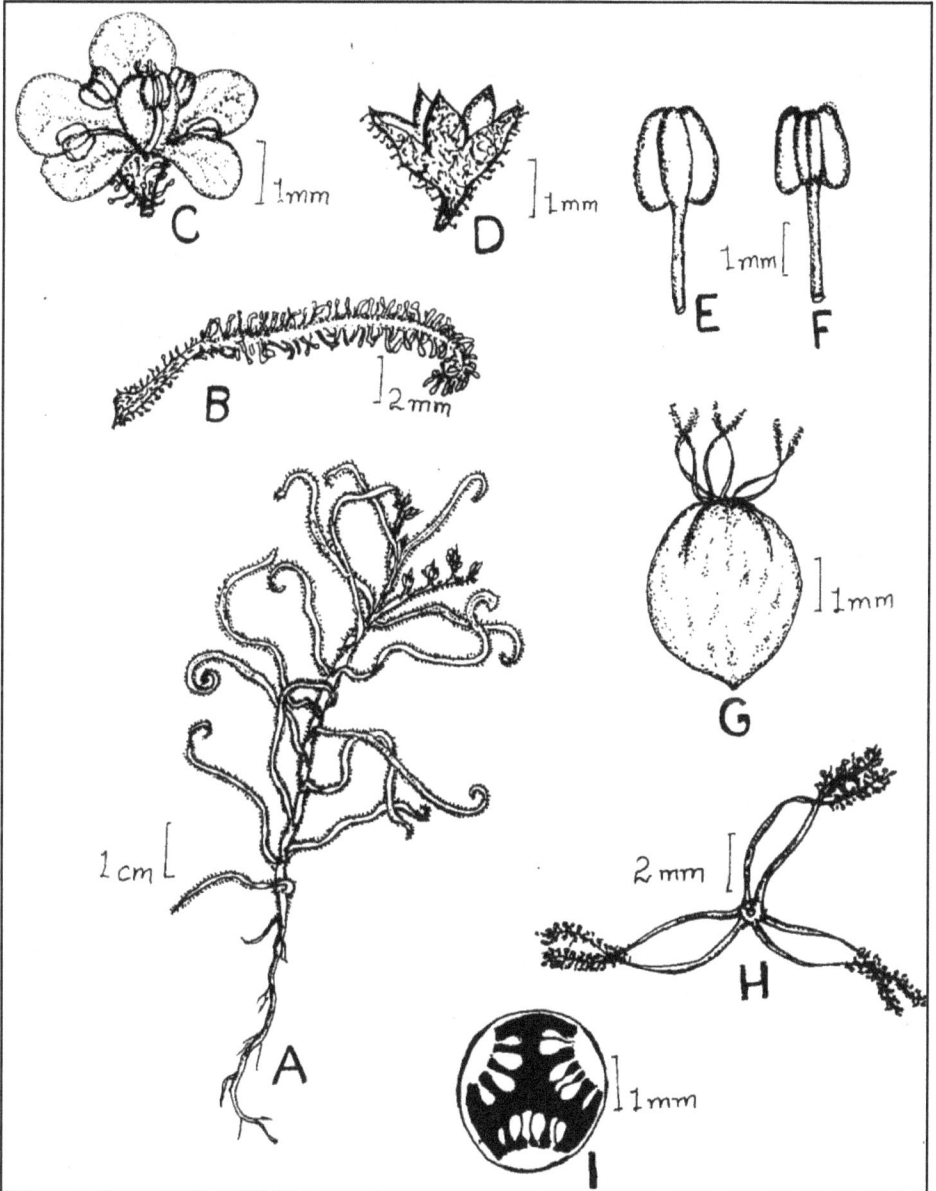

Figure 33: Drosera indica L.

A. Habit, B. Leaf, C. Flower-entire, D. Calyx, E&F. Stamens, G. Pistil, H. Stigma, I. Ovary c.s.

Anogeissus acuminata (Roxb. ex DC.) Guill. & Perr., Fl. Seneg. Tent. 7: 280. 1832; C.B.Clarke in Hook. f., Fl. Brit. India 2: 330. 1957; Gamble, Fl. Madras 1: 466. 1919. *Conocarpus acuminata* Roxb. ex DC., Prodr. 3: 16. 1828. *Anogeissus acuminata* (Roxb. ex DC.) Guill. & Perr. var. *lanceolata* Wall. ex C.B. Clarke in Hook. f., Fl. Brit. India 2: 451. 1878.

Large deciduous trees, up to 30 m high; bark dark grey, rough; branches drooping, blackish brown or coppery, glabrous. Leaves elliptic-lanceolate, 2.5-4 × 1.5-2 cm, mostly sub-opposite, base obtuse, margin entire, apex acute, glabrous above, puberulous beneath, secondary nerves to 6 pairs, slender and distant; petiole to 6 mm, pubescent. Flowers greenish yellow, in globose heads, usually solitary on bracteate peduncles, peduncles 0.5–2 cm long, hairy, bracts 1-2 pairs, sub foliaceous on peduncles, bracteoles spathulate; calyx tube with ovary 3-5 mm long, ferrugineous, expanded portion shortly campanulate, glabrous, teeth 5; stamens 10, 2-3 mm long in 2 series, style 3-7 mm long. Fruits 6.8 mm in diameter, broader than long with upper part pubescent and beak much shorter than diameter, wings 2, membranous, dentate, ending in persistent calyx tube; seeds ovoid.

Common in Adilabad and Karimnagar districts usually on river banks.

Fl.: August – March; Fr.: September-May.

Bheemaram (ADB), *GO* & *DAM* 5172; Tekupalli (KMM), *PVS* 84054 (BSID).

Anogeissus latifolia (Roxb. ex DC.) Wall. ex Bedd., Fl. Sylv. S. India 1: 15. 1869; C.B. Clarke in Hook. f., Fl. Brit. India 2: 450. 1878; Gamble, Fl. Madras 1: 466. 1919; Scott, Kew Bull. 33: 560. 1979. *Conocarpus latifolia* Roxb. ex DC., Prodr. 3: 16. 1828.

Vern.: *Chiru-manu, Elama.*

Deciduous trees, up to 25 m high, bark greyish to whitish, smooth, peeling of in thin flakes, branchlets grey-pubescent. Leaves alternate, sub-opposite, elliptic or suborbicular, 3-10 × 2.5-6 cm, chartaceous, nerves 7-12 pairs, glabrous above, rusty below, base round or cordate, margin entire, apex obtuse-rotund; secondary nerves 8-10 pairs, conspicuous beneath; petiole 6-8 mm. Flowers pale yellow, axillary, aggregated in cymose heads; calyx tube 3 mm, pubescent, teeth 5, short, triangular, 1.5 mm, persistent; petals 0; stamens 10 in 2 series, anthers small, filaments to 1.5 mm long; ovary to 2 mm, 1- celled, densely tomentose, ovules 2, pendulous, styles to 1 mm, thickened at base, villous. Fruits clustered into globose head, ca 1 × 1.5 cm, compressed, 2-winged ending in a persistent calyx tube; seed solitary, pale brown, ellipsoid, glabrous.

Common; it forms one of the major components of the moist and dry deciduous forests of all districts.

Fl.: April – September; Fr.: September – April.

Bendala RF (ADB), *GO* 4483; Mosra RF (NZB), *TP* & *BR* 6105; Sardanha (MDK), *BR* & *CP* 6122; Kothapalli (RR), *MSM* 10447; Osmangunta (NLG), *BVRS* 424; Rechapalli (KRN), *GVS* 22249 (MH); Irlapenta (MBNR), *BSS* & *SKB* 32368; Bhadrachalam (KMM), *RCS* 102441 (BSID); way to Farahabad (MBNR), *SRS* 104588 (BSID); Tadwai RF (WGL), *RKP* 110804 (BSID).

CALYCOPTERIS Lam.

Calycopteris floribunda (Roxb.) Lam. ex Poir. Encycl. Suppl. 2: 41. 1811; C.B. Clarke in Hook. f., Fl. Brit. India 2: 449. 1878; Gamble, Fl. Madras 1: 467. 1919. *Getonia floribunda* Roxb., Pl. Coromandel 1: 61. t. 87. 1798.

Vern. : *Bontha, Adavi jama, Bandi murugdu.*

Scandent shrubs, up to 3 m high; branches drooping, rusty brown-puberulous. Leaves alternate-subopposite, oblong or elliptic-oblanceolate, 3.8-13.8 × 2.3-5.4 cm, softly pilose above, rusty tomentose beneath or glabrous, prominently gland-dotted beneath, base obtuse or cuneate, margin entire, apex acuminate-caudate; secondary nerves 4-8 pairs, prominent beneath, petiole to 6 mm. Flowers greenish white, 2 cm long, in terminal and/or axillary, large fulvous panicles; bracts foliaceous, 1 cm long, bracteoles lanceolate, 3 mm; calyx hairy, tube 8 mm long, apically campanulate, lobes 6 mm, recurved, accrescent in fruit, scarious, linear–oblong, 3-nerved, veined; petals 0; stamens 10, biseriate, shorter than lobes, filaments subulate, 1.8 – 2 mm, anthers didymous; ovary 1.5 mm, densely villous out side, 1-celled, 3-ovuled, pendulous. Fruits ovoid-ellipsoid, 5-7 cm long, 5- ribbed, indehiscent, crowned by the persistent calyx.

Common in Adilabad, Warangal, Nalgonda, Karimnagar and Mahabubnagar districts in dry deciduous forests.

Fl. & Fr.: February-May.

Kotpalli RF (ADB) *GO* & *DAM* 5144; Neelvai; RF (ADB), *GO* & *DAM* 5147, 5162; Appapur (MBNR), *SKB* & *BSS* 30701; Saleswaram (MBNR), *BR* & *SKB* 30727; Pakhal (WGL), *ANH* 15912 (MH), *RKP* 108249 (BSID); Kambalpally (NLG), *BVRS* 313 (NU); Nellippakka RF (KMM), *RCS* 99028 (BSID).

COMBRETUM Loefl. *nom. cons.*

1. Calyx tube produced more than 2.5 cm beyond the ovary .... **C. indicum**
1. Calyx tube not or only shortly produced beyond the ovary:
    2. Bracteoles small; calyx and corolla lobes 4 each;
        fruit with 4 wings, globose:
        3. Leaves pale and thin when dry; calyx tube above the
            ovary widely campanulate; fruit 1.8 cm diam .... **C. albidum**
        3. Leaves greenish and shining when dry; calyx tube
            above the ovary funnel-shaped .... **C. latifolium**
    2. Bractoles linear, as long as flowers; calyx and corolla lobes
        5 each; fruit with 5 wings, oblong .... **C. album**

Combretum albidum G. Don, Trans. Linn. Soc. London 15: 429. 1827. *C. ovalifolium* Roxb., Fl. Ind. 2 : 226. 1832; C.B.Clarke in Hook. f., Fl. Brit. India 2: 458. 1878; Gamble, Fl. Madras 1: 469. 1919.

Vern. : *Yadatheega, Putangi, Shirtalboddi, Bandikota, Bandi kattu tige.*

Large climbing shrubs, to 3 m long; branchlets greyish, terete, densely yellowish to brownish, lepidote when young, glabrous later. Leaves opposite decussate, ovate, ovate-elliptic, 3.7-11.5 × 3.8-7 cm, thin coriaceous, 7-9-nerved, almost parallel, glabrous, base cuneate- subacute, margin entire, apex acuminate or emarginate; secondary nerves 3-5 pairs, reticulations prominent beneath, green or pale brownish after drying; petiole to 0.7 –2.2 cm, glabrous, channeled. Flowers pale white, in axillary or terminal, 3-8 cm long panicles of spikes; bracteoles minute, to 1 mm; calyx tube glabrous, 4 mm, lobes 4, 1 mm; reflexed; petals 4, oblong, 1.5 × 0.5 mm; stamens 8, biseriate incurved, filaments subulate, 2.5 mm, anthers 0.25 mm; ovary 1-celled, 2 mm, ovules 2-5, pendulous, style subulate. Berries 1.5-2 × 1.5-2 cm, globose, 4 or 5-angular, 4-winged, wings chartaceous; seed solitary.

Common in deciduous forests of most districts.

Fl.: February- March; Fr.: May onwards.

Bheemaram (ADB), *GO* 5169; Laxmapur (NZB), *TP* & *BR* 6274; Gowraram RF (NZB), *TP* & *BR* 6356; Gangapur RF (MDK), *BR* & *CP* 11432; Mannanur (MBNR), *TDCK* 13696; Pakhal (WGL), *ANH* 15967 (MH); Mahadevpur to Kataram (KRN), *SLK* 70811 (NBG); Kacharam (KMM), *PVS* 84051 (BSID); Kinnerasani Wild life sanctuary (KMM), *RR* 112680 (BSID).

Combretum album Pers., Syn. Pl. 1: 411. 1805; Gangopadhyay & Chakrabarty, J. Econ. Taxon. Bot. 14: 225. 1997. *C. roxburghii* Spreng., Syst. Veg. 2: 331. 1825. *C. decandrum* Roxb., Pl. Coromandel 1 : 49. 59. 1795; C.B. Clarke in Hook. f., Fl. Brit. India 2: 452. 1878; Gamble, Fl. Madras 1: 468. 1919.

Large climbing shrubs, to 2.5 m long, with many pendent branches; branchlets pale yellow. Leaves opposite, oblong, 4-12 x 2.5-6.5 cm, glabrous on both surfaces, base cuneate, margin entire, apex suddenly acuminate; secondary nerves 5-8 pairs; petiole to 6 mm. Flowers white, polygamous, in panicled spikes; bracts foliaceous; calyx tube 1.5 – 2.5 mm long, villous, expanded portion cupular, 2-2.5 × 2-4 mm, tawny villous, lobes 5; petals 5, ovate – oblong or oblong – lanceolate, 1-1.5 mm long; stamens 3-4 mm long; style 2-3 mm long. Fruit oblong to elliptic, 1.5- 3.5× 0.5-2 cm, subquadrate, nearly smooth; sessile wings 5(-4), 1.5 – 3.5 cm, thinly chartaceous, brown to reddish – brown or blackish, glabrous.

Abundant in open forests of Karimnagar, Warangal, Khammam and Hyderabad districts.

Fl.: September- March; Fr.: November - July.

Aklaspur (KRN), *GVS* 22504 (MH); Pakhal (WGL), *ANH* 15929 (MH), *RKP* 108250 (BSID); Tadwai (WGL), *RKP* 110802 (BSID); Sathiyampet RF (KMM), *RR* 112630 (BSID); Poosugundi (KMM), *PVS* 84046 (BSID).

Combretum indicum (L.) DeFilipps, Useful Pl. Dominica 277. 1998. *Quisqualis indica* L., Sp. Pl. ed. 2: 556. 1762; C.B.Clarke in Hook.f., Fl. Brit. India 2: 459. 1878; Gamble, Fl. Madras 1: 469. 1919.

Vern.: *Radhamadhav, Rangoon malle.*

Large woody climbing shrub; young branchlets rusty-tomentose. Leaves broadly elliptic or elliptic-oblong, 5-7 x 3-4 cm; rounded or subcordate at base; acuminate or often sub-caudate at apex, margins entire; petioles 5 mm long. Flowers showy, fragrant, pendulous, borne in terminal and axillary spikes in opposite leaves, 5-10 cm long; bracts lanceolate, 5 mm long; calyx tube 3.5 cm long; lobes shortly triangular, 1.5-2 mm long; corolla lobes oblong, sparsely pubescent, 6 x 1.5 cm long; filaments 4 mm long; gynoecium 4.5 cm long. Fruit narrowly ellipsoid, appressed-pubescent, 2-3 cm long, 5-angled.

Grown in gardens and house yards as an ornamental.

Fl. & Fr .: Major parts of the year, but profuse in June-October.

Bhadrachalam (KMM), *RCS* 99038 (BSID).

Note: Petioles are persistent and transformed into recurved spines after falling of lamina. Flowers are pale yellow at first, then turning orange and finally red.

Combretum latifolium Blume, Bijdr. Fl. Ned. Ind. 641. 1825. *C. extensum* Roxb. ex G. Don, Trans. Linn. Soc. London 15: 414. 1827; Hook. f., Fl. Brit. India 2: 458. 1878; Gamble, Fl. Madras 1: 469. 1919.

Large, scandent, straggling shrubs, branches glabrous. Leaves ovate-orbicular, 5-15 × 4-6.5 cm, elliptic-oblanceolate, glabrescent, shortly acuminate at apex, rounded at base. Flowers yellowish-green in axillary racemes, calyx lobes 4, triangular, acute, corolla lobes 4, equaling calyx lobes, oblong, slightly emarginate. Fruits ovoid, up to 3.5 × 3.2 cm, wings 4, membranous, reddish, horizontal striations.

Occasional in dry deciduous forest in Warangal district.

Fl. & Fr.: December – May.

Mallur (WGL), *CSR* 1547 (KU).

TERMINALIA L. *nom. cons.*

1.   Fruit not winged, ovoid or subcompressed :

   2.   Leaves clustered at the ends of branchlets, alternate; flowers in simple spikes :

      3.   Fruit 2-ridged when dry, ellipsoid .... **T. catappa**

      3.   Fruit faintly 5-ridged, ovoid .... **T. bellirica**

   2.   Leaves not clustered at the ends of branches, opposite or sub- opposite; flowers in simple spikes or in panicles .... **T. chebula**

1.   Fruit winged :

   4.   Fruit shortly angled or winged; glands on the petiole close to the base of leaf blade .... **T. arjuna**

   4.   Fruit always winged, not angled; glands inserted at the base of leaf blade :

5. Leaves with 2 stalked glands on the midrib, at base of
   the leaf blade; fruits glabrous .... **T. tomentosa**

5. Leaves with 2 sessile glands on the midrib, at the base of
   the leaf blade; fruits velvety .... **T. coriacea**

Terminalia arjuna (Roxb. ex DC.) Wight & Arn., Prodr. Fl. Ind. Orient. 314. 1834; C.B. Clarke in Hook. f., Fl. Brit. India 2: 447. 1878; Gamble, Fl. Madras 1: 465. 1919.

Vern.: *Tellamaddi, Nirumaddi.*

Large deciduous trees, up to 30 m high, often buttressed; outer bark stiff, flaking off in pieces, inner bark white, smooth, branchlets pubescent. Leaves alternate to sub-opposite, oblong or obovate-oblanceolate, 8-15 × 5-7 cm, thick coriaceous, nerves 14-16 pairs, parallel, glabrous, base obtuse-subcordate, margin crenate-serrate, apex obtuse; petiole to 1.5 cm, 2 glands on the petiole, close to the base of lamina. Flowers dull yellow in panicles of spikes with linear bracteoles; bracts 2 mm; calyx 3 mm long, hairy within; stamens 10 in 2 series, 1-2 mm; style 3 mm. Fruit drupe, 5-angled, 5-winged, wings equal, woody, dark brown to blackish when dry.

Common in deciduous forests in all districts on the banks of rivers and streams; often planted.

Fl.: April-July; Fr.: August – December.

Kundaram RF (ADB), *TP & GO* 5458; Ibrahimpet RF (NZB), *TP & BR* 6122; Kudichintalabailu (MBNR), *TDCK* 15386; Narsapur RF (MDK), *TP & CP* 14047;, *KMS* 7991 (CAL); Lothuvagu (RR), *MSM & KH* 10294; Bhupatipur (KRN), *GVS* 22299 (MH); Bhadrachalam RF (KMM), *RCS* 102444 (BSID).

Terminalia bellirica (Gaertn.) Roxb., Pl. Coromandel t. 198. 1805; " *bellerica*"; C.B. Clarke in Hook. f., Fl. Brit. India 2: 445. 1878; Gamble, Fl. Madras 1: 463. 1919. *Myrobalanus bellirica* Gaertn., Fruct. 2: 90. t. 97. 6f. 1791 " *bellirina*".

Vern.: *Tandra, Tadi, Tani.*

Large deciduous trees, up to 30 m high, bark bluish-grey with vertical cracks; branchlets warty, with persistent leaf scars. Leaves densely clustered at the tips of branchlets, alternate, broadly elliptic, 12-15 × 6-10 cm, coriaceous, 4-6-nerved, glabrous, finely pellucid-dotted above, base cuneate, oblique, margin entire-subcrenulate, apex rotund, emarginate; petiole to 10 cm. Flowers pale-white in axillary spikes; peduncle 1.2 cm long; calyx 2 mm, tube cupular, 1 mm long, densely villous inside, lobes triangular; stamens 10, filaments 3 mm long, anthers 1 mm; ovary 2 mm long, style 4 mm long, base villous. Drupe ovoid, ca 3 × 2.3 cm, softly tomentose, obscurely 5-ridged, wings 0.

Common in deciduous forests of all districts.

Fl.: April – May; Fr.: June – November.

Boath RF (ADB), *GO & PVP* 4633; Mamidipally RF (NZB), *BR & KH* 9016; Narsapur RF (MDK), *TP & CP* 14035; Forest south of the lake, Narsapur (MDK), *KMS* 7978 (CAL); Anantasagar RF (RR), *MSM & NV* 10391; Kudichintalabailu (MBNR),

TDCK 13606; Saleswaram (MBNR), *SKB* & *BSS* 30585; Pakhal (WGL), *SSR* & *KSM* 18522; Kambalpally (NLG), *BVRS* 606 (NU); Near Mahadevpur (KRN), *SLK* 70801 (NBG).

Terminalia catappa L., Mant. Pl. 128. 1767; C.B. Clarke in Hook. f., Fl. Brit. India 2: 444. 1878; Gamble, Fl. Madras 1: 463. 1919.

Vern : *Badam.*

Deciduous trees, up to 25 m high; crown spreading, trunk buttressed; branchlets with persistent leaf scars, rusty tomentose. Leaves densely clustered at the ends of branchlets, alternate, obovate- 15-25 × 10-12 cm, thin coriaceous, 10-12-nerved, glabrous above, tomentose below, base cuneate, margin entire, apex obtuse-rotund, retuse, two glands at base of the leaf blade, petiole 1-2 cm. Flowers whitish-yellow in axillary spikes; calyx tube 3-7 x 1.1.3 mm, pubescent to glabrous, expanded portion cupular, 3.5 – 5 × 3-5 mm; stamens 10 in two series, 3-4 mm long; disc villous; style 2-4 mm long. Drupe ellipsoid, 2- ridged, when dry glabrous, 4-6 cm long, wings 0.

Frequently planted in residential colonies and house yards.

Fl. & Fr.: September- June.

Amangal to Shadnagar (MBNR), *SRS* 104558 (BSID); Ramanagutta (KMM), *RR* 107998 (BSID).

Terminalia chebula Retz., Observ. Bot. 5 : 31. 1789. incl. forma. *tomentosa;* C.B. Clarke in Hook. f., Fl. Brit. India 2: 446. 1878 *p.p.;* Gamble, Fl. Madras 1: 464. 1919. *Myrobalanus chebula* (Retz.) Gaertn., Fruct. Sem. Pl. 2: 91. t. 97. 1791.

Vern. : *Karakkai.*

Deciduous trees, up to 20 m high, bark brown or grey or blackish, shallowly or deeply fissured; branchlets rusty-villous or glabrescent. Leaves opposite, ovate or elliptic-obovate, 7-12 × 5-7 cm, thin coriaceous, nerves 7-9 pairs, densely woolly below, glabrous at length, base rounded, margin entire, apex subacute- obtuse, apiculate; glands 2 at the base of the leaf blade; secondary nerves 6-10 pairs, more or less prominent above and beneath, arcuate – round; petiole 1.5 – 2.5 cm long; sulcate above, puberulous. Flowers dull white, in simple or branched axillary spikes; bracteoles conspicuous, linear, acute, hairy, caducous; calyx campanulate, 3 mm long, glabrous outside, villous within, lobes 5; stamens 10, biseriate; ovary inferior, 2 mm long, style 3.2 mm long, carpel 1, ovules 1 or 2. Drupes obovoid or ellipsoid, 4 × 2.5 cm, glossy, glabrous, faintly angled, wings 0.

Frequent in deciduous forests in Adilabad, Karimnagar, Ranga Reddi, Hyderabad and Mahabubnagar districts.

Fl. & Fr.: May-July.

Jaipur RF (ADB), *GO* 4487, 4967; Aklaspur (KRN), *GVS* 20221 (MH); Anantasagar RF (RR), *MSM* & *NV* 10410; Farahabad (MBNR), *TDCK* 15381; Mannanur (MBNR), *SKB* & *BSS* 30718 (BSID).

Terminalia coriacea (Roxb.) Wight & Arn., Prodr. Fl. Ind. Orient. 315. 1834; Gamble,
Fl. Madras 1: 465. 1919. *Pentaptera coriacea* Roxb., Fl. Ind. 2 : 438. 1832. *Terminalia tomentosa* (Roxb. ex DC.) Wight & Arn. var. *coriacea* (Roxb.) C.B. Clarke in Hook. f.,
Fl. Brit. India 2: 448. 1878.

Deciduous trees, up to 15 m high, bark deeply cracked; branchlets velvety-pubescent, with persistent leaf-scars. Leaves alternate or decussate, elliptic or oblong-obovate, 10-15 × 5-8 cm, thick coriaceous, nerves 10-12 pairs, punctate, glabrous above, velvety-pubescent and glaucous below, base obtuse-cordate, oblique, margin entire-subcrenate, apex obtuse, emarginate, glands 2, sessile at the base of leaf blade. Flowers dull yellow, in axillary or terminal, 14 cm long panicles; calyx 5 – lobed; stamens 10 in 2 series, filaments 4.5 mm long, anthers 0.5 mm; ovary 2.5 mm, style 5.5 mm. Drupes yellowish-velvety, 5-angled, 5-winged, equal, thin with horizontal lines, apex rounded.

Common in deciduous forests in Adilabad, Karimnagar, Medak, Ranga Reddi and Mahabubnagar districts.

Fl.: June-July; Fr.: August- October.

Jaipur RF (ADB), *GO* & *PVP* 4967; Anantagiri (RR), *SSR & KSM* 18518; Kodimial (KRN), *GVS* 20143 (MH); Narsapur RF (MDK), *TP* & *CP* 14015; Pagarikutta-Narsapur (MDK), *KMS* 6648 (CAL); Kothapalli (RR), *MSM* 10449; Domalapenta (MBNR), *VBH* 84949 (BSID); Bandi RF (KMM), *RCS* 104293 (BSID).

Terminalia tomentosa (Roxb. ex DC.) Wight & Arn., Prodr. Fl. Ind. Orient. 314. 1834;
C.B. Clarke in Hook. f., Fl. Brit. India 2: 447. 1878; Gamble, Fl. Madras 1: 465. 1919. *Pentaptera tomentosa* Roxb. ex DC., Prodr. 3: 14. 1828.

Vern.: *Nallamaddi, Inamaddi, Maddi.*

Deciduous trees, up to 25 m high, young parts villous, bark rough, black with deep vertical fissures and transverse cracks. Leaves alternate or opposite, oblong or elliptic, 6-15 × 4.5-7 cm, thick coriaceous, nerves about 15 pairs, parallel, glabrous above, tomentose below, base and apex obtuse, margin crenulate, glands turbinate, on the midnerve at the base of leaf-blade; secondary nerves 18-21 pairs, articulate; petiole to 2 cm. Flowers light yellow, in simple spikes in axillary panicles; calyx 5-lobed, tube 1.5 mm long, villous, triangular; stamens 10, filaments 3-4 mm; ovary terete, 2 mm, 1-celled; ovules 2 or 3, pendulous; style to 2 mm. Drupe ovoid, brown, ca 4 cm in diam., glabrous, with 5 equal wings running horizontally from the axis.

Common member of deciduous forests in all districts.

Fl.: May-July; Fr.: October –December.

Bheemaram RF (ADB), *GO* & *PVP* 4494; Manchippa RF (NZB), *BR* & *PSPB* 9614; Sirnapally RF (NZB), *BR* & *GO* 9086; Mothugudem (KMM), *SSR* 15246; Madikonakutta-Narsapur RF (MDK), *TP* & *CP* 11984, *KMS* 6727 (MH); Central University campus (RR), *MSM* 15162; Kodimial (KRN), *GVS* 20143 (MH); Pakhal (WGL), *KMS* 13159 (MH).

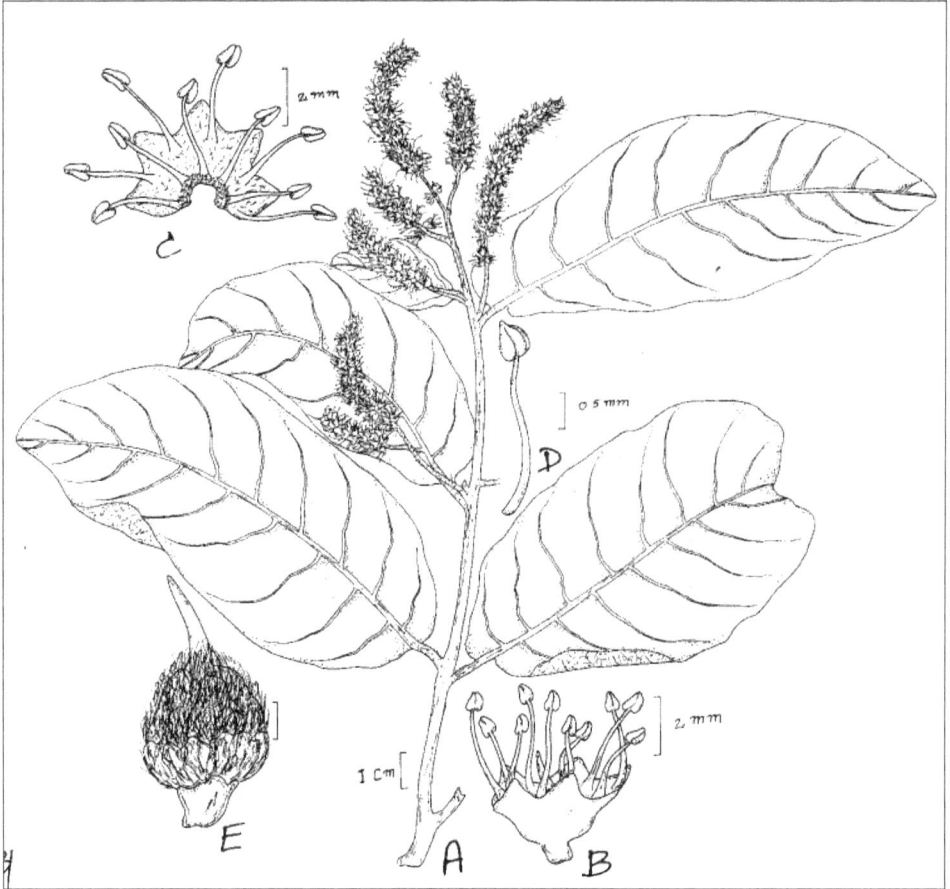

Figure 34: Terminalia coriacea (Roxb.) Wight & Arn.
A. A habit, B. Flower, C. Calyx, split open, D. Stamen, E. Pistil.

MYRTACEAE

1. Fruit berry:
   2. Ovary 2-celled; seeds one or a few .... **Syzygium**
   2. Ovary many-celled; seeds numerous .... **Psidium**
1. Fruit capsule:
   3. Flowers bright red or crimson in elongate spikes; petals
      distinct, free .... **Callistemon**
   3. Flowers cream-coloured in capitate umbels; petals fused and
      form a calyptra as a lid .... **Eucalyptus**

CALLISTEMON R.Br.

Callistemon citrinus (Curtis) Skeels, U.S.D.A. Bur. Pl. Industr. Bull. 282: 49. 1913.
Metrosideros citrina Curtis, Bot. Mag. t. 260. 1794.

Evergreen tree, 3-6 m high; branches spreading, silky-hairy when young. Leaves linear-lanceolate, 3-8 × 0.4-0.9 cm, rigid, apex acute, margin entire, base narrowed; midrib and secondary nerves prominent; glands lateral. Flowers bright red or crimson in 5 – 10 cm long, terminal spikes. Capsule woody, depressed globose, 4 mm long, glabrous, contracted at summit.

Cultivated in gardens and parks.

Fl. & Fr.: Throughout the year.

EUCALYPTUS L'Herit.

1. Peduncles flattend .... **E. globulus**
1. Peduncles not flattend .... **E. camaldulensis**

Eucalyptus camaldulensis Dehn., Cat. Horti Camald. ed. 2. 20. 1832. *E. rostrata* Schl., Linnaea 20: 365. 1922.

Tall tree, branches drooping; bark smooth, white or ashy, decorticating in rounded plates. Leaves lanceolate, falcate, 10-20 × 1-2 cm. Flowers whitish in axillary 3-12-flowered umbels. Fruit 4-6 mm in diam., rim convex, valves exserted.

Planted extensively in Bhadrachalam area for its pulp wood, occasional elsewhere.

Fl. & Fr.: November – January.

Eucalyptus globulus Labill. Rel. Voy. rech. Perouse 1 : 153. 1799; Hook. f. Fl. Brit. India 2 : 462. 1878; Gamble, Fl. Madras 1 : 486. 1919.

Large tree, 20-60 m high; bark greyish-or bluish-white. Leaves broad-lanceolate, or lanceolate-falcate, 10-30 x 3-4 cm, acuminate, dark-green and lustrous, sessile. Flowers to 4 cm across, usually solitary, rarely 2-3-nate, axillary; peduncle much reduced; calyx quadrangular, verrucose; operculum umbonate, verrucose, shorter than calyx-tube; stamens 1-1.5 cm long; anthers obovate, versatile, 0.08-0.15 cm long.

Fruit sessile, depressed-globose to broadly turbinate, 4-ribbed, verrucose, 1-1.5 x 1.5-3 cm, with rim and depressed or convex crown.

Planted on a large scale on waste lands under Social forestry programmes.

Fl. & Fr .: February-May.

Many more species of *Eucalyptus* are being grown by the forest department and ITC Bhadrachalam

PSIDIUM L.

Psidium guajava L., Sp. Pl. 470. 1753; Duthie in Hook. f., Fl. Brit. India 2: 468. 1878, *"guyava"*; Gamble, Fl. Madras 1: 472. 1919.

Vern.: *Jama.*

Trees, up to 9 m high, bark smooth, pinkish brown, exfoliating in small flakes; branchlets appressed tomentose. Leaves decussate, oblong, ovate-lanceolate, 6-12 × 2.5-5.5 cm, coriaceous, secondary nerves 10-16 pairs, prominent below, submarginal, sparsely tomentsoe above, densely tomentose beneath, pellucid punctate, base obtuse-cuneate, margin entire, apex acute; petiole 7-10 mm long. Flowers white, 3 cm across, axillary, solitary or in pairs; calyx urceolate, lower portion adnate to the ovary, upper portion free and irregularly lobed; petals free, deciduous; stamens many, filaments 1.5-2 cm long; ovary 4-5-celled, ovules many. Berries globose or pyriform varying in size and shape, ca 4 × 3.5 cm; seeds numerous, embedded in pleasantly flavoured sweet pulp.

Cultivated and run wild in most districts.

Fl. & Fr.: Nearly all round the year.

Appapur (MBNR), *TDCK* 16812; Bhadrachalam (KMM), *P.A.N.Reddy* & *AJR* 20472.

SYZYGIUM Gaertn. *nom. cons.*

1. Shrubs or small trees, leaves narrow, oblong; berries crowned with the cup-like calyx limb .... **S. salicifolium**
1. Large trees, leaves elliptic-oblong; berries not crowned with calyx limb .... **S. cumini**

Syzygium cumini (L.) Skeels in U.S.D.A. Bur. Pl. Industr. Bull. 248: 25. 1912. *Myrtus cumini* L., Sp. Pl. 471. 1753. *Eugenia jambolana* Lam., Encycl. 3: 198. 1789; Duthie in Hook. f., Fl. Brit. India 2: 499. 1879. *Syzygium jambolanum* (Lam.) DC., Prodr. 3. 259. 1828; Gamble, Fl. Madras 1: 481. 1919.

Vern.: *Neredu.*

Evergreen, densely foliaceous trees, up to 30 m high, with ashy-brown smooth bark. Leaves opposite decussate, elliptic-oblong or lanceolate, 8-15 × 3-5.5 cm, coriaceous, glabrous, glossy above, base cuneate-obtuse, margin entire, apex acuminate; secondary nerves numerous, prominent on the lower surface; petiole to 1.8 cm, reddish. Flowers pale greenish white, sessile, axillary or terminal in trichotomous panicles, sweet scented; bracteole triangular, 9 mm, gland dotted, pedicel to 3.8 mm long; calyx 3 x 2.25 mm, tube turbinate, lobes 4; petals 4, cream, united and form a calyptra, 1 mm broad; stamens numerous, filaments 4 mm long, anthers 0.75 mm; ovary 1.2 – 1.8 mm, style filiform, 6 mm long. Berries ovate or oblong, glabrous, 1-2.5 × 1-1.5 cm, shining bright to dark-purple; seed solitary.

Common in all forest districts. Planted along roads and in gardens.

Fl.: February – May; Fr.: April- August.

Janakpur (ADB), *PVP* 9492; Mamidipally (NZB), *BR* & *KH* 9006; Akkannapet (MDK), *BR* & *CP* 11563; Gacchibowli (RR), *MSM* 15194; Appapur (MBNR), *TDCK* 15327;

Syzygium salicifolium (Wight) J. Graham, Cat. Pl. Bombay 73: 1839. *S. heyneanum* (Duthie) Wall. ex Gamble, Fl. Madras 1: 482. 1919. *Eugenia heyneana* Duthie in Hook. f., Fl. Brit. India 2: 500. 1878.

Vern.: *Chinna neredu.*

Large shrubs or small trees, up to 6 m high; bark smooth, greyish; branchlets greenish white, glabrous. Leaves opposite, elliptic-oblong or lanceolate, 8-12 × 2-5 cm, base narrow, margin entire, apex usually acuminate, pelucid punctate, glabrous; secondary nerves numerous, slender, with reticulate intermediate nerves, petiole to 1 cm. Flowers white, lateral from the scars of fallen leaves, rarely axillary. Berries 1-2.5 cm long, oblong-ellipsoidal, pale-purple, glabrous, crowned with the cup-like calyx-limb.

Rare in Nizamabad and Khammam districts in river beds and along streams.

Fl. & Fr.: January-May.

Jalalpur RF (NZB), *BR* & *KH* 9659; Perantapalli (KMM), *PVS* 84064 (BSID).

Several species and hybrids of *Eucalyptus* are extensively planted on the hill slopes of various forest ranges by the State Forest Department under Afforestation scheme. They are also planted in dry zones of the state under Social forestry in wastelands, gardens, and cultivated fields.

LECYTHIDACEAE

Key 1 (Vegetative)

1. Leaves puberulous beneath; secondary nerves up to 7 pairs, irregular, apex obtuse or subacute .... **Barringtonia**

1. Leaves glabrous on both surfaces; secondary nerves regular, apex shortly acuminate-long acuminate, acute or obtuse:

    2. Leaves thin; secondary nerves curved upwards, angle of divergence narrow; margin entire; base cuneate or acute; cultivated .... **Couroupita**

    2. Leaves thick; secondary nerves curved towards margins, angle of divergence broad, margin crenulate-denticulate, base attenuate; wild .... **Careya**

Key 2 (Flowering)

1. Flowers on the trunk or old wood; stamens fused into 2 uneqal bundles and forming a hood over ovary; cultivated .... **Couroupita**

1. Flowers in the axils of leaves; stamens free:

   2. Flowers yellowish-white, 6 cm across on short spikes, all staminal filaments antheriferous .... **Careya**

   2. Flowers pale pink, 1-3 cm across in axillary pendulous racemes; few staminal filaments antheriferous .... **Barringtonia**

Key 3 (Fruiting)

1. Berry small, ovoid, angular; seed 1 .... **Barringtonia**

1. Berry large, globose, rounded; seeds many, embedded in pulp:

   2. Fruit not foetid, crowned by persistent calyx lobes .... **Careya**

   2. Fruit foetid, not crowned by calyx lobes .... **Couroupita**

BARRINGTONIA J.R. Forster & J.G.A. Forster *nom. cons.*

Barringtonia acutangula (L.) Gaertn., Fruct. Sem. Pl. 2: 97. t. 101. 1791. subsp. **acutangula;** Bedd., Fl. Sylv. t. 204. 1872; C.B. Clarke in Hook. f., Fl. Brit. India 2: 508. 1879; Gamble, Fl. Madras 1: 487. 1919; Payens; Blumea 15: 226. 1967. *Eugenia acutangula* L., Sp. Pl. 471. 1753.

Evergreen trees, up to 12 m high; bark thick, dark grey, distinctly furrowed on old tree; young branches densely tomentose. Leaves oblong-obovate, oblanceolate, 8-25 × 4-15 cm, coriaceous, glossy above, nerves 10-13 pairs, base attenuate, margin serrulate, apex obtuse-round or subacute, petiole 5 mm. Flowers pale red, 1 cm long in pendulous, axillary, about 40 cm long spicate racemes; sepals ovate or oblong, 2.5 x 2 mm, pubescent, denticulate; petals pale pink or red, 5 mm long, caducous; stamens numerous, curved, filaments dark pink; ovary usually 4-celled; style to 5mm. Berries 3 x 1 cm, obtusely quadrangular, fibrous, crowned with persistent calyx tube; seed solitary, ovoid.

Common in Adilabad, Karimnagar and Warangal districts along streams and on swampy land.

Fl.: March-June; Fr.: July – December.

Jodevagu-Bheemaram RF (ADB), *TP* & *GO* 5453; Bheemaram (ADB), *GO* & *DAM* 5166; Vempalli (KRN), *GVS* 20184 (MH); Pakhal lake (WGL), *KMS* 13110 (MH); Pilutla RF (MDK), *RG* 106793 (BSID); Appanapally RF (MBNR), *SRS* 109781 (BSID); Gunavaram (KMM), *RCS* 98943 (BSID); Near Ramappa lake (WGL), *RKP* 105354 (BSID).

CAREYA Roxb. *nom. cons.*

Careya arborea Roxb., Pl. Coromandel t. 218. 1819; C.B. Clarke in Hook. f., Fl. Brit. India 2: 511. 1879; Gamble, Fl. Madras 1: 488. 1919.

Deciduous trees, 10-15 m high; bark thick, dark brown, smooth, with exfoliating scales; young branches gnarled with persistent leaf scars, pubescent. Leaves apically clustered, broadly oblong-ovate, base attenuate, margin crenulate-denticulate, apex shortly acuminate; secondary nerves 16-18 pairs, petiole to 1 cm. Flowers yellowish-white in terminal spikes or racemes on naked branches; bract lanceolate, 8 × 4 mm; calyx tube campanulate or funnel – shaped, 2 × 1.5 cm, glabrous, limb 4-lobed, lobes imbricate, 7 × 9 mm, persistent; petals 4, 3.8 x 2 cm, oblanceolate; stamens many in several rows, slightly connate at base, staminal column 6 mm long, filaments 2-5 cm long, anthers 2 mm, inner and outer rows of filaments without anthers; ovary 4-5-celled, 6.2 cm long, style to 4 cm. Berry ca 5 × 5 cm, globose, bright-green, fibrous, rind thick, crowned with the limb of calyx and remains of style; seeds many in fleshy pulp.

Occasional in all districts in deciduous forests.

Fl.: February-April; Fr.: May- July.

Bheemaram (ADB), *GO* & *DAM* 5168; Laxmapur RF (NZB), *TP* & *BR* 6285; Gowraram RF (NZB), *TP* & *BR* 6355; Pocharam RF (MDK), *BR* & *CP* 11605; Lothvagu (RR), *MSM* & *KH* 10297; Appapur (MBNR), *TDCK* 13687; Farahabad (MBNR), *SSR* 15537, *KP, VSL* & *BR* 38812; Kudurpalli (KRN), *SLK* 70715 (NBG).

COUROUPITA Aubl.

Couroupita guianensis Aubl., Hist. Pl. Guine 708. t. 282. 1775; Curtis, Bot. Mag. t. 3158 & 3159. 1832.

Vern.: *Nagalingam.* English: Cannon ball tree.

Deciduous tree, up to 20 m high; bark brownish-grey, smooth; branchlets glabrous. Leaves clustered at the ends of branchlets, thin coriaceous, oblanceolate or oblong-obovate, 14 – 40 × 8 -17 cm, glabrous, pubescent on the nerves beneath, apex acute or obtuse, margin entire or obscurely serrulate, base acute or cuneate, often oblique; secondary nerves 15-20 pairs, prominent beneath; petiole to 1.8 cm. Flowers reddish or tinged yellow on the outside, 10 cm across, cauliflorous; stamens fused into 2 unequal bundles and forming a hood over ovary. Fruit pale reddish-brown, woody, pendulous, 15 cm in diam., inner pulp ill smelling.

Introduced. Cultivated in gardens and temples.

Fl. & Fr.: Throughout the year.

Perantapalli (KMM), *RCS* 98991 (BSID).

MELASTOMATACEAE

1. Leaves not ribbed, but penninerved with faint intramarginal nerves; ovary 1-celled, seed 1 .... **Memecylon**

1. Leaves 3 or more-ribbed from the base; ovary 4-5-celled, seeds many:

    2. Stamens equal; fruit opening by pores at apex .... **Osbeckia**

    2. Stamens unequal; fruit bursting irregularly .... **Melastoma**

MELASTOMA L.

Melastoma malabathricum L., Sp. Pl. 390. 1753, "*malabathrica*"; C.B. Clarke in Hook. f., Fl. Brit. India 2: 523. 1879; Gamble,Fl. Madras 1: 450. 1919.

Evergreen shrubs or small trees, up to 3 m high, branchlets 4-angled, branchlets, petioles and peduncles densely clothed with acute or acuminate, often serrulate scales. Leaves ovate-oblong or lanceolate-oblong, 5-10 × 3-4 cm, base usually rounded, margin entire, apex acute, strigose above with appressed thick hairs, tomentose beneath, 3-5-nerved; petiole 7.5 - 12 mm. Flowers bright purple in terminal corymbose 1-5-flowered panicles. Fruit broadly ovoid, truncate 1- 1.2 cm long; seeds numerous.

Rare in wet places of forests in Warangal district (C.S.Reddy, 2001).

Fl.: September – March.

Mallur (WGL), *CSR* 1642 (KU).

MEMECYLON L.

1. Leaves ovate, 8-10 pairs of nerves family visible, when dry
   upper surface greenish .... **M. edule**
1. Leaves elliptic or elliptic-ovate, nerves not visible, yellowish
   when dry .... **M. umbellatum**

Memecylon edule Roxb., Pl. Coromandel t. 82. 1798; C.B. Clarke in Hook. f., Fl. Brit. India 2: 563. 1879. *p.p.;* Gamble, Fl. Madras 1: 504. 1919.

Evergreen trees, up to 10 m high; bark deep-brown, closely and regularly fluted. Leaves ovate, 3-7 × 1-3 cm, coriaceous, glabrous, base subacute, margin entire, apex acuminate, retuse; secondary nerves 8-10 pairs; intramarginal vein and nerves inconspicuous; petiole to 4 mm. Flowers blue, rarely white, in copious, diffuse umbels; peduncle to 3 cm, pedicel to 4 mm; calyx tube campanulate, to 1.5 mm, lobes 4, suborbicular, 1 mm; petals 4, blue, broadly ovate, 3 mm; stamens 8, filaments to 5 mm; ovary to 1.5 mm, style to 6 mm. Berries globose, 7 × 5 mm, black-purple, seeds up to 5 mm.

Common on dry evergreen forests in Khammam district.

Fl.: April – June; Fr.: July.-August.

Bandirevu RF (KMM), *RCS* 98929 (BSID); on the bank of Thellavaku river (KMM), *RR* 112633 (BSID).

Memecylon umbellatum Burm. f., Fl. Ind. 87. 1768; Gamble,Fl. Madras 1: 504. 1919.

Vern.: *Alli, Peddalli, Kukka alli.*

Shrubs or small trees, up to 4 m high; branches terete, glabrous, glaucous; bark corky, fluted. Leaves elliptic or elliptic-ovate, 6 × 4 cm, coriaceous, glabrous, obtuse, base attenuate, margin slightly revolute, apex emarginated; secondary nerves 6-9, obscure, petiole to 5 mm. Flowers reddish-purple in umbellate cymes; peduncle almost reduced or 0, branchlets to 2 mm, pedicel to 2.5 mm; calyx tube to 1 mm; lobes 4; petals 4, blue, orbicular, to 2 mm; stamens 8, filaments to 4.5 mm; ovary globose to 1.5 mm, 1-

celled; style to 1 cm. Berries 1 × 1 cm, globose, reddish-tinged, ripening black, with persistent calyx-tubes, seed 1.

Occasional along streams of forests in Warangal district (C.S. Reddy, 2001)

Fl.: April – June; Fr.: July-August.

Tadwai (WGL), *CSR*1034 (KU).

OSBECKIA L.

1. Anthers not beaked .... **O. muralis**

1. Anthers beaked .... **O. zeylanica**

Osbeckia muralis Naudin, Ann. Sci. Nat. Bot. Ser. 3: 14 : 56. 1850. *O. truncata* D. Don ex Wight & Arn., Prodr. Fl. Ind. Orient. 332. 1834. *p.p.; nom. illeg;* C.B. Clarke in Hook. f., Fl. Brit. India 2: 514. 1879; Gamble, Fl. Madras 1: 494. 1919.

Erect herbs, up to 30 cm high, branches quadrangular, hirtellous. Leaves ovate-lanceolate or elliptic, 3-5 × 1.5-2.5 cm, hirsute on both the surfaces, base obtuse, margin entire, apex subacute, nerves 3 from base. Flowers pink with bluish tinge, axillary, solitary, often clustered at ends of branches; bracts narrowly triangular, pedicels to 3 mm long; calyx broadly triangular, ca 1.5 mm long, cuspidate, persistent, ciliated, apex stellate with long bristles; petals pale purple, 2-3.5 mm long, broadly elliptic to obovate; stamens 8, anthers not beaked. Capsules 1 × 0.6 cm, ovoid, bristly at top; seeds numerous, minute, muriculate.

Rare in the moist deciduous forests of Mahabubnagar district.

Fl. & Fr.: August- November

Osbeckia zeylanica L. f., Suppl. Pl. 215. 1781; C.B. Clarke in Hook. f., Fl. Brit. India 2: 516. 1879; Gamble, Fl. Madras 1: 494. 1919.

Herbs, up to 30 cm high, branches quadrangular, striate, hirtellous along angles. Leaves oblong or lanceolate, 2-2.5 × 0.5-1 cm, membranous, 3-nerved, hirsute, base subacute-cuneate, margin entire-serrulate, apex acute, petiole to 3.5 mm. Flowers pinkish-purple, axillary, solitary, often 3-5 clustered at the ends of branches in leafy capitate heads; calyx-tube with both simple and stellate bristles, lobes ciliate, the apices stellate-bristly; petals 4, pink to light purple, broadly obovate, 8 × 3 mm, ciliate at apex; stamens 8, filaments to 4 mm, anthers beaked, beak to 1mm; ovary 2.5 mm, style to 1 cm. Capsules ribbed, 0.8 x 0.6 cm, ovoid, seeds numerous, minute, finely muriculate.

Occasional in Medak, Karimnagar and Hyderabad districts.

Fl. & Fr.: November-March

Pocharam tank (MDK), *TP* & *CP* 12030; Pandivagu, Narsapur RF (MDK), *BR* 11201; Aklaspur (KRN), *GVS* 22485 (MH).; Bandirevu RF (KMM), *RCS* 98924 (BSID).

LYTHRACEAE

1. Herbs; flower(s) solitary, in cymes or racemes:

    2. Flowers in axillary cymes:

        3. Stamens inserted at the base of the calyx tube; seeds hollowed on one side .... **Nesaea**

        3. Stamens inserted in the middle of the calyx tube; seeds flat .... **Ammannia**

    2. Flower(s) solitary in leaf axils or in racemes .... **Rotala**

1. Trees or shrubs; flowers in panicles :

    4. Flowers not zygomorphic; calyx-tube straight; leaves not black dotted beneath:

        5. Flowers tetramerous; stamens 8; capsule irregularly dehiscent; seeds pyramidal, wingless .... **Lawsonia**

        5. Flowers hexamerous; stamens numerous; capsule regularly dehiscent; seeds flat, winged .... **Lagerstroemia**

    4. Flowers zygomorphic; calyx tube bent; leaves black-dotted beneath .... **Woodfordia**

AMMANNIA L.

1. Leaves tapering at base; seldom basal lobes auricled; petals absent:

    2. Herbs; leaf base cuneate; flowers pedicellate at anthesis .... **A. baccifera**

    2. Sub-shrubs; leaf base sub-cordate; flowers sub sessile .... **A. aegyptiaca**

1. Leaves cordate at base; basal lobes auriculate; petals present .... **A. multiflora**

Ammannia aegyptiaca Willd., Hort. Berol. t. 6. 1809. *A. baccifera* L., subsp. *aegyptica* (Willd.) Koehne, Bot. Jahrb. Syst. 1: 260. 1880; C.B. Clarke in Hook. f., Fl. Brit. India 2: 569. 1879; Gamble, Fl. Madras 1: 510. 1919. *A. salicifolia* sensu C.B. Clarke in Hook. f., Fl. Brit. India 2: 569. 1879. non Blume 1856.

Erect or decumbent herbs, up to 50 cm high, usually much branched, branches quadrangular. Leaves opposite, sessile-subsessile, linear-elliptic, oblanceolate, 1-4 × 0.4-1 cm, becoming smaller towards the apex, base cuneate, margin entire, apex acute, glabrous. Flowers green with reddish-tinge in axillary clusters, often forming whorls, pedicels at centre, rarely 1 mm long, sepal tube broadly bell-shaped or obpyramidal, 1-2.5 mm long, 1-2.5 mm in diameter, with 8 longitudinal ribs; sepal appendages absent; petals minute or absent; stamens 4, included, filaments red, anthers yellow. Capsules 5 × 0.5 cm, linear, sparsely pilose; seeds numerous, minute with prominent raphes, bright red.

Common in all districts, in wet places.

Fl. & Fr.: March- December.

Satnella river (ADB), *PVP* & *GO* 4230; Nirmal – Adilabad way (ADB), *MHR* 14487; Ghanapur RF (NZB), *BR* & *KH* 6495; Sarvapur (NZB), *TP* & *BR* 6369; Patancheru (MDK), *BR* 11304, 11305; Kodimial (KRN), *GVS* 21846 (MH); Karkalpadu RF (MBNR), *SRS* 104553 (BSID); Panaparthi RF (KMM), *RR* 112504 (BSID).

Ammannia baccifera L., Sp. Pl. 120. 1753. C.B. Clarke in Hook. f., Fl. Brit. India 2: 569. 1879; Gamble, Fl. Madras 1: 510. 1919.

Erect annual herb, 15-40 cm high, stems and branches slender. Leaves 1.5-4. × 0.3-1.3 cm, linear-oblong or lanceolate, base attenuate – cuneate. Flowers reddish, in axillary clusters, forming whorls; calyx campanulate, 4-lobed; petals absent; stamens 4; ovary 4-celled, style ca 0.5 mm. Capsules depressed-globose, red.

Occasional in Adilabad, Nizamabad, Medak, Ranga Reddi, Karimnagar and Nalgonda districts.

Fl. & Fr.: September – March.

Potchera stream (ADB), *GO* & *DAM* 5062; Rajakampet (NZB), *BR* & *KH* 9054; Pochammaralu (MDK), *BR* & *CP* 11492; Patancheru (MDK), *BR* & *CP* 11304, 11305; Anantasagar RF (RR), *MSM* & *KH* 11007; Nagarjunasagar (NLG), *KMS* 19158 (MH).

Ammannia multiflora Roxb., Fl. Ind. 1: 426. 1820; C.B. Clarke in Hook. f., Fl. Brit. India 2: 570. 1879; Gamble, Fl. Madras 1: 509. 1919.

Erect glabrous herbs, up to 65 cm high, with tetragonous, slightly winged branches. Leaves sessile, linear to linear-oblong or elliptic-oblong, 1-2 × 0.2-0.4 cm, chartaceous, base auriculate, subcordate, margin entire, apex obtuse. Flowers minute, red in axillary cymes; pedicels up to 2 mm long, but some flowers sessile or nealry so; calyx tube broadly campanulate, 1-2 mm in diameter, 1-1.5 mm long, 8 - ribbed at anthesis; petals 4, pinkish, up to 0.5 mm long; stamens 4. Capsules globose, 1-2 mm broad, exceeding calyx-tube; seeds numerous, minute, red, smooth, hemispherical.

Common in Adilabad, Nizamabad, Hyderabad, Karimnagar and Mahabubnagar districts in wet places.

Fl. & Fr.: August- April.

Jannaram (ADB), *TP* & *PVP* 4202; Kunthala way (ADB), *MHR* 13271; Nirmal fort (ADB), *MHR* 14422; Gandhari RF (NZB), *TP* & *BR* 7227, 7235; Gummadidala (MDK), *BR* & *CP* 14116; Kammarapally (NZB), *TP* & *BR* 6234.

LAGERSTROEMIA L.

1. Shrubs; calyx with a distinct ring demarcating the lobes from the tube; flowers pinkish purple, or white .... **L. indica**

1. Trees; calyx not as above:

    2. Flowers white; calyx-tube smooth in capsule:

        3. Leaf apex obtuse-retuse .... **L. parviflora**

        3. Leaf apex acute to acuminate .... **L. microcarpa**

    2. Flowers purplish-lilac; calyx tube ribbed in capsule .... **L. speciosa**

Lagerstroemia indica L., Amoen. Acad. 4 : 137. 1759, Syst. Pl. (ed. 10) 1076. 1759 & Sp. Pl. (ed.2) 734. 1762; Hook.f. Fl. Brit. India 2 : 575. 1879; Gamble, Fl. Madras 1 : 513. 1919.

Shrub with spreading or ascending branches, 2-3 m high. Leaves oblong, 4-7.5 x 2-4.5 cm. rounded or subcordate at base, obtuse, mucronate; petiole 2-3 mm long. Flowers in short axillary and terminal, 2-3-flowered paniculate cymes; peduncles to 1 cm long, pilose; pedicels 3.5 mm long; calyx campanulate, 8-10 mm long, reddish green, teeth triangular-acute, 3-4 mm long, erect in fruit; petals white or purple, broadly orbicular with crisped margins. Capsules globose, 1-1.5 to cm in diam. acute at apex; seeds compressed, winged at one end, brown.

Grown in gardens and house yards as an ornametal shrub for showy flowers.

Fl. & Fr .: May-July.

Note: Two varities are under cultivation, one with flowers white and other with flowers pinkish purple.

Lagerstroemia microcarpa Hance, J. Bot. 16: 107. 1878.

Deciduous trees, to 30 m high, bark 6-8 mm thick, greyish or greyish-white, smooth, peeling off in thin long and broad flakes. Leaves opposite, distichous, 4.5-10 x 3.7-6.5 cm, elliptic, ovate, elliptic-lanceolate or ovate-lanceolate, base attenuate or acute, margin entire, apex acute or acuminate, glabrous and shining above, velvety pubescent beneath, chartaceous. Flowers white, in axillary and terminal panicles; calyx smooth, hoary, without a ring; lobes 6, triangular, persistent, reflexed; petals 6, 3 mm long, obovate, white, clawed; stamens numerous, inserted at the base of calyx tube; filaments long, exserted; ovary half inferior, sessile, glabrous, 4-6-celled, ovules many; style long, curved; stigma capitate. Fruit a capsule, 8-12 mm long, ovoid, 4-6 valved, dehiscent; seeds many, elongate, falcately winged, brownish.

Occasional in deciduous forests of Medak district.

Fl. & Fr.: June-February

Hamsanpalli Extension (MDK), *RG* 104067 (BSID); Medak (MDK), *RG* 104080 (BSID).

Lagerstroemia parviflora Roxb., Pl. Coromandel t. 66. 1796; C.B. Clarke in Hook. f., Fl. Brit. India 2: 575. 1879; Gamble, Fl. Madras 1: 513. 1919.

Vern.: *Chennangi, Gullakaraka, Chinnagi.*

Deciduous trees, up to 20 m high, bark ash-coloured, peeling in thin vertical flakes; branchlets puberulous. Leaves opposite decussate, lanceolate or elliptic, coriaceous, glabrous and shining above, pale or greyish, some times hoary–tomentose beneath, base cuneate, subacute, margin entire, apex obtuse- retuse, secondary nerves 6-8 pairs, arcuate, prominent beneath, petiole to 4 mm. Flowers white in terminal and axillary, 8-12 cm long panicles, fragrant; calyx tube 7 mm, with a ring inside; lobes 6, to 3 mm; petals 6, white, ovate–suborbicular, to 6 mm; stamens many, filaments to 1 cm; ovary 2.5 mm, style to 1 cm. Capsule ellipsoid, brown, 1-3 × 1-1.5 cm, dehiscent by 3 or 4 valves, seeds many, flattened, winged at tips.

Common in deciduous forests.

Fl. & Fr.: May-November.

Satrugada (ADB), *GO* & *PVP* 4253; Nasarullabad RF (NZB), *TP* & *BR* 6166; Nadipally (NZB), *BR* & *KH* 6452; Nathyanipalli (MDK), *BR* & *CP* 11251; Mohammadabad RF (RR), *MSM* & *KH* 10214; Mothugudem (KMM), *SSR* 15243; Kambalpally (NLG), *BVRS* 622; Kodimial (KRN), *GVS* 20072 (MH); Near Vatavarlapalli (MBNR), *VBH* 84958 (BSID).

Lagerstroemia speciosa Pers., Syn. Pl. 2: 72. 1807. *Munchausia speciosa* L., Hausvater 5: 257. 1770. *Lagerstroemia reginae* Roxb., Pl. Coromandel t. 65. 1796. *L. flos-reginae* Retz., Observ. Bot. 5. 25. 1788. *p.p.;* C.B. Clarke in Hook. f., Fl. Brit. India 2: 577. 1879; Gamble, Fl. Madras 1: 513. 1919.

Vern.: *Varagogu.*

Large deciduous trees, up to 15 m high, bark pale, smooth, flaking of in regular pieces. Leaves opposite, oblong-lanceolate or elliptic, 10-20 ✕ 3-7.5 cm, base acute or rounded, margin entire, apex acuminate, glabrous, finely reticulate on both surfaces, green beneath, secondary nerves 10- 15 pairs, conspicuous, curving upwards and forming loops; petiole to 5 mm. Flowers purplish-lilac, in large terminal panicles; sometimes reaching 30 cm even; bracts 2; bracteoles 2 on the pedicel; calyx tube ribbed outside, without any ring inside, 0.8 cm long (in fruit), lobes persistent; petals 2.5 cm or more long; stamens subequal. Capsule broadly ovoid, 1.5-2.5 cm long, the lower third enclosed in the persistent calyx; seeds winged, glabrous, pale brown.

Cultivated as an ornamental tree, sometimes for timber.

Fl. & Fr.: April-August.

On the way to Bhadrachalam (KMM), *RCS* 98955 (BSID).

LAWSONIA L.

Lawsonia inermis L., Sp. Pl. 340. 1753; Gamble, Fl. Madras 1: 514. 1919. *L. alba* Lam., Encycl. 3. 106. 1789; C.B. Clarke in Hook. f., Fl. Brit. India 2: 573. 1879.

Vern.: *Gorinta.*

Bushy shrubs, up to 3 m high, branchlets spine-tipped, with greyish-white smooth bark. Leaves opposite decussate, obovate-elliptic, 1.5-2.5 ✕ 0.5-1 cm, base attenuate, margin entire, apex acute, subsessile. Flowers white or creamy-white in terminal panicle, fragrant; pedicel 3 mm; calyx-tube cupular, lobes 4, spreading; petals 4, orbicular or obovate; stamens 8, inserted in pairs on the rim of calyx tube; ovary 4-celled, style 5mm. Capsules globose, glabrous, smooth; seeds minute, brown, smooth, glabrous.

Cultivated as hedge plant, in all plains districts, at times an escpe.

Fl. & Fr.: February-June.

Kalapalli (ADB), *GO* & *DAM* 5146; Laxmapur (NZB), *TP* & *BR* 6271; Pochammaralu (MDK), *BR* & *CP* 11495; Anantasagar RF (RR), *MSM* & *NV* 10386;

Appapur (MBNR), *TDCK* 16843; Nelkal (NLG), *BVRS* 457 (NU); Kinnerasani Wild Life Sanctuary (KMM), *RR* 112696 (BSID); Mehabubghat (WGL), *RKP* 11506 (BSID).

NESAEA Commers. ex Kunth

Nesaea lanceolata (B. Heyne ex C.B. Clarke) Koehne in Engler Bot. Jahrb. Syst. 3: 325. 1882; Gamble, Fl. Madras 1: 510. 1919. *Ammannia lanceolata* B. Heyne ex C.B. Clarke in Hook. f., Fl. Brit. India 2: 570. 1879.

Erect or decumbent herbs, up to 40 cm high. Leaves opposite decussate, oblong-elliptic, 1-3.5 ✕ 0.4-1 cm, base attenuate, margin entire, apex obtuse. Flowers pink, tetramerous, axillary, solitary or dichasial cymes; bracts lanceolate to almost oblong, as long as or longer than sepals; sepal tube bell–shaped, 4 or 5-angular, 2.5 – 3 mm long, glabrous or hairy, sepal lobes 4 or 5, as long as or shorter than the tube, sepal appendanges shorter or longer than the lobes, usually with hairs at the tips; petals 4 or 5, pink, obovate to sub-orbicular, half as long as or shorter than sepal lobes; stamens 4 or 5; anthers 0.5 mm long. Capsule globose, or nearly so, 2 mm in diameter; seeds 35-40, compressed, brown to reddish- brown,.

Occasional in moist places in Nizamabad and Hyderabad districts.

Fl. & Fr.: August-November.

Gandhari RF (NZB), *BR* 7230; Sirnapalli RF (NZB), *BR* & *PSPB* 9574.

ROTALA L.

1. Capsule dehiscing by 2 valves :

　　2. Flowers in terminal, long racemes .... **R. serpyllifolia**

　　2. Flowers solitary, axillary .... **R. indica**

1. Capsule dehiscing by more than 2 valves :

　　3. Capsule dehiscing by 4 valves :

　　　　4. Flowers axillary, solitary, petals minute, subulate .... **R. illecebroides**

　　　　4. Flowers in close terminal spikes; petals large, broadly obovate .... **R. rotundifolia**

　　3. Capsule dehiscing by 3 valves :

　　　　5. Leaf base cuneate .... **R. mexicana**

　　　　5. Leaf base cordate or auriculate :

　　　　　　6. Capsule as long as calyx-tube; seeds numerous .... **R. densiflora**

　　　　　　6. Capsule exceeding calyx tube; seeds 10-12 .... **R. fimbriata**

Rotala densiflora (Roth ex Roem. & Schult.) Koehne in Engler, Bot. Jahrb. Syst. 1: 164. 1880; Gamble, Fl. Madras 1: 508. 1919. *Ammannia densiflora* Roth ex Roem. & Schult. in L., Syst. Veg. 3 : 304. 1818.

Erect, simple or branched, slender, flaccid glabrous herbs, 10-40 cm high, sometimes creeping below, stems 4-angled, winged. Leaves decussate, linear-

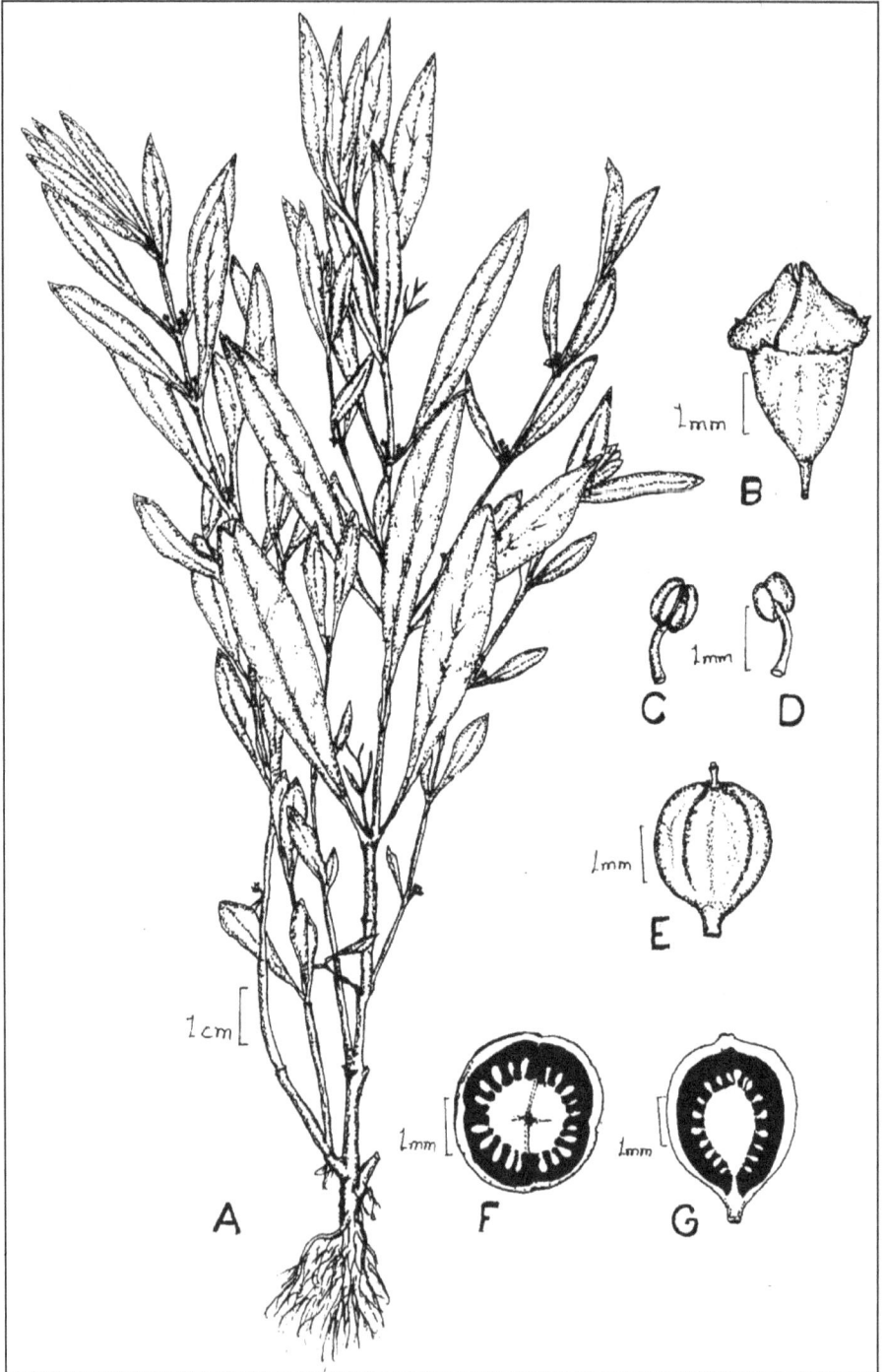

Figure 35: Nesaea lanceolata (B. Heyne ex C.B. Clarke) Koehne

A. Habit, B. Flower-entire, C&D. Stamens, E. Pistil, F. Ovary c.s., G. Ovary l.s.

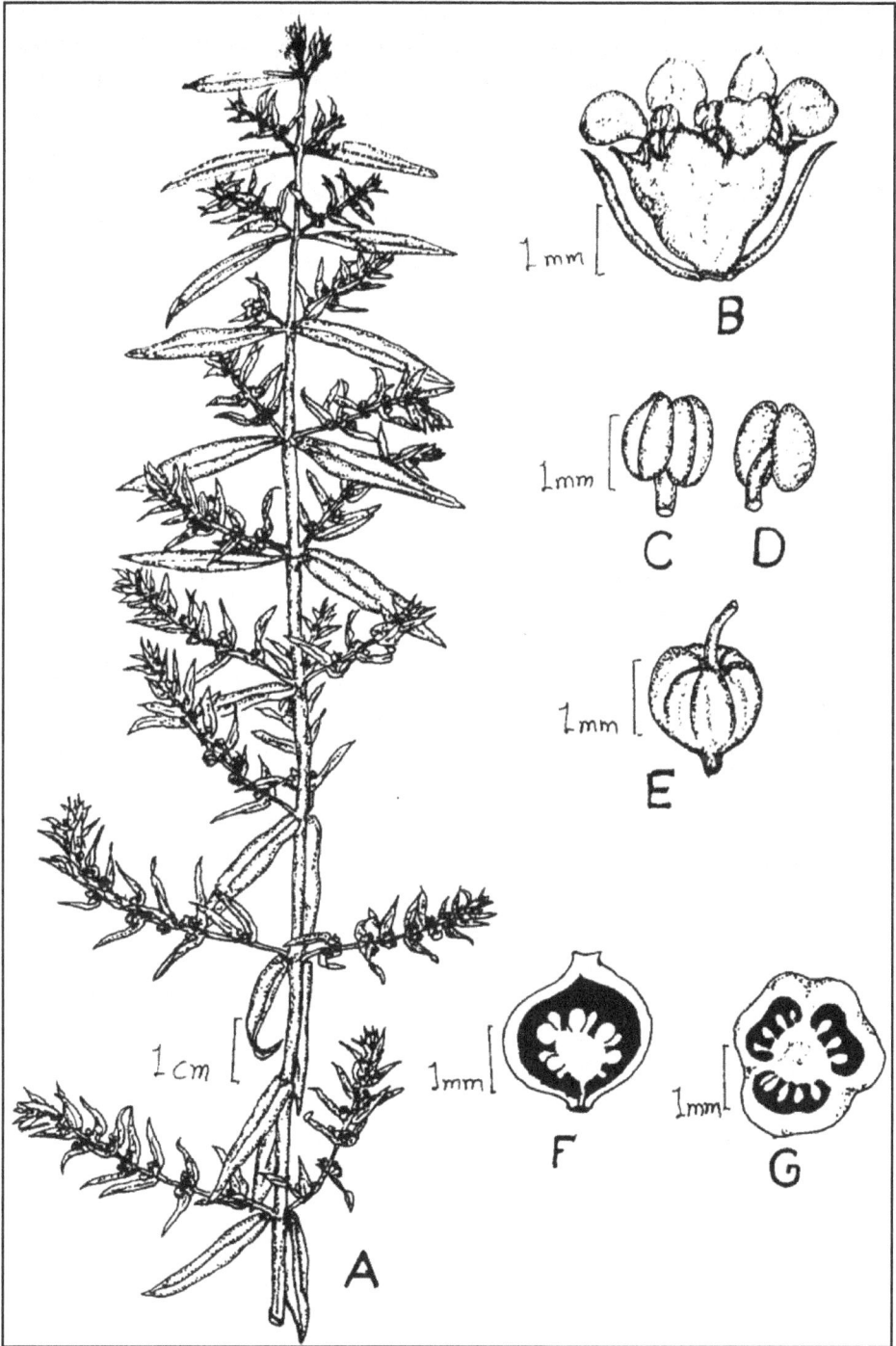

Figure 36: Rotala densiflora (Roth ex Roem. & Schult.) Koehne

A. Twig, B. Flower-entire, C&D. Stamens, E. Pistil, F. Ovary l.s., G. Ovary c.s.

lanceolate, 0.5-2 × 0.1- 0.4 cm, base subcordate-obtuse, margin entire, apex subacute or obtuse. Flowers pink, solitary, axillary; floral bracts almost dimorphic on main stem and lower branches like foliar; on upper branches bracts like and secondary exceeding the flowers; bracteoles longer than sepals with distinct – midribs; calyx tube campanulate, 1 mm long, lobes 5 or rarely 4, triangular to shallowly triangular, 0.5 mm long, sepal appendages linear or capillary, as long as or up to 2 times longer than sepal lobes or very rarely rudimentary; petals 5 or rarely 4, bright pink or rarely white, 0.5 - 1 mm long, at least as long as the sepal lobes; stamens 5 or rarely less, inverted on the lower half of the sepal tube; anthers inserted. Capsule globose, as long as calyx tube, dehiscing by three valves; seeds numerous, reddish-pale brown, glabrous.

Common in most of the plains districts in moist places.

Fl. & Fr.: September-October.

Annaram (NZB), *BR* & *CPR* 717; Maddigunta (NZB), *BR* & *GO* 9560, 9562; Kothapalli RF (RR), *MSM* & *KH* 10562; Bhupatipur (KRN), *GVS* 22476 (MH); Pakhal (WGL), *ANH* 15914 (MH); Hyderabad (HYD), *Campbel* 1834 (DD).

Rotala fimbriata Wight, Icon. Pl. Ind. Orient. t. 217. 1839; Gamble, Fl. Madras 1: 508. 1919. *Ammannia pentandra* Roxb. var. *fimbriata* (Wight) C.B. Clarke in Hook. f., Fl. Brit. India 2: 569. 1879.

Erect herbs, up to 60 cm high, creeping below, stem weakly 4-angled. Leaves decussate, lanceolate, 1.5 × 0.3 cm, lower ones cuneate to cordate at base, upper ones cordate or auriculate at base, margin entire, apex obtuse-acute. Flowers yellow, solitary, sessile, axillary; floral bracts like upper leaves, usually some what auriculate at base, bracteoles acute, triangular to subulate, rarely more than 1mm long; calyx tube campanulate, 2-2.5 mm long, sepal lobes 5 or rarely 6, dentate, 0.5 mm long, sepal appendanges absent; petals 5 or rarely 6, bright pink, pinnately divided into linear segments, 2-3 mm long; stamens 5 or rarely 6, inserted at the base of the sepal tube; anthers exserted. Capsule 4 mm, dehiscing by 3-valves, exceeding calyx tube; seeds 10-12, compressed, granular.

Rare in Adilabad, Warangal, Nizamabad, Medak, Karimnagar, Mahabubnagar and Hyderabad districts.

Fl. & Fr.: January –March.

Nirmal – Adilabad road (ADB), *MHR* 14447; Mingaram RF (NZB), *BR* 9291; Kondagutta (KRN), *GVS* 25687 (MH); Medak to Bodhan (MDK), *RG* 113529 (BSID).

Rotala illecebroides (Arn. ex C. B. Clarke) Koehne in Engler, Bot. Jahrb. Syst. 1: 161. 1881; Gamble, Fl. Madras 1: 508. 1919. *Ammannia pentandra* Roxb. var. *illecebroides* Arn. ex C.B. Clarke in Hook. f., Fl. Brit. India 2: 569. 1879.

Erect, annual herbs up to 10 cm high; stem creeping below; branches slender, 4-angled or 4-winged. Leaves broadly ovate–cordate, 2-4 x 1-3 mm, semi-amplexicaul at base, acute. Flowers sessile, solitary in the axils of bracts; calyx campanulate; tube less than 1 mm long; teeth 4, triangular-acuminate, 1 mm long, apiculate at apex;

appendages 4, linear, about as long as the calyx teeth; petals pale pink, obovate, sometimes absent. Capsules globose, 1 mm in diam. 3-4-valved; seed ellipsoidal, 0.3 mm long, brown.

Rare in Warangal, Medak and Hyderabad districts.

Fl. & Fr.: December- January.

Edupayalu RF (MDK), *RG* 113450 (BSID).

Rotala indica (Willd.) Koehne in Engler, Bot. Jahrb. Syst. 1: 172. 1880; Gamble, Fl. Madras 1: 508. 1919. *Peplis indica* Willd., Sp. Pl. 2 : 244. 1799. *Ammannia peploides* Spreng., Syst. Veg. 1: 444. 1825; C.B. Clarke in Hook. f., Fl. Brit. India 2: 566. 1879.

Prostrate or suberect herbs, up to 30 cm high, rooting at lower nodes, stem weakly 4-angled or terete. Leaves decussate, obovate, spathulate, suborbicular, or elliptic-lanceoalte, 6-12 × 3-8 mm, base cuneate or obutse, rarely cordate, margin entire, cartilaginous, apex rotund or emarginate. Flowers small, bright pink or red, axillary, solitary, sessile; bracts polymorphic, leaf like, bracteole solitary in axils of bracts; calyx tube 1.5 mm, appendages 0, lobes 4, lanceolate, 1mm, petals 4, rose to pink, obovate, 0.5 mm; stamens 4, filaments 1 mm, ovary 0.8 mm; style 0.5 mm. Capsule minute, ellipsoidal, smooth, glabrous; seeds very minute, narrowly ellipsoid, pale-brown or reddish-brown, glabrous.

Common in Warangal and Hyderabad districts in rice fields and other wet places.

Fl. & Fr.: September-January.

Banks of Pakhal lake (WGL), *KMS* 11678 (MH); Pakhal (WGL), *ANH* 15913 (MH).

Rotala mexicana Cham. & Schltdl., Linnaea 5: 67. 1830. *R. rosea* (Poir.) C. D.K. Cook. & H.Hara, Enum. Fl. Pl. Nepal 2: 173. 1979. *Ammannia rosea* Poir. in Lam., Encycl. (Suppl.) 329. 1810. *Rotala leptopetala* (Blume) Koehne in Engler Bot. Jahrb. Syst. 1: 162. 1880; Gamble, Fl. Madras 1: 508. 1919. *Ammannia leptopetala* Blume, Mus. Bot. 2: 134. 1856. *A. pentandra* Roxb., Fl. Ind. 1: 427. 1832; C.B. Clarke in Hook. f., Fl. Brit. India 2: 568. 1879. *Rotala pentandra* (Roxb.) Blatter & Hallb., J. Bombay Nat. Hist. Soc. 25 : 707. 1918. *p.p.*

Erect herbs, up to 30 cm high, sometimes creeping at base, simple or branched from below. Leaves decussate, lanceolate, 0.5-2 × 0.2-0.6 cm, upper floral leaves becoming bractiform, oblong, lower leaves broader; base cuneate, margin entire, apex obtuse. Flower solitary, axillary, sessile, pink in colour; floral bracts like foliage leaves and exceeding the flowers, bracteoles shorter than calyx tube, with out a midrib; calyx tube campanulate, lobes 5, rarely 4, greenish or pinkish, shallowly triangular, 0.25 mm long, sepal appendages subulate, longer, rarely shorter than the sepal lobes or sometimes rudimentary; petals 5 or rarely 4 or rudimentary, pink, distinctly longer than wide, 0.25 mm long, as long as or shorter than the sepal lobes; stamens 5 or less; anthers included with in the sepal tube. Capsule globose, 1.5 mm, dehiscing by 3-valves, exceeding calyx tube, seeds 20-25, compressed, 0.5 mm, yellow to straw – coloured.

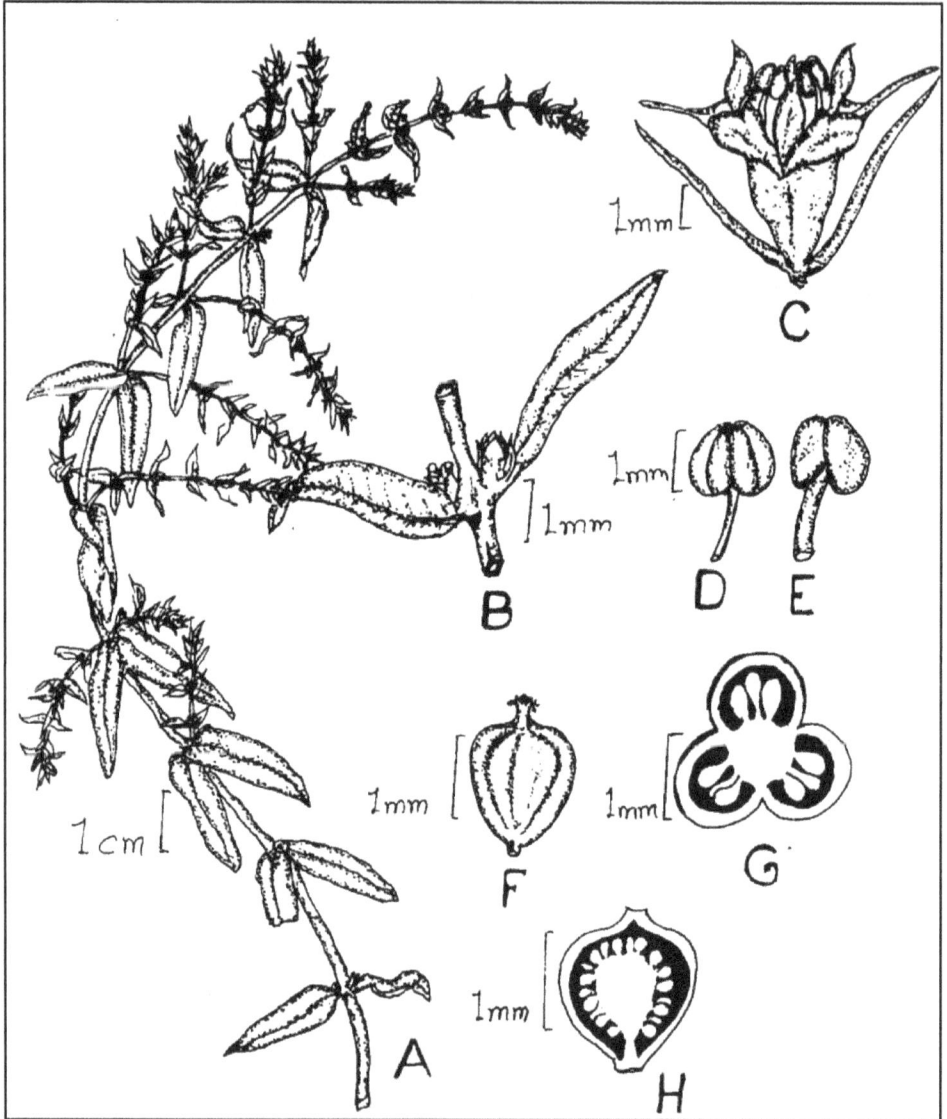

Figure 37: Rotala mexicana Cham. & Schltdl [=Syn. *Rotala rosea* (Poir.) C. D.K.Cook]
A. Twig, B. Flowers-enlarged, C. Flower-entire, D&E. Stamens, F. Pistil, G. Ovary c.s., H. Ovary l.s.

In moist places, ponds and river banks in Adilabad, Nizamabad, Medak and Hyderabad districts.

Fl. & Fr.: December- April.

Jannarum (ADB), *TP* & *PVP* 4197; Kubgall RF (NZB), *TP* & *BR* 6202; Ooracherau lake, Narsapur (MDK), *KMS* 6756 (MH); Parnasala RF (KMM), *RCS* 102469 (BSID).

Rotala rotundifolia (Buch.-Ham. ex Roxb.) Koehne in Engler, Bot. Jahrb. Syst. 1: 175. 1880; Gamble, Fl. Madras 1: 509. 1919. *Ammannia rotundifolia* Buch.-Ham. ex Roxb., Fl. Ind. 1: 446. 1820; C.B. Clarke in Hook. f., Fl. Brit. India 2: 566. 1879 *p.p.*

Creeping or decumbent herbs, up to 20 cm long, thick, branched stems at base. Leaves (aerial) opposite, sessile or shortly petioled, orbicular-ovate, 5-9 × 3.5-8 mm, base subcordate, margin entire, apex rotund; submerged leaves linear to orbicular. Flowers rose-coloured in dense terminal spikes; bracts ovate, 1-4 mm long; bractoles to 1.5 mm, equal to calyx tube; calyx tube campanulate, to 1.5 mm long, lobes 4, 1 mm; petals 4, pink, ovoid – suborbicular, to 1.5 mm; stamens 4, filaments 1.5 mm; ovary to 0.5 mm, style 0.2 mm. Capsule ellipsoid, a little longer than broad, equal to calyx tube, dehiscing by 4 valves; seeds 15-20, brown-yellow, ellipsoid, peltate.

Common in Ranga Reddi district, in rice fields and other wet places.

Fl. & Fr.: April-July.

Rotala serpyllifolia (Roth) Bremek., Acta Bot. Neerland 3 : 149. 1953. *Micranthus serpyllifolia* Roth, Nov. Pl. Sp. 282. 1821. *Ammannia tenuis* C.B. Clarke in Hook. f., Fl. Brit. India 2: 567. 1879.

Very slender, erect, 10-30 cm high, glabrous herbs, flowering shoots red and shining. Leaves opposite, subsessile, broadly ovate-elliptic, 2-15 mm long, glabrous, obtuse or acute, base cuneate, tip acute or obtuse. Flowers minute, bright-pink or red, in terminal, slender, 2-8 cm long racemes, on filiform peduncles; floral bracts green or tinged with red, smaller than foliage leaves, rarely exceeding 4 mm long, lanceolate to oblong; bracteoles caducous, linear-lanceolate, 1.5 mm long, sepal tubular like, constricted at the tip, 2.5 – 3 mm long, sepal lobes 4, widely deltate, 0.5 mm long, sepal appendages absent; petals 4, obovate, somewhat longer than sepal lobes; stamens 4, anthers included within the calyx tube. Capsule narrowly ellipsoid, much longer than broad, red, glabrous, smooth, 2-valved; seeds numerous, very minute, narrowly ellipsoid, pink or yellow.

Rare in moist localities in Adilabad, Nizamabad and Medak districts.

Fl. & Fr.: October-January.

Potchera water falls (ADB), *GO* & *DAM* 5071; Sirnapalli river (NZB), *BR* & *GO* 9075; Pochammaralu (MDK), *BR* & *CP* 11491.

WOODFORDIA Salisb.

Woodfordia fruticosa (L.) Kurz., J. Asiat. Soc. Bengal, Pt. 2, Nat. Hist. 40: 56. 1871; Gamble, Fl. Madras 1: 511. 1919. *Lythrum fruticosum* L., Sp. Pl. 641. 1762. *Woodfordia floribunda* Salisb., Parad. Lond. t. 42. 1806; C.B. Clarke in Hook. f., Fl. Brit. India 2: 572. 1879.

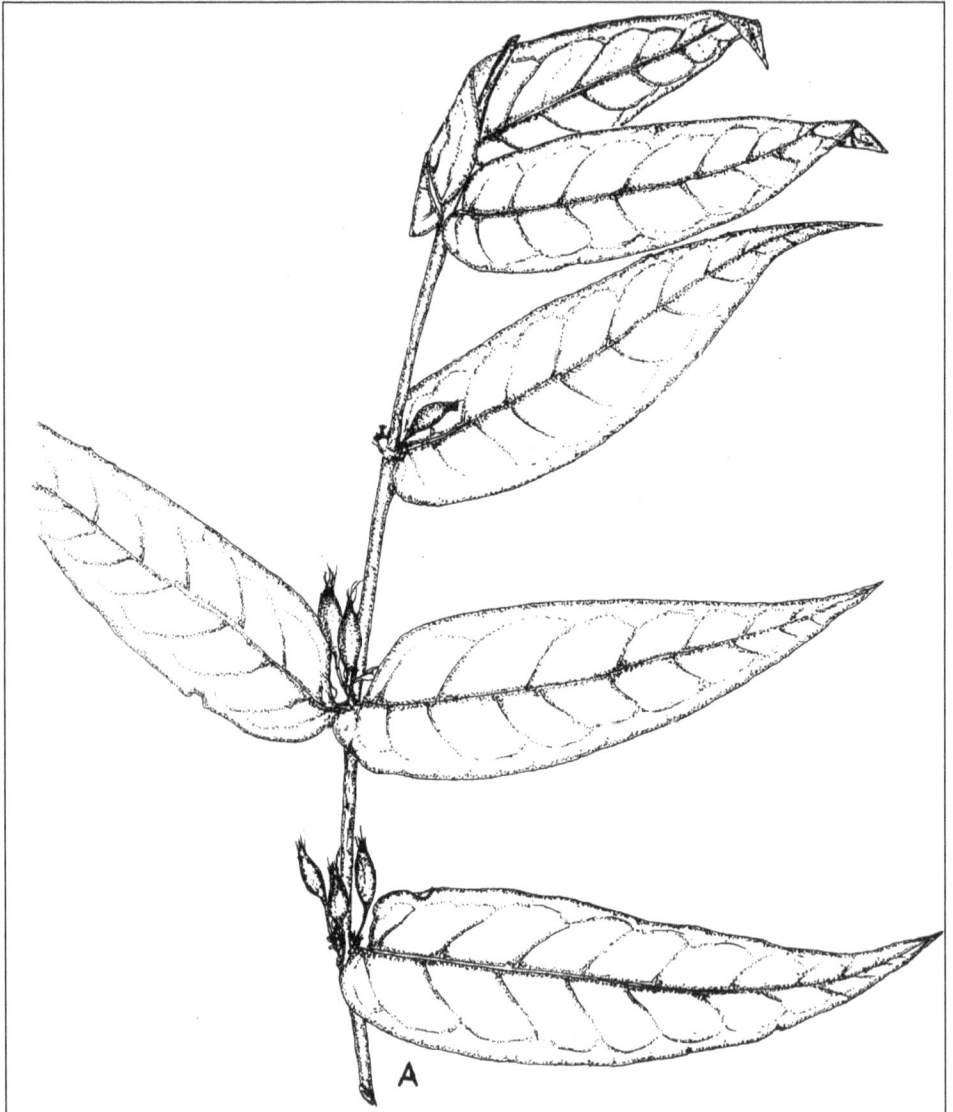

Figure 38: Woodfordia fruticosa (L.) Kurz.
A. Twig.

Large woody shrubs, up to 4 m high; branches terete, bark peeling off in threads. Leaves opposite, lanceolate, 4-8 × 1-2.5 cm, base rounded or cordate, margin entire, apex acuminate, strigose above, tomentose and black-glandular beneath, laterals join to form intramarginal vein. Flowers greenish red, in axillary panicled cymes; calyx 1.5 cm long, striate, covered with glandular dots, with a small campanulate base and a long slightly curved teeth ca 2.5 mm long, triangular, acute; petals greenish red, slightly longer than the calyx teeth, narrowly linear, produced at the apex to a long fine point. Capsules oblong or ellipsoid; seeds minute, pale- brown, numerous, smooth, excavated on one side.

Common in most districts in deciduous forests, on dry hill sides and rocky places.

Fl. & Fr.: November – May.

Satnella Dam (ADB), *GO* & *PVP* 4226; Laxmapur (NZB), *TP* & *BR* 6295; Pochammaralu (MDK), *BR* & *CP* 11475; Kinnerasani RF (KMM), *RR* 112693 (BSID); Ramanagutta (KMM), *RR* 108007; Killah (MDK), *RG* 104002 (BSID); Veerapuram (WGL), *PVS* & *NRR* 76916 (BSID).

PUNICACEAE

PUNICA L.

Punica granatum L., Sp. pl. 472. 1753; Hook.f., Fl. Brit. India 2: 581. 1879; Backer in Steenis, Fl. Males. I. 4: 226. 1951.

Vern.: *Danimma.* English: Pomegranate.

A large shrub; branchlets often spine-tipped. Leaves mostly decussate, simple; stipules 0. Flower bisexual, terminal, solitary or in clusters; calyx tubular, adnate to the ovary, lobes 5-7, valvate; petals 5-7, imbricate, and crumpled; stamens numerous, filaments free; ovary inferior, 3-7-celled, cells superposed in 2-series, ovules many, the upper ones with parietal placentae, style slender, stigma capitate. Berry large, globose or ellipsoid, crowned by clayx-lobes, many-seeded.

Cultivated for fruits.

Fl. & Fr.: March – May.

ONAGRACEAE

LUDWIGIA L.

1. Plants aquatic, floating; flowers pentamerous; capsule thick-walled .... **L. adscendens**

1. Plants marshy, erect, flowers tetramerous; capsule thin-walled:

  2. Stamens 4 .... **L. perennis**

  2. Stamens 8:

   3. Flowers less than 1 cm across; seeds apically 1-seriate, embedded in endocarp at base .... **L. hyssopifolia**

   3. Flowers more than 2 cm across; seeds multiseriate throughout, not embedded in endocarp .... **L. octovalvis**

Ludwigia adscendens (L.) Hara, J. Jap. Bot. 28: 291. 1953. *Jussiaea adscendens* L., Mant. Pl. 1: 69. 1767. *J. repens* L., Sp. Pl. 1: 69. 1753; non *Ludwigia repens* Forst. 1771; C.B. Clarke in Hook. f., Fl. Brit. India 2: 587. 1879; Gamble, Fl. Madras 1: 516. 1919.

Floating aquatic herbs, with white spongy floats, branchlets ribbed, glabrescent. Leaves alternate, variable, oblong-elliptic, 3-3.5 × 0.5-2 cm, glossy above, glabrous, base attenuate, margin entire, apex obtuse, petiole to 2 cm. Flowers yellow, axillary, solitary; peduncle to 3 cm; pedicel 3-5 cm; calyx 5, deltoid lobes, apex acuminate; petals 5, cream, with yellow blotches inside, obovate, base cuneate, apex emarginate or rounded; stamens 10, filaments sub-equal, anthers oblong, 2 mm; ovary 5-celled, ovules many, style 6 mm, hairy at base, stigma globose, obscurely lobed. Capsule 1-4 cm long, cylindric, thick-walled, 10-ribbed, glabrous; seeds smooth, reticulated, embedded in endocarp.

Frequent along moist ditches and streams in Nizamabad, Medak and Karimnagar districts.

Fl. & Fr.: June-January.

Pothangal (NZB), *BR* & *KH* 9646; Medak (MDK), *BR* & *CP* 11567; Indira park (HYD), *KI* & *NVN* 18741; Manuguru (KMM), *RCS* 104216 (BSID); Gunavaram (KMM), *RCS* 98941 (BSID).

Ludwigia hyssopifolia (G. Don) Exell, Garcia de Orta 5: 471. 1957. *Jussiaea hyssopifolia* G. Don, Gen. Syst. 2: 693. 1832. *J. linifolia* Vahl, Eclog. Amer. 2: 32. 1798, non *Ludwigia linifolia* Poir. 1813; Gamble, Fl. Madras 3: 1875. 1936.

Herbs or shrubs, up to 1 m high, branchlets glabrous. Leaves oblong-elliptic, 2-6 × 0.3-1.2 cm, with 10-12 pairs of lateral nerves, base attenuate, margin entire, apex acuminate, petiole to 1cm. Flowers yellow, solitary, pedicel to 4 mm; bracteoles deltoid, 1mm, at the base of calyx-tube; calyx-tube 1cm, lobes 4, lanceolate, 4mm, apex acuminate; petals 4, yellow, obovate, 3.5 mm, base cuneate, apex round-subtruncate; stamens 8, filaments 2 mm, short ones epipetalous; ovary 1 cm, 4-celled, dilated at apex, ovules many, style 2 mm, stigma globose or 4-lobed. Capsule subterete, 2 × 0.2 cm, thin-walled, 8-ribbed, puberulous, apex smooth, base torulose, seeds multiseriate, free and 1-seriate at the apex.

Common in forests in Karimnagar Khammam and Warangal districts.

Fl. & Fr.: December-January.

Manuguru RF (KMM), *RCS* 104217; Parnasala RF (KMM), *RCS* 99052 (BSID); Dhaniyayigudem (KMM), *RR* 108521 (BSID); 10 Km towards north from Pasra (WGL), *RKP* 108196 (BSID).

Ludwigia octovalvis (Jacq.) P.H.Raven, Kew Bull 15: 476.1962. subsp. **sessiliflora** (Mich.) P.H.Raven, Kew Bull. 15: 476. 1962. *Jussiaea octonervia* forma *sessiliflora* Mich. in Masters, Fl. Bras. 13 (2): 171. 1875. *J. suffruticosa* L., Sp. Pl. 388. 1753; C.B. Clarke in Hook. f., Fl. Brit. India 2: 587. 1879; Gamble, Fl. Madras 1: 516. 1919. *J. villosa* Lam., Encycl. 3. 331. 1789; Gamble, Fl. Madras 3: 1297. 1935.

Erect undershrubs, up to 3 m high, branchlets densely tomentose. Leaves oblong-elliptic, 2-6 × 0.3-1.2 cm, 10-12 pairs of lateral nerves, base attenuate, margin entire, apex acuminate, petiole 1.5 mm. Flowers axillary, solitary, tetramerous; pedicel 1.5 m; bracteoles subulate, at the base of calyx-tube; calyx-tube 2 cm, pubescent, lobes 4, lanceolate, 8 × 4 mm; petals 4, yellow, obovate, 1 × 0.8 cm, base cuneate, apex round; stamens 8, filaments to 3mm, short ones epipetalous, anthers oblong, 2 mm; ovary 2 cm, 4-celled, ovules many, style 3 mm, stigma subglobose. Capsule terete, 4 × 0.6 cm, thin-walled, 8-ribbed, pubescent; seeds multiseriate, free, ovoid.

Common along water courses and on rocky crevices of the streams in Adilabad, Nizamabad, Medak and Ranga Reddi districts.

Fl. & Fr.: August-March.

Birsaipet (ADB), *GO* 4602; Chandoor (NZB), *BR* & *GO* 9211; Gangapur RF (MDK), *TP* & *CP* 12092; Malkalchruvu (RR), *MSM* & *KH* 10226; Pullaechelma (MBNR), *SRS* 109710 (BSID).

Ludwigia perennis L., Sp. Pl. 119. 1753; excl. *follis oppositis. L. parviflora* Roxb., Fl. Ind. 1 : 440. 1820; C.B. Clarke in Hook. f., Fl. Brit. India 2: 888. 1879; Gamble, Fl. Madras 2: 517. 1919.

Herbs or subshrubs, up to 60 cm high, branchlets puberulous at apices. Leaves narrowly elliptic-lanceolate, 2-8.5 × 0.5-1.5 cm, chartaceous, glabrous, base attenuate, margin entire, apex acuminate, petiole to 1 cm. Flowers yellow, axillary, solitary; bracteoles subulate, 1 mm; pedicel 2 mm; calyx-tube 1cm, lobes 4, lanceolate, 4mm; petals 4, yellow, elliptic, to 4 mm, base cuneate, apex obtuse; stamens 4, filaments 2mm, anthers oblong, 6 mm; ovary 1 cm, 4-celled, ovules numerous, style 1.5 mm, stigma globose. Capsule pale-brown or straw-coloured; seeds numerous, minute, brown, smooth, glabrous.

Common almost in all districts in rice fields and other wet places.

Fl. & Fr.: August-January.

Jannaram (ADB), *TP* & *PVP* 4188; Kammarapally (NZB), *TP* & *BR* 6231; Yedapally (NZB), *BR* & *KH* 6472; Narsapur RF (MDK), *BR* & *CP* 11276; Sardhana (MDK), *CP* 11775; Mohammadabad (RR), *MSM* 10442; Pakhal (WGL), *KMS* 71671 (MH); Kambalpally (NLG), *BVRS* 232; Rechapally (KRN), *GVS* 22273; Aklaspur (KRN), *GVS* 25631 (MH); Poochavaram RF (KMM), *RCS* 102489 (BSID); Borapuram (MBNR), *BSS* & *SKB* 32366.

TRAPACEAE

TRAPA L.

Trapa natans L., Sp. Pl. 120. 1753. var. **bispinosa** (Roxb.) Makino, Bot. Mag. (Tokyo) 11: 283. 1897. *T. bispinosa* Roxb., Pl. Coromandel t. 234. 1815; C.B. Clarke in Hook. f., Fl. Brit. India 2: 590. 1879; Gamble,Fl. Madras 1: 518. 1919.

Aquatic floating herbs. Leaves dimorphic, submerged leaves opposite, root-like, pinnatipartite, with filiform segments, floating leaves broadly ovate-rhomboid, 2-4 ×

2.7-6 cm, dark green above, tomentose beneath, glabrous above, subcoriaceous; petiole 10 – 18 cm long, the inflated portion ca 2.5 cm long, hairy or woolly. Flowers white, above water surface, axillary, solitary. Nut 1.5- 2.5 cm long and as much as broad, obscurely 4-angled, glabrous, 2 or all angles spinescent, shortly beaked, 1-seeded.

In Central University campus in Ranga Reddi district (Seshagiri Rao, 2012).

Fl. & Fr.: September-December.

TURNERACEAE

TURNERA L.

1. Leaf-apex obtuse-rounded or subacute; bracteoles linear- subulate, 1-nerved, entire, without glands; pedicels adnate to petiole throughout .... **T. subulata**

1. Leaf-apex tapering, acuminate; bracteoles lanceolate, penninerved, serrate, with 2 large glands; pedicels adnate to petiole only at base, but free above .... **T. ulmifolia**

Turnera subulata Sm. in Rees, Cycl. 36. no.2. 1817. *T. ulmifolia* L. var. *elegans* (Otto) Urb., Monog. Turn. 139. 1883; Gamble, Fl. Madras 1: 523. 1919. *T. elegans* Otto in Nees, Hor. Phys. Berol 36. 1820.

Herbs, up to 80 cm high, branchlets densely downy-pubescent. Leaves ovate-oblong, 2-5 × 1-3 cm, chartaceous, 7-9-nerved, pubescent, base cuneate, slightly decurrent, margin crenate-dentate, apex obtuse-rounded or subacute; petiole up to 1 cm, with diskoid foliar glands. Flowers pale yellow, solitary, axillary; bracteoles linear-subulate, 1-nerved, 1 cm long, entire, without glands; pedicel and petiole adnate throughout; calyx tube to 7 mm; lobes linear, equal to calyx tube; petals 5, 1.5 × 1 cm. Capsules globose, to 5 mm, pubescent.

Occasional in urban areas on roadsides in Karimnagar district (Naqvi, 2001).

Fl. & Fr.: October-January.

Turnera ulmifolia L., Sp. Pl. 271. 1753. *T. ulmifolia* L. var. *angustifolia* (Mill.) Willd. ex Urb., Monog. Turn. 141. 1883; Gamble, Fl. Madras 1: 523. 1919. *T. angustifolia* Mill., Gard. Dict. ed. 8. n. 2. 1768.

Erect, much branched herbs or shrubs, up to 1 m high, branchlets glandular-pubescent. Leaves alternate, oblong or lanceolate-elliptic, 1.5 - 4.5 × 1.5-3.5 cm, chartaceous, 15-nerved, appressed pubescent, base cuneate, decurrent, margin 2-serrate, apex tapering, acuminate; petiole 1-3 cm. Flowers yellow, in axillary, few-flowered cymes; pedicel adnate to petiole only at base, but free above; calyx tube 1 cm, pubescent, lobes linear, to 1.5 cm; petals 5, yellow; stamens 5, filaments to 2.5 cm; ovary ovoid, pubescent, to 3 cm, styles 3, 1.5 – 2 cm, pubescent. Capsules globose, to 8 mm; seeds subterete, 2.5 mm, pitted.

Common weed of road sides, and waste ground.

Fl. & Fr.: June- November.

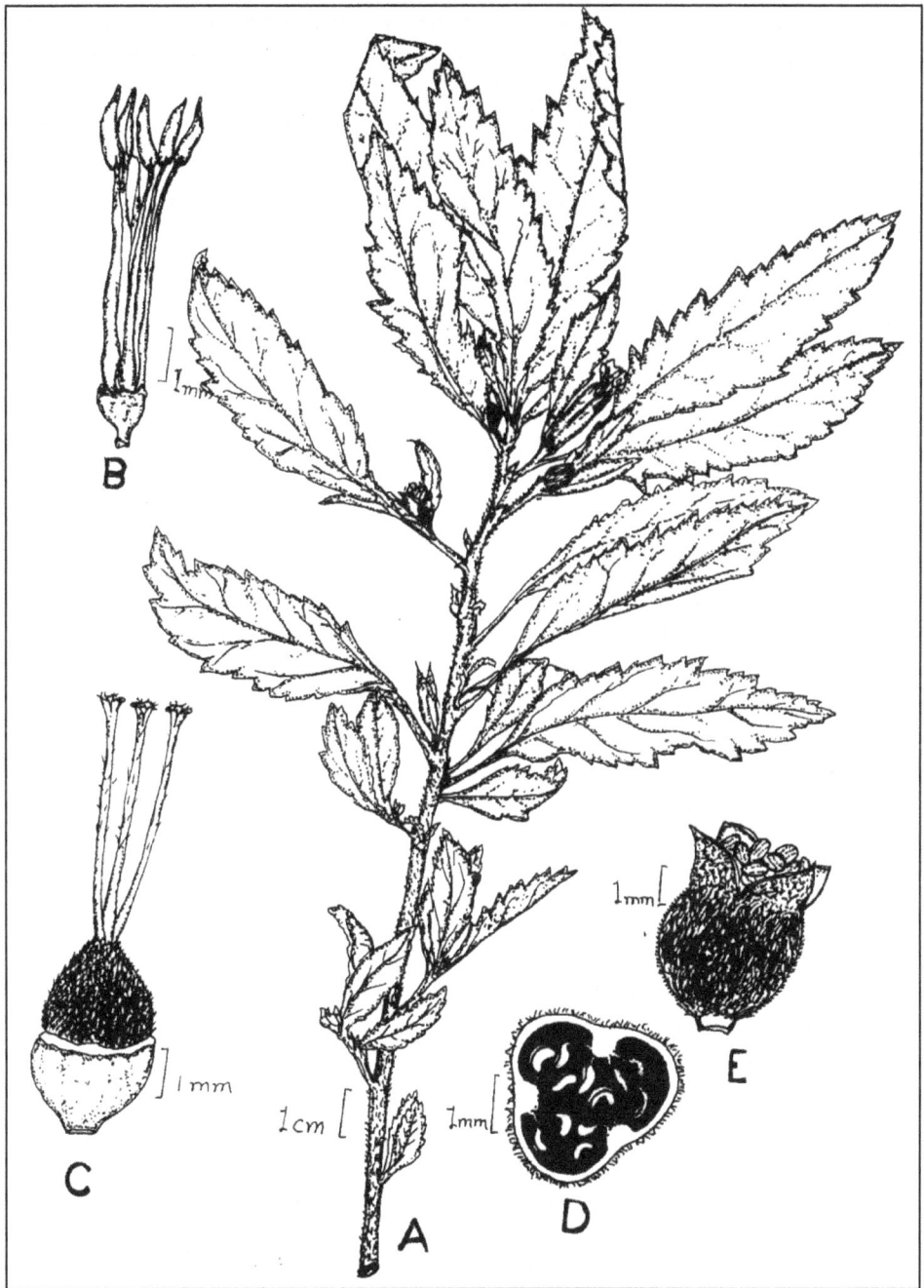

Figure 39: Turnera ulmifolia L.

A. Twig, B. Pistil & stamens, C. Pistil, E. Capsule, D. Ovary c.s.

Nirmal (ADB), *GO* & *BR* 9406; Lingampet (NZB), *BR* & *CPR* 9239; Medak (MDK), *CP* 11722; Venkat Reddipalli (RR), *MSM* & *KH* 10309; Rechapalli (KRN), *GVS* 22280 (MH); Perantapally RF (KMM), *RCS* 98996 (BSID); Medak (MDK), *RG* 104028 (BSID).

PASSIFLORACEAE

PASSIFLORA L.

1. Wild plants; stems and leaves pubescent, foetid .... **P. foetida**

1. Cultigens; stems and leaves not as above .... **P. edulis**

Passiflora edulis Sims, Curtis Bot. Mag. t. 1989.1818; Gamble, Fl. Madras 1 : 524. 1919; Bailey, Man. Cult. Pl. 691.1949.

Large herbaceous climber; stems often angled; tendrils simple. Leaves deeply palmately 3-lobed, 5-10 cm long, rounded or subcordate at base, serrate; petiole 2 cm long with 2 glands at apex; stipules linear. Flowers solitary, axillary, on long peduncles; sepals lanceolate, green outside, white within, horned on the back, 2-2.5 cm long; petals white, lanceolate, 3 cm long; coronal filaments double ring of bluish purple threads, about as long as the petals; stamens united into a tube around the gynophore. Berry oblong-ovoid, fleshy, yellow.

Cultivated in gardens as covering trellises, archways, and hedge plant, for its showy flowers and edible fruits. Common garden ornametal climber (Naqvi, 2001; Ramana, 2010).

Fl. & Fr.: Round the year.

Passiflora foetida L., Sp. Pl. 959. 1753; Masters in Hook. f., Fl. Brit. India 2: 599. 1879; Gamble, Fl. Madras 1: 524. 1919.

Climbing herbs; up to 1.5 m long, branches faintly angular, densely hispid, tendrils axillary. Stipules subreniform, 0.5 – 1 cm, deeply cleft into glandular processes. Leaves palmately 3-lobed to halfway, 5-10 × 4- 7 cm, chartaceous, appressed glandular-pubescent, base cordate, margin subentire to ciliate, apex acute. Flowers white, axillary, solitary; involucral bracts bipinnatifid; peduncle 3-4 cm; bracts and bracteoles 2-3 cm, deeply pinnatisect, glandular pubescent; calyx tube short, lobes ovate – lanceolate, 1.5 cm, apically spurred; petals slightly shorter than calyx –lobes, outer corona hairs to 1 cm and inner one to 2 mm; disc prominent, annular; androgynophore 4-6 mm, filaments to 5 mm, flat, anthers to 3.5 mm; ovary globose, 1 × 0.8 mm, style to 4 mm. Berries subglobose, glabrous, orange-red when ripe; seeds oblong, tetragonous, deeply pitted on faces, glabrous, pale to dark-brown.

Commonly run wild near towns and along road sides in many places in the plains.

Fl. & Fr.: June-December

Sirpur RF (ADB), *GO* & *PVP* 4927; Mosra (NZB), *TP* & *BR* 6059; Medak (MDK), *CP* 11733; Nagarjunakonda (NLG), *BVRS* 117 (NU); Pakhal (WGL), *ANH* 15930 (MH), *KMS* 11685 (MH); Subedari (WGL), *RKP* 105253 (BSID); Pakhal (WGL), *RKP* 111566 (BSID); Wyra (KMM), *RR* 108070(BSID); Gunavaram (KMM), *RCS* 98945(BSID).

CARICACEAE

CARICA L.

Carica papaya L., Sp. Pl. 1036. 1753.

Vern.: *Boppayi.* English: Papaya.

Dioecious tree; latex milky; leaf scars on stem. Leaves scattered, crowding towards stem apex; blade to 60 cm long, deeply divided and subdivided into several lobes with acute apex; petiole to 1 m, to 3 cm wide. Male flowers in racemes, to 1 cm, pendulous; calyx 5-lobed, obscure; corolla-tube to 6 cm, 5-lobed, twisted in bud, lobes 1 × 0.5 cm, cream; stamens 5+5, outer shortly stalked, inner sessile. Female flowers 2-4 per axil, peduncle to 2 cm; bracts fleshy, to 2 cm, caducous; calyx-lobes 5, to 8 mm, lobes acute; petals 5, lanceolate, obtuse, to 6.5 × 1.8 cm; ovary 4 × 1.8 cm, stigmatic lobes fimbriate. Berry spherical to pyriform, 30 × 15 cm, orange or yellow when ripe; seeds wrinkled, black, ovoid, 2 mm wide, enclosed in a gelatinous membrane.

Cultivated in households for its edible fruits.

Fl. & Fr.: Throughout the year.

Errapenta (MBNR), *TDCK* 16832.

Native of Tropical America, introduced in India during 16th century by Portuguese (Mehra, 1965). Commonly cultivated in gardens for its edible fruits. Fruits besides being edible has good medicinal properties as carminative and diuretic.

CUCURBITACEAE

Note: Renner and Pandey (2012) recently gave an account on accepted names, synonyms, geographical distribution of Cucurbitaceae of India. This treatment is followed in this book.

1. Petals free:
   2. Petiole with a pair of glands below the lamina; all flowers solitary; fruit polymorphic, woody, indehiscent .... **Lagenaria**
   2. Petiole without glands below the lamina; male flowers racemose, female solitary; fruit oblong or cylindric, fibrous, dehiscing by a terminal operculum or irregularly .... **Luffa**
1. Petals united:
   3. Tendrils branched or rarely simple :
      4. Fruits spinous-tuberculate .... **Momordica**
      4. Fruits smooth :
         5. Petals fimbriate on margins .... **Tichosanthes** *p.p.*
         5. Petals not fimbriate on margins
            6. Leaves deeply lobed :
               7. Male flowers solitary:

8. Leaves pinnately lobed .... **Citrullus**

8. Leaves palmately lobed or entire:

    9. Calyx lobes entire; corolla lobes recurved at the apex; anthers united into an oblong head .... **Cucurbita**

    9. Calyx lobes serrate; corolla lobes not recurved or reflexed at the apex; anthers free, exserted .... **Benincasa**

7. Male flowers clustered .... **Diplocyclos**

6. Leaves angular or shallowly lobed .... **Trichosanthes** *p.p.*

3. Tendrils simple :

10. Bracts ciliate .... **Blastania**

10. Bracts glabrous :

    11. Male flowers solitary:

        12. Petals white; anthers united .... **Coccinia**

        12. Petals yellow; anthers free .... **Citrullus** *p.p.*

    11. Male flowers in clusters or racemes :

        13. All stamens 2-celled .... **Zehneria**

        13. All stamens not 2-celled :

            14. Connectives apically crested .... **Cucumis** *p.p.*

            14. Connectives apically uncrested:

                15. Ovary not rostrate .... **Cucumis** *p.p.*

                15. Ovary rostrate at the apex:

                    16. Stamens inserted at the base of calyx tube, dioecious .... **Solena**

                    16. Stamens inserted in the mouth of the calyx tube; monoecious .... **Corallocarpus**

BENINCASA Savi

Benincasa hispida (Thunb.) Cogn. in DC., Monog. Phan. 3: 513. 1881; Renner & Pandey, Phytokeys 20: 60. 2013. *Cucurbita hispida* Thunb., Fl. Jap. 322. 1784. *Benincasa cerifera* Savi, Bibl. Ital. 9: 158. 1818; C.B.Clarke in Hook. f., Fl. Brit. India 2: 616. 1879.

Vern.: *Budida gummadi.*

Annual climbing herb. Petioles hirsute, 5-20 cm long. Leaves reniform-rotund, 10-25 × 10-25 cm, base deeply cordate, margin sinuate, dentate, upper surface scabrous, lower shortly hispid, 5-7-lobed. Tendrils slender, 2-3-fid. Peduncles hirsute, male 5-15 cm long, female 2-4 cm long. Male flowers: calyx tube broadly campanulate, lobes 5, sub-foliaceous, 8-12 × 3-5 mm; corolla rotate, 5-partite, lobes obovate; stamens 3, free, inserted at the calyx tube, filaments hispid, 2-3 mm long; ovary ovoid or cylindric, softly hairy, style 2-3 mm long. Fruits fleshy, hairy when young, waxy bloom when mature; seeds compressed, ovoid, yellowish white.

Cultivated in many parts of the State for the fruit vegetable.

Fl. & Fr.: June – October.

BLASTANIA Kotschy & Peyr.

Blastania garcinii (Burm. f.) Cogn. in A. & C. DC., Monog. Phan. 3: 629. 1881; Gamble, Fl. Madras 1: 540. 1919; Chakravarthy in Fasc. Fl. Ind. 11: 13. 1982; Renner & Pandey, Phytokeys 20: 61. 2013. *Sicyos garcinii* Burm. f., Fl. Ind. 211. 1768. *Ctenolepis garcinii* (Burm. f.) C.B. Clarke in Hook. f., Fl. Brit. India 2: 629. 1879; Jeffrey, Kew Bull. 34: 793. 1980; Pullaiah & Chennaiah, Fl. Andhra Pradesh 1: 407. 1997.

Slender annual climbers, to 5 m long, tendrils simple, branches subfiliform, branchlets scabrid. Leaves 2.5-5 cm diameter, membranous, deeply 5-lobed, both surfaces hirsute when young, scabrid with white spots at length, deeply 3-5-lobed, lobes unequal, obovate, mostly acute, denticulate, petiole to 5 cm. Plants monoecious. Male flowers yellowish-white, peduncle usually capillary, puberulous, subcapitately 3-4-flowered, peduncle filiform, flowers 4 mm wide; pedicel to 1.5 mm; calyx-tube campanulate, 1mm, lobes 5, linear, 0.5 mm; corolla rotate, greenish, petals 5, ovate, 1.5 mm, pubescent; stamens 3, filaments free, inserted at the base of the calyx tube, anthers ovate, straight, one 1-celled, others 2-celled, connective not produced; pistillode absent. Female flowers solitary, dull white; peduncle 3mm; flowers 1.5mm wide; calyx-tube 0.5 mm; petals 1mm; ovary ovoid; placenta 1 or 2, ovule 1 per cell, horizontal, style 0.5 mm; stylar disc absent, stigma obscurely 3-lobed, obtuse, papillose; staminodes absent. Fruits reddish, inversely sub-reniform, smooth, 6-8 × 3-4 mm, glabrous; seeds 1-2, oblong. 7 × 4 mm, smooth, sulcate on one face, convex on the other.

Common in most of the districts in waste lands and open forests.

Fl. & Fr.: August-December

Birsaipet (ADB), *GO* & *PVP* 4590; Gowraram (NZB), *TP* & *BR* 6391; Choutkoor (MDK), *TP* & *CP* 11927; Molachintapally (MBNR), *TDCK* 15334; Kodimial (KRN), *GVS* 21830 (MH); Mallepally (NLG), *BVRS* 535 (NU); Ondikudisai (KMM), *RR* 113838 (BSID); Gangaram (WGL), *PVS* & *NRR* 76931 (BSID).

CITRULLUS Schrad. *nom. cons.*

1. Tendrils simple; leaf lobes narrow; bracts subulate;
   fruits below 8 cm across .... **C. colocynthis**

1. Tendrils 2-fid; leaf lobes broad; bracts spathulate;
   fruits above 8 cm across .... **C. lanatus**

Citrullus colocynthis (L.) Schrad., Linnaea 12 : 414. 1838; C.B. Clarke in Hook. f., Fl. Brit. India 2: 620. 1879; Gamble, Fl. Madras 1: 536. 1919; Chakravarthy in Fasc. Fl. Ind. 11: 20. 1982; Renner & Pandey, Phytokeys 20: 62. 2013. *Cucumis colocynthis* L., Sp. Pl. 1011. 1753.

Trailing scabrid herbs, tendrils slender, short, sparsely villose- hirsute, simple or 2-3-fid, stems angular. Leaves ovate or narrowly triangular, deeply 3-7-lobed, lobes pinnatifid-sinuous, 6-12 × 4-6 cm, rigid, densely villous hirsute below lobes, terminal lobes large, lobes obtuse or acute, margin crisped. Flowers monoecious, yellow, both male and female solitary. Fruits globose, slightly depressed, 5-7 × 4.5-5 cm, striped green and white when young, yellow when ripe; seeds ovate-oblong, yellowish-brown, not margined.

Common in open sandy or stony lands of Adilabad, Karimnagar and Nalgonda districts.

Fl. & Fr.: May-October.

Kowtal (ADB), *GO* & *PVP* 4833; Hajipur hills (MBNR), *TDCK* 13618; Godavary (KRN), *KMS* 9715 (MH); Bhupatipur (KRN), *GVS* 22478 (MH); Rekularam (NLG), *BVRS* 625 (NU).

Citrullus lanatus (Thunb.) Matsum. & Nakai, Cat. Sem. & Spor. Hort. Bot. Univ. Imp. Tokyo 1916: 30. 1920 ("1916'); Renner & Pandey, Phytokeys 20: 62. 2013. *Momordica lanata* Thunb., Prodr. Pl. Cap. 13. 1794. *Citrullus vulgaris* Schrad. ex Eckl. & Zeyh., Enum. Pl. Aqfric. Austral. 2:279. 1836; Gamble, Fl. Madras 2: 536. 1919.

Vern.: *Kalingara, Puchkai, Karbuja.* English: Watermelon.

Trailing herb, stems angular, villose. Leaves triangular-ovate, 5-15 × 2.5-12 cm, deeply trifid, segments pinnatifid, obovate-lanceolate, softly pubescent. Flowers 1.5-2 cm across, globose or ellipsoid, green or variegated; seeds numerous, obovate-oblong, black, embedded in red pulp.

Usually cultivated on sandy soil along the river beds and streams, at times found as an escape.

Fl. & Fr.: April – July.

Billyguttalu (KMM), *RR* 105973

COCCINIA Wight & Arn.

Coccinia grandis (L.) Voigt., Hort. Suburb. Calc. 59. 1845; Chakravarthy in Fasc. Fl. Ind. 11: 24. 1982; Renner & Pandey, Phytokeys 20: 60. 2013. *Bryonia grandis* L., Mant. Pl. 126. 1767. *Coccinia indica* Wight & Arn., Prod. Fl. Ind. Orient. 1. 347. 1834. *Cephalandra indica* (Wight & Arn.) Naudin, Ann. Sci. Nat. Bot. Ser. 5. 5: 16. 1866; C.B. Clarke in Hook. f., Fl. Brit. India 2: 621. 1879. excl. syn.

Perennial herbaceous climbers; tendrils slender, simple, branchlets apically pubescent, glabrous at base. Leaves simple, entire to palmately 3-5-lobed, 3-6.5 × 3-6 cm, chartaceous, glabrous, rarely scaly, punctate above, glandular below, basal sinus subrotund-cordate, margin denticulate, apex obtuse, mucronate. Flowers large, dioecious; male flowers white, solitary; female flowers also white, solitary. Male flower 1.5 cm wide; corolla campanulate, white; petals 1.5 × 1 cm; stamens inserted at the base of the calyx tube, filaments connate into a column, anthers connate, triplicate, flexuous, one 1-celled, others 2-celled, connectives narrow, pistillode absent. Female flower 2.5cm wide, petals 1.2 × 1 cm; ovary oblong, ovules many, horizontal, staminodes 3, subulate. Berries ovoid- oblong, sub-glabrous, round at both ends; red when mature, seeds ovoid-oblong, compressed, granular.

Common in most plains districts on hedges and bushes.

Fl. & Fr.: Through out the year.

Satnella dam (ADB), *GO* & *PVP* 4212; Ankole (NZB), *TP* & *BR* 6185; Yadapally – Bodhan Road (NZB), *BR* & *KH* 6477; Ponnala (MDK), *CP* 11365; ICRISAT (MDK), *LJGV* 2375 (CAL); Naskal (RR), *MSM* & *PRSR* 15112; Appapur (MBNR), *TDCK* 16847; Peddavuru (NLG), *BVRS* 508 (NU); Kodimial (KRN), *GVS* 20057 (MH); Pakhal (WGL), *KMS* 11575 (MH); C.T.Padu (MBNR), *BR* & *MV* 41380; Koraguddalu (KMM), *RR* 105943 (BSID); Kajipet (WGL), *RKP* 111563 (BSID).

CORALLOCARPUS Welwitsch

Corallocarpus epigaeus (Rottl. & Willd.) C.B. Clarke in Hook. f., Brit. India 2: 628. 1879; Gamble, Fl. Madras 1: 541. 1919; Chakravarthy in Fasc. Fl. Ind. 11: 28. 1982; Jeffery, Kew Bull. 34. 792. 1980; Renner & Pandey, Phytokeys 20: 64. 2013. *Bryonia epigaea* Rottl. & Willd. in Ges. Naturf. Freunde Berlin Neue Schriften. 4: 223. 1803. *Aechamandra epigaea* (Rottl. & Willd.) Arn., J. Bot. (Hooker) 3 : 274. 1841. *Corallocarpus gracillipes* (Naudin) Cogn. in DC., Monog. Phan. 3: 650. 1881; Gamble, Fl. Madras 1: 541. 1919. *Rhynchocarpa epigaeus* (Rottl. ex Willd.) C.B. Clarke var. *gracilipes* Naudin, Ann. Sci. Nat. Bot. Scr. 4. 16: 179. 1862.

Thick stemmed climbers; rootstock large, tuberous; branches glabrous, angular-acute. Leaves broadly suborbicular, angled or deeply 3-5-lobed, 2.5- 4.5 × 2.5-5 cm, densely villose, lobes obovate, lobulate, base cordate, margin irregularly denticulate, apex obtuse, mucronate, tendrils simple. Flowers monoecious, staminate flowers in crowded racemes on long axillary peduncles; pistillate flowers usually solitary on short peduncles from the same axis. Male flowers: corolla campanulate, greenish yellow; stamens free, inserted into the middle of calyx-tube; anthers: one 1-celled, others 2-celled, connectives green, faintly bifid, apically produced, pistillode obscure. Female flower 2mm wide; corolla yellow; ovary oblong-ovoid, beaked, ovules 8-10, horizontal. Berries ellipsoid, stalked, base attenuate, glabrous, apex beaked; seeds 6-9, ovoid, or subglobose, smooth, rigid.

Occasional in Mahabubnagar district.

Fl. & Fr.: September-March.

Molachintapally (MBNR), *TDCK* 15319.

CUCUMIS L.

1. Connectives apically crested:
   2. Perennials; leaves deeply 5-7-lobed .... **C. callosus**
   2. Annuals; leaves angular or lobed :
      3. Ovary softly hairy; young fruits not tuberculate .... **C. melo**
      3. Ovary hispidulous; young fruits tuberculate .... **C. sativus**

1. Connectives apically uncrested:

    4. Leaves densely, softly villous beneath; seeds smooth .... **C. leiospermus**

    4. Leaves shortly hirsute or scabrous beneath;
        seeds rugose .... **C. maderaspatanus**

Cucumis callosus (Rottl.) Cogn. in Engler, Das. Pflanzenr. 4. 275. 2: 129. 1924; Chakravarthy in Fasc. Fl. Ind. 1: 31. 1982. *Bryonia callosa* Rottl. in Ges. Naturf. Freunde Berlin Neue Schriften 4: 210. 1803. *Cucumis trigonus* Roxb., Fl. Ind. 2 : 722. 1824; C.B. Clarke in Hook. f., Fl. Brit. India 2: 619. 1879; Gamble, Fl. Madras 1: 535. 1919.

Perennial, weak climbing herbs, branches grooved, hispid, almost prickly, tendrils simple, Leaves deeply palmately 5-7-lobed, lobes round or ovate-oblong, often narrowed at base, base cordate, apex round, dentate or lobulate. Plants monoecious, flowers yellow, male axillary, solitary, sometimes 2-3 clustered, female flowers solitary. Fruits obovoid, striped white, seeds many, oblong, flattened, smooth.

Common in Adilabad and Ranga Reddi districts.

Fl. & Fr.: July-February.

Itkial (ADB), *TP & PVP* 4233; Malkalcheruvu (RR), *MSM & KH* 10277.

Note: Sequences representing *C. callosus, C. pubescens,* and *C. trigonus* all cluster with *C. melo* (Sebastian *et al.,* 2010) and likely present wild progenitors of domesticated *C. melo.* Jeffrey (1980) preferred to list *C. trigonus* as a separate species, and Chakravarty (1982) treats *Cucumis trigonous* as synonym of *Cucumis callosus* and separated from *Cucumis melo.* Renner and Pandey (2012) treat *Cucumis callosus* and *Cucumis trigonous* as synonyms of *Cucumis melo.* Till the matter is resolved we treat *Cucumis callosus* and *Cucumis melo* as separate species.

Cucumis leiospermus (Wight & Arn.) Ghebretinsae & Thulin, Novon 17(2): 77. 2007; Renner & Pandey, Phytokeys 20: 67. 2013. *Bryonia leiosperma* Wight & Arn., Prodr. Fl. Ind. Orient. 1: 345. 1834. *Mukia leiosperma* (Wight & Arn.) Arn., Madras J. Lit. Sci. 12: 50. 1840; C.B. Clarke in Hook. f., Fl. Brit. India 2: 623. 1879; Pullaiah & Chennaiah, Fl. Andhra Pradesh 1: 414. 1997. *Melothria leiosperma* (Wight & Arn.) Cogn. in A. & C. DC., Monogr. Phan. 3: 622. 1881; Gamble, Fl. Madras 1: 539. 1919; Chakravarthy in Fasc. Fl. Ind. 11: 81. 1982.

A small hispid climber, stems robust, angular, sulcate, villose. Leaves broadly ovate, 6-8 × 4-6 cm, 5-angled or 3-5-lobed, scabrid above, densely villous below, base cordate, margin denticulate, apex acute, mucronate. Flowers yellow, both male and female fasciculate, pedicels very short. Male flowers 5 mm across; calyx tube 4 mm, villous, lobes linear, 3 mm; corolla yellow, petals ovate, 2mm, villous; stamens 3, filaments 1mm, free, anthers oblong, 2 mm, free, erect, ciliate, connectives shortly produced at apex; pistillode 0.5 mm. Female flowers 6 mm across; ovary ovoid, 2 mm, villous, style 1.2 mm. Fruit globose, 1 × 0.5 cm, glabrous, smooth.

Occasional on hedges and on shrubs in Medak district.

Fl. & Fr.: July-September.

Choutkoor (MDK), *TP* & *CP* 11908.

Cucumis maderaspatanus L., Sp. Pl. 1012. 1753; Renner & Pandey, Phytokeys 20: 67. 2013. *Mukia maderaspatana* (L.) Roem., Syn. Monogr. 2: 47. 1846; Pullaiah & Chennaiah, Fl. Andhra Pradesh 1: 414. 1997. *Melothria maderaspatana* (L.) Cogn. in DC., Mon. Phan. 3: 623. 1881; Gamble, Fl. Madras 1: 539. 1919; Chakravarthy in Fasc. Fl. Ind. 11: 83. 1982. *Bryonia scabrella* L. f., Suppl. 424. 1781. *Mukia scabrella* (L.f.) Arn., Hook. J. Bot. 3: 276. 1841; C.B.Clarke in Hook. f., Fl. Brit. India 2: 623. 1879 (excl. syn.).

Prostrate or climbing, scabridly hairy, monoecious annuals. Leaves ovate-deltoid, angular or 3-5-lobed, 4-6.5 × 4-6 cm, scabrid above, shortly hispid below, base cordate, margin denticulate, apex acuminate, mucronate; petiole 2-4 cm. Flowers pale-to-bright-yellow, male and female in same axils, male pedicillate, female sessile. Male flowers 6 mm across; calyx-tube 3 mm, villous, lobes linear, 2 mm; corolla yellow, petals ovate, 3 mm, villous without; stamens 3, filaments 1.5 mm, anthers oblong, 2 mm; pistillode 0.5 mm. Female flowers solitary or in cluster about male flowers; ovary 3 mm, villous; style 2.5 mm. Fruits 0.5-1 cm broad, spherical, glabrous, smooth; seeds grey-light black, ovoid-oblong, compressed, somewhat wrinkled, scorbiculate, 4 × 2.5 mm, pitted.

Common on hedges or on low herbs or shrubs.

Fl. & Fr.: July-October.

Surzapur (ADB), *TP* & *GO* 4119; Gowraram (NZB), *TP* & *BR* 6392; Nagulabanda (MDK), *CP* 11335; Mohammadabad (RR), *MSM* 10498; Pakhal (WGL), *KMS* 13153 (MH); Bhongir (NLG), *BVRS* 551 (NU); Kodimial (KRN), *GVS* 20058; Aklaspur (KRN), *GVS* 25627 (MH); Pasra (WGL), *RKP* 108139 (BSID); Yellandu – Karepally (KMM), *RR* 105950 (BSID); C.T.Padu (MBNR), *MV* & *BR* 41371.

Note: *Melothria* in its modern circumscription is confined to the New World and does not occur in India. Its two Indian species have been moved to *Cucumis* and *Solena*.

Cucumis melo L., Sp. Pl. 1011. 1753; var. **melo** C.B. Clarke in Hook. f., Fl. Brit. India 2: 620. 1879; Gamble, Fl. Madras 1: 535. 1919; Chakravarthy in Fasc. Fl. Ind. 11: 32. 1982; Renner & Pandey, Phytokeys 20: 68. 2013. *C. melo* L. var. *cultis* Kurz., J. Asiat. Soc. Bengal 46: 102. 1877. *C. melo* L. var. *agrestis* (Naudin) Greb., Kulturpflanze Beih 2 : 424. 1959. *C. melo* L. var. *agrestis* Naudin, Ann. Sci. Pl. 4: 614. 1805; Gamble, Fl. Madras 1: 535. 1919.

Perennial, prostrate or slender, monoecious herbs, with thick root stock. Leaves suborbicular, 9 × 8 cm, palmately 3-7-lobed, lobes obtuse, suborbicular, denticulate, base cordate, villose or subhirsute. Male flowers fasciculate, peduncles slender, 0.5-3 cm; calyx-tube narrow, campanulate, villose, 6-8 mm long, lobes subulate; corolla ca 2 cm long, lobes ovate-oblong, acute; staminal filaments very short, anthers 3-4 mm long; pistillode ca 1 mm long. Female flowers solitary; corolla yellow; ovary softly hairy, style 1-2 mm long, stigma connivent. Berries rounded or ellipsoid, obscurely trigonous, smooth, yellow with age, with green stripes when young; seed obovoid, rounded at apex.

Largely cultivated on the sandy beds of rivers when the stream has subsided to its hot season channel.

Fl. & Fr.: January-June.

Ankusapuram (ADB), *GO* 4332; Ibrahimpet RF (NZB), *TP* & *BR* 6121; Banjapally (NZB), *BR* & *CPR* 9250; Lingareddypally (MDK), *CP* 11716; Peddavura (NLG), *BVRS* 520 (NU); Pakhal (WGL), *KMS* 13108 (MH); Janakalancha RF (WGL), *RKP* 108191 (BSID).

Cucumis sativus L., Sp. Pl. ed. 1: 1072. 1753; C.B. Clarke in Hook. f., Fl. Brit. India 2: 630. 1879; Gamble, Fl. Madras 1: 535. 1919; Chakravarthy in Fasc. Fl. Ind. 11: 35. 1982; Renner & Pandey, Phytokeys 20: 69. 2013.

Annual monoecious herbs, trailing or climbing, stems hirsute. Leaves 12-18 cm across, almost as much as broad, broadly cordate-ovate, pedately 5-7-nerved, palmately 3-5-lobed, densely hairy. Male flowers fasciculate, peduncles slender; calyx-tube narrow, campanulate, villose-hirsute, 8-10 cm long, lobes subulate, spreading; corolla 2-3 cm long, lobes acute; staminal filaments very short; anthers 3-4 mm long. Female flowers solitary or fasciculate, peduncle robust, 1-2 cm long; ovary fusiform, muricate, aculeate, hairy. Fruits of various shapes, narrowly or broadly oblong, pale to dark-green or pale yellow, glabrous; seeds whitish, oblong, 8-10 × 3-5 mm, both ends subacute.

Cultivated, also as an escape.

Fl. & Fr.: Throughout the year, profuse in monsoon.

Wankidi (ADB), *GO* 4477; Vempalli (ADB), *GO* & *PVP* 4758; Gacchibowli (RR), *MSM* 15193; Appapur (MBNR), *TDCK* 16849.

CUCURBITA L.

1. Calyx segments linear, apex leaflike; fruiting pedicel conspicuously enlarged at apex .... **C. moschata**

1. Calyx segments linear or linear lanceolate, apex not leaflike; fruiting pedicel not strongly enlarged at apex:

    2. Leaf blade triangular or ovate-triangular, irregularly 5-7-lobed; calyx segments linear-lanceolate; fruiting pedicel angular-sulcate, slightly thickened at apex .... **C. pepo**

    2. Leaf blade reniform or orbicular, almost entire or minutely dentate; calyx segments lanceolate; fruiting pedicel not angular-sulcate, not thickened at apex .... **C. maxima**

Cucurbita maxima Duch. ex Lam., Encycl. 2: 151. 1786; C.B.Clarke in Hook.f., Fl. Brit. India 2: 622. 1879; Gamble, Fl. Madras 1: 543. 1919; Renner & Pandey, Phytokeys 20: 71. 2013.

Vern.: *Erra gummadi.*

Annual, prostrate herbs; stems cylindric, grooved. Petioles 5-19 cm long, hairy. Leaves reniform or orbicular with 5-rounded shallow lobes, 6-19 × 7-30 cm, coarsely hairy, margin entire or minutely dentate. Tendrils 2-6-fid. Male flowers: peduncles 10-17 cm long; calyx-tube club-shaped, 5-10 mm long, lobes 5, lanceolate, hairy; corolla 4-7 cm long, campanulate, lobes 5, reflexed, yellow; stamens 3, filaments thick. Female flowers: solitary, peduncles 5-7 cm long. Fruit of various forms, usually large; seeds ovate, white or yellowish.

Cultivated for its fruit which is used as a vegetable.

Fl. & Fr.: March – August.

Ramanagutta (KMM), *RR* 107984 (BSID).

Cucurbita moschata (Duchesne ex Lam.) Duchesne ex Poir. in F.Curier, Dict. Sci. Nat. 11: 234. 1818; C. Jeffrey, Kew Bull. 34: 799. 1980; Chakravarthy in Fasc. Fl. India 11: 41-42. 1982; Renner & Pandey, Phytokeys 20: 72. 2013. *C. pepo* L. var. *moschata* Duchesne ex Lam., Encycl, 2: 152. 1786.

Vern.: *Karbuja.* English: Muskmelon.

Stout climber; branchlets scabrid. Leaves large, cordate, ovate, 5-angular or lobed; tendrils 2 or more. Flowers axillary, solitary, monoecious, large, yellow. Male flower: calyx-tube campanulate, lobes 5, linear, apex leaf-like; corolla campanulate, petals 5, apically recurved; stamens 3, inserted at the base of the calyx-tube; filaments free, anthers linear, confluent, one 1-celled, others 2-celled, cells sigmoid-flexuous; connectives not produced. Female flower: ovary oblong, placentae 3, ovules many, horizontal; style short, thick, stigmas 3-5, 2-lobed; staminodes 3. Fruit large, indehiscent, fruiting pedicel conspicuously enlarged at apex; seed margin thickened.

Common cultivated species in S. India.

Fl. & Fr.: March-October.

Cucurbita pepo L., Sp. Pl. 1010. 1753; C.B.Clarke in Hook. f., Fl. Brit. India 2: 622. 1879; Renner & Pandey, Phytokeys 20: 72. 2013.

Vern.: *Gummadi*; English: Pumpkin.

Annual, prostrate herb. Petioles 6-9 cm long, with rigid pungent hairs below. Leaves 5-lobed, lobes obtuse or acute, sinus obtuse; base cordate, margin dentate; upper surface scabrous, lower with glandular club-shaped hairs. Tendrils branched. Monoecious, flowers solitary. Male flowers: peduncles 3-6 cm long, 5-angular; calx-tube obscurely 5-angled, segments linear-lanceolate; corolla campanulate, lobes erect, acute, yellow; stamens 3, filaments swollen below. Female flowers: calyx and corolla as in male flowers; staminodes 3; ovary ovoid, 1-locular. Fruits variable, small or large, pulp fibrous; seeds whitish yellow.

Cultivated as a vegetable.

Fl. & Fr.: August – January.

DIPLOCYCLOS (Endlicher) Post & Kuntze

Diplocyclos palmatus (L.) Jeffrey, Kew Bull. 15: 132. 1962; Chakravarthy in Fasc. Fl. Ind. 11: 48. 1982; Renner & Pandey, Phytokeys 20: 73. 2013. *Bryonia palmata* L., Sp. Pl. 1012. 1753. *B. laciniosa* L., Sp. Pl. 1013. 1753 *p.p.,* excl. typus; C.B. Clarke in Hook. f., Fl. Brit. India 2: 622. 1879. *Bryonopsis laciniosa* sensu Naudin, Ann. Sci. Nat. Bot. Ser. 4. 12. 141. 1859; Gamble, Fl. Madras 1: 534. 1919.

Annual, slender, climbing herbs, stems elongate, tendrils 2-fid. Leaves 8-12 cm in diameter, orbicular, deeply palmately 5-lobed, upper scabrous, lower smooth, margin denticulate or undulate or subcrenulate; petiole to 5 cm. Male flowers greenish-yellow in axillary fascicles of 3-6; pedicel 1 cm; flowers 1.5 cm wide; calyx-tube broadly campanulate, 3 mm, lobes 5, linear, 0.5 mm; corolla greenish-yellow, broadly campanulate, petals 5, ovate, 5 mm, pubescent; stamens 3, filaments free, 1mm, inserted in the middle of calyx-tube, anthers ovate, 2 mm, one 1-celled, others 2-celled, cells linear, slightly flexuous, connectives broad, apically not produced; pistillode absent. Female flowers solitary or few in fascicles, ovary globose, 4 mm, puberulous; placentae 3, ovules a few, horizontal, style 3 mm, glabrous, stylar disc absent, stigma 3-fid, papillose. Fruits spherical, yellowish-green when young, brick-red when ripe with white vertical streaks; seeds grey belted, narrowly acute attenuate with raised projections of both faces.

Common in Nizamabad, Karimnagar, Warangal, Medak, Ranga Reddi, Hyderabad and Mahabubnagar districts in hedges and on bushes.

Fl. & Fr.: April-December.

Gowraram (NZB), TP & BR 6392; Pathur (MDK), *TP & CP* 12122; Narsapur town (MDK), *BR* 11233, *KMS* 6671 (CAL); Pargi (RR), *MSM* & *PRSR* 12795; Laxmapur (MBNR), *TDCK* 13679; Rechapalli (KRN), *GVS* 21899 (MH); Nathyanipally (MDK), *BR* 11249; Rangapur RF (MBNR), *AJR* 20479; Yelandu – Karepally (KMM), *RR* 105954 (BSID); Jhakharam (WGL), *RKP* 105310 (BSID).

Note: The Linnaean epithet *'laciniosa'* has been erroneously applied to the widely distributed plant of the old world tropics, in the combination *Bryonopsis laciniosa* (L.) Naudin. Actually this is an American plant of which the correct name is *Cayaponia laciniosa* (L.) Jeffrey and that of old world species is *Diplocyclos palmatus* (L.) Jeffrey.

LAGENARIA Ser.

Lagenaria siceraria (Molina) Standl., Publ. Field Mus. Nat. Hist. Chicago, Bor. Ser. 3: 435. 1930; Renner & Pandey, Phytokeys 20: 78. 2013. *Cucurbita siceraria* Molina, Sagg. Storia Nat. Chile 133. 1782. *Lagenaria vulgaris* Seringe, Mem. Soc. Phys. Hist. Nat. Geneve 3(1): 25. 1825; C.B.Clarke in Hook. f., Fl. Brit. India 3: 613. 1879.

Vern.: *Sorakai, Anapakai*; English: Bottle gourd.

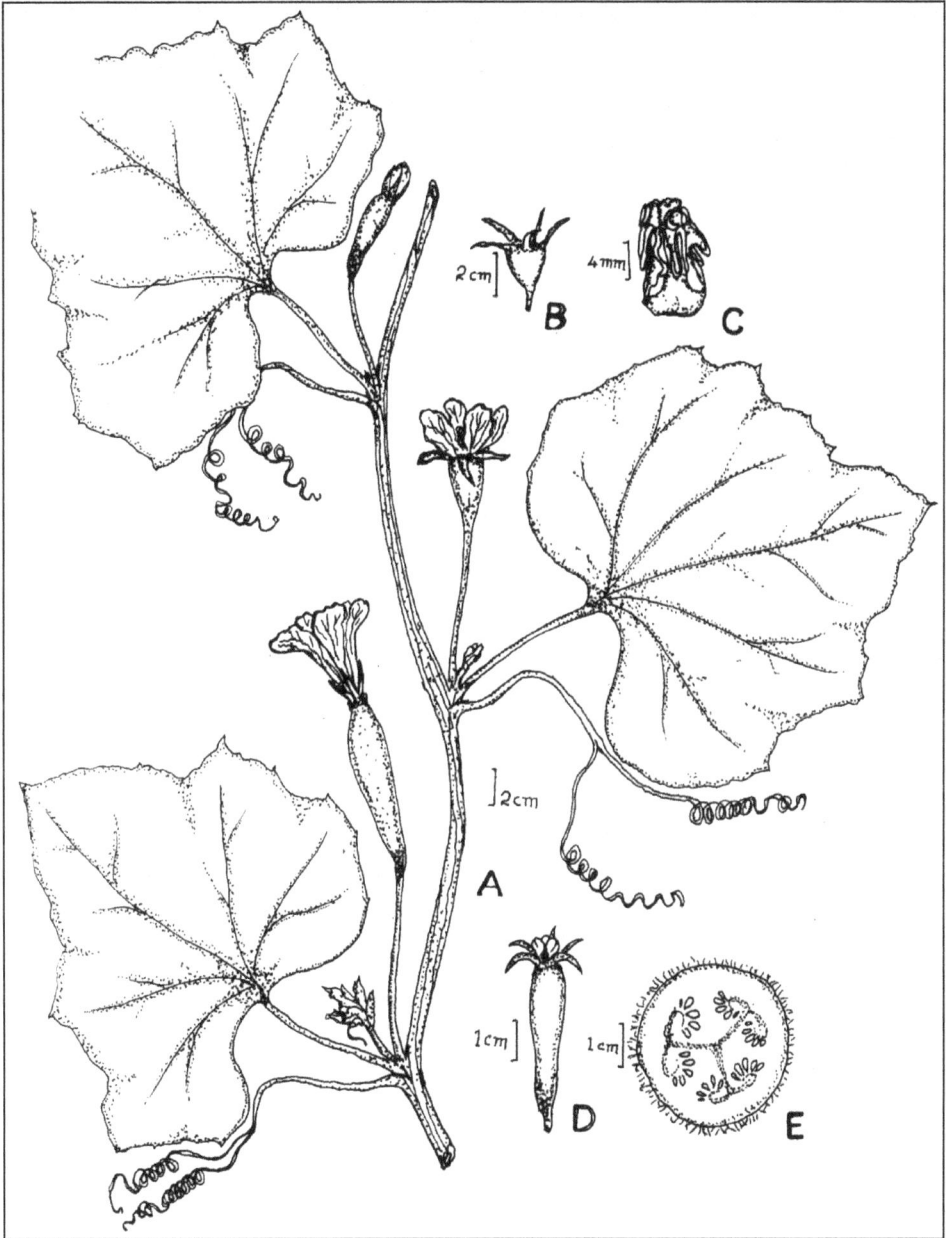

Figure 40: Lagenaria siceraria (Molina) Standl.
A. Twig, B. Calyx, C. Stamens, D. Pistil, E. Ovary c.s.

Annual, monoecious herbs, softly pubescent. Leaves suborbicular-cordate, angular or shortly trilobed, apex acute or shortly acuminate, 5-7-nerved; petiole 5-30 cm long, thick, subcylindric, often hollow, biglandular at apex; tendril bifid. Flowers solitary, white. Male flowers: long pedunculate; calyx tube narrow, campanulate or infundibuliform, sepals 5, small; petals 5, free, oblong-obovate, retuse; stamens 3, inserted at the calyx-tube, filaments free, anthers included, free or slightly coalascent, one 1-locular, two 2-locular. Female flowers: calyx-tube cup-shaped, calyx and corolla like the male; staminodes 3, obsolete; ovary ovoid or cylindric, style thick, stigmas 3, bilobed. Fruits indehiscent, very fleshy, variously shaped, greenish yellow; seeds many, obovate, compressed.

Cultivated throughout the state.

Fl. & Fr.: July – February.

Kotapalli (ADB), *GO* 4529.

LUFFA P. Miller

1. Roots tuberous .... **L. tuberosa**
1. Roots not tuberous :
    2. Stamens 5 .... **L. cylindrica**
    2. Stamens 3 .... **L. acutangula**

Luffa acutangula (L.) Roxb., Fl. Ind. 3 : 713. 1832 incl. var. *amara;* C.B. Clarke in Hook. f., Fl. Brit. India 2: 615. 1879; Chakravarthy in Fasc. Fl. Ind. 11: 67. 1982; Gamble, Fl. Madras 1: 533. 1919; Renner & Pandey, Phytokeys 20: 78. 2013. *Cucumis acutangula* L., Sp. Pl. 1011. 1753.

Vern.: *Beerakaya.*

Annual climbers, sparingly scabrous, tendrils bifid. Leaves 15-20 cm long and as broad, palmately 5-7-angled or sublobate, base cordate, margin entire or obscurely 5-angular or deeply 5-lobed, apex acute, tendrils sub-hispid, often trifid. Flowers pale-yellow, male and female in the same axil, male peduncle many-fowered, female solitary. Fruit 10-angled, apex obtuse or slightly acute, not warty.

There are two varieties, one cultivated another wild. The fruit of the wild is bitter and smaller, whereas cultivated fruits are larger and edible, used as vegetable. Extensively cultivated for fruit vegetable and wild variety found wild in all districts.

Fl. & Fr.: June-October.

Khammam to Dhaniyayigudem (KMM), *RR* 108519 (BSID); Krishna river bank (NLG), *KMS* 9808 (CAL).

Luffa cylindrica (L.) M. Roem., Syn. Monog. 2: 63. 1846; Jeffrey, Kew Bull. 15: 355. 1962; Chakravarthy in Fasc. Fl. Ind. 11: 70. 1982; Renner & Pandey, Phytokeys 20: 79. 2013. *Momordica cylindrica* L., Sp. Pl. 1009; 1753. *Luffa aegyptiaca* Mill., Gard. Dict. ed. 4. 1768; C.B.Clarke in Hook. f., Fl. Brit. India 2: 614. 1879; Gamble, Fl. Madras 1: 532. 1919.

Vern.: *Thikka beera* (Wild); *Netibeera* (cultivated).

Monoecious, slender climbing, glabrous herbs; branchlets glabrous. Leaves 5-7-lobed, 5-10 × 6-8 cm, lobes ovate, thin coriaceous, gland-dotted, scabrid, base cordate, margin sharply serrate, apex acute or acuminate, tendrils 3-fid, puberulous. Flowers monoecious, bright yellow, males in short racemes, female solitary. Fruits 10-30 cm long, cylindric, oblong, pale to dark olivaceous green with dark green longitudinal stripes; seeds flat, narrowly winged, glabrous.

There are two varieties of this species, the cultivated and the wild. The fruit of the wild is bitter and smaller, whereas cultivated fruits are larger and edible, used as vegetable. The wild one grows occasionally in hedges and bushes.

Fl. & Fr.: Throughout the year.

Appapur (MBNR), *BSS* & *SKB* 30657.

Luffa tuberosa Roxb., Fl. Ind. 3 : 717. 1832; Chakravarthy in Rec. Bot. Surv. Ind. 17: 81. 1959 & in Fasc. Fl. Ind. 11: 74. 1982. *Momordica tuberosa* Cogn. in DC., Mon Phan. 3: 454. 1881; Gamble, Fl. Madras 1: 532. 1919. *M. cymbalaria* Fenzl ex Naudin, Ann. Sci. Nat., Bot., Sér. 4, 12: 134. 1859; C.B.Clarke in Hook. f., Fl. Brit. India 2: 618. 1879; Renner & Pandey, Phytokeys 20: 82. 2013.

Perennial climbers, roots tuberous, stems slender, striate. Leaves orbicular, reniform, 2-5 × 2-4 cm, shallowly 5-7-lobed, hirsute, basal sinus broad, margin irregularly dentate, apex acute. Flowers white, monoecious, male flowers 2-5 in racemes, female solitary. Fruits angular, pyriform with 8 sharp ridges, 2- 4 × 1.5 cm, attenuated at base and apex; seeds broadly ovoid, 7 × 4.5 mm, not margined, smooth, blackish grey.

Weed in cultivated fields in Medak and Mahabubnagar districts.

Fl. & Fr.: July-October.

ICRISAT (MDK), *LJGV* 2171 (CAL).

Note: This species has been removed to *Momordica* under *M. cymbalaria* Fenzl. by C.B. Clarke in Hooker's Flora of British India (2: 618. 1879). Cogniaux as referred above named it as *M. tuberosa* (Roxb.) Cogn., based on Roxburgh's *Lufa tuberosa*. The fruit is a specific character in *Luffa* and there is no reason to shift this species to *Momordica* which has either muriculate or echinate fruits but never angular. In his original description, Roxburgh stated that the fruit is exactly like *L. amara* Roxb. (*L. acutangula* var. *amara*) but without the stopple. Moreover, the leaves of all the species of *Momordica* contain true cystoliths on the lower surface. Cystoliths are absent in this species. Foliaceous bracts which are a common feature in *Momordica* are also absent in this species. So *Luffa tuberosa* Roxb. is retained against *Momordica cymbalaria* Fenzl.

MOMORDICA L.

1. Plants monoecious; fruits narrowed at the ends, ribbed
    and tubercled .... **M. charantia**

1. Plants dioecious; fruits obtuse at the ends, not ribbed, but softly spinous:

    2. Petioles eglandular; flowers yellow; fruits muricate .... **M. dioica**

    2. Petioles with 2-5 glands; flowers white; fruit not
        as above .... **M. cochinchinensis**

Momordica charantia L., Sp. Pl. 1009. 1753; C.B.Clarke in Hook. f., Fl. Brit. India 2: 616. 1879; Gamble, Fl. Madras 1: 532. 1919; Chakravarthy Fasc. Fl. Ind. 11: 89. 1982; Renner & Pandey, Phytokeys 20: 81. 2013.

Vern.: *Kakara.* English: Bitter gourd.

Climbing herbs; branches slender, grooved, sparingly puberulous, tendrils simple. Leaves about 6 × 6 cm, palmately 3-9-lobed, denticulate, cordate at base, sinus large, scabrous; petioles 4 cm long, faintly hairy. Flowers unisexual, monoecious; bracts basal on the peduncle, orbicular or reniform. Male flowers yellow, corolla campanulate, 1.2-1.5 cm across, basal scales 3, lobes 5, ovate; stamens 3, inserted at the mouth of calyx-tube, filaments short, free, anthers cohering, free at length, one 1-celled, others 2-celled, cells triplicate, flexuous, connectives apically not produced; pistillode glanduliform or absent. Female flowers solitary; ovary oblong, fusiform, placentae 3, ovules many, horzontal, style short, slender, stylar disc absent, stigma trifid, papillose. Fruits up to 15 × 6 cm, profusely tuberculate; seeds several, flattened, faces sculptured, margins corrugated.

Very commonly cultivated speices for its edible bitter fruits, often found wild.

Fl. & Fr.: August-December.

Velutla RF (NZB), *BR* 9528; Medak (MDK), *CP* 11723; Mombojipally (MDK), *BR* & *CP* 11642; Mohammadabad (RR), *MSM* 19237; Laxmapur (MBNR), *TDCK* 15341; On the way to Bhadrachalam (KMM), *RCS* 98963 (BSID); Thupakulagudem (WGL), *PVS* & *NRR* 76982 (BSID).

Momordica cochinchinensis (Lour.) Spreng. Syst. Veg. 3 : 14. 1826; Hook.f. Fl. Brit. India 2 : 618. 1879; Chakrav. in Rec. Bot. Surv. India 17 : 95. 1959; Jeffrey, Kew Bull. 34 : 790 & 801. 1980.

Dioecious perennial climber; roots tuberous; stems robust, angular; tendrils simple. Leaves sub-orbicular in outline, cordate at base, divided to middle or almost to base into 3-5 lobes, with 2-5 glands on margins near base; petioles 5-6 cm long, glandular at middle and apex. Male flowers solitary on 6-15 cm long peduncles; calyx lobes narrow, 1.5-1.6 cm long; corolla white, tinged with yellow, pubescent; segements obovate to oblong; filaments with black and white marks. Female flowers on 5-6 cm long peduncles. Fruits ovoid, bright red, fleshy, densely aculeate, 10-15 cm long; seeds ovoid, compressed, black, sculptured.

Cultivated in kitchen gardens for fruits used as vegetable.

Fl. & Fr.: August-December.

Momordica dioica Roxb. ex Willd., Sp. Pl. 4 : 605. 1805; C.B.Clarke in Hook. f., Fl. Brit. India 2: 617. 1879 (excl. syn.); Gamble, Fl. Madras 1: 532. 1919; Chakravarthy in Fasc. Fl. Ind. 11: 94. 1982; Jeffrey, Kew Bull. 34: 790. 1980; Renner & Pandey, Phytokeys 20: 82. 2013.

Perennial, trailing or climbing glabrous herbs with tuberous roots. Leaves broadly ovate, deeply 3-5-lobed, 4-10 × 3.5-8 cm, glabrous, base cordate, margin denticulate, apex obtuse-acute. Flowers unisexual, dioecious, solitary, yellow. Male flowers peduncle 5-15 cm long, with persistent spathaceous deeply concave orbicular bracts wrapping around the buds and enclosing the base of flowers; calyx-tube campanulate, 1cm long, lobes oblong-lanceolate or linear, as long as the tube. Female flowers peduncle shorter than the male flowers, with small bract inserted at the middle; calyx-tube lobes linear-lanceolate, 5-7.5 mm long; ovary densely hirsute-fimbriate. Fruits 3-5 × 2-3.5 cm, ovate or obovate, softly echinate; seeds ovate, flat, smooth, pale yellow or red.

Occasional in cultivated fields and open forests.

Fl. & Fr.: July-November.

Annaram RF (NZB), *BR* & *CPR* 7170; Kusumasamudram (RR), *MSM* & *NV* 10347; Kodimal (KRN), *GVS* 20033 (MH); Gangaram (WGL), *PVS* 84018 (BSID); Guttalagangaram (WGL), *PVS* 76979 (BSID).

SOLENA Lour.

Solena amplexicaulis (Lam.) Gandhi in Saldanha & Nicolson, Fl. Hasan 179. 1976; Jeffrey, Kew Bull. 34: 793. 1980; Renner & Pandey, Phytokeys 20: 86. 2013. *Bryonia amplexicaulis* Lam., Encycl. 1: 496. 1785. *Melothria amplexicaulis* (Lam.) Cogn., Mon. Phan. 3: 621. 1881; Chakravarthy in Fasc. Fl. Ind. 11: 71. 1982. *M. heterophylla* (Lour.) Cogn. in DC. *loc.cit.* 3 : 618. 1881; Gamble, Fl. Madras 1: 539. 1919. *Zehneria umbellata* Thwaites, Enum. Pl. Zeyl. 125. 1858; C.B.Clarke in Hook. f., Fl. Brit. India 2: 625. 1879.

Dioecious, climbing herbs, stems strongly grooved, glabrous, roots tuberous, tendrils simple. Leaves highly variable in shape, deltoid-ovate, ovate-oblong or suborbicular, 4-8 × 2.5-4.5 cm, 3-5-lobed, hastate, sometimes denticulate, cordate at base, often amplexicaul, sinus small or big, scabrous. Flowers pale yellow to creamy-white, axillary, subsessile. Staminate flowers in a short corymb; 6 cm across; peduncle 2.5 cm; bracts linear, 1.5 mm; pedicel 2 mm; calyx-tube 4 mm, glabrous, lobes 5, triangular, 1mm; corolla campanulate, petals 5, white, oblong-ovate, 3 mm, pubescent without, triangular; stamens 3, inserted at the base of calyx-tube, connective papillose on the top; pistillode 3-lobed. Pistillate flowers solitary, short-peduncled from the same axil as the male flowers; ovary narrow. Fruits ellipsoid or subglobose, reddish-yellow or orange-coloured.

Occasional in Adilabad, Mahabubnagar and Karimnagar districts.

Fl. & Fr.: May-October

Neelwai (ADB), *TP* & *GO* 5429 and 5444; Jaipur RF (ADB), *GO* & *PVP* 4953; Aklaspur (KRN), *GVS* 25628 (MH); Mannanur RF (MBNR), *VBH* 86618 (BSID).

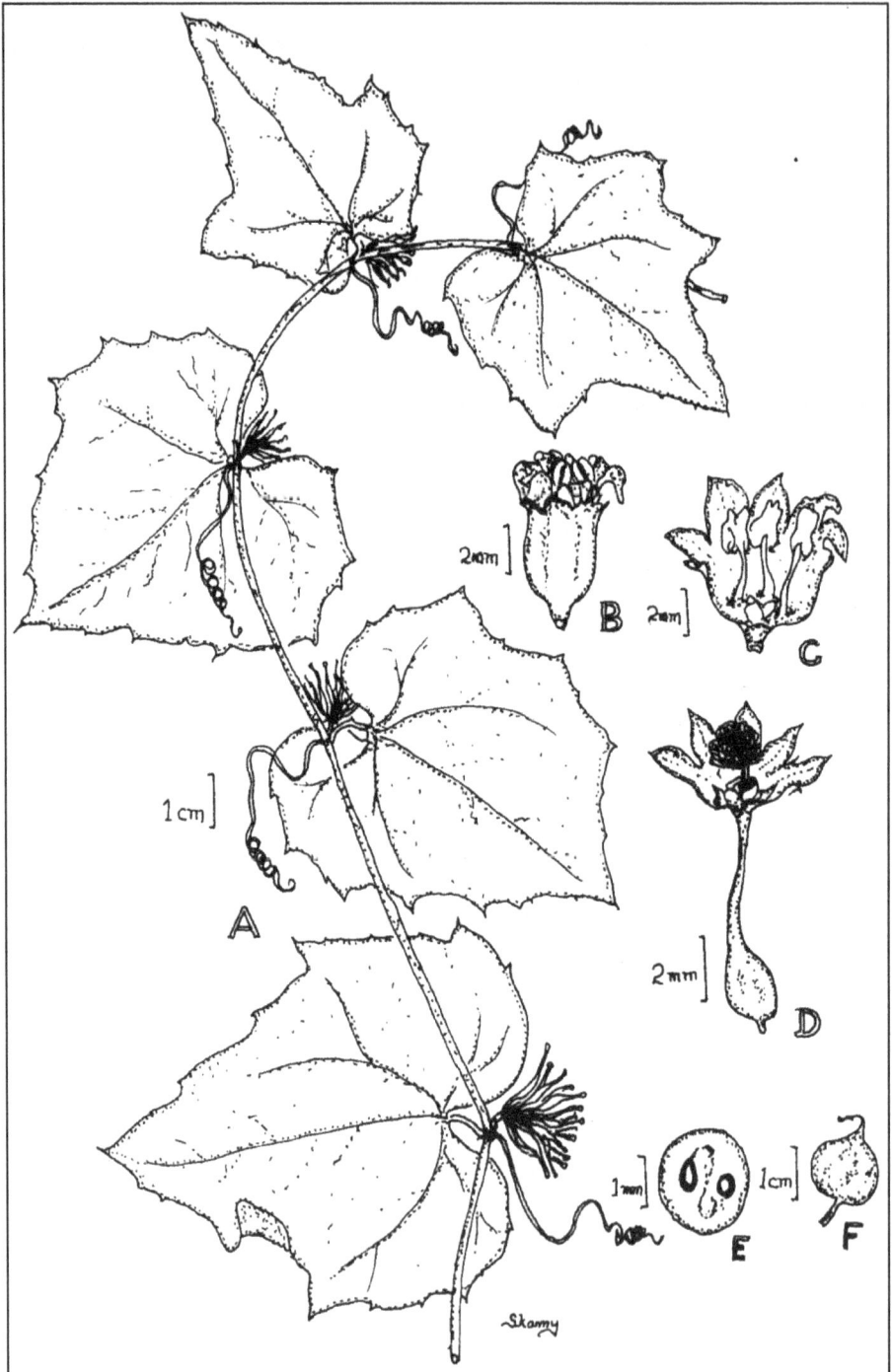

Figure 41: Solena amplexicaulis (Lam.) Gandhi

A. Twig, B. Male flower, C. Male flower open, D. Female flower, E. Ovary c.s., F. Fruit.

TRICHOSANTHES L.

1. Male racemes minutely bracteate or ebracteate .... **T. cucumerina**

1. Male racemes with large bracts :

    2. Female flowers in racemes, bracteate .... **T. anaimalaiensis**

    2. Female flowers solitary, ebracteate .... **T. tricuspidata**

Trichosanthes anaimalaiensis Bedd., Pres. Madras J. Lit. Sci. Ser. 3, 1: 47. 1864; Chakravarthy, Fasc. Fl. Ind. 11: 108. 1982; Jeffrey, Kew Bull. 34: 798. 1980; Renner & Pandey, Phytokeys 20: 88. 2013. *T. annamalayana* Bedd., Trans. Linn. Soc. 25: 217. 1865; Gamble, Fl. Madras 1: 530. 1919.

Large climbers, stem sulcate, slightly hirsute. Leaves 3-7-lobed, irregularly and deeply serrate; upper surface very scabrous, pubescent beneath; tendrils 2-3-fid; petiole to 4 cm. Flowers unisexual. Male in racemes, bracts large; calyx tube oblong, apically dilated, glabrous, lobes 5, lanceolate, entire; corolla rotate, white, petals 5, fimbriate; stamens 3, inserted at the base of calyx-tube, filaments short, anthers connate, one 1-celled, others 2-celled, cells conduplicate, connective narrow, not produced; pistillodes 3, minute. Female flowers usually in racemes, bracts large, laciniate; calyx-tube laciniate; ovary globose, glabrescent; ovules many, horizontal, style elongate, slender, disc absent, stigma 3-lobed. Berry globose, beaked, smooth.

Rare in forests in Mahabubnagar district (Raghava Rao, 1989).

Fl. & Fr.: April – September.

Trichosanthes cucumerina L., Sp. Pl. 1008. 1753; C.B.Clarke in Hook. f., Fl. Brit. India 2: 609. 1879; Gamble, Fl. Madras 1: 529. 1919; Chakravarthy in Fasc. Fl. Ind. 11: 112. 1982; Renner & Pandey, Phytokeys 20: 91. 2013. *T. anguina* L., Sp. Pl. 1008. 1753; C.B. Clarke in Hook. f., Fl. Brit. India 2: 610. 1919.

Vern.: *Potlakai;* English: Snake gourd.

Monoecious, climbing herbs, branches sulcate, sparsely puberulous, tendrils slender, branched. Leaves orbicular-reniform or broadly ovate, angular or 5-7-palmately lobed, lobes obovate, 7-10 × 9.5-10 cm, membranous; densely glandular-pubescent below, base cordate, margin distantly denticulate, apex acute. Flowers white, male flowers in racemes, bracts absent, female flowers solitary, male and female flowers in same or different axils. Fruits 3-7 × 2.5-5 cm, glabrous, green and striped when fresh, scarlet or orange when ripe; seeds many, flattened.

Common in most of the districts.

Fl. & Fr.: June-August

Rechapalli (KRN), *GVS* 25693 (MH); Pakhal RF (WGL), *KMS* 13137 (MH).

Trichosanthes tricuspidata Lour., Fl. Cochinch. 589. 1790; var. **tricuspidata.** Renner & Pandey, Phytokeys 20: 97. 2013. *T. palmata* Roxb., Fl. Ind. 3 : 704. 1832; C.B. Clarke in Hook. f., Fl. Brit. India 2: 606; Gamble, Fl. Madras 1: 529. 1919. *T. lepiniana* (Naudin) Cogn. in DC. Monog. Phan 3: 377. 1881; Gamble, Fl. Madras

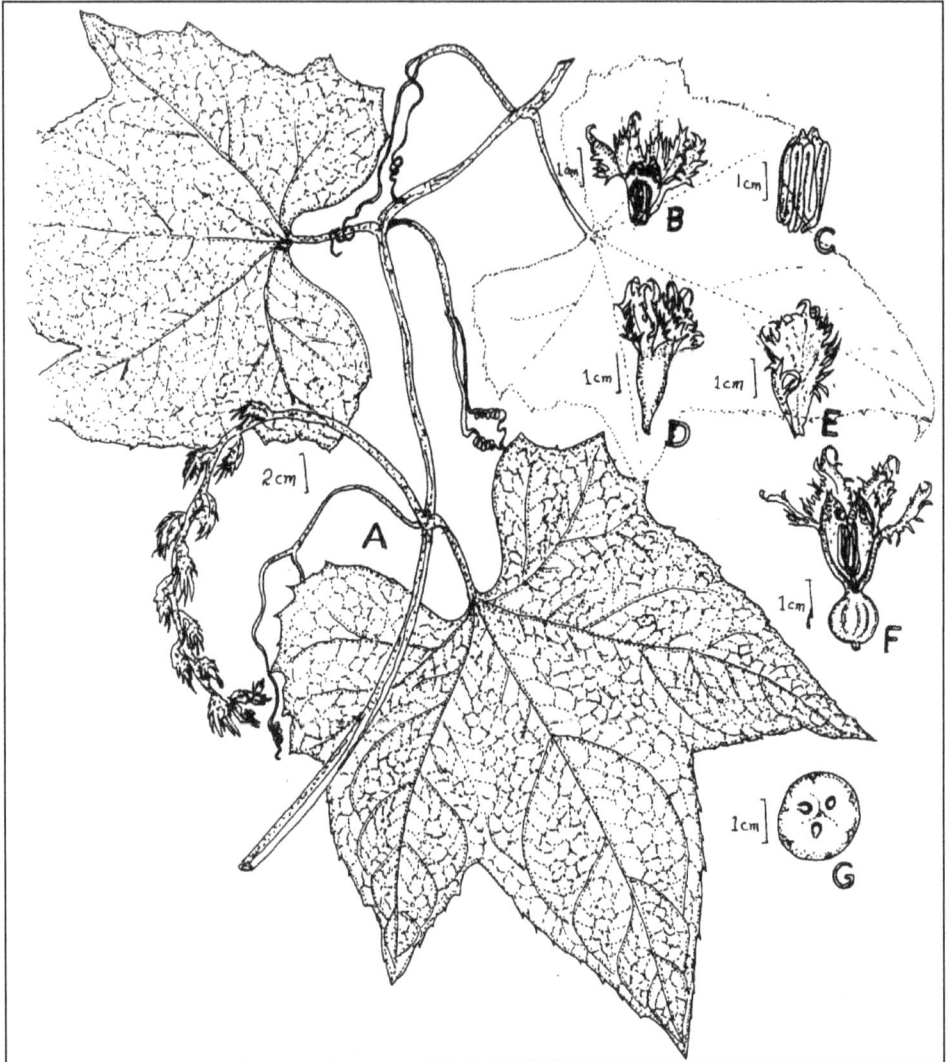

Figure 42: Trichosanthes tricuspidata Lour.

A. Twig, B. Male flower-open, C. Stamens, D. Female flower, E. Petal, F. Female flower-open, G. Ovary c.s.

1: 529. 1919; Chakravarthy in Fasc. Fl. Ind. 11: 116. 1982. *Involucraria lepiniana* Naudin, Huber. Oat. 11. 1868.

Climbing herbs, stems robust, branched sulcate, tendrils branched, 3-fid. Leaves polymorphous, deeply 3-7-lobed, lobes oblong or ovate- elliptic, sometimes lobulate, 10-18 × 8-15 cm, thin, coriaceous, base scabrid, cordate, margin entire, distantly denticulate, apex acuminate; petiole 2-5 cm. Flowers white, axillary. Male 5-10-flowered racemes; peduncle to 10 cm; bract ovate or obovate, 1-2 × 0.5-1 cm, deeply laciniate; pedicel to 5mm, calyx-tube. 3 cm, glabrous; lobes lanceolate, 1.5 × 0.5 cm; corolla white; petals ovate, 2 × 1 cm, long-laciniate, stamens 3; filaments 2 mm, anthers 1 cm, ciliate, connective slightly produced at the apex. Female solitary; ebracteate; calyx-tube 1.5 cm, lobes lanceolate, 2 × 1 cm, laciniate; petals involute, 2 cm; ovary globose, 1.5 cm, puberulous; style 1.5 cm; staminode 1cm. Fruits ca 7 × 4 cm, ovoid globose, red with orange stripes when ripe; seeds many, ellipsoid.

Occasional near cultivated lands in Adilabad, Nizamabad, Medak, Ranga Reddi and Mahabubnagar, districts.

Fl. & Fr.: June-December.

Bijjur (ADB), *GO* & *PVP* 4874; Mamidipally RF (NZB), *BR* & *SPB* 9602; Medak (MDK), *CP* 11736; Mohammdabad RF (RR), *MSM* 10477; Bhadrachalam (KMM), *AJR* & *PAN Reddy* 20464; Medak (MDK), *CP* 11736 (BSID).

Note: Jeffrey (1980) and Lu & *al.* (2011) treat *T. bracteata* as a synonym of *T. tricuspidata*, while Chakravarty (1982) recognized *T. bracteata* with two varieties; while Renner and Pandey (2012) treated them as separate species.

ZEHNERIA Endlicher

1. Plants monoecious; leaves angular and rarely lobed, ovate- subdeltoid .... **Z. maysorensis**

1. Plants dioecious; leaves often lobed, rarely angular, ovate-suborbicular .... **Z. perpusilla**

Zehneria maysorensis (Wight & Arn.) Arn., J. Bot. (Hooker) 3 : 275. 1841; Jeffrey, Kew Bull. 15: 366. 1962 & 34: 795. 1980; Renner & Pandey, Phytokeys 20: 102. 2013. *Bryonia maysorensis* Wight & Arn., Prodr. Fl. Ind. Orient. 345; 1834. *Melothria perpusilla* var. *subtruncata* Cogn. in DC., Monogr. Phan 3: 606. 1881; Gamble, Fl. Madras 1: 528. 1919. *M. perpusilla* auct. non (Blume) Cogn.; Chakravarthy in Fasc. Fl. Ind. 11 : 85. 1982.

Slender, monoecious scandent herbs. Tendrils simple, rarely bifid. Leaves membranous, ovate-subdeltoid, 5-10 × 6-8 cm, simple or 3-7-lobed, often hispid beneath, base cordate, margin remotely serrate; petiole to 5 cm. Flowers yellow, small dioecious. Male flowers more than 10 in a cluster, peduncle to 4 cm, pedicel to 4 mm; calyx-tube 2.5 mm, lobes ovate, 0.5 mm; corolla cream to yellowish, petals ovate, 3 mm; stamens 3, filaments to 3 mm, anthers 2 mm; pistillode 2 mm, obovoid. Female flowers solitary or 2-3 in a cluster; pedicel 3-6 mm, ovary 2 mm, globose-oblong; style 1.5 mm. Fruit a berry, globose, reddish when ripe.

Rare in Nizamabad district.

Fl. & Fr.: June – December.

Kamareddy (NZB), *BR* & *CPR* 9231.

Zehneria perpusilla (Blume) Bole & M.R. Almeida, J. Bombay Nat. Hist. Soc. 79(2): 315. 1983; Renner & Pandey, Phytokeys 20: 104. 2013. *Melothria perpusilla* (Blume) Cogn. in DC., Monog. Phan. 3: 607. 1881; Gamble, Fl. Madras 1: 538. 1921. *Cucurbita perpusilla* Blume, Cat. Hort. Buitenz. 105. 1823. *Zehneria hookeriana* Arn., Hook. J. Bot. 3: 275. 1841; C.B.Clarke in Hook. f., Fl. Brit. India 2: 624. 1879. *Z. scabra* auct; Pullaiah & Chennaiah, 1: 418. 1997.

Slender climbers. Leaves ovate-suborbicular, angular or 3-5-lobed, 3.5-5 × 2.5-3.5 cm, chartaceous, base truncate, sub-cordate, margin denticulate, apex acuminate, pubescent below; petiole to 1.5 cm. Flowers greenish-white, dioecious. Male flowers in apical umbellate racemes, 15-flowered; peduncle to 2.5 cm, pedicel to 1 mm, filiform; calyx-tube 3.5 mm, lobes linear, 0.5 mm; corolla greenish to white, petals ovate, 2.5 mm; stamens 3, filaments to 2 mm, anthers 2 mm, ovate; pistillode 1 mm. Female flowers solitary and/or 3-5 in a cluster; pedicel 1.5 cm; ovary 3 mm, globose, style 2 mm. Fruits ovoid, finely reticulate, red when ripe; seeds ovate-oblong, base slightly attenuate, smooth, distinctly marginate, brown, compressed.

Occasional in semievergreen forests in Adilabad district.

Fl. & Fr.: June-December

Rebbana (ADB), *GO* & *PVP* 4848.

BEGONIACEAE
BEGONIA L.

Begonia trichocarpa Dalzell, Hooker's J. Bot. Kew Gard, Misc, 3: 230. 1851; Gamble, Fl. Madras 1: 546. 1919.

Large, fleshy, caulescent tuberous herbs. Leaves ovate, sparsely pubescent, base cordate, margin serrate, apex acute, tapals 4 in male, 5 in female. Capsule wings subequal.

Rare in Narsapur forest of Medak district (Narasimha Rao 1986).

Fl. & Fr.: August-November.

Narsapur (MDK), *GNR* 4299 (MH).

CACTACEAE

1. Stem not jointed, angular, areoles without glochids; flowers more than 15 cm long, perianth funnel-like, united; stamens inserted on perianth .... **Cereus**
1. Stem jointed, flattend, areoles with glochids; flowers less than 10 cm long; perianth rotate, free; stamens inserted on the receptacle:
   2. Unarmed, stamens exserted .... **Nopalaea**
   2. Armed; stamens included .... **Opuntia**

CEREUS P. Miller

Cereus pterogonus Lem., Cact. Gen. Sp. Nov. 59. 1839; Curtis, Bot. Mag. t. 5360. 1863; Burkhill, Rec. Bot. Surv. India 4: 294. 1911.

Shrubs, up to 6 m high, stems columnar, 3-9-winged or angled, wings to 4 cm across; areoles echinate without glochids, lanate; spines 15, unequal. Leaves 0. Flowers lateral, sessile, 15 cm across, perianth tube terete at base, funnel-shaped with few scales below, outer ones greenish-white, inner ones white; stamens many, inserted at various levels on the perianth, filaments slender, anthers oblong, to 2.5 mm; ovary subterete, angular, 2.5 –3 × 0.8 – 1 cm, with scales and areoles; ovules many on several parietal placentae, style to 10 cm, stigmatic lobes 10-15, to 1 cm. Fruits seldom seen.

Common in Nizamabad and Mahabubnagar districts.

Fl. & Fr.: March-July.

Anandanagar (NZB), *BR* & *KH* 9651; Nagaram (NZB), *BR* 9712.

NOPALAEA Salm-Dyck

Nopalaea cochenillifera (L.) Salm-Dyck, Cact. Hort. Dyck. 64. 1850; Backer & Bakh. f., Fl. Java 1: 316. 1963. *Opuntia cochenillifera* (L.) Mill., Gard. Dict. ed. 8. Opuntia No. 6. 1768; Gamble, Fl. Madras 1: 548. 1919. (*coccinellifera*). *Cactus cochenillifer* L., Sp. Pl. 468. 1753.

Unarmed, erect, much branched shrub, joints flat, obovate-oblong, 8-40 × 5-7.5 cm, thick, margin entire, base and apex rounded. Spines usually absent, when present 1-3 per areole. Leaves conic, early deciduous. Flowers rose-coloured, up to 5.5 cm long; sepaloids with brilliant red or green midrib, largest ones ovate-deltoid; petaloids bright red, ovate to obovate; stamens exserted. Fruit red, ellipsoid; seeds grey or tannish.

A Mexican species, it was introduced at Machilipatnam in Krishna district in Andhra Pradesh in 1788 with the object of acclimatizing the cochineal insect, for which it is suited.

Occasional as an escape in Mahabubnagar district.

Fl. & Fr.: January-August

OPUNTIA L.

1. Spines 3-7 per areole, entirely translucent yellow :
    2. Spines thick, the largest often flattened and curved; fruit purplish .... **O. stricta**
    2. Spines slender, all straight; fruit reddish .... **O. elatior**
1. Spines 1-3 per areole, brown tipped .... **O. ficus-indica**

Opuntia elatior Mill., Gard. Dict. ed. 8. n. 4. 1768; Gamble, Fl. Madras 1: 548. 1919.

Erect shrubs, up to 2 m high, with jointed stems, joints flat, oblong-obovate, 5-30 × 3-15 cm, succulent, glaucous-green, with pale-yellowish spines. Leaves small, subulate, caducous. Areoles with 7-13, straight, slender, tawny or purplish-black spines. Flowers at first yellow, changing to red or reddish-purple, axillary, solitary, on free margin of joints; petals 3-3.5 cm long; stamens purple, included. Berries obovoid, dark red, prickly, fleshy, edible, with glochidiate and spiny areoles.

Raised as a hedge along cultivated fields; in scrub forest as an escape.

Fl. & Fr.: December-May.

Opuntia ficus-indica (L.) Mill., Gard. Dict. Ed. 8. No. 2. 1868. *O. vulgaris* Mill., Gard. Dict. ed. 8. n. 1. 1768. *O. mona-cantha* (Willd.) Haw., Suppl. Pl. Succ 81. 1819; Gamble, Fl. Madras 1: 548. 1919. *Cactus monacanthos* Willd. Enum. Pl. Suppl. 33. 1814.

Erect subshrubs, up to 2 m high; areoles densely woolly, brownish, glochidiate, spines 1-3 per areole, very unequal, brown tipped. Leaves subulate, lanceolate, to 4 cm long, early deciduous. Flowers to 8 cm across, perianth numerous, outer ones yellow, shaded pink, inner-ones yellow; stamens numerous, filaments 0.6 to 2 cm long, anthers to 2 mm; ovary 2.5 x 1 cm, glochidiate; pistil longer than the stamens. Berry obovoid.

Common in Mahabubnagar and Medak districts.

Fl. & Fr.: Through the year.

Mingaram (NZB), *TP* & *BR* 6317; Nagulabanda (MDK), *CP* 11328; Dummadugu (MDK), *GNR* 4617 (HH).

Opuntia stricta (Haw.) Haw., Syn. Pl. Succ. 191. 1812. var. **dillenii** (Ker-Gawl.) Benson, Cact. & Succ. J. America 41: 126. 1969. *Cactus strictus* Haw., Misc. Nat. 188. 1803. *Opuntia dillenii* (Ker-Gawl.) Haw., Suppl. Pl. Succ. 79. 1819; C.B. Clarke in Hook. f., Fl. Brit. India 2: 657. 1879; Gamble, Fl. Madras 1: 548. 1919. *Cactus dillenii* Ker-Gawl. in Bot. Reg. 3 : t. 255. 1818.

Erect branching, xerophytic shrubs, up to 5 m high; joints flat, obovate, 15-20 × 8-10 cm, joints fleshy with many areoles bearing flattened yellowish spines and with woolly hairs and tufts of numerous bristles. Flowers yellow, solitary near the margins of joints; perianth many, outer tepals to 1.5 cm, inner ones obovate, 3 × 2 cm, mucronate; stamens many, included, filaments to 2 cm; ovary 2.5 cm, glochidiate, style apically branched. Berry globose, fleshy, areolar and glochidiate.

Common in scrub jungles, roadsides and neglected lands.

Fl. & Fr.: January-August,

Potchera (ADB), *GO* & *DAM* 5094; Nagulabanda (MDK), *CP* 11328; Sivanoor (MDK), *BR* & *CP* 11548; Naskal (RR), *MSM* & *PRSR* 15120; Hyderabad, *J.H. Burkill* 25027 (CAL).

MOLLUGINACEAE

1. Ripe carpels free; one ovule in each locule .... **Gisekia**

1. Ripe carpels united; many ovules in each locule :

    2. Staminodes present :

        3. Seeds strophiolate .... **Glinus**

        3. Seeds estrophiolate .... **Corbichonia**

    2. Staminodes absent .... **Mollugo**

CORBICHONIA Scop.

Corbichonia decumbens (Forssk.) Exell, J. Bot. 73: 80. 1935. *Orygia decumbens* Forssk., Fl. Aeg.-Arab. 103. 1775; C.B. Clarke in Hook. f., Fl. Brit. India 2: 661. 1879; Gamble,Fl. Madras 1: 551. 1919.

Diffusely branched glabrous herbs. Leaves decussate and alternate, pairs unequal, obovate-suborbicular, 1.5-4.5 × 1-2.5 cm, fleshy, glabrous, base cuneate or attenuate, margin entire, apex rotund-obtuse, mucronate; petiole to 0.5 cm long, with membranous wing. Flowers pink, leaf-opposed or terminal, in cymes or lax racemes, up to 7 cm; peduncle to 15 cm long; bracts lanceolate, to 0.3 cm long; sepals ovate, apiculate, green; stamens and staminodes many. Capsule 0.4-0.6 cm long, pale-yellow, smooth, glabrous, 5-valved; seeds many, brownish black, reniform, minutely tuberculate.

Occasional in sandy soils in Warangal district (C.S. Reddy, 2001).

Fl. & Fr.: November-May.

Madikonda (WGL), *CSR* 1113.

GISEKIA L.

Gisekia pharnaceoides L., Mant. Pl. 562. 1771; C.B. Clarke in Hook. f., Fl. Brit. India 2: 661. 1879; Gamble, Fl. Madras 1: 553. 1919.

Diffuse or suberect herbs. Leaves opposite, oblong or obovate-oblanceolate, pairs unequal, 1-2 × 0.5-1 cm, fleshy, pubescent, base cuneate, margin entire, apex obtuse, mucronate. Flowers white or pink on filiform peduncles in umbellate cymes; sepals ovate, persistent; stamens 5, alternating with the sepals, less than 1 mm long, filaments dilated at base; carpels 5, unicellular, ovule solitary, basal, styles 5, free, lateral. Capsule 5-celled, papillose, indehiscent; seed 1 per cell, rounded, minutely pitted.

Common on wastelands in Medak, Hyderabad and Mahabubnagar districts.

Fl. & Fr.: July-September.

Choutkoor (MDK), *TP* & *CP* 11939; Amrabad (MBNR), *BR* & *BSS* 32572.

GLINUS L.

1. Plants densely tomentose; pedicels less than 0.6 cm long; stamens 10; capsule 5-valved, densely hairy, seeds faintly tuberculate .... **G. lotoides**

1. Plants glabrous; pedicels more than 1 cm long; stamens 5; capsule 3-valved, glabrous; seeds tuberculate .... **G. oppositifolius**

Glinus lotoides L., Sp. Pl. 463. 1753. *Mollugo lotoides* (L.) Kuntze, Revis Gen. Pl. 214. 1891; Gamble, Fl. Madras 1: 552. 1919. *M. hirta* Thunb., Prodr. Fl. Cap. 24. 1794; C.B. Clarke in Hook. f., Fl. Brit. India 2: 662. 1879.

Radically spreading or suberect herb, branches densely stellate-tomentose. Leaves opposite-whorled, obovate, suborbicular or spathulate, unequal, 1-1.5 × 0.5-2 cm, stellate tomentose, narrowed at base, margin entire, apex obtuse-rotund. Flowers greenish-white, 1-6 in axillary fascicles; sepals greenish-white, ovate, densely stellate-pubescent on dorsal surface; stamens 10, staminodes usually 10; ovary 5-locular, styles 5. Capsules ovoid, enclosed in persistent calyx, 5-valved, densely hairy; seeds many, reniform, faintly tuberculate, dark brown or nearly black.

Common in dry sandy places and on wastelands of most districts.

Fl. & Fr.: March-July.

Potchera (ADB), *GO* & *DAM* 5078; Laxmapur (NZB), *TP* & *BR* 6277; NSF (MDK), *BR* & *CP* 11672; Mohammadabad (RR), *MSM* & *KH* 10227; Hussainsagr-Hyderabad (HYD), *KMS* 5956 (MH); Kodimial (KRN), *GVS* 20092 (MH); Narsapur (MDK), *KMS* 7969; Nagarjunasagar (NLG), *BVRS* 128 (NU); Ramanagutta (KMM), *RR* 107964 (BSID); Farahabad to Vatavarlapally (MBNR), *SRS* 104593 (BSID); Pocharam Wild Life Sanctuary (MDK), *RG* 116418 (BSID).

Glinus oppositifolius (L.) A. DC., Bull. Herb. Boissier ser. 2. 1: 559. 1901. *Mollugo oppositifolia* L., Sp. Pl. 89. 1753; Gamble, Fl. Madras 1: 552. 1919. *M. spergula* L., Syst. Nat. ed. 10. 881. 1759; C.B. Clarke in Hook. f., Fl. Brit. India 2: 662. 1879.

Prostrate, diffusely branched, glabrous herbs. Leaves 3-5 in a whorl, oblong-oblanceolate or elliptic-obovate, unequal, 0.5-3 × 0.5-1.5 cm, base cuneate, margin entire, apex obtuse; petiole to 3 mm. Flowers white, axillary, on 7 mm long, slender pedicels, in 5-7-flowered cluster; sepals 5, green, oblong-lanceolate, 5 mm, glabrous, margin membranous; stamens 5, filaments 2.5 mm, anthers oblong, 1 mm, staminodes bifid, 2.5 mm; ovary 3-lobed, 2.5 mm, 3-celled, ovules numerous, style 3, 1 mm long. Capsule ellipsoid, enclosed in persistent calyx, 3-valved, glabrous; seeds minute, many, subreniform, dark-brown, tuberculate.

Common weed in wastelands in most districts.

Fl. & Fr.: Throughout the year.

Alampalli (ADB), *TP* & *PVP* 4176; Laxmapur (NZB), *TP* & *BR* 6276; Pocharam (MDK), *TP* & *CP* 12064; Mohammadabad (RR), *MSM* & *KH* 10228; Pakhal lake (WGL), *KMS* 13147 (CAL); Kodimial (KRN), *GVS* 20101 (MH); Bhadrachalam RF (KMM), *RCS* 104284 (BSID); Rayagir (NLG), *BVRS* 17 (NU).

MOLLUGO L.

1. Leaves all radical .... **M. nudicaulis**
1. Leaves cauline:
   2. Leaves narrow-linear or acicular; stem terete; flowers in umbelliform, simple cymes .... **M. cerviana**

2. Leaves other than linear or acicular; stem angular; flowers in dichasial cymes :

    3. Inflorescence rachis straight; pedicels curved; seeds prominently tuberculate .... **M. pentaphylla**

    3. Inflorescence rachis zigzag; pedicels straight; seeds smooth or obscurely granular .... **M. disticha**

Mollugo cerviana (L.) Ser. in DC., Prodr. 1: 392. 1824; C.B. Clarke in Hook. f., Fl. Brit. India 2: 663. 1879; Gamble, Fl. Madras 1: 553. 1919. *Pharnaceum cerviana* L., Sp. Pl. 272. 1753.

Annual glabrous herbs, up to 12 cm high, branching from base. Radical leaves rosulate, narrow, cauline leaves 5-7 in a whorl at nodes, narrowly linear or acicular, 1-2 × 0.2 cm, sessile or subsessile, glaucous below, base attenuate, margin entire, apex obtuse. Flowers white in terminal and axillary, trichotomous, umbellate cymes; peduncle short, 1.5 cm, slender; bract subulate, 0.5 mm; pedicel to 5 mm; sepals 5, subequal, elliptic, 1-5-2 mm, obtuse; stamens 5-8; filaments 1 mm; ovary 3-lobed, 3-celled, styles 3, to 0.2 mm. Capsule broadly ellipsoid, seeds minute, reniform, blackish brown.

Occasional in sandy places in Medak, Hyderabad, Mahabubnagar and Nalgonda districts.

Fl. & Fr.: June-November.

Rampur (MBNR), *BR* & *BSS* 32630; Peddavura (NLG), *BVRS* 509 (NU).

Mollugo disticha Ser. in DC., Prodr. 1: 392. 1824; C.B. Clarke in Hook. f., Fl. Brit. India 2: 663. 1879; Gamble, Fl. Madras 1: 553. 1919.

Erect annual herbs, branchlets slightly pubescent. Leaves 3-5 in fascicles at nodes; obovate-oblanceolate or elliptic, 0.5-1.5 × 0.2-0.4 cm, pubescent, base cuneate, margin entire, apex acute. Flowers white, shortly pedicellate in leaf-opposed and/or terminal cymes, branches often prolonged as racemes; stamens 5. Capsules ovoid, seeds many, pale pink, obscurely granular.

Common weed along road sides and open places in Medak, Ranga Reddi and Nalgonda districts.

Fl. & Fr.: June-October.

Mekavaripalli (RR), *TP* & *MSM* 11045; Medak (MDK), *RG* 104150 (BSID); Mallepally (NLG), *BVRS* 536 (NU).

Mollugo nudicaulis Lam., Encycl. 4: 234. 1797; C.B. Clarke in Hook. f., Fl. Brit. India 2: 664. 1879; Gamble, Fl. Madras 1: 553. 1919.

Erect, annual herbs, up to 15 cm high. Leaves all radical, spathulate, 2-4.5 × 1-2 cm, glabrous, base attenuate, margin entire, apex obtuse-retuse; petiole 1 cm. Flowers white in many-peduncled, trichotomously branched racemes, arising from the base, branches leafless, angular, bearing membranous bracts at the nodes; bracts 1mm;

pedicel to 4 mm; sepals 5, white, elliptic, 3 mm, hooded; stamens 3-5; filaments 1.5-2 mm; ovary oblong, 3-lobed, 3-celled, styles 3. Capsules ellipsoid or nearly globose, 2mm, brown; seeds many, black, reniform, granular.

Common in dry places, open fields and along roadsides in all districts.

Fl. & Fr.: June-December.

Vempalli (ADB), *GO* 4787; Gannaram (NZB), *TP* & *BR* 6406; Poosanpalle (MDK), *TP* & *CP* 11953; Dharmapuram (RR), *MSM* & *KH* 10284; Rampur (MBNR), *BR* & *BSS* 34482; Kodimial (KRN), *GVS* 20064 (MH); Ramanagutta (KMM), *RR* 10800 (BSID); Yenkur (KMM), *RR* 107901 (BSID); Vijayapuri (NLG), *KMS* 9844 (MH); Nagarjunasagar (NLG), *BVRS* 61 (NU).

Mollugo pentaphylla L., Sp. Pl. 89. 1753; Gamble, Fl. Madras 1: 553. 1919. *M. stricta* sensu C.B. Clarke in Hook. f., Fl. Brit. India 2: 663. 1879. *p.p.* non. L. 1762.

Erect to ascending annual herbs, up to 20 cm high, branchlets glabrous, dichotomously branched. Leaves radical as well as cauline, radical leaves oblanceolate, spathulate, cauline leaves 3-5 in a whorl at each node, obovate-elliptic, 1-2.5 × 0.4-1 cm, base cuneate-attenuate, margin entire, apex obtuse-subacute, petioles 2 mm. Flowers white in terminal or leaf-opposed lax peduncled cymes; peduncle 8 cm; bracts 1 mm; pedicels curved, 5-8 mm; sepals 5, white, ovate-oblong, 2 mm; stamens 3, filaments 1.5 mm; ovary ellipsoid, 1.8 mm, 3-celled, styles 3. Capsules ellipsoid, faintly 3-lobed; seeds many, minute, reniform, dark-brown, prominently tuberculate.

Common weed in open places of all districts.

Fl. & Fr.: Throughout the year, predominantly June-November.

Bijjur (ADB), *GO* & *PVP* 4880; Laxmapur (NZB), *TP* & *BR* 6287; Chedmal (NZB), *BR* & *CPR* 9261; Sangareddy tank (MDK), *BR* & *CP* 11293; Kusumasamudram RF (RR), *MSM* & *NV* 10330; Amrabad (MBNR), *BR* & *BSS* 32574; Rampur (MBNR), *GS* 18224; Koriguttalu (KMM), *RR* 106028 (BSID); Nellippakka RF (KMM), *RCS* 99089 (BSID); Nagarjunasagar (NLG), *BVRS* 60; Mahadevpur (KRN), *SLK* 70905 (NBG); Pakhal (WGL), *KMS* 11622 (MH); Pasra RF (WGL), *RKP* 10815 (BSID).

AIZOACEAE

1. Styles 1 .... **Trianthema**
1. Styles 2-5 .... **Zaleya**

TRIANTHEMA L.

1. Leaves orbicular-obovate or oblong; stamens 10-20;
    seeds 3-many .... **T. portulacastrum**
1. Leaves linear-lanceolate to elliptic-lanceolate;
    stamens 5; seeds 2 .... **T. triquetra**

Trianthema portulacastrum L., Sp. Pl. 223. 1753; Gamble, Fl. Madras 1: 550. 1919. *T. monogyna* L., Mant. Pl. 69. 1767; C.B. Clarke in Hook. f., Fl. Brit. India 2: 660. 1879.

Prostrate or ascending herbs, up to 50 cm long, branchlets glandular-tomentose or subglabrous, thickened and flattened at the nodes. Stipules inserted on the pouch, triangular, to 2.5 mm, membranous. Leaves subopposite, very unequal, one of the lower pair much smaller than the other, leaf blade obovate-orbicular, or oblong, 1.5-3.5 × 1-3 cm, subsucculent, purplish on margins, base cuneate, margin entire, apex obtuse, apiculate; petiole to 2.5 cm, base membranous, connate with leaf base into a pouch. Flowers white or bright pink, axillary, solitary in pouch or between forks of branches; bracteoles lanceolate, to 3.5 mm, sessile; calyx tube 1.5 mm, lobes 5, oblong – lanceolate, to 5 mm; stamens 10-20, filaments to 4.5 mm, ovary 3.5 mm, glabrous, style 1, to 3 mm. Capsules circumscissile, glabrous, partly concealed in the petiolar hood; seeds 3-many, with concentric muricate lines.

Common in most of the districts as a weed of roadside and wastelands.

Fl. & Fr.: July-November.

Kammarpally (NZB), *TP* & *BR* 6189; Toopran (MDK), *BR* & *CP* 11281; ICRISAT (MDK), *LJGV* 3193 (CAL); Dharmapur (RR), *MSM* & *KH* 10286; Kodimial (KRN), *GVS* 20105 (MH); Moosi River (HYD), *KMS* 5976 (MH); Khammam hills (KMM), *RR* 113814 (BSID); Rekularam (NLG), *BVRS* 626 (NU).

Trianthema triquetra Rottler ex Willd. in Ges. Naturf. Freunde Berlin Neue Schriften 4: 181. 1803; var. **triquetra;** Gamble, Fl. Madras 1: 551. 1919. *T. crystallina* auct non Vahl 1790; C.B.Clarke in Hook. f., Fl. Brit. India 2: 660. 1879.

Prostrate herbs, up to 50 cm long, branchlets radiating from a rootstock, stems reddish, studded with minute papillae. Stipules triangular. Leaves linear-lanceolate to elliptic-lanceolate, succulent, 2-5 × 1.5-2 cm, base subacute, shining, margin entire, apex obtuse, petiole 4 mm, slightly sheathing at base. Flowers pink or white, axillary, solitary or clustered; calyx tube 1.5 mm, lobes 0.5 mm, triangular; stamens 5, not exceeding calyx-lobes, ovary obconical, style 1, 0.5 mm. Capsules with a solitary seed in the operculum and basal part; seeds 2, black, reniform with concentric rings.

Occasional in waste lands and open forests in Hyderabad, Mahabubnagar and Nalgonda districts.

Fl. & Fr.: August-September.

Moosi River (HYD), *KMS* 5981 (MH); Nelikal (NLG), *BVRS* 459 (NU).

ZALEYA Burm.f.

Zaleya decandra (L.) Burm. f., Fl. Indica 110. t. 31. f. 3. 1768. *Trianthema decandra* L., Mant. Pl. 7. 1767; C.B. Clarke in Hook. f., Fl. Brit. India 2: 661. 1879; Gamble, Fl. Madras 1: 551. 1919.

Prostrate herbs, up to 40 cm long, branchlets glabrescent. Leaves estipulate, elliptic-lanceolate, elliptic-oblong, 2-3 × 1-1.5 cm, subsucculent, papillose, base acute, margin entire, apex obtuse or rounded, petiole to 7 mm, dilated, amplexicaul. Flowers pinkish in dense axillary or terminal sub-umbellate cymes or fascicles; bracteoles linear, 1.5 mm; pedicel to 2 mm, not enclosed in aeriole; calyx tube 2mm, sepals 5, 4.5 mm; stamens 10, unequal, filaments to 2 mm; ovary terete, 3 mm, styles 2-5, filiform, to 2 mm. Capsule truncate separating into 2 parts, each with 2 seeds; seeds orbicular or reniform, compressed, rugulose.

Common in most of the districts on dry lands and also as road side weed.

Fl. & Fr.: June-November.

Tandur (ADB), *TP* & *GO* 5481; Kammarapally (NZB), *TP* & *BR* 6188; Chevella (RR), *TP* & *MSM* 19255; Jalapenta (MBNR), *BSS* & *SKB* 32616 (BSID); Vijayapuri (NLG), *KMS* 19302 (MH).

APIACEAE (*nom. alter.* UMBELLIFERAE)

1. Umbels simple .... **Centella**
1. Umbels compound :
  2. Leaves simple .... **Eryngium**
  2. Leaves usually pinnate or decompound, flowers white:
    3. Ovary and fruit bristly or scaly hairy:
      4. Involucres pinnatifid; rays 15-30 .... **Daucu**s
      4. Involucres not pinnatifid; rays 2-9:
        5. Fruit laterally compressed; mericarps not winged on margins:
          6. Calyx teeth distinct, subulate; vittae solitary
            under each secondary ridge .... **Cuminum**
          6. Calyx teeth inconspicuous or absent:
            7. Leaves simple or pinnate with simple leaflets;
              furrow on mericaprs 2-3-vittate .... **Pimpinella**
            7. Leaves pinnate with divided leaflets;
              furrows on merciarps 1- vittate .... **Trachyspermum**
        5. Fruit not laterally compressed; primary
          ridges prominent .... **Seseli**
  2. Ovary and fruit entirely glabrous .... **Coriandrum**

CENTELLA L.

Centella asiatica (L.) Urb. in Mart., Fl. Bras. 11: 287. t. 78. f. 1. 1879; Gamble, Fl. Madras 1: 556. 1919. *Hydrocotyle asiatica* L., Sp. Pl. 234. 1753; C.B. Clarke in Hook. f., Fl. Brit. India 2: 669. 1879.

Prostrate herbs, up to 75 cm, stems creeping with long stolons, rooting at nodes, grooved, sparsely hairy. Leaves simple, orbicular-reniform, 1-2.5 × 1-5.3 cm, puberulous, base cordate, margin crenate-dentate, apex rounded, petiole up to 20 cm, sheathing at base. Flowers pink or red in axillary fasciculate umbels, central flowers sessile, laterals pedicellate, bracteate; involucral bracts 2; peduncle to 6 mm, pedicel to 0.5 mm; calyx lobes 5, triangular, to 0.5 mm; petals 5, purplish red, to 1.5 mm, ovate, acute to obtuse; stamens red, to 1 mm, filaments 0.5 mm; ovary 2-celled, styles bifid, distant, free from the disc, stigmas 2, simple. Fruit 0.4-0.5 cm long, ovoid, mericarp 7-9-ribbed, prominently reticulate.

Common in plains of most districts, in wet places.

Fl. & Fr.: Throughout the year.

Kadam dam (ADB), *GO* & *BR* 9415; Alisagar (NZB), *BR* 7284; Medak (MDK), *BR* & *CP* 11466; Rudraram RF (RR), *TP* & *MSM* 12723; Kudichintalabailu (MBNR), *TDCK* 15352; Pakhal (WGL), *RKP* 108268 (BSID).

CORIANDRUM L.

Coriandrum sativum L., Sp. Pl. 256. 1753; Hook.f. Fl. Brit. India 2 : 717. 1879; Gamble, Fl. Madras 1 : 566. 1919.

Vern.: *Dhaniyalu, Kothimeera.* English: Coriander.

Erect, annual herb, to 20 cm long; stems slender, branched, terete, striate. Lower leaves palmately-partite; middle ones pinnate; upper ones pinnate-bipinnate; ultimately segments linear-laneolate, 0.5 mm long. Flowers in umbels, terminal and axillary, compound, 3-5 rayed; peduncles 2-10 cm long; involucres 3-5, linear, 4-5 mm long; calyx tube 1.5 mm, lobes 5, triangular, upper ones lanceolate, lower ones triangular; petals white to pale pink, radiating; other petals of rays 4-4.5 mm long, bipartite, with inflexed tips. Fruits nearly globose, 3-3.5 x 2-2.5 mm. Mericarps hollow inside, with dorsal primary ribs strongest and undulated, and secondary ones filiform and obscure.

Cultivated for its leaves used as flavouring material and fruits used as spice.

Fl. & Fr.: July-September, December-March.

Kadthala (ADB), *GO* & *DAM* 5000.

CUMINUM L.

Cuminum cyminum L. Sp. Pl. 254. 1753; Hook.f. Fl. Brit. India 2 : 718. 1879; Buwalda, Blumea 2 : 178. 1936 & in Stennis, Fl. Males. ser I. 4 : 131. 1949.

Vern.: *Jilakara.*

Annual herb; stems 15-30 cm long, erect, strongly divergently branched from base. Leaves bipinnate; short petioled or sessile on a sheath, to 1 x 0.5 cm, with membranaceous white margins, auriculate at apex or tapering into petiole; lamina 4-10 cm long, segments to 1.5 mm broad, linear. Flowers compound umbels opposite to the leaves or terminal; peduncles 2-4 cm long; rays 4-6, 1-1.5 cm long; calyx–teeth 1-1.5 mm long, linear–subulate, persistent; petals obcordate, with inflexed tips, white to reddish, 1 x 0.5 mm. Mericarps laterally flattened, 5-7 x 2.5 -3 mm; main ribs filiform, bristly; ridges with stellate-hairy line.

Cultivated in small scale in kitchen gardens.

Fl. & Fr.: March-June.

Kadthala (ADB), *GO* & *DAM* 5009.

DAUCUS L.

Daucus carota L., Sp. Pl. 242. 1753; Hook.f., Fl. Brit. India 2 : 718. 1879; Buwalda, Blumea 2 : 208. 1936 & in Steenis, Fl. Males Ser. I. 4 :140. 1949.

Vern.: *Carrot.* English: Carrot.

Biennial herbs; stems 15-120 cm high, solitary, glabrous or papillate hispid. Leaves oblong in outline, 5-15 x 2-7 cm, 2-3 pinnate; ultimate divisions linear-lanceolate, acute and mucronate at apex, glabrous or hispid on veins and margins, 3-12 x 0.5-2 mm. petioles 4-10 cm long. Flowers in few-flowered terminal umbels; calyx teeth triangular, acute; petals unequal, white. Fruit ovoid, 3-4 x 1.8-2 mm; secondary ribs with prominent spines, glochidiate.

Cultivated in smale scale in kitchen gardens.

Fl. & Fr.: January–March.

Note: There are two varieties under cultivation; one *D. carota* var. *sativa,* cultivated for its fleshy edible roots and other *D. carota* var. *carota* cultivated as an ornamental in gardens.

ERYNGIUM L.

Eryngium foetidum L., Sp. Pl. 232. 1753; Buwalda in Steenis, Fl. Males. 1: 4: 126, 1949; Krahulik & Theob. in Dassan. & Fosberg., Rev. Handb. Fl. Ceylon 3: 487.1981; C.S.Reddy & V.S.Raju, J. Econ. Taxon. Bot. 26: 195-198. 2002.

Vern.: *Brahma kottimeera.*

Erect, perennial, aromatic herb, up to 50 cm high, branchlets dichotomously branched; roots long and fusiform. Leaves sessile, oblanceolate – oblong, 4-12 × 1-3 cm, spinosely – serrate, obtuse, glabrous, base cuneate and more or less sheathing. Bracts of inflorescence 1-6 cm, often palmatilobate to partite, with spiny tips and teeth, strong nerved, lowermost often like normal leaves, peduncles to 1 cm, heads to 1 cm, ovoid or cylindrical; involucral bracts 5-7, spreading, nearly lanceolate, with few spiny teeth. Flowers sessile in the axils of narrow membranous – marginated bracts, 1.25 - 1.5 mm long; sepals (4) 5, lanceolate, acute, margin membranous; petals whitish, obovate or oblanceolate, to 7.5 mm long; stamens 5, filaments inflexed in bud. Mericarps ovoid, 1-1.5 mm long, densely papillose.

Rare, occurs in dampy soils close to stream banks in Warangal district (C.S.Reddy and V.S.Raju, 2002).

Fl. & Fr.: April-August.

Raghunathapally (WGL), *CSR* 1253 (KU).

PIMPINELLA L.

Pimipinella heyneana (Wall. ex DC.) Kurz, J. Asiat. Soc. Bengal 46: 115. 1877; C.B.Clarke in Hook. f., Fl. Brit. India 2: 684. 1879; Gamble, Fl. Madras 1: 560. 1919. *Helosciadium heyneanum* Wall. ex DC., Prodr. 4: 106. 1830.

Erect annual herb, up to 1 m high, branchlets glabrous. Leaves 3-foliolate, 7 cm long, leaflets oblong-lanceolate or partite, 3-5 × 1.5-2.5 cm, pubescent or glabrous, base obtuse, margin serrate, apex acuminate, petiole up to 10 cm, sheath 2.5 cm. Flowers white or pale-pink, minute in compound umbels; peduncles 3.5 – 10 cm, rays 4-20, to 2.5 cm, glabrous; flowers ca 15 per ray, bracts 1 or 2, lanceolate, caducous; bracteole 1-3, acicular 1.5 cm, caducous; pedicel 0.5 –1 cm, flowers white, 1.5 mm across; sepals 0; petals 5, ovate, 0.5 mm, stylopodium flat; stamens 5; ovary 5 mm; base puberulous, style 0.5 mm. Cremocarp ovoid, mericarps glabrous, carpophore bifid; seeds nearly terete.

Very rare in hill forests of Khammam and Mahabubnagar districts.

Fl. & Fr.: December- March.

Sukumamidi near Mothugudem (KMM), *EC* 7455; Farahabad (MBNR), *MVS* & *MB* 38005; Jalapenta (MBNR), *BSS* & *SKB* 32609.

SESELI L.

Seseli diffusum (Roxb. ex Smith) Santapau & Wagh, Bull. Bot. Surv. India 5: 108. 1963. *Ligusticum diffusum* Roxb. ex Smith in Rees. Cyclop 21: 11. 1812. *Seseli indicum* Wight & Arn., Prodr. Fl. Ind. Orient. 371. 1834; C.B.Clarke in Hook. f., Fl. Brit. India 2: 693. 1879; Gamble, Fl. Madras 1: 561. 1919.

Erect or diffuse annual herbs, 5-15 cm high. Leaves 2-3-pinnate, 2-4 × 1-2.5 cm, hispidly pubescent. Bracts 4-5, minutely pubescent, bracteoles many, smaller than bracts. Flowers white or pink in compound umbels; calyx 0; petals 5, unequal with medium folds, stamens 5; disc epigynous; styles 2. Fruit oblong-obovate or subglobose, densely glochidiate, not laterally compressed; seed semiterete, inner surface concave.

Rare in Nalgonda and Nizamabad districts.

Fl. & Fr.: August-December

Kandakurthy-Godavari banks (NZB), *BR* 7258; Kambalapalli (NLG), *BVRS* 227 (NU).

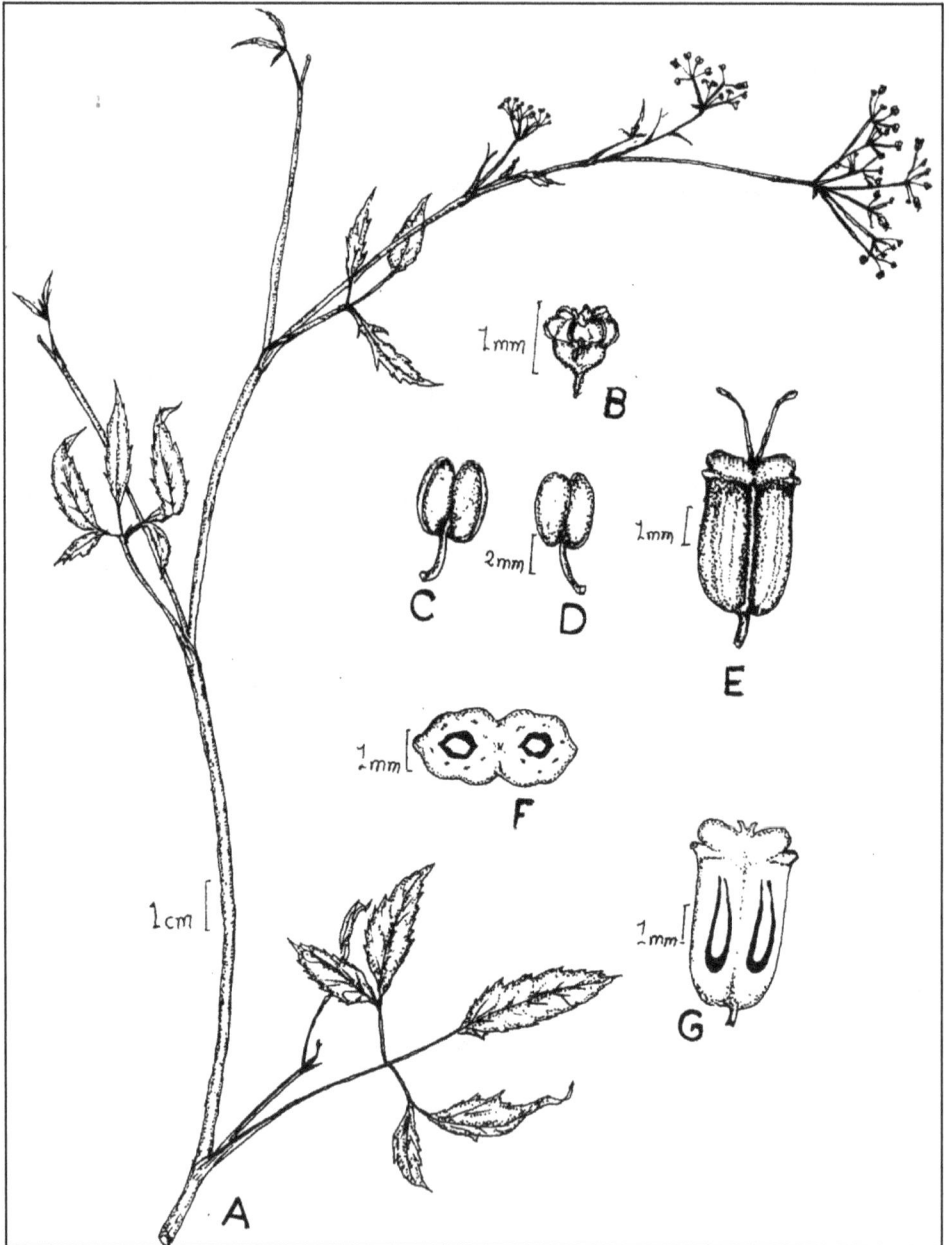

Figure 43: Pimipinella heyneana (Wall. ex DC.) Kurz
A. Twig, B. Flower-entire, C&D. Stamens, E. Pistil, F. Ovary c.s., G. Ovary l.s.

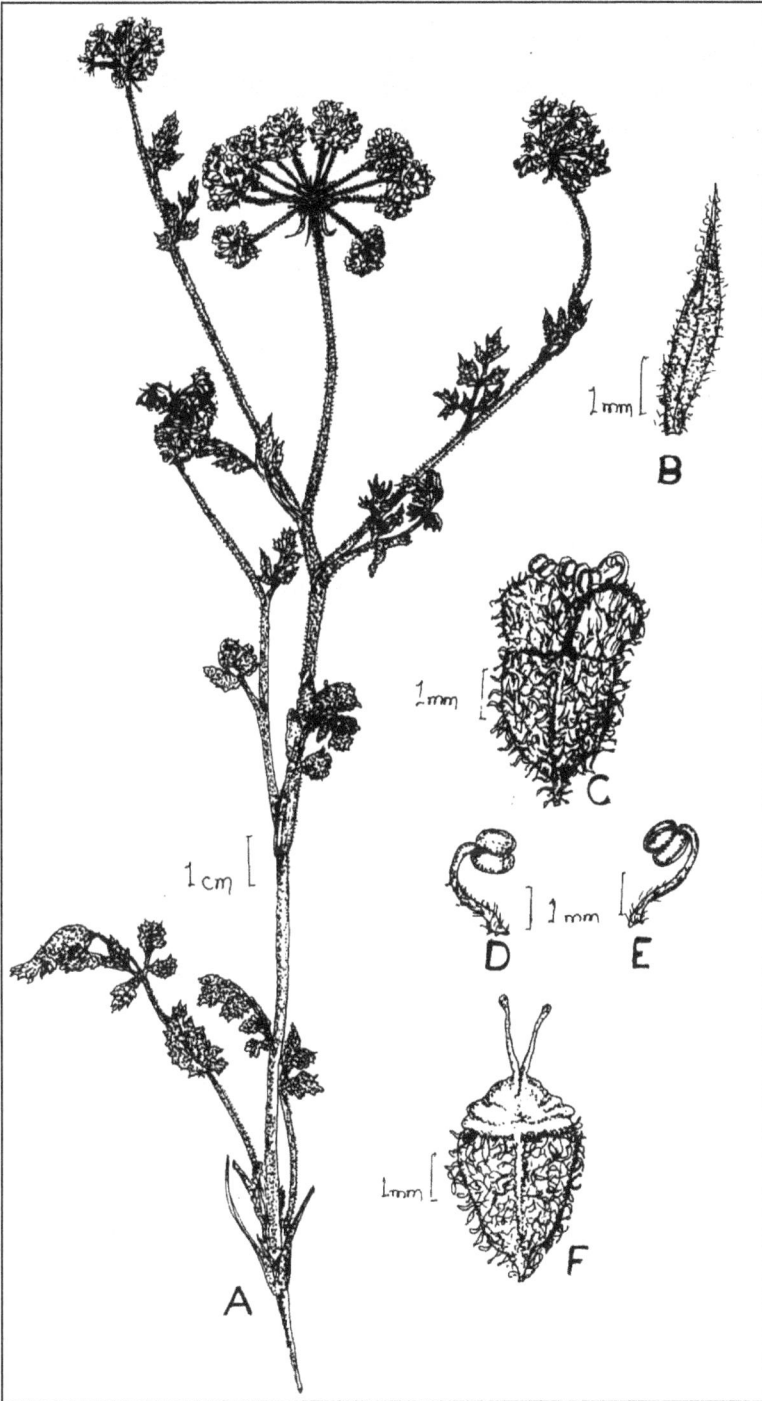

Figure 44: Seseli diffusum (Roxb. ex Sm.) Santapau & Wagh
A. Twig, B. Bract, C. Flower-entire, D&E. Stamens, F. Pistil.

TRACHYSPERMUM Link *nom. cons.*

Trachyspermum ammi (L.) Sprague, Bull. Misc. Inform. Kew 1929: 228. 1929. *Sison ammi* L., Sp. Pl. 252. 1753. *Carum copticum* Benth. ex Hiern. in Oliver, Fl. Trop. Africa 3: 12. 1871; C.B. Clarke in Hook. f., Fl. Brit. India 2: 682. 1879.

Vern.: *Vamu.* English: Bishop's weed.

Slender herbs, up to 60 cm high, minutely pubescent or glabrescent, branched. Leaves pinnately decompound, ultimate segments narrowly linear, glabrous. Flowers whitish in compound umbels. Fruit ovoid, carpels subpentagonous, dorsally compressed, ridges usually distinct.

Cultivated, often an escape.

Fl. & Fr.: December-March.

Gannaram (NZB), *TP* & *BR* 6401.

ALANGIACEAE

ALANGIUM Lamarck *nom. cons.*

Alangium salvifolium (L.f.) Wangerin in Engler, Pflanzenr. 4 (220 b), Heft. 41: 9. 1910; Gamble, Fl. Madras 1: 572. 1919. subsp. **salvifolium.** *Grewia salvifolia* L.f., Suppl. 409. 1782. *Alangium lamarckii* Thwaites, Enum. Pl. Zeyl. 133. 1859; C.B. Clarke in Hook. f., Fl. Brit. India 2: 741. 1879. *A. salvifolium* (L.) Wangerin subsp. *decapetalum* (Lam.) Wangerin in Engler, Pflanzenr 4. 220 b. Heft. 41: 11. 1910. *A. decapetalum* Lam., Encycl. 1: 174. 1783. *A. salvifolium* (L.f.) subsp. *hexapetalum* (Lam.) Wangerin in Engler, Pflanzenr. 4. 220 b, Heft. 41: 9. 1910; quoud monem, excl. deser. *Alangium hexapetalum* Lam., Encycl. 1: 175. 183.

Deciduous shrubs or trees, up to 12 m high, trunk often with numerous holes; bark grey, orange-yellow when young, deeply fissured; branchlets appressed tomentose. Leaves oblong- lanceolate, 8-12 × 2.5-4.5 cm (new foliage following the flowers measure less), glabrous above, glabrescent or puberulous below; apex subacute to acuminate, margin entire, base oblique, 3-nerved from base; secondary nerves 6 pairs, irregular; petiole to 1 cm, tomentose. Flowers to 2.5 × 1.5 cm, scented, in irregular axillary fascicles; calyx tube cupular, 2.5 mm, tomentose, lobes ca 10, triangular-ovate, 0.5 mm; petals 10, white, 2.5 × 0.2 cm, linear oblong, reflexed; stamens ca 20, filaments to 1 cm with a fleshy and villous base, sub connate; ovary turbinate, to 2 mm, style to 2 mm, glabrous, stigma capitate. Berry ellipsoid, 2 × 1 cm, red or blackish purple, pubescent, succulent with bony endocarp, crowned by calyx lobes; seed solitary, ovoid, 0.8 × 0.4 cm.

Common in open habitats in Adialabad, Warangal, Nizamabad, Hyderabad, Medak, Ranga Reddi, Khamamam and Mahabubnagar districts.

Fl. & Fr.: February-July.

Kadam RF (ADB), *GO* & *BR* 9409; Rampur (ADB), *SSR* 19311; Mudheli RF (NZB), *TP* & *BR* 6359; Gangapur RF (MDK), *TP* & *CP* 12067, 11435; Mohammadabad (RR), *MSM* & *KH* 10211; Farahabad (MBNR), *MVS* & *MB* 38640; Vatvarlapally (MBNR), *TDCK* 13602; ICRISAT (MDK), *LJGV* 3795 (CAL); Koida (KMM), *PVS* 84067 (BSID), Pocharam Wild Life Sanctuary (MDK), *RG* 11359 A (BSID); Tadwai (WGL), *RKP* 110801 (BSID).

www.ingramcontent.com/pod-product-compliance
Lightning Source LLC
Chambersburg PA
CBHW020216290326
41948CB00001B/64

.